AF179455

# Contemporary Ergonomics 2005

# Contemporary Ergonomics 2005

Editors

**Philip D. Bust**

*Loughborough University*

**Paul T. McCabe**

*Atkins, Warrington, UK*

Taylor & Francis

The Ergonomics society

© 2005 Taylor & Francis

Did 16th Century Pagoda builders base their designs on human neck architecture?
*Bernard Masters*
© B. Master 2005

The effect of handle angle on fencing sport
*Chih-Lon Lin and Chih-Lin Chang*
© Crown Copyright 2005/ROC

Baggage handling in narrow-bodied aircraft: identification and assessment of musculoskeletal injury risk factors
*Sarah Tapley and David Riley*
© Crown Copyright 2005/HSL

Reducing the risk of musculoskeletal disorders in construction workers
*Margaret Hanson and Taran Barrington*
© Crown Copyright 2005/HU-TECH

Passenger behaviour on amusement rides
*Edmund Milnes*
© Crown Copyright 2005/HSL

Confessions of an expert witness
*Alan Jackson*
© Crown Copyright 2005/HSE

Usability evaluation methods for child participants
*Helen Edwards and Rachel Benedyk*
© Crown Copyright 2005/UCLIC

Human factors in air traffic management: Making a difference
*Barry Kirwan, Mark Rodgers and Dirk Schaefer*
© Crown Copyright 2005

Distributed situational awareness
*Neville Stanton, Chris Baber, Guy Walker, Paul Salmon and Damian Green*
*Rob Houghton, Richard McMasters and Huw Gibson*
© Crown Copyright 2005/HFI-DTC

Impacts of automation revisited
*Elizabeth Carver*
© Crown Copyright 2005/BAE-SYSTEMS

Emergency egress from multi-crew military aircraft
*D Harrison and F Sturrock*
© Crown Copyright 2005/BAE-SYSTEMS

Succeeding at Human Factors Integration (HFI)
*Monica Sen Gupta*
© Crown Copyright 2005/BAE-SYSTEMS

Assessment of the use of carry chairs by UK ambulance staff
*Jeremy Ferreira and Leanne Stanley*
© Crown Copyright 2005/HSL

Web accessibility for all
*Ben Dawson, Bill Wessel, Freyja Lockwood and Caroline Gowing*
© Crown Copyright 2005/BAE-SYSTEMS

Event Analysis Of Systemic Teamwork (EAST); A novel integration of ergonomics methods to analyse C4i activity
*Guy H Walker, Huw Gibson, Neville A. Stanton,*
*Chris Baber, Paul Salmon and Damian Green.*
© Crown Copyright 2005/UNIVERSITY OF BIRMINGHAM

The Event Analysis Of Systemic Teamwork (EAST): Methodology and analyses of railway maintenance tasks
*Huw Gibson, Guy H Walker, Chris Baber, Neville A. Stanton, and Paul Salmon.*
© Crown Copyright 2005/HFI-DTC

Integrating human factors methods: The HF methods matrix
*Paul Salmon , Neville Stanton, Guy Walker , Damian Green and Chris Baber*
© Crown Copyright 2005/HFI-DTC

Floor cleaning – Drop that mop!
*Robert Shaw, Kevin Hallas, Paul Lemon and Steve Thorpe*
© Crown Copyright 2005/HSL

Tactile paving – right or wrong?
*Marianne Loo-Morrey and Steve Thorpe*
© Crown Copyright 2005/HSL

Roller coaster slip tests putting slip testing back on the rails!
*Kevin Hallas, Robert Shaw, Paul Lemon and Steve Thorpe*
© Crown Copyright 2005/HSL

Managing fatigue risks
*Claire Dickinson*
© Crown Copyright 2005/HSE

The effect of signal infrastructure on train driver visual behaviour on signal approaches
*T Luke, N Brook-Carter, R Smith, A M Parkes, A Mills and E Grimes*
© TRL 2005

Measuring pilot workload in single and multi-crew military aircraft
*Fiona Sturrock and Christian Fairburn*
© Crown Copyright 2005/BAE-SYSTEMS

Reproduced with permission of the Controller of her Britannic Majesty's Stationary Office. The views expressed are those of the Author and do not necessarily reflect the views or policy of the Controller or any government department. The Controller, any government and Taylor & Francis Ltd accept no responsibility for the accuracy of any recipe, formula instruction contained within this population.

Printed and bound in Great Britain.

All rights reserved. No part of this book may be reprinted or reproduced or utilised in any form or by any electronic, mechanical, or other means, now known or hereafter invented, including photocopying and recording, or in any information storage or retrieval system, without permission in writing from the publishers.

Every effort has been made to ensure that the advice and information in this book is true and accurate at the time of going to press. However, neither the publisher nor the authors can accept any legal responsibility or liability for any errors or omissions that may be made. In the case of drug administration, any medical procedure or the use of technical equipment mentioned within this book, you are strongly advised to consult the manufacturer's guidelines.

ISBN 0 415 37448 0

# CONTENTS

## ENVIRONMENTAL ERGONOMICS



## ERGONOMICS 4 SCHOOLS



## HOSPITAL ERGONOMICS



## INCLUSIVE DESIGN

## SLIPS, TRIPS AND FALLS – BIOMECHANICS



## SLIPS, TRIPS AND FALLS – INJURY PREVENTIONS

## SLIPS, TRIPS AND FALLS – TRIBOLOGY



## TRANSPORT

## GENERAL ERGONOMICS

# PREFACE

*Contemporary Ergonomics 2005* is the proceedings of the Annual Conference of the Ergonomics Society, held in April 2005 at Hatfield University, Hertfordshire, UK. The conference is a major international event for Ergonomists and Human Factors Specialists, and attracts contributions from around the world.

Papers are chosen by a selection panel from abstracts submitted in the autumn of the previous year and the selected papers are published in *Contemporary Ergonomics*. Papers are submitted as camera ready copy prior to the conference. Each author is responsible for the presentation of their paper. Details of the submission procedure may be obtained from the Ergonomics Society.

The Ergonomics Society is the professional body for ergonomists and human factors specialists based in the United Kingdom. It also attracts members throughout the world and is affiliated to the International Ergonomics Association. It provides recognition of competence of its members through its Professional Register. For further details contact:


The Ergonomics Society
Elms Court
Elms Grove
Loughborough
Leicestershire
LE11 1RG
UK

Tel: (+44) 1509 234 904
Fax: (+44) 1509 235 666

Email: ergsoc@ergonomics.org.uk
Web page: http://www.ergonomics.org.uk

# APPLICATIONS OF ERGONOMICS

# CONSTRUCTING NEW HARNESS FIT CHARTS USING 3D ANTHROPOMETRIC INFORMATION

**Hongwei Hsiao, Richard Whisler, Tsui-Ying Kau, Joyce Zwiener, Jinhua Guan & James Spahr**

*National Institute for Occupational Safety and Health, 1095 Willowdale Road, Morgantown, West Virginia 26505, USA*

The harness manufacturing industry has a pressing need to update harness fit charts to accommodate diverse workforces in the construction industry as well as in supporting new roles for women in the workforce. This paper presents an improved harness sizing system for vest-type harness-design applications. Three dimensional (3D) torso-scan data from 108 women and 108 men were obtained using a 3D whole-body scanning system. Human-harness interfaces, both in static and suspended forms, were digitally captured. A combination of weight, height, gender, and 3D information for upper and lower torso regions yielded a logistic regression model that correctly classified 96% of the participants to their best fit size in a validation test. The outcomes suggested the replacement of the current 4-size system with a new system that contains 2 sizes for females and 3 sizes for males. Two approaches to determine the adjustment range for each critical harness component for each harness size were proposed for harness designers to consider.

## Introduction

A review of the historical anthropometric sources for commercial harness sizing revealed that they are based on dimensions which were derived from U.S. military population databases collected in the 1970's and 1980's (Bradtmiller et al., 2000). Population anthropometry evolves over time (Bodzsar, 2000); large variations exist over the last decades in body dimensions among the U.S. civilian population (HHS, 2001). In addition, diverse workforces in the construction industry, as well as new roles for women in the workforce, make it likely that more construction occupations will show a greater variation in their range of body dimensions and shapes as compared to that in the 70s and 80s. The harness manufacturing industry has a pressing need to reassess the current sizing schemes. The goals of this study were to identify anthropometric characteristics that affect the fit of vest type fall-protection harnesses to workers and to correspondingly develop an enhanced sizing scheme. The results should have a direct impact on practical design consideration for harnesses and have a potential impact on the development of standards for harness sizing systems.

## Method

### *Participants*

One hundred and eight males and 108 females participated in this study. Their ages ranged from 18 to 56. The average height was 175.7 cm (6.5 cm S.D.) and mass was

85.5 kg (14.2 kg S.D.) for males, and 162.6 cm (6.6 cm S.D.) and 66.2 kg (12.4 kg S.D.) for females.

*Independent variable*

*Harness fit rating*
Harnesses were tested while the participant was both standing and suspended, and a pass or fail rating was assigned based on 4 criteria. For the standing condition, the pass criterion was that the harness-back D-ring was positioned between the inferior and superior borders of the scapula. For the suspended condition, 3 criteria needed to be met to receive a pass rating: the suspension angle had to be equal to or less than 35 degrees; the chest strap should not make contact with the neck; and the center of gravity (COG) of body had to be behind the hip rings. These criteria were evaluated by examining the interface between the harness and the human torso through 3D scan images. A harness fit rating of pass was assigned if all 4 criteria were met.

*Body dimensions*
The maximum breadth, depth, and height of the upper-torso volume space and the lower-torso volume space were extracted from 3D scan images. The breadth measure refers to side-to-side distance, depth to front-to-back, and height to top-to-bottom distance of the two torso regions (also known as "torso bounding boxes"). The 10th rib landmark was used to separate the torso images into the upper and lower torso regions. These dimensions were used to explore the interaction of the general body shape and dimensions with harness fit, based on the "bounding box" concept in the computer graphics field. Body weight, stature, and 17 other body dimensions that may be associated with harness sizing were also measured.

*Dependent variable*

*Best-fit size*
Four available harness sizes (XSM, STD, XLG, and SXL) were fitted for each participant. The investigator adjusted the harness to achieve the best possible fit and asked the participant about the comfort/discomfort of the harness and recorded the best-fit size.

*Apparatus*

A Cyberware WB4 3-D full-body scanner (Cyberware Inc., Monterey, CA) was used to register the interface between the harness and participant during normal standing, as well as to quantify the changes in the harness fit when the participant was suspended. The accuracy of the scanning system was tested to an average error of 2.9 mm. A Toledo scale (Mettler-Toledo Inc., Worthington, OH) was used to measure body weight, and a custom-made suspension system was developed to suspend the participants safely.

*Procedure*

Upon arrival, the participant viewed a web page, which described the details of the study and the tasks that the participant was to perform. Each participant signed an informed consent form and then changed from street clothes into bicycle shorts for men and bicycle shorts with halter top for women. The participant tried on 4 sizes of a vest-style harness. One investigator performed the fit test with feedback from the participant and recorded the fitting result for each size. The "best-fit" size was then selected for a subsequent fit assessment. The participant was scanned with the Cyberaware full-body scanner in an erect standing posture with a harness (referred to as A scan, Figure 1). The scan took 17 seconds. After the first scan, an investigator attached the steel cable of the suspension system to the back D-ring of the harness, and

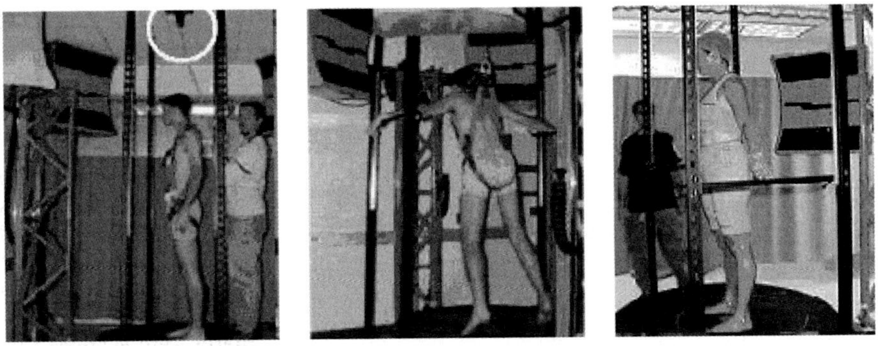

**Figure 1. A standing posture with harness donned (A scan; 1a), a suspended position (B scan; 1b), and a standing posture without a harness (C scan; 1c) were used to evaluate human-harness interface.**

raised the participant until the participant's toes were off the floor to simulate a post-fall condition. The participant was scanned a second time (referred to as B scan). The participant was lowered to the floor and was then scanned the third time without a harness (referred to as C scan). Finally, the participant changed back into street clothes and was compensated for his/her time, and then was dismissed.

## Data extraction

Body landmarks from surface scans with the harness donned (A scans) were sometimes hidden by the placement of the harness. Alignment of the A and C scans allowed the hidden landmarks in A scans to be copied from C scans so that the evaluations of the interface between the harness components and the body were possible. One of the harness-fit tests was to determine whether the body's COG fell behind the harness hip rings during suspension. To assign the COG within the suspended conditions (B scans), an alignment of A and B scans was performed based on crotch and right and left anterior superior iliac spines.

## Results

A multinomial logistic regression with forward hierarchical approach was used to search for the most influential factors on sizes. The dependent variable was the best-fit size of the harness, and potential independent variables included gender, height, weight, traditional anthropometry, and torso bounding box data. Only those participants who received a "pass" fit rating were used to construct the new sizing chart. These participants were randomly assigned to either the estimation or validation subset of data. The estimation subset was used to create a harness-sizing-classification model while the validation subset gave a post estimate (i.e., validation) of the percentage of the user population that could be fit correctly using the estimation model. Upper and lower torso box breadth and depth, gender, body height, and weight were found to be most influential. Four logistic regression models were derived to predict the best fit size. The "Standard Size (STD)" was set as the reference size and the resulting probability factor fell between 0 and 1, with 0.51 or above as a measure of favor of tested size over the reference size. The probability of wearing small size (XSM) instead of STD for males is $1/(1 + \exp(-681.075 + 3.894 * \text{BreadthUp} - 3.66 * \text{DepthLow} + 3.164 * \text{DepthUp} + 1.854 * \text{Stature} + 2.441 * \text{Weight} + 2.001 * \text{BreadthLow}))$ … [Equation 1]. For females, the equation is: $1/(1 + \exp(-657.348 + 3.894 * \text{BreadthUp} - 3.66 * \text{DepthLow} + 3.164 * \text{DepthUp} + 1.854 * \text{Stature} + 2.441 * \text{Weight} + 2.001 * \text{BreadthLow}))$ … [Equation 2]. The probability equation of wearing XLG instead of STD for males is: $1/(1 + \exp(79.220 - 0.386 * \text{BreadthUp} - 2.464 * \text{DepthLow} + 1.77 * \text{DepthUp} + 0.035 * \text{Stature} - 0.409 * \text{Weight} - 0.038 * \text{BreadthLow}))$ … [Equation 3]; and for females the equation is $1/(1 + \exp(81.315 - 0.386 * \text{BreadthUp} - 2.464 * \text{DepthLow} + 1.77 * \text{DepthUp} + 0.035 * \text{Stature} - 0.409 * \text{Weight} - 0.038 * \text{BreadthLow}))$ … [Equation 4]. These models accurately classified 96% of the validation subset and thus provided unique information for harness designers to restructure current harness sizing plans to accommodate an estimated 96% of the population. For practical purposes, any size selection chart must be based on dimensions that are readily known or easily measured by the consumer. To explore the possibility that there could be simpler models for harness end-users, only gender, weight, and stature were involved in generating logistic regression models. These models accurately classified 93% of the classification subset and 96% of the validation subset. The simplified models are illustrated in two charts in Figure 2.

With the new sizing scheme, harness manufacturers would be interested in knowing the adjustment range of each harness component for harness redesign use (e.g., in determining the cut length of each harness component). There are several ways to determine the adjustment range. First, all the participants (including those in the FAIL group of harness fit test) can be classified into their predicted best-fit sizes, using the prediction equations 1–4. Using descriptive statistics, the adjustment range of a harness component for each harness size for each gender can be determined. For instance, 66 female participants were predicted to use a Female-XSM harness. Their Front Strap Length measurements have a population-weighted mean of 57.5 cm with a standard deviation of 2.9 cm. The adjustment range of the front strap for 98% of the population in this body-height-weight category can be estimated as 50.9 cm to 64.2 cm. The second approach would also classify all the participants into predicted best-fit size groups, using equations 1–4. A principal component analysis (PCA) (Meindl et al., 1993) would identify representative body models that consider multi-harness component dimensions at the same time. This approach would accommodate individuals with unique ratios among torso dimensions, such as a short torso with large belly depth as well as a long torso with wide chest width and median belly depth. Therefore, the adjustment

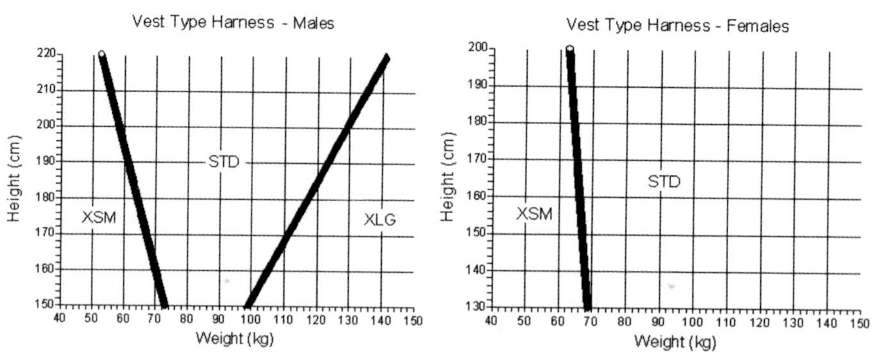


**Figure 2. Proposed sizing charts for males (left) and females (right) for vest type harnesses; the possibility for males to use a SXL size and females to use a XLG or SXL size is remote.**

range for each harness component is likely to be slightly smaller and more close to the reality than that computed from the first approach.

## Conclusion

Updated harness sizing systems are needed to accommodate diverse workforces and new roles for women in the construction industry. This study developed a logistic regression model that correctly classified 96% of the participants to their best fit sizes, i.e., 2 sizes for females and 3 sizes for males. The model is a combination of weight, height, gender, and 3D information for upper and lower torso regions. The new sizing charts were graphed by gender, body weight and body height for harness manufacturers to consider for redesigning harnesses as well as for harness users to consider for selecting an optimal size of harness.

## Acknowledgments

The authors gratefully acknowledge the support of Jennifer Whitestone, Chris Lafferty, Joseph Feldstein, Paul Keane, and Dr. Al Amendola.

## References

Bodzsar E. B. 2000, A review of Hungarian studies on growth and physique of children, *Acta Biological Szegediensis*, **44**(1–4):139–153.

Bradtmiller B., Whitestone J., Feldstein J., Hsaio H., Snyder K. 2000, Improving fall protection harness safety: contributions of 3-D scanning, *Proceedings of Scanning 2000: The European Meeting Point for Scanning* (Paris, France: Numerisation 3D), 117–128.

HHS [U.S. Department of Health and Human Services] 2001, The surgeon general's call to action to prevent and decrease overweight and obesity, U.S. Government Printing Office: Washington D.C., VI–VII.

Meindl R. S., Hudson J. A., Zehner G. F. 1993, *A multivariate anthropometric method for crew station design* (AL-TR-1993-0054), U.S. Air Force Armstrong Laboratory, Wright-Patterson AFB, Ohio.

# DID 16TH CENTURY PAGODA BUILDERS BASE THEIR DESIGNS ON HUMAN NECK ARCHITECTURE?§


**Bernard Masters**

*College of Chiropractors, 106 London Street, Reading, Berkshire RG1 4SJ*



Ancient Japanese pagodas have withstood earthquakes for centuries due to their unusual architecture. The monocoque stories are damped by a central pole and oscillate individually, inhibiting forced resonance, and energy is dissipated throughout the building. The pagodas are seen to mimic cervical vertebral movement (evidence from cine-x-radiographs), and the "dens" of the atlanto-axial joint is similar in shape to the central peg mechanism (shinbashira). The human vertebrae have small raised joints (unco-vertebral joints) abutting the disc, these translate lateral movement into pendular displacement similar to the action of the pagoda floors – *a chain of interconnected pendulums*. Did the ancient Japanese architects seek inspiration from human neck bones? Do these design features give an insight into acceleration/deceleration injuries?


## Introduction

Ward, the anatomist, observed in 1838 that the upper femoral architecture of the hip joint looked like a shelf bracket (Trueta, 1968, a), and in 1886, the engineer Culmann noted that the femur looked like a crane he was designing (Thompson, 1917, a). From these observations, anatomist Meyer and Culmann theorized stress loading patterns in bone (Trueta, b). Goodwin (1994) explored shape changes from complex non-linear constraints and Vogel (1988) explored functional anatomy in terms of fabricated structures.

Gravity is part of the complex fields of force that influence the form of organisms (Thompson, 1917, b) and architectural design. Tall structures, subject to ground movements, are limited by column mechanics (Gordon, 1988) but their activities can be mathematically modeled as "lumped mass" (Boswell and d'Mello, 1993). A build-up of moderate vibration can produce excessive loading that can destroy structures – as in the Tacoma Narrows Bridge disaster (Hopkins, 1970).

The human frame can also be viewed as an upright beam where balance to keep the eyes level is paramount and where the vertebrae are an integral "engine" in a complex locomotive system (Gracovetsky, 1988). Tall structures are inherently unstable and are stiffened for safety. However, pagoda design and the human spine appear to counteract gravitational forces by absorbing unwanted vibrations in unexpected ways.

## The pagoda design

Pagodas of 100 metres in height have successfully withstood earthquakes for hundreds of years (Ueda, 1997). A fabricated wooden central pole, like a ship's mast, is driven deep into ground and extends upwards through the building. Around the pole, modular


§ My grateful thanks to Alice Roberts, Department of Anatomy, Bristol Medical School, for her assistance.

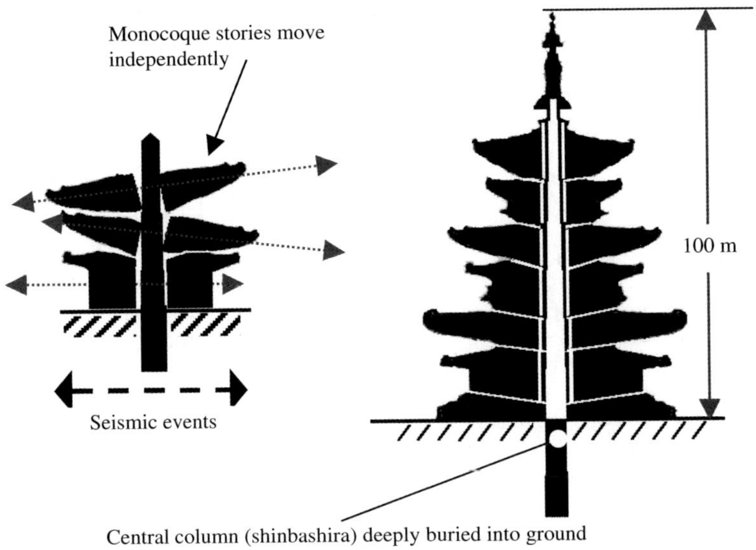


**Figure 1. The principle of the earthquake resistant pagoda (after Ueda, 1997). The floors are free to move but are arrested by the central column. The different shape of the floors may also dissipate resonance.**

monocoque stories are piled up on one another so that in a seismic event the floors can move independently, but movement is arrested by the central column, counteracting the effect of ground disturbances. Some of the floors have extended roof structures and Isheda (1998) observed that these extensions may effectively lower the centre of gravity like the tightrope walker's pole, thus increasing stability (see fig. 1). Non-uniform floors may also dissipate movement in the same way as tuned mass dampers (TMD), by moving out of phase with the ground movements and the other floors.

## Behaviour of the spine as a tall structure

All structures have their own natural frequency of vibration, (Burton, 1958) and are subject to physical laws (McNeill Alexander, 1983). The effects of harmonics on tall axial modal structures can produce unpredicted resonance as in tall ship mast failure (David, 1995). To impede destructive resonance, anti-vibrational devices can absorb the unwanted energy. Masters & Sugiyama (1980) employed stop-frame analysis cine-radiography of human neck movement and reported erratic vertebral motion or "ragged movement" by individual vertebra driven by spinal muscles – sometimes going in the opposite direction to the overall spinal movement. The phenomenon was again reported by Breen (1988) tracking the lumbar vertebral movements. Moroney (1992) described multiple centers of movement for each neck vertebrae. The human frame is a particularly flexible system and is constantly undergoing postural changes to maintain balance (Masters, 1999) not unlike the pagodas response to ground disturbance. The pagoda floors and the vertebral bodies in the neck may both behave like a line of vertical pendulums (see fig. 2) resisting external forces (Gordon, 1991). The unco-vertebral joints at the side of the discs may arrest movement by chaotic out-of-phase activity (see figs 1 & 2) acting as "un-tuned" mass dampers.

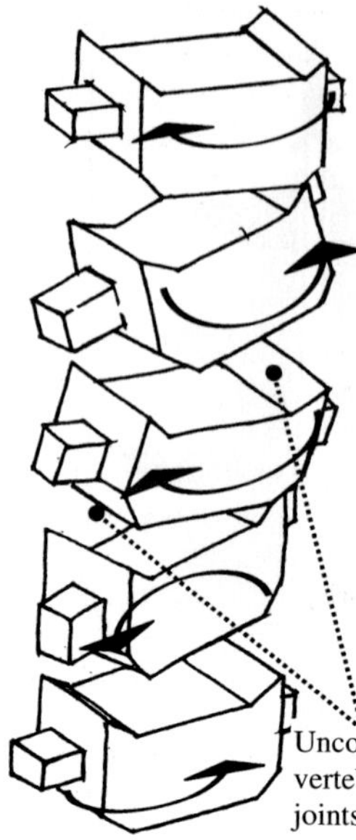


**Figure 2. Neck bones as a chain of pendulums. Lipped unco-vertebral joints restrain the lateral movement of the vertebrae (adapted from Kapandji, 1974).**

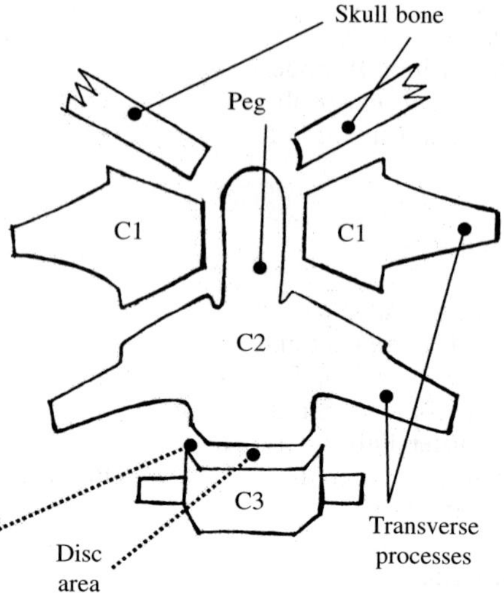


**Figure 3. Peg of C2 resembles the shinbashira restraining C1, likened to monocoque floors (adapted from Kapandji, 1974 p. 185) – see fig. 1.**

## Vibration resistant structures and Buckminster-Fuller tensegrity masts

Figure 3 shows a transverse section of the skull sitting on the ring of bone (C1), pierced by a peg-like structure of C2 (Kapandji, 1974). The peg of C2 resembles the shinbashira or central column inhibiting movement of C1. While C3 to C7 share the a similar shape and range of movement they have no shinbashira damping but rely on an interrelationship between vertebrae similar to that of a tensegrity mast (Buckminster-Fuller, 1975) where the loading on the column is counteracted by tensional forces, not compression, where spinal muscles hold individual vertebrae in juxtaposed appropinquity.

## Conclusion

These concepts threaten our image of the spinal column as *a stack of blocks subject only to compression* (Robbie, 1977). Combining Kapandji's models (see figs 2 & 3) with Masters, & Sugiyama's, Breen's and Moroney's cine-radiographic studies may give an insight into the complex activity of the spine where the head is held in the Frankfort plane keeping the eyes and ears level. The constant "irregular patterned" motion of the vertebrae would prevent forced resonance or unwanted behaviour in the interconnected chain of vertebral pendulums. Study of the harmonic movements of tall structures may give an understanding of human spinal behaviour in acceleration/deceleration injuries.

Where *form follows function and function follows form*, all mechanisms are subject to the constraints of physics, and the study of fabricated mechanisms can lead us to a better understanding of nature's solutions. Cervical vertebrae appear to parallel the pagoda's extended roofs as an anti-vibrational device and the peg of axis C2 mimics the shinbashira. If modern architects and engineers are returning to nature for inspiration, it seems reasonable to assume that the 16th Century pagoda builders could also have sought inspiration from the human spine.

## References


Boswell, F. and d'Mello, C. 1993, *The dynamics of structural systems* (Blackwell Scientific Publications, London), 4

Breen, A.C., Allen, R. and Morris, A. 1989, Spine kinematics: a digital video-fluoroscopic technique, *Journal of Biomedical Engineering*, **11**, May, 224–228

Buckminster Fuller, R. and Applewhite, E.J. 1975, *Synergetics, Explorations in the Geometry of Thinking* (Macmillan Publishing, New York)

Burton, R. 1958, *Vibration and Impact* (Addison Wesley, Reading, Massachusetts), 10

David, A. 1995, in Murrant J., *Boating Bible: An Essential Handbook* (Collins/Angus, Australia)

Goodwin, B. 1994, *How the Leopard Changed Its Spots* (Phoenix Giants, London), 134

Gordon, J.E. 1998, *Science of structure and materials* (Scientific American Library), 43

Gordon, J.E. 1991, *Structures* (Penguin Books, London), 229

Gracovetsky, S. 1988, *The Spinal Engine* (Springer-Verlag Wein, New York), 7

Hopkins, H.J. 1970, *A Span of Bridges* (David & Charles, Newton Abbot), 231

Isheda, S. 1998, Earthquake resistant buildings, *Tomorrow's World*, BBC TV

Kapandji, I.A. 1974, *The Physiology of the Joints* (Churchill Livingstone, Edinburgh)

Masters, B. and Sugiyama, I. 1980, A cineradiographic study of the kinetic relationship between cervical vertebrae, *Bulletin of the European Chiropractic Union*, **28**, no.1, 11–33

Masters, B. 1999, *An examination of the neuro-musculo-skeletal health of the of the horsewoman in relation to saddle design*, PhD Thesis (University of Exeter, Devon), 65

McNeill Alexander, R. 1983, *Animal Mechanics* (Blackwell, Oxford)

Moroney, S.P. 1992, Biomechanics of the Cervical Spine, in Scott Haldeman (ed.) *Principles and Practice of Chiropractic* (Appleton & Lange, Connecticut), 111

Thompson, D'Arcy, W. 1917 & 1992, *On Growth and Form*, this edition 1992 (Canto, Cambridge University Press), (a) 323 (b) 288

Robbie, D.L. 1977, Tensional Forces in the Human Body, *Orthopaedic Review*, **VI**, No. 11, November, 45–48

Trueta, J. 1968, *Studies of the Development and Decay of the Human Frame* (Heinemann, London), (a) 37 (b) 36

Ueda, A. 1997, Why Five story-storied Pagodas do not fall down, http://www.kansai.gr.jp/culture_e/build/archi/balance.htm

Vogel, S. 1988, *Life's Devices, The physical world of animals and plants* (Princeton University Press, Princeton), 269

# THE EFFECT OF HANDLE ANGLE ON FENCING SPORT

**Chih-Lon Lin[1] & Chih-Lin Chang[2]**

[1]*National Kaohsiung Marine University, Kaohsiung, Taiwan,*
[2]*Hsiuping Institute of Technology, Taichung, Taiwan*

A suitable blade angle could not only reduce ulna deviation to avoid wrist injury, but can also enhance the sports performance of fencing. Improperly designed hand tools devices have many undesired injuries and accidents. This study found that angle of 18 and 21 provides the best performance in fencing. The results are useful for the man-machine design of fencing and other sports equipments. We suggest that academic circles and ergonomics specialist have to pay more attention to the growing trend in sports industry – especially in sports performance and sports injury.

## Introduction

Practiced skill and tactics could improve the athlete's multiple strategy to win the game. Strict training is necessary to accomplish successfully use of this tactics. The main requirement for training includes muscular strength, movement velocity, durability, body softness, and sensitivity, especially in the arms, legs and waist (Fencing, 1996). There are four main tactics of fencing that including Active-attacks, simultaneous-attacks, parry-ripostes, and counter-attacks. Chang & Lin (2001a) indicated that offensive defense strategy includes parry-ripostes and counter-attacks. The parry-ripostes is the basis of defense tactics, thus no difference on hit performance between winner and loser. Chang & Lin also found that the key point for winning the game is using counter-attack tactics properly. Since tactics is the key factor to victory and skill is the basis to perform tactics, coaches frequently make change skill training to accommodate changes in tactics.

If coaches ignore the athlete's physiological characteristics, exercise excessively or adopt incorrect training methods that go against the principles of ergonomics during the training period, the athlete may suffer serious injuries (Nigg, Denoth & Neukomm, 1981. Fink (1993) indicated the main regions where a fencer frequently suffers injuries are the joints of the ankles, knees, and hips. Injuries affect the effect of training seriously, thus the athlete can't make continuous progress in skill.

Indeed, many sports tools are not designed for the efficient and safe operation by humans – especially in repetitive activity. Improper designed hand tools and devices have many undesired consequences, including accidents and injuries (Sanders & McCormick, 1993). For example, Aghazadeh and Mital (1987) indicated that there were over 260,000 hand tools-related injuries in the United States each year and that the associated medical costs alone come to some 400 million dollars. The human factors study of fencing is necessary to prevent these sports injuries and to enhance sports performances (Shi, 1991).

Today, many hand tools are crafted for uncounted specific applications as well as for general-purpose activities (Sanders & McCormick, 1993), thus man–machine interface design of fencing is also very important and a well-designed interface that should be considered (Shi, 1991). For example, Lin & Chang (2001a, 2001b) found that a suitable downward blade angle could contribute to the accomplishment of effective hit, accuracy and subjective satisfaction rating. A suitable downward angle could also reduce ulna deviation and prevent wrist injury. Lin & Chang (2001a) conducted an experiment of blade angle from 3 to 15, who found that 9 downward blade angle had the highest performance on the effective hit and accuracy, the next is the 15 blade, but the 15 blade had the highest subjective satisfaction rating also.

The idea of bending the tool, not the wrist, has been applied to other things by John Bennett (Emanuel, Mills & Bennett, 1980), who patented the idea of bent handles (19 ± 5) for all tools and sports equipment. According to Bennett's idea, the experimental design of ideal blade angle should be in the range of 14 to 24 degree. Thus the larger blade angle is, the more comfortable it is to the wrist and fewer injuries would occur.

The main purpose of this study is to know the fencing performance with different blade angle in the range of 14 to 24 degrees. We conducted an experiment to analyze the sports performance with several different blade angular magnitudes and use the foil (a type of blade with attack-right) as experimental sword.

## Experimental design

An experiment with several different blade angles is conducted to analyze the effect of fencing performance. The Subjects are eight male fencers representing the university fencing team aged 22 ± 3.

### *Experimental variable*

The independent variable is downward blade angle that includes 5 levels, they are 9, 15, 18, 21 and 24. The dependent variables are effective hit, accuracy and subjective satisfaction rating. The analysis of variance (ANOVA) method was adopted for inference statistics.

### *Experimental equipment*

Eight France Lames sword for international competition, and a target for shooting, and a fencing outfit with a 20 × 16 cm copper induced current metal.

### *Experimental action*

"Attack hit action". All attack actions are the same as Chang's experiment (Chang and Lin, 2001b). The distance between coach and subject has to be kept constant. The coach was always on the initiative position. The subject has to follow the coaches commands for required actions, such as 6th or 4th counter parries, 8th or 2nd fling hits. The subjects' replacements always keep the same distance to the coach by advancing or by retreating step. The fling hit may use lunges or retreating steps that depends on the distance from the coach. Each subject has 10 fling hits for each handle angle, for a total of 50 hits per subject in this experiment.

"Accuracy hit action". To avoid experimental errors to occur as in Lin & Chang's experiment of 3-points target (Lin & Chang, 2001), this study uses 5-point targets ranged from 5 points to 1 point. The target circle diameter is 3 cm, 5 cm, 7 cm, 9 cm and 11 cm, respectively. Any hit outside the area or non-touch hitting is considered as 0 point. The distance from the start line to the target is the length of a straight-arm with a weapon plus the length of two steps in retreat. Each subject has 10 trials to lunge with each handle angle, for a total of 50 hits per subject in this experiment.

## Results and discussions

### *Analysis of the effective hit*

There are five types of handle angles; each angle had ten fling-hit trials, a total of 50 fling-hits for each subject. Subjects hit scores are shown in table 1. The five blade angles' average score were 5.88, 7.63, 7.75, 9.38 and 8.50, respectively. The effective hit scores were significantly affected by using different handle angles ($p < 0.01$), see table 2. The result of Scheffe's multiple comparisons of post-hoc tests indicated that subset 3, include 18, 21 and 24 downward angles, have the highest hit score (table 3). But there is no significant difference on subjective satisfaction rating.

### *Analysis of accuracy*

The five types of handle angles are the same as with the fling hit and each angle had ten lunge trials for a total of 50 lunges-hit for each subject. The subjects' scores are shown in table 4, the five angle handle's average scores are 28, 33.5, 33.25, 32.75 and 29.63, respectively. There is no significant difference on subjective satisfaction rating. But accuracy is significantly affected by using the different handle angles ($p < 0.001$), see

**Table 1. ANOVA table of effective hit.**

| Dependent var. | Effective hit | SS | dof | MS | F | P |
|---|---|---|---|---|---|---|
| Model | | 53.650 | 4 | 13.41 | 11.69 | .000* |
| Error | | 40.125 | 35 | 1.14 | | |
| Corrected total | | 93.775 | 39 | | | |

* p 0.05.

**Table 2. Scheffe's multiple comparisons of post hoc tests on effective hit.**

| Downward angle degree | N | Alpha .05 subset | | |
|---|---|---|---|---|
| | | 1 | 2 | 3 |
| 9 | 8 | 5.88 | | |
| 15 | 8 | | 7.63 | |
| 18 | 8 | | 7.75 | 7.75 |
| 24 | 8 | | 8.50 | 8.50 |
| 21 | 8 | | | 9.38 |

**Table 3. Accuracy score of each subject.**

| Downward angle degree | Subject no. | | | | | | | | | |
|---|---|---|---|---|---|---|---|---|---|---|
| | A | B | C | D | E | F | G | H | Average | S.D. |
| 9 | 26 | 28 | 33 | 34 | 26 | 19 | 30 | 28 | 28.00 | 4.69 |
| 15 | 34 | 27 | 34 | 37 | 34 | 31 | 38 | 33 | 33.50 | 3.42 |
| 18 | 36 | 29 | 26 | 28 | 38 | 34 | 41 | 34 | 33.25 | 5.20 |
| 21 | 23 | 32 | 34 | 42 | 35 | 27 | 39 | 30 | 32.75 | 6.18 |
| 24 | 36 | 29 | 33 | 36 | 33 | 18 | 31 | 21 | 29.63 | 7.42 |

**Table 4. ANOVA table of accuracy.**

| Dependant var. | Accuracy hit | SS | dof | MS | F | P |
|---|---|---|---|---|---|---|
| Model | | 564.90 | 4 | 141.225 | 6.445 | .001* |
| Error | | 766.87 | 35 | 21.911 | | |
| Corrected total | | 1331.77 | 39 | | | |

* p 0.05.

**Table 5. Scheffe's multiple comparisons of post hoc tests on accuracy.**

| Downward angle degree | N | Alpha .05 subset 1 | 2 | 3 |
|---|---|---|---|---|
| 9 | 8 | 26.75 | | |
| 24 | 8 | 28.00 | 28.00 | |
| 15 | 8 | 33.50 | 33.50 | 33.50 |
| 21 | 8 | | 34.63 | 34.63 |
| 18 | 8 | | | 36.25 |

table 5. The result of Scheffe's multiple comparisons of post-hoc tests indicated that subset 3, include 15, 18 and 21 downward angles, had the highest accuracy (table 6)

*Experiment summary*

The important finding of this experiment is that 18 and 21 downward angle has the best overall performance. Because of the 18 and 21 both downward angles have the highest effective hits and accuracy with the same rating of subjective satisfaction.

## Conclusions

This study found that the range of handle angle are 19.5 ± 1.5, 18 and 21 and the best range in overall performance of fencing sport. This finding is same as Bennett's idea of bending handles (19 ± 5) for all tools and sports equipment, not the wrist (Emanuel, Mills & Bennett, 1980). But the range is more narrow and smaller than in Bennett's finding. The reason may be that fencing has highly dynamic movement and it requires extreme accuracy (Cavanagh & Lafortune, 1980), so such an optimal angle range should be smaller. Finally, this study suggests that blade angle design should be considered with the ergonomics theory. Because a suitable interface design of fencing could not only reduce ulna deviation to avoid wrist injuries, but can also enhance sports performances, including accuracy and the effective hit.

## References

Aghazadeh, F. and Mital, A. 1987. Injuries due to hand tools. *Applied Ergonomics*, **18**(4): 273–278

Cavanagh, P. and Lafortune, M. 1980. Ground reaction forces in distance running. *Journal of Biomechanics*, **13**: 397–406

Chang, C.L. and Lin, F.T. 2001. The analysis of injuries on fencing. *Journal of Hsiu-Ping*, Taiwan, **7**(3): 239–250

Chang, C.L. and Lin, F.T. 2001a. Statistical analysis of the influence on fencing tactics in competition. *Journal of National Taiwan College of physical education*, **8**: 239–251

Chang, C.L. and Lin, F.T. 2001b. Effects of foil downward angle on fling hit. *Journal of National Taiwan College of physical education*, **9**: 591–607

Emanuel, J., Mills, S., and Benett, J. 1980. In search of a better handle. *Proceedings of the Symposium: Human Factors and Industrial Design in Consumer Products*. Medford, MA: Tufts University

*Fencing*, 1996. Fencing in Beijing College of physical education, Beijing, Mainland Chain

Fink, P. 1993. *Force in the forward leg during a fencing lunge*. Unpublished doctoral thesis of Purdue University

Lin, F.T. and Chang, C.L. 2001a. An application for research of human factors on foil blade angle. *Occupational Ergonomics*, pp. 351–354

Lin, F.T. and Chang, C.L. 2001b. Human factors design and application of foil blade angle. *Proceeding of 21st FISU Congress*, Beijing, pp. 211–212

Nigg, B.M., Denoth, J. and Neukomm, P. A. 1981. Quantifying the load on the human body: Problems and some possible solutions. In R.C. Nelson (Eds.), Biomechanics VIB, Baltimore: University Park Press, pp. 88–99

Sanders M.S. and McCormick E.J. 1993. *Human factors in engineering and design*. Singapore Mcgraw-Hill, pp. 383–409

Shi, S.K. 1991. Human Factors and sports. *Journal of culture and physical education*, Taipei, **18**: 48–51

# BAGGAGE HANDLING IN NARROW-BODIED AIRCRAFT: IDENTIFICATION AND ASSESSMENT OF MUSCULOSKELETAL INJURY RISK FACTORS

**Sarah Tapley & David Riley**

*Health & Safety Executive, Ergonomcs Team, Magdalen House, Trinity Road, Bootle L20 3QZ*

The manual loading and unloading of passenger baggage in narrow-bodied aircraft is a strenuous manual handling operation. It is often subject to tight time constraints, involves working in the confined space of the baggage holds in awkward postures, and handling loads of unknown weight, up to 32 kg. To date there appears to have been little in the way of a formal identification and assessment of the musculoskeletal injury risk factors. This paper reviews the relevant literature and reports the findings from a series of visits observing the apron baggage handling operations for Boeing 737 series aircraft at two UK airports. Recommendations are made for possible measures to reduce the risk of musculoskeletal injury to the baggage handling crews.

## Introduction

Musculoskeletal Disorders (MSDs) are responsible for 40% of all reported incidents at airports, of this figure the majority are reported by ground handlers. The loading and unloading of baggage in narrow-bodied passenger aircraft has been of concern to the air industry, HSE and other European enforcement agencies for many years.

There are a variety of methods used to load and unload these aircraft. However there has been little evidence that risk assessments for the tasks have considered all the relevant risk factors. Many assessments have concentrated on the aircraft hold sill height to decide on whether mechanical assistance should be provided.

## Project aim and methodology

This study aimed to identify the MSD risk factors present during the manual loading and unloading of 737 series aircraft. This will provide information to ensure best working practices to reduce the risk of manual handling injuries to staff when loading and unloading to a level as low as reasonably practicable.

An observation based approach was used to gather data using video, as well as measurements and interviews with staff during fieldwork carried out at 2 UK airports, observing complete turnarounds (off-load and on-load) of 737 series aircraft.

## The apron baggage handling operation

The baggage handling operation is conducted by teams of 4 or 5 workers. Usually two operate inside the hold and 2 or 3 operate outside on the apron. There are important differences between on- and off-loading. For on-loading, 2 or 3 workers unload baggage from carts into the hold doorway, either direct or via powered conveyor, to two workers in the hold (in reduced headroom ~1165 mm), usually knelt or stood stooped. These two work "in series", one near the hold door passes baggage deeper into the hold for the other worker to stack. For off-loading, the two hold workers pass baggage out to their co-workers either direct, or via a powered conveyor onto flatbed trucks. A team will typically deal with 5 turn-arounds per shift, each lasting 25 minutes, handling up to around 150 baggage items in and out for each. The size and shape of baggage is variable and may be awkward, for example bicycle boxes, golf clubs, skis and pushchairs. Carts and conveyors may be moved manually requiring forces in excess of 400 N. The workers usually rotate between tasks. Baggage item weight is difficult to quantify (not routinely recorded) but the average item weight is probably around 15–20 kg.

## Literature review

The MSD injury risk for baggage handlers working in the holds of narrow-bodied aircraft has been recognised as a significant problem worldwide for 20 years. Annual prevalence rates for low back pain and knee pain are high compared with industrial workers as well heavy manual occupations such as Lumberjacks, Construction Workers and Railway Maintenance Workers (Andersson et al., 1984, Foundation for Occupational and Environmental Medical Research and Development, 1989, Stålhammar et al., 1986, Dell, 1998, Dickinson, 1998, Riley, 2004).

Physical activities associated with an increase risk of back disorder are: heavy physical work; lifting and handling of loads; and awkward postures, e.g. bending and twisting (De Beck and Hermans, 2000).

There is a body of research that has investigated the demands of lifting in reduced headroom, much in relation to mining. However, the relative risks associated with kneeling versus standing severely stooped do not appear clear. In particular the findings of Ridd (1985) are at odds with those of Gallagher and co-workers: Ridd reports a greater loss of lifting capacity associated with standing stooped than when kneeling; Gallagher et al. (1994) found that kneeling produced a higher peak compression force on the spine than stooping, indicating that handling 25 kg stooped was equivalent to handling 15 kg kneeling. Graveling et al. (2003) estimate that this means that kneeling may present 22% greater risk than stooping, and generally advocate standing stooped rather than kneeling.

Overall, kneeling to perform a manual lifting operation must be viewed as presenting a greater risk of injury than when standing upright. It compromises stability, and restricts movement. Unfortunately, in the context of baggage handling, and in terms of risk of injury, the distinction between kneeling and standing stooped is less clear.

From observation of baggage handlers standing stooped in the hold, it is considered that one of the principles for good manual handling (Graveling et al., 2003), that of "Keep the load close to your waist" (with the aim of minimising the horizontal distance between the load and the low back) may be impractical for the workers to adopt, and in this context, the degree of departure from this principle is greater when stood stooped than when kneeling. The differences between the two in-hold tasks are such that stooping for the stacking/unstacking worker would present a greater risk and would also be impractical due to the often further reduced headroom for this task, while it is a viable alternative for the door/transfer worker. When handling while stood stooped the upper arm is held almost vertical, and the moment at the shoulder will be low. When kneeling, the upper arm is used in a position away from the vertical, and moments at the shoulder will consequently be increased (Gallagher and Unger, 1990).

Generally, the Maximum Acceptable Weight of Lift when kneeling has been shown to decrease in comparison to stooped and standing postures. A reduction of around

15% is considered to represent a reasonable consensus from the literature (Boocock, 1997), and this was incorporated in our assessments for kneeling work.

## The main high risk aspects identified

### *The load*

The weight varies up to 32 kg, often unlabelled, has poor handles and awkward shapes.

### *The individual*

Body size differences.
Time pressure to complete within 25-minute turn-around between flights.

### *The environment*

18 hour per day operation, all year round. Handlers are exposed to weather conditions, noise from aircraft, tight space constraints within holds and inconsistent lighting conditions.

### *The task*

External – On-load:

- Items are handled at a rate of up to one every 6 seconds, the risk of injury ranges from low if the item is below 10 kg to very high if the item is above 26 kg.
- Items are moved outside acceptable parameters if loaded directly into the hold, hands are above shoulder level for every transfer.
- Items are handled within acceptable parameters if a conveyor is used to transfer the items from cart to hold.

External – Off-load:

- Handlers stand on the ground between the aircraft and truck – handling between over shoulder level and waist level.
- Handlers stand on the flatbed – the handler stands stooped due to reduced headroom in the hold doorway.
- For both the above, the co-worker on the flatbed carries items from the back and stacks them at the front, involving stooping to pick items from below knee level.
- Some handlers use a conveyor to present items at a suitable level while stood on the flatbed. Items were carried and stacked.

Internal – For both on-load and off-load operations:

- Handlers were observed to regularly be at the extremes of reach when grasping and placing items.
- Handlers were handling in the zone between knuckle level (at floor level) and head level (hold ceiling height).
- Handlers typically handled from one side of the body to the other. A 45-degree twist to either side is typical, greater twisting was observed. Twist and sideways bending was observed to be routine.

Internal off-load:

- The handling frequency was observed to be higher than for on-load, at slightly faster than one item every 4 seconds.
- Items were pulled from the stack, although some were lifted/lowered.

**Table 1. Key risk areas.**

| Internal | External |
|---|---|
| Very high handling frequency. | Loading/unloading directly to/from hold: |
| Space constraints: | – Handling above shoulder level. |
| – Kneeling and stooping. | – Handling below knee level. |
| Reaching, twisting, sideways bending. | Pushing/pulling carts and equipment. |

**Table 2. Specific risk reduction measures.**

| Internal | External |
|---|---|
| Avoid squatting and prolonged kneeling | Avoid direct-to-hold loading/unloading. |
| Avoid standing stooped for stack-ing/unstacking. | Use powered conveyors to unload to flat bed trucks. |
| Alternate between kneeling and sitting according to stack level. | Use powered conveyors to load from cart to hold. |
| Alternate between standing stooped, kneeling and sitting for the hold door to stacker transfer task. | Use tugs to move carts and equipment. |
| Use both hands to lifting | |
| If lifting with one hand, use other hand for additional support. | |

**Table 3. General risk reduction measures.**

| |
|---|
| Reduce handling rate. |
| Reduce baggage weights. |
| Baggage weight labelling. |
| Task specific training for handlers and check-in staff. |
| Passenger education about weight and size of luggage at ticket point of sale. |
| Formal job rotation to control exposure to the highest risk operations. |

Internal on-load:

- The handling frequency was slower at around one item per 5 to 6 seconds.
- Handlers lifted and placed items into the stack.

## Summary

There has been relatively little research specifically concerning the apron based aspects of baggage handling for narrow-bodied jets, though handling whilst inside the hold and in limited headroom has received attention. What literature there is, and our observations, indicate that this is a high-risk activity. There are options available to reduce the risks, however, these need to be adopted throughout the industry to be fully effective, by reviewing, researching and developing alternative methods, including the design of vehicles and equipment.

## References

Andersson, G., Biering-Sørensen, F., Hermansen, L., Jonsson, B., Jørgensen, K., Kilbom, Å., Kuorinka, I., and Vinterberg, H. (1984) Nordiska frägeformulär för kartläggning av yrkesrelaterade muskuloskeletala besvär. *Nordisk Medicine*, **99**, 54–55.

Foundation for Occupational and Environmental Medical Research and Development, Orebro (1989) Reference data for the YMK form. FHV 002D 003D 007D. FHV version March 1985. Clinic of Occupational Medicine. Regional Hospital, 701 85 Orebro. HSE Translation 89/90.

Stålhammar, H.R., Leskinen, T.P.J., Kourinka, I.A.A., Gautreau, M.H.J., and Troup, J.D.G. (1986) Postural, epidemiological and biomechanical analysis of luggage handling in an aircraft luggage compartment. *Applied Ergonomics*, **17.3**, 177–183.

Dickinson, C.E. (1998) interpreting the extent of musculoskeletal disorders. In Hanson M.A. (Ed) Contemporary Ergonomics 1998, pages 36–40 London: Taylor and Francis.

Riley, D. (2004) Manual Handling in the Rail Sector in South Wales. HSL Internal Report ERG/04/10.

Boocock, M. (1997) Manual materials in a kneeling posture, HSE Internal Report (EWP/97/16).

De Beck, O. and Hermans, V. (2000) Research on work-related low back disorders. Bulbar: European Agency for Safety and Health at Work.

Dell, G. (1998) Airline baggage handler back injury: a survey of baggage handler opinion on causes and prevention. Safety Science Monitor. Volume 2, Issue 2, Article 6.

Ridd J.E. (1985) Spatial restraints and intra-abdominal pressure. Ergonomics; **28**, 149–166.

Gallagher, S., Hamrick, C., Love, A.C., and Marras, W.S., (1994) Dynamic biomechanical modelling of symmetric and asymmetric lifting tasks in restricted postures. Ergonomics; **37**, 1289–1310.

Gallagher, S. and Unger R.L. (1990) Lifting in four restricted lifting conditions: psychophysical, physiological and biomechanical effects of lifting in stooped and kneeling postures, *Applied Ergonomics*, **21**, 237–245.

Graveling, R.A., Melrose, A.S., and Hanson, M.A., (2003) The principles of good manual handling: Achieving a consensus. Prepared by the Institute of Occupational Medicine for the Health and Safety Executive RESEARCH REPORT 097.

# ERGONOMIC INTERVENTION ON MUSCULOSKELETAL DISCOMFORT AMONG CRANE OPERATORS AT WASTE-TO-ENERGY-PLANTS

**Dirk M. Ditchen, Rolf P. Ellegast, Christian A. Herda & Ulrike M. Hoehne-Hueckstaedt**

*Berufsgenossenschaftliches Institut für Arbeitsschutz (BGIA), Alte Heerstr. 111, 53754 Sankt Augustin, Germany*

The paper presents an investigation of crane operator workplaces in the German waste management in regard to musculoskeletal discomfort. As crane operators complain about pains in the neck and shoulders, their working postures have been checked using a person-centred measuring system called CUELA. As a result of these measurements the factors for the complaints were identified and possible improvements for the redesigning of the workplace have been recommended. After several design modifications of the workplaces – especially the crane operators' seats – further measurements have been carried out to evaluate the impact of the reconstruction. The results of both measurements have been assessed separately and finally compared with each other. The main findings of this investigation are outlined in this paper.

## Introduction

The statutory accident insurance institutions or "Berufsgenossenschaften" (BG) of the industrial sector in Germany are running a research and testing institute, the BG Institute for Occupational Safety and Health (BGIA). The BGIA supports the accident insurance institutions in their statutory function to prevent accidents and occupational dieseases at work e.g. work-related musculoskeletal disorders.

The point of departure for this project was the complaints of employees in a waste-to-energy facility about pains in the neck, shoulders and arms. These employees were operating a clamshell crane which was used to pick up waste out of an enclosed receiving pit (waste bunker). As these complaints were only expressed by crane operators at one waste-to-energy facility (F1) and not by their counterparts at another, identically designed facility (F2), the statutory accident insurance institution for the gas, district heating and water industry (BGFW) initiated a study into a comparison of the two workplaces using modern measurement techniques.

The aims of this study were as follows: Firstly, investigation of the existing workplaces to ascertain the factors that gave rise to the aforementioned complaints. Secondly, taking appropriate ergonomic preventative measures to improve the working situation. At last, conducting further investigations to evaluate the impact of the ergonomic modifications of the workplace.

## Methods

The first measurements in both facilities took place in 2001. The crane operator workplaces both in F1 and F2 were investigated using a measuring system called CUELA

**Figure 1. Application of the CUELA system.**

**Table 1. Measured body angles and measuring sensors of CUELA ("Sitting").**

| Joint/body region | Dimension | Sensors |
|---|---|---|
| Head | Flexion/extension | Inclinometer |
| Cervical vertebrae | Flexion/extension | Calculated |
| Trunk | Flexion/extension in lumbar spine and upper thoracic spine; torsion, lateral flexion | Inclinometer, gyroscope |
| Pelvis | Inclination | Inclinometer, gyroscope |
| Hip joint | Flexion/extension | Potentiometer |
| Knee joint | Flexion/extension | Potentiometer |

(computer-assisted recording and long-term analysis of musculoskeletal load) that has been developed in the BGIA. This system enables the continous recording and analysis of physical load factors of the musculoskeletal system to be carried out at the workplace itself (Ellegast and Kupfer, 2000).

The CUELA measuring system consists of sensors and a miniature computer that can be attached to the person's working clothes (see Figure 1). The system can be adjusted to body size and height. The employees wearing the measuring system can go about their work in the usual way.

The load data (joint movements, applied forces) are measured with high time resolution (50 Hz) throughout a shift and stored on a memory card. As the system will run on a commercially available battery, the measuring system can also be used in the field at mobile workplaces. In this study, a special version of the system adapted for sitting workplaces was used. Table 1 shows the different body/joint movements that are continuously recorded.

The measurements were additionally documented on video. By synchronising the video recording with the measured data the load readings can be matched with the actual work situation.

Immediately after measurement the data can be placed into specially developed software and displayed. Using this software it is possible to display body posture at any

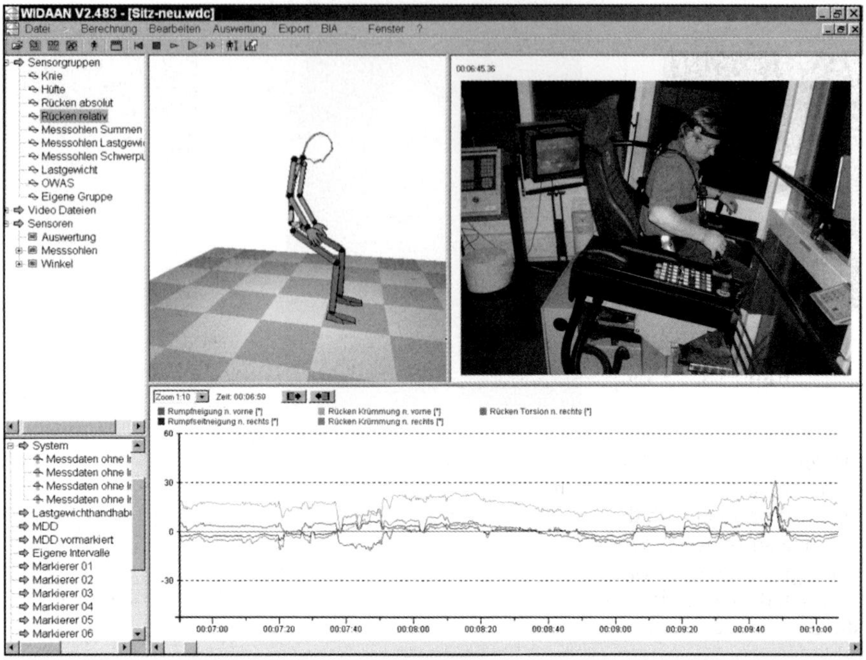


**Figure 2. CUELA software: measuring data, animated figure, video.**

given point with the aid of a three-dimensional computer-animated figure, a time dependent graph of the measured body and joint angles and a video sequence showing the associated work situation (see Figure 2).

The CUELA software automatically issues a series of statistical evaluations to give a quick impression of extreme and static body postures. The body angle positions are analysed with reference to the literature and the relevant standards e.g. extreme angle positions, asymmetrical posture patterns (assessed in accordance with ISO/CD 11226, prEN 1005-4, Drury 1987) or static postures (assessed in accordance with EN 1005-1).

Two employees per power station have been measured on two different working shifts each.

## Results

### *2001 – power stations 1 & 2 (F1 & F2)*

The initial measurements in 2001 on the test persons revealed significantly worse head, upper body and arm postures for crane operators at F1 than their counterparts at F2. Operators of F1 in fact spent a large proportion of their working hours with their heads strongly inclined forward and with extremely unfavourable static arm positions. These significantly worse postures could be found for head, neck and trunk. Figure 3 shows the typical working posture of the crane operators in F1 in 2001 ("old seat"). In F2 these postures have not been found in that extreme way. Finally, the cause of these differences in posture was found in the relative positions of the crane operators' cabs within the waste bunker of the power station. At F2 the cab was at the face side, so that

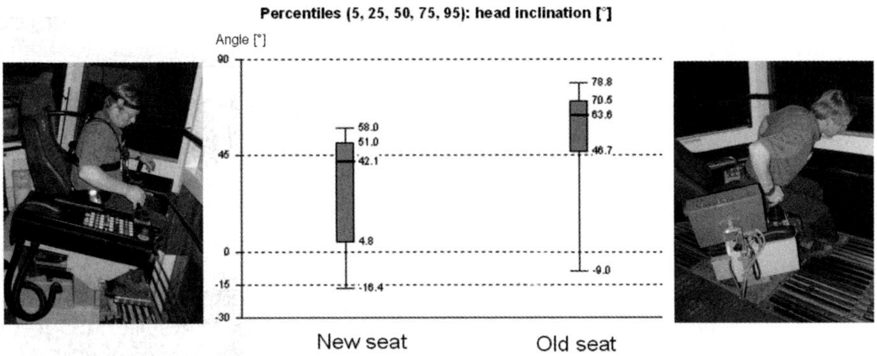


**Figure 3. Comparison of new and old seat in power station 1 (F1).**

the operator's field of vision and working area was in front of or slightly below the crane operator. In contrast to that, the main working area of crane operators in F1 is for the most part beneath them because their cab was on the longside of the waste bunker.

Since subsequent conversion of the crane operators' cab at F1 was not possible, a redesign of the seat was recommended as a means of improving body posture. It was postulated that the ideal seat and associated instruments in these working conditions should tilt forward steplessly and be individually adjustable.

*2004 – power station 1 (F1)*

In 2004 a new designed seat showing most of the recommended features has been installed in the crane operators' cab at F1. Further measurements with the CUELA system took place to evaluate the impact of the new installed seat. The new seat, the arm rests and the instruments were adjustable steplessly in height and inclination (see Figure 3, "New seat"). Each crane operator could adjust his own ideal position and save it to a memory in the seat.

The new measurements yielded significantly improved head, upper body and arm positions for crane operators. As an example, Figure 3 shows the distributions of the angles of the head inclination for the old and the newly designed seat. There was a significant reduction in extreme inclination of more than 45°. Similar results could be found for flexion and bending of the trunk and of the cervical spine.

The arm positions have not been measured with this version of CUELA but have been documentated by video. These extreme postures have obviously improved with the new seat, as well.

## Discussion

At first sight, the regarded workplaces in both power stations seemed to be almost identical, the work organisation (e.g. shift duration and shift schedule) and work task were the same, too. Nevertheless, complaints about musculoskeletal discomfort occured only by crane operators in F1, even if these employees were working alternately in both plants.

The complaints have been affirmed by the results of the body posture measurements with the CUELA system: in F1, crane operators were working most of their time in more extreme unfavourable postures of head, neck, arms and shoulders than their colleagues in F2 did. The difference between both facilities was found in the position of the crane operators' cab within the waste bunker. This positon of the cab was the cause whether the crane operators' main area of operation was in front of them or straight underneath the cab. If their area of vision was underneath their cab, they had to incline their heads and necks extremely. Furthermore, the installed non-adjustable seat and its instruments have been designed for working in upright position, so that working in extreme inclined positon led to extreme unfavourable static arm postures.

As subsequent redesign of the existent cab in F1 was beyond all question, other possible measures of prevention had to be found. Apart from changing work organisation (e.g. job rotation) and individual prevention (e.g. physiotherapeutic exercises) technical measures were recommended (redesign of the seat).

The seat has been adapted to the real working conditions in F1 and after its installation and startup there was a sharp decline in unfavourable body postures. This was documented by further measurements with the CUELA system which yielded significantly improved working postures of the crane operators. A survey of the workforce also showed a sharp decline in musculoskeletal disorders. The new seat was given a very positive rating by the crane operators.

## Conclusion

In the planning phase of new power stations or similar workplaces the constructors should consider the right position of the crane operators' cab. If for certain civil engeneering reasons the cab has to be located on an unfavourable position as in F1, an adequate crane operator's seat must be installed.

The new adjustable seat in this case wasn't a big financial investment – actually it wasn't much more expensive than the ordinary seat. For that reason this kind of seat should be the favourite choice both for new constructions and for already existing facilities.

## References

DIN EN 1005-1 2002, Menschliche körperliche Leistung Teil 1: Begriffe (Beuth, Berlin)

Drury, C. G. 1987, A Biomechanical Evaluation of the Repetitive Motion Injury Potential of Industrial Jobs. Seminars in Occupational Medicine, **2**, 41–49

Ellegast, R. and Kupfer, J. 2000, Portable posture and motion measuring system for use in ergomomic field analysis. In: Landau Ergonomic Software Tools in Product and Workplace Design (Ergon, Stuttgart )

ISO/CD 11226 1995, Ergonomics – Evaluation of working posture, International Organization of Standardization

prEN 1005-4 2002, Menschliche körperliche Leistung Teil 4: Bewertung von Körperhaltungen und Bewegungen bei der Arbeit an Maschinen (Beuth, Berlin)

# REDUCING THE RISK OF MUSCULOSKELETAL DISORDERS IN CONSTRUCTION WORKERS

**Margaret Hanson & Taran Barrington**

*Hu-Tech Ergonomics, 91 Hanover Street, Edinburgh, EH2 1DJ, www.hu-tech.co.uk*

Construction is in the top three industrial sectors for prevalence of work related musculoskeletal disorders (WRMSDs) in the UK. The Health and Safety Executive (HSE) estimates suggest that up to 30% of the construction workforce are significantly affected by WRMSD injuries. HSE commissioned this research to identify practical solutions that could be applied to reduce the risk of musculoskeletal injury occurring within the trades of bricklayers and plasterers – two trades that are identified as having a particularly high prevalence of WRMSDs. A series of case studies illustrating practices that reduce the MSD risks are presented.

## Introduction

The HSE estimates that up to 30% of the construction workforce are significantly affected by WRMSD injuries. This has an obvious cost to the industry as well as to the individual. The UK construction industry has an aging workforce, with fewer young people entering the industry. Skilled trades people are in short supply, and the industry has some difficulty in recruiting, partly due to it being seen as a heavy industry to work in, with people at risk of experiencing discomfort.

Recent research has indicated that MSDs are particularly prevalent in the construction trades of bricklayers and plasterers (Reid et al. 2001). Tasks in the construction industry can involve handling of heavy components, repetitive movement, awkward postures, working under pressure and in poor weather (cold, wet); all these are known to be risk factors for MSDs. Problems are particularly common in the lower back, shoulders and knees for plasterers, and in the lower back, neck, shoulders, elbows, wrists and hands for bricklayers.

Because of the high prevalence of MSDs, HSE has identified construction as one of their target industries for reducing prevalence of musculoskeletal disorders (as part of their Priority Programme). They are also focussing on MSDs generally within this programme. With this attention on MSDs and construction, the HSE required practical guidance for the industry on what could be done to reduce the risk of these problems.

The study reported here, commissioned by HSE, sought to identify practical ways to reduce the risks of MSDs in these trades. All stages of the construction process were considered, from design and planning, to site management and the trades people's tasks. The aim was to develop a series of practical examples and case studies which could be disseminated to HSE Inspectors in the construction industry, and ultimately, to the public.

## Methods

Internet searches and literature reviews were undertaken to identify tools and equipment available that could reduce the risk of injury. Site visits, to a range of construction sites (small, medium, large), were undertaken to observe the tasks, identify good working practices, and discuss the feasibility of implementing risk reduction measures on other sites. Discussions were also held with architects and designers to understand how these risks could be avoided through appropriate design. Most risk reduction solutions identified were available and in use in the UK. A small number of solutions were identified as best practice in other countries; these were included where they would be practicable in the UK.

## Risk reduction solutions

The case studies and examples that were developed were categorised into the Pre-construction phase and the Construction phase. Separate case studies were developed for bricklayers and plasterers. Each case study contains a summary, a description of where the solution could be used (e.g. small, medium or large sites, new-build or refurbishment), a description of the problem and the solution, a testimonial from users where available, where the solution could be obtained, and an indication of the cost.

A sample of the solutions are presented below.

### *Pre-construction phase: ensuring good knowledge and communication*

Many of the musculoskeletal risks identified on construction sites arise due to a lack of consideration of the trades people who will actually build the building. If due thought is given to the construction process, and if different stakeholders are aware of the implications of their decisions, it is possible to reduce the risk in the design.

Duties are placed on the client, designer, principal contractors and subcontractors under the Construction (Design and Management) Regulations (1994) to reduce the risk of injury in the construction, maintenance and demolition of the building. Manual handling and musculoskeletal risks need to be considered, assessed and reduced through appropriate planning, work organisation and specification of materials. Awareness training for designers in their duties and of different risk reduction options can be helpful, as can ensuring that there is good communication between different stakeholders in the construction process. This may be achieved through regular (e.g. weekly) team meetings where the client, principal contractor and all subcontractors meet together and identify solutions to reduce any risks that have been identified in the construction process.

### *Pre-construction phase: considering the materials specified*

The materials that are specified will have an impact on the people who handle them. Weights can be reduced, and materials made easier to handle if this is considered when the materials are specified. Examples are given below.

Standard plasterboard is 1.2 m wide, and 2.4 m long. It typically weighs over 25 kg (depending on thickness). Its width makes it awkward to handle (arms have to be abducted if holding it vertically), and awkward to see past. Narrower plasterboard (0.9 m wide) is available. This is 25% lighter than the same length/thickness of 1.2 m wide board, and is therefore typically less than 20 kg. It is also easier to see past.

Foundations can be specified to be concrete poured rather than block laid. This eliminates the risk of musculoskeletal injury that arises from laying blocks in narrow trenches with limited access, where twisted and stooped postures are often required.

It should be ensured that materials are not over-specified in relation to the structural strength required for the building. Materials of a higher structural strength specification are usually heavier and may not be required.

Metal lintels can be specified instead of concrete lintels as these typically weigh about 3–4 times as much as metal lintels for the same length / structural strength.

Hollow or cut-away concrete doorsteps should be specified rather than solid ones, as this will reduce the weight significantly (approximately by 50%). Hollow or cut-away steps usually still provide the structural strength required.

Blocks with handhold cut-outs are available that facilitate the manual handling of blocks.

Special thin-joint mortar is available that requires a 2 mm layer of mortar between blocks rather than the standard 10 mm. This will reduce the amount of mortar that has to be handled on site, and the weight of the mortar that has to be placed on the blocks. However, it is acknowledged that more blocks may have to be handled during the rise of a wall.

Prefabricated units (e.g. of toilet facilities) or use of wall panels (as has occurred with some of the newer structures, e.g. the Swiss Re building in London) eliminates the need for bricks and blocks to be laid; materials can be mechanically handled into place and attached to the super-structure. This eliminates the risks associated with brick and block laying.

### *Pre-construction phase: planning the construction*

The design of the building and methods and equipment that will be used to construct it can have a significant impact on the risk of musculoskeletal injury.

Consideration should be given to the delivery of materials to ensure that maximum use is made of mechanical handling devices available on site. This may include laying roadways into a development prior to digging foundations, so that tele-handlers can be driven along these roadways to deliver materials mechanically. The position of loading bays should also be considered in relation to the movement of materials within a building, such that the distance to be travelled internally is reduced. The sequence of delivery of materials should also be considered, so that items can be mechanically handled into place before internal structures prevent this.

Buildings should be designed with no change in floor level for each storey so that trucks and trolleys can be used to deliver materials internally during the construction.

Mast climbers can be used to lay bricks from. They provide a working platform whose height can be adjusted precisely; brick can then be laid at waist height all the way up the rise of the wall.

Scaffolding is available with hop-ups (2 boards wide); these provide a working platform 500 mm below the main run of the scaffolding (5 boards wide), and can then be raised to the scaffolding level as the wall is built.

### *Construction phase: site management*

Much can be done in the construction phase of a project to reduce the risk of musculoskeletal injury. Many unforeseen issues often arise during construction, and the site manager has a key role to play not only in planning the construction site, but in making appropriate decisions during the course of the construction.

Materials should be delivered to an appropriate point on the site so that the requirement to manual handle items is reduced. This may include scheduling the delivery of materials so that maximum use can be made of mechanical handling devices.

Materials should be kept dry so that their weight is not increased. A wet block can weigh approximately 3 kg more than a dry block. Having an appropriate materials storage area (e.g. covered) can facilitate this.

Scaffolding can be used at ground level to raise the height of the bricks and mortar worked on, in order to reduce the stooping required when accessing them.

*Construction phase: tools and equipment*

A number of tools and range of equipment are available to facilitate both bricklayers and plasterers in their tasks. Although many of these tools can be useful, they should be seen as the lowest form of control after eliminating and reducing the risk (through appropriate design and site management).

The mortarboard is usually placed on two blocks, meaning it is about 200 mm above ground level, and bricklayers have to repetitively stoop to it. A lightweight platform for the mortarboard is available which raises it to a more appropriate height (approximately 650 mm) and thus reduces the bending required to access the mortar.

Mechanical lifts are available that can be used for raising lintels into position above a doorway.

Pallet trucks with pneumatic tyres allow pallets of materials to be manually moved within buildings without the need to carry them. Note that buildings have to be designed with no internal steps for this option to be able to be used.

Sack trucks can be used rather than wheelbarrows for moving blocks. The centre of gravity of the load can be held directly over the wheels of the sack truck, whereas the forward position of the wheel on the wheelbarrow in relation to the load means more loading is taken on the body compared with a sack truck.

Brick tongs allow 6–8 bricks to be picked up at one time. They are very simple to use, and provide a handle for lifting the bricks. This prevents the extended finger position that can occur when picking them up manually. However, with the brick tongs the weight of the bricks will still be carried, which may still present risks.

A trellis platform system is available that provides a working platform within a building. This can raise the tradesperson to the right height for working without the need for scaffolding or mobile towers.

Plasterers often have to access small areas above head height; they will often stretch or stand in inappropriate surfaces (e.g. upturned tubs). Proprietary hop-ups are available that will raise them to an appropriate height and provide a secure footing.

A range of plasterboard handling trolleys and hoists are available which can eliminate the need to manually handle boards. Some of these integrate with the delivery of plasterboard on a pallet from a truck. Some convert from being a trolley into a workbench, presenting the plasterboard at about waist height, so it can be handled and trimmed to size as required at an appropriate height. A range of hoists are available that raise the plasterboard up to ceiling height. However, they can be slow to operate, and may not be practical in small developments.

Some simple plasterboard handles are available that facilitate the handling of single sheets of plasterboard. Plasterboard can be difficult to handle due to its thickness (12–15 mm) typically requiring a pinch grip to hold it, which can contribute to discomfort. These tools provide a handle that enables a power grip to be used. However, the board still has to be manually handled.

A plasterboard clip is available for assisting with fixing the plasterboard to the ceiling, a task that is particularly difficult and forces the plasterer to adopt awkward postures, with them typically supporting the load with their head and stretching to attach nails through the edges. The clip holds one end of the plasterboard to the joists, while the plasterer is able to move to the other end of the board and attach it.

A tool is available that holds a vertical sheet of plasterboard against a wall, allowing the plasterer to move freely to position the board to the wall, and thus eliminating the need for awkward postures which occur when holding the board in place with one hand and securing it with the other.

A mechanical tool is available for mixing wet plaster; mixing by hand (using a shovel) places a significant load on the body and can be tiring. The powered tool reduces this loading.

## Dissemination of solutions

The risk reduction measures have been presented as a series of case studies written specifically for (1) bricklayers and (2) plasterers. These will be placed on the HSE intranet site, available for HSE Inspectors in the Construction Industry to use. Ultimately, they will be placed on the HSE's external website to be available for public access. The findings have been presented to inspectors and industry groups and have been favourably received.

Please note that the case study examples presented represent good practice in the authors' view, and do not constitute an endorsement of a system or product by HSE.

## Acknowledgements

The authors would like to thank HSE for funding the study, and the many companies and organisations who contributed to the development of the case studies of solutions.

## References

Reid A., Pinder A. and Monnington S. 2001, *Musculoskeletal problems in bricklayers, carpenters and plasterers: Literature review and results of site visits.* ERG/01/01 Health and Safety Laboratory, Sheffield.

# TURNING BANKING ON ITS HEAD – THE ERGONOMICS CHALLENGES

**Andy Nicholson[1], Dave Moore[2] & Helen Gagg[1]**

[1]*Hu-Tech Ergonomics, 29 Marefair, Northampton NN1 1SR, www.hu-tech.co.uk*
[2]*Head of Occupational Health and Safety, Abbey, 201 Grafton Gate East, Milton Keynes MK9 1AN*

As a key element of its recovery plan, Abbey embarked upon a new retail branch concept to match its re-branding strategy of "turning banking on its head". Ergonomists from Hu-Tech were brought into the team tasked with developing the branch concept and assisting in the process. The ergonomists were consulted at each key phase of the development. Many of the concepts were new and relatively untried in a retail bank environment.

The paper will present the successes and the compromises that were made during the programme to achieve a desirable outcome, which was to make the branch network inclusive in design and, as far as possible, to meet ergonomics good practice.

## Introduction

*"We've changed our name, introduced a bright new look, simplified our product range, rewritten all our letters and restructured our organisation"*, so said the Abbey website (www.aboutabbey.com, October 2004).

To introduce the "bright new look" to the retail banking network, architects were appointed by the Chief Executive to turn strategy into reality. At a very early stage, a limited number of key staff in Abbey were asked to act as internal advisers to the project team. One such player was the Occupational Health and Safety Department. Since this was such a large scale project and involved substantial capital investment, Hu-Tech were appointed as ergonomics advisers to support the project team through the health and safety function with particular reference to ensuring compliance with health and safety regulations and good ergonomics practice.

The Abbey branch network had been somewhat neglected over the previous 10 years. It reflected an old style format that not only presented a poor image in the marketplace, but could also be giving rise to a risk of musculoskeletal injury for staff working at the branch counter. It was hoped this could be reduced by the appropriate and judicious application of ergonomics in the re-design process. At the same time the business was undertaking a thorough review of its "people and processes" with a view to improving service delivery and profitability of the network. This would have a direct effect on work organisation within the branches – a factor that also needed to be considered in the design process.

### *Design philosophy*

Abbey wanted its branches to be distinctive and to stand out in the marketplace, to be open and welcoming, easy to use, and to be built around customer service. This meant that traditional counters were seen as an unnecessary barrier, typical banking moulds would have to be broken, and that an entirely new design concept would be required.

*Design team*

The design team constituted key individuals from external suppliers and stakeholder departments including, design consultants (design lead), architects, operational managers and staff, and internal subject matter specialists. These various vested interests and viewpoints introduced complexities into the design process which, in turn, imposed further constraints on what could be achieved by the ergonomists. Examples included the choice of PCs and IT supplier which meant we had to work with equipment of set dimensions; and the decision not to purchase proprietary furniture but to design these primarily to meet aesthetic requirements.

*Design concept*

The concept proposed by the design consultants/architects to meet this philosophy was:

- XOpen-plan working
- Queue management system (ticket machine)
- No barriers between customers and staff – avoid rising screen protection/glass screens but provide adequate security protection for Abbey staff
- Staff to work standing up and free to move around the banking hall
- Visible equipment and paperwork reduced to minimum
- Cash management through Teller Cash Recycling Units (TCRs)

## Ergonomics challenges

The design philosophy and concept presented some considerable challenges, but also some potential "wins", from an ergonomics perspective. On the one hand the lack of barriers between customers and staff removed constraints on reach to an "exchange point" (typically cash, card or passbook). Similarly the requirement to have staff standing simplified some of the design process due to the presence of chairs e.g. getting in the way when staff stand up to access other workstations – an event that was becoming increasingly necessary with limited "back office" functions to support "frontline" staff. However, having staff standing for a significant part of their working day was a new concept with new challenges – how to prevent the onset of leg and low back pain and discomfort.

An additional requirement was the need for branch counter workstations to be accessible by all. It was considered essential for the new design (that is likely to be in place for several years) to be, as far as possible, inclusive in design providing high levels of DDA compliance. It should be noted that an audit of the software for DDA purposes was not within the ergonomists' brief.

The constraints imposed by the various interests and viewpoints of the internal stakeholders introduced some considerable challenges. Examples included the choice of PCs and IT supplier which meant we had to work with equipment of set dimensions; and the decision not to purchase proprietary furniture but to design these primarily to meet aesthetic requirements. This latter decision was particularly difficult to resolve, but the argument for compliance with regulations and British, European and International Standards was persuasive and eventually drove the purchasing decision.

## Constraints

Hu-Tech was involved at all key design stages. Often this involved negotiating with other key functions with vested interests – not least the architects who were the custodians of the design brief. Every component of the "front-office" retail banking furniture and facilities was to be re-designed. The time frame imposed was approximately 8 months from concept to opening the first new branch.

Due to the extreme time limitations imposed it quickly became apparent that design decisions would have to be taken without fully testing each component at the usual mock-up user testing stages. In these circumstances, it was necessary to amalgamate feedback from the first user trial with our experience gained from similar scenarios in other industry sectors. In this way we had to extrapolate the likely results and success of each design.

Wherever possible the process of (i) development of an ergonomics specification, (ii) mock-up, (iii) fitting/user trial, redesign and (iv) final fitting/user trial before (v) sign-off, was adopted, and was put forward to the design team as the correct way to proceed. It was not uncommon to find that shopfitters were unable to build a final mock up for a second user evaluation within the time available before the furniture was scheduled for delivery and installation in the flagship branch. However, for key pieces of furniture the full design process was followed, as described below.

## Case studies

### *"Kiosk"*

The purpose of the Kiosk was to be a means of Abbey being where people assemble, e.g. large leisure facilities. In this instance a design concept had been developed and built by the designers/architects before Hu-Tech were involved. Thus our involvement was simply to provide comment on the design, its functionality and usability. At a later stage the brief was changed to include accessibility for disabled customers. The use of the facility by disabled staff was considered separately through the "People and Process" initiative.

It quickly became apparent that the initial design failed to meet many basic ergonomics requirements. Space availability was woefully inadequate, functionality was poor, and much of the equipment failed to meet the requirements of the Health and Safety (Display Screen Equipment) Regulations. For this reason the design team agreed that a second mock up should be built based on design criteria provided by the ergonomists. Thus a design specification was drawn up and used by the architects as the basis for the revision. The specification was extended when it was agreed that the Kiosk must be physically accessible to all.

A modification of the Kiosk called the "Sales" or "Talk Pod" was also introduced. The function of this design was as an informal interview area within the main banking hall. While some of the design issues could be transferred from the Kiosk, others had to be considered and developed separately. Most notable was the wider range of functions required by staff operating the Pod.

The ergonomics specification provided for both the Kiosk and the Sales Pod included the following:

- Width (diameter) of the table to meet the needs of staff and customers
- Overall internal width (diameter) of the interview space
- Table height and thickness
- Full range of adjustment required for the visual display screen

- Customer and staff reach and space requirements
- Position of associated equipment on the desk and other surfaces
- Positioning of table support mechanisms
- Footrest dimensions
- Height of customer seating and how to make the access and egress more usable

*Branch counters*

From the outset, and before the involvement of the health and safety or ergonomics expertise, it was stated that the counter must be operated in a standing posture. This is unusual, but not unique in the banking environment, and created a particular issue that had to be addressed. It was accepted that it was reasonably foreseeable that standing at a counter and serving customers for long periods of time could lead to lower back and leg discomfort. Thus, as far as possible, the design included features that permitted staff to alter their posture while standing, for example the inclusion of a low foot rest built into the rear of the counter which would enable staff to occasionally rest one leg and alter their hip orientation. It also included specially chosen anti-fatigue matting and sit-stand stools. These were only to be used when staff were not serving, and customers were not waiting to use the counter.

It was stressed to the design team that while these measures should diminish the potential for discomfort, it was essential that a management process be introduced to ensure staff were not serving at the counter for more than about 2 hours without a seated break. It was agreed that this could mean moving to a seated workstation (such as the Sales Pod or interview room). A policy of managing pregnant workers and serving at the counter was developed by the occupational health function.

Since the counter was introduced later in the design process it was possible for two mock-ups were made – the second incorporating minor changes determined by a review of the first.

The ergonomics specification provided for the counter included the following specific aspects:

- Counter dimensions
- Full range of adjustment required for the visual display screen
- Customer and staff reach requirements
- Position of till and keyboard and use of a transfer point between customer and staff
- Position of associated equipment under the desk
- Position of the automated teller unit relative to the staff positions
- Design for disabled
- Features of the sit/stand stool
- Storage facilities

At a very late stage it was decided to introduce a 5-position counter with rising screen protection for one location. Timescales were such that no mock-up could be built. Ergonomics advice was sought and modifications based on the standard two position counter were provided based on drawings provided by the architects. While this was an unsatisfactory position to be in, it was agreed that a reasonable opportunity to improve the ergonomic design was better than none; a decision that was later justified when the drawings were reviewed.

The Occupational Health and Safety Manager and lead ergonomist prepared training material for the Customer Service Advisers to make sure they knew how to work safely, including good practice when working at the Kiosk, Sales Pod, and counter. These were incorporated within a training manual provided to all staff on induction to the new branch as each one was opened.

## Outcome

Whilst staff and their union representatives initially expressed concern regarding some of the new concepts, their fears were quickly resolved when they worked with the new facilities.

The adoption of a consultative approach to workstation design and the provision of clear guidelines, training and support helped staff understand and adapt to their new working environment with minimal concern.

In particular, on-site guidance and advice on the sensible, practical measures staff can take to avoid the early onset of fatigue when working standing up was greatly appreciated, as was clear guidelines on safe working practices regarding job rotation in the branch.

It was very pleasing to see staff with many years of experience working in a traditionally designed branch quickly adapting to their new workstations and valuing the intrinsic benefits and flexibility working standing up brings rather than having to adopt the rigid, constrained working postures demanded by a seated counter position.

## References

Health & Safety (Display Screen Equipment) Regulations 1992 as amended by the Health and Safety (Miscellaneous Amendments) Regulations 2002, HSE Books, L26, ISBN 0 7176 2582 6

# PASSENGER BEHAVIOUR ON AMUSEMENT RIDES

**Edmund Milnes**

*Health & Safety Laboratory (HSL), Harpur Hill, Buxton, Derbyshire SK17 9JN*

Passengers' behaviour on Amusement rides has been identified as a significant contributing factor in certain incidents. Concerns have also been raised by the amusement industry that certain groups of passengers are engaging in more risk taking behaviours than has historically been the case. A field study of passenger behaviours was carried out to catalogue behaviours and establish behaviour frequency benchmarks. Filming took place at 15 venues incorporating 51 discrete rides and 25 ride types. This paper provides an overview of the frequencies of particular types of behaviour and their associations with passenger age, gender and perceived motives. Additional observations include a low number of exceptional passenger behaviours (i.e. very low frequency/high risk behaviours).

## Introduction

For several years there has been increasing concern within the amusement industry and the Health & Safety Executive (HSE) regarding the behaviour of passengers on amusement rides. In certain incidents passengers' behaviour is believed to have been a significant causal factor. For example reaching out of a moving ride and striking an arm against a stationary object. There are also concerns within the amusement industry that certain groups of passengers now engage in more risk taking/thrill enhancement behaviours than previously (e.g. compared to 10 or 15 years ago). If ride operators are aware of potential passenger behaviours on specific rides they may be more able to take steps to safely accommodate those behaviours or prevent them occurring. This could involve design modifications, increased alertness to behaviours and/or enhancing passenger management systems (announcement/cctv facilities and information signs). Due to the concerns about passenger behaviour and the scope for improving safety using improved knowledge in the area, HSE funded a research project with the agreement of co-operation with data collection from the UK Showmans Guild and BALPPA (British Association of Leisure Parks, Piers and Attractions). The research project involved two main stages; a) a review of literature on thrill-seeking and risk taking behaviours and b) a field study to develop a catalogue of passenger behaviours and benchmark their frequencies and associations with variables such as passenger age and gender.

This paper presents the key field study findings rather than the literature review. The literature review is a detailed piece of research in its own right (Wright, 2004). The overall conclusions were that personality traits are inextricably linked to risk taking behaviours and children typically have more sensation seeking tendencies and lower risk perception coupled with a lower ability to deploy knowledge to avoid risky situations developing.

## Field study method

### *Data collection*

Passengers' behaviours were recorded by filming rides being used at a range of venues. A total of 15 venues were visited; 4 Theme parks, 4 Park Fairs, 3 Street Fairs, 1 Indoor

**Table 1. Behaviour categories and their repetition rates.**

| Behaviour categories | Repetition rates |
|---|---|
| One or Two hand waving | |
| Turning head/trunk | One-off behaviour |
| Turning around in seat | |
| Turning and kneeling on seat | |
| Leaning out of containment | Repeated dynamic action |
| Reaching out of containment (arm or leg) | |
| Extended reaching out of containment | |
| Interfere with containment device | Single sustained action |
| Inappropriate exiting of ride | |
| Violent behaviour | Prolonged action (all through ride cycle) |
| Boisterous behaviour | |
| Standing up | |
| Pointing | |
| Active interact with containment | Sustained and repeated action |
| Kicking | |

Fair, 1 Music Festival and 2 Pleasure Beaches. Passengers were not informed that they were being filmed and a key aim was to avoid them becoming aware of being filmed. This was achieved by filming from crowds or vantage points which were unlikely to draw passengers' attention. Clear and explicit permission was obtained from key stakeholders; HSE, Park owners/operators, Showmans Guild/BALPPA and in some cases Local Authorities and the Police. All visits were arranged in advance by HSL.

Rides were selected for filming based on their type in order to obtain behaviour data from a wide and representative range of different ride types. Passenger density, time of day and general demographics of people attending each venue were also considered when selecting rides. Rides were filmed for a minimum of 20 minutes. This period was extended if the ride was busy and a good filming vantage point was being used.

### *Data analysis*

The observed behaviours were transcribed by categorising them in a spreadsheet. Each observed behaviour was assigned into one of the categories in Table 1 along with a repetition rate (categories of repetition are also in Table 1). "Behaviour" refers to physical action, not movement caused by ride motion or actions to brace against ride motion.

Passengers' approximate/estimated age and gender were recorded alongside their behaviour. Passengers were assigned into one of the following age groups; 0–10, 11–15, 16–21, 22–30, 30–50, >50 yrs.

An estimate was also made of passengers' motives for their behaviour, for each behaviour a primary motive was assigned. Motivation is a complex issue and people may often have more than one factor influencing their behaviour, however in many cases a clear primary motivation could be seen, for example people waving to friends (communication) or children turning and kneeling in their seats to maintain visual contact with their parents. Table 2 contains the list of motives which were assigned to behaviours.

## Overview of field study scope

A total of 7096 passengers were observed on rides from whom a total of 2872 behaviours were observed. The gender split of the passengers who exhibited behaviours was

**Table 2. Potential motives for behaviours.**

| | | |
|---|---|---|
| Communication | Copying | Peer pressure |
| Aggression/competition | Exhibitionism | Encouragement |
| Curiosity | Boredom/Natural pose | Impatience |
| Lack of info./knowledge | Illness/nausea | Fear |
| Thrill enhancement | Discomfort | Excitement |

**Table 3. Types of ride that were observed (& number of each ride type observed, & total number of passengers observed on ride type).**

| | | |
|---|---|---|
| Traditional Carousel (5, 1166) | Bench (5, 437) | Jets (1, 287) |
| Log Flume (4, 654) | Twister (5, 443) | Waltzer (3, 406) |
| Pirate Ship (2, 545) | Jumper (3, 327) | Superbob (2, 399) |
| Superbowl (2, 126) | Junior Autodrome (2, 113) | Mini Jet (2, 121) |
| Juvenile coaster (1, 13) | Spinning coaster (1, 239) | Coaster (3, 739) |
| Tagada (1, 121) | Paratrooper (1, 151) | Juvenile Track (1, 173) |
| Adventurer Wave Surf (1, 43) | Toyset Roundabout (1, 40) | Pendulum (1, 164) |
| Superspin (1, 72) | Breakdance (1, 112) | Chairlift (1, 108) |
| | Orbiter (1, 289) | |

59.4% female to 40.6% male. A total of 51 discrete rides were observed which comprised of 25 ride types. Table 3 lists the ride types which were observed with additional information on number of each type observed and total passenger numbers.

## Field study results

### *Overview statistics*

Approximately 92% of observed behaviours were exhibited by passengers estimated to be aged 21 yrs. or younger. This broadly reflects the overall age demographic of all observed passengers. The 3:2/female:male gender ratio of passengers who exhibited behaviour is also believed to reflect the overall gender ratio of all observed passengers.

The most common behaviours were one and two handed waving, turning head/trunk and pointing. All 4 of these behaviours occurred on nearly all types of ride and accounted for 85% of all behaviours (comprising 24%, 46%, 10% and 5% respectively).

The four most common motives, associated with 82% of behaviours were communication, exhibitionism, curiosity and thrill enhancement (comprising 29%, 8%, 7% and 38% respectively).

### *Passenger age vs. behaviours*

Behaviour of young passengers (estimated ages <=10 yrs) appeared to be motivated by communication and curiosity more than behaviours shown by passengers aged 11–21 yrs. 40% of 0–10 yr old behaviours were associated with communication compared with an average of 17.5% in the 11–21 yrs age groups and the percentages for behaviours associated with curiosity were 12% (0–10 yrs) to an average of 3% (11–21 yrs). Passengers in the 11–21 yrs age groups showed an average of 61% of all their behaviours related to exhibitionism and thrill enhancement combined, compared to 29% in passengers aged 0–10 yrs. However, the total number of behaviours motivated by communication were found to increase again to an average of 62% amongst passengers aged >22 yrs whilst behaviours motivated by thrill enhancement and exhibitionism drop to average levels of 11% and 2% respectively in these older age groups.

Young childrens' tendency to communicate results in high levels of one handed waving amongst the 0–10 yrs age group (which accounted for 29% of all observed behaviours for 0–10 yr olds). They also frequently turn and move in their seats to maintain visual fixation on items off the ride as they rotate which appeared to be mainly motivated by curiosity and communication. Turning head/trunk and turning around in seat comprised 21.5% of all 0–10 yrs behaviours compared with just 0.9% and 0.04% of 11–15 yrs and 16–21 yrs behaviours respectively.

*Standing up during ride cycles*

Standing up accounted for 4.4% of all observed behaviours (126 behaviours) and was the fifth most common behaviour. The risk associated with standing, the motives for standing and the physical activity varies widely between different types/designs of ride. For example it may represent a significant risk on a Log Flume or Twist ride where there may be a risk of ejection or impact against parts of the ride, and a lesser risk on Traditional Carousels which move more slowly and are designed to allow standing.

*Associations between gender and specific behaviours*

No evidence was found of a strong gender bias for any of the behaviours. However, females' behaviours were more frequently associated with communication than were males' behaviours (31% compared with 25% for males). Females also showed lower levels of behaviour associated with curiosity (6% compared with 10% for males). Overall females showed fewer behaviours associated with motives other than the main 4 compared to males (15% compared to 21%). Males' tendency to engage in more curiosity driven behaviours is reflected by a notably higher proportion of males than females reaching their arm out of containment, kneeling in their seat (younger passengers) and interfering with containment systems/devices. However, the relatively low number of occurrences of these behaviours (just 3 and 4 observations for kneeling and interfering with containment respectively) means it is not possible to associate these behaviours with males significantly more than females.

## Additional observations

Only a small number of "exceptional" behaviours were observed which, in combination with the ride design, were believed to contribute to a significant risk to the passenger. These behaviours were as follows:

- Reaching leg out of containment on Pirate ship
- Lying back/reclining on horses/climbing between horses on Traditional Carousels
- Standing in rear foot well whilst reaching to front console on Mini-jets
- Reaching out of side of log flume while it is rocking
- Standing on log flume or coaster whilst in motion
- Swinging on Paratrooper chair before ride starts (risk to passengers embarking)

If a ride stops at an unscheduled point (e.g. during an emergency stop or power failure) it is likely that some passengers will begin to move around (e.g. kneel on seats, stand up) which may result in their moving into positions of danger. This typically will occur on juvenile rides when children are unaware that the ride will restart.

Passengers appeared to act on the advice of signs to secure their belongings. This was seen as being due to passengers interpreting the consequences of not doing so (material loss) and acting to prevent those undesirable consequences. However, it was not possible within the field study design to establish whether signs had a significant effect on behaviour *during* rides (e.g. signs to maintain a grip on the handrail etc.). It is possible that if passengers can be helped by signs to appreciate potential consequences of not following behavioural instructions, those signs may become more effective.

Behaviour related to exhibitionism and thrill-enhancement (two handed waving/ exceptional behaviours) are considered likely to increase amongst passengers who have been drinking alcohol although this was not specifically noted in observations. Inhibition-reducing effects of alcohol may also lead to increased incidences of behaviour due to increased copying and stronger peer pressure effects. Alcohol and other drugs (prescription or otherwise) may also contribute to passengers becoming dizzy/ disoriented during fast/spinning rides, which could cause falls during egress from rides and ride platforms.

## Summary

This paper provides an insight into the scale and design of HSL's Passenger Behaviour field study and is based on the key points in the project report which is available to the public (Milnes, 2004). For a copy of the full report contact the HSE Information Centre.

## References

Milnes, E. 2004, Passenger Behaviour on Amusement Rides: Field Study Report. HSL Report ERG/04/24. Available to the public.

Wright, S. 2004, Passenger Behaviour at Fairgrounds: A Literature Review. HSL Report. File EPPR/41. Available to the public.

# THE HEALTH AND SAFETY EXECUTIVE MANUAL HANDLING ASSESSMENT CHART (MAC) BENEFITS

**Sophie Mawle**

*Occupational Health, Coca-Cola Enterprises Ltd, Uxbridge, Middlesex*

It is well reported that musculoskeletal disorders (MSDs) are the most common form of work related ill health in Great Britain (HSE, 2002). This company is no different and in 2002 when the focus was on tracking and trending long term absence, MSDs were one of the main causes. Research concerning manual handling has primarily focused on training; it is understood that teaching of manual handling does not work alone (reported at the Ergonomic Society Conference, 2003). The MAC tool was promoted at the Ergonomic Society Conference in 2003; from here it was decided this was to be incorporated into our ergonomics programme. The MAC was trialed at a distribution site which consisted of warehouse operatives and drivers. During the trial the MAC promoted workplace design; correct use of equipment and correct individual manual handling technique. It promoted team work between employees, managers and occupational health. The outcome has been positive, with an overall decrease in MSDs and an increase in employee attendance.

## Introduction

Manual handling is every employers' concern under the Guidance on Regulations for the Manual Handling Operations Regulations (1992) and the implications of incorrect handling possibly resulting in: reduced attendance; productivity; performance and output (Coztug and Buckle, 2004). Statistics estimate 1.1 million people a year are affected with an associated musculoskeletal disorder, which as a result leads to a total of 12.3 million work days lost (HSE, 2002). Poor manual handling practices have primarily been blamed as the main cause in work related musculoskeletal disorders (WRMSDs). WRMSDs are described as inflammatory and degenerative conditions which result in pain and functional impairment and may effect the neck, shoulder, elbows, forearms, wrists and hands (Buckle and Devereux, 2002). The MAC was developed by the HSE to assist in the identification, assessment and evaluation of significant risk factors for manual handling injury. Three separate flow charts were developed for a quick assessment of single person lifting operations, carrying operations and team handling. The charts use a traffic light system to indicate level of risk with a numerical indication also provided (Quarrie, 2003). This study focuses on the advantages the MAC has to assisting with solutions concerning workplace design, the correct use of manual handling equipment and individual manual handling behaviour; while attempting to demonstrate the importance of in-house team work as the way forward for manual handling solutions.

## Method

The ergonomics team at the site consisted of a transport manager, a warehouse coordinator, a safety advisor, a warehouse operative and an occupational health advisor.

The team then went through an education/training session on the MAC by the occupational health team. The trial commenced in the warehouse, utilizing 19 employees. The warehouse operatives were briefed that the trial would be commenced by the warehouse co-ordinator; in order to provide confidence that the assessment would be undertaken in a realistic fashion by the operatives. The observational team was split into pairs before assessing individuals' working practices in the warehouse (this was aimed to promote reliability and validity in the results recorded). A similar situation occurred with the drivers (N = 32) and a simulation of three main phases of work (perceived as worst case scenarios) were conducted at the site. The three phases consisted of: unloading at the side of a vehicle; team handling between two drivers and carrying stock. Each individual was assessed during each phase of the simulation procedure. Every driver was scored during each phase of the simulation. Constructive two way feedback was then given between the manager and driver with input from occupational health and on two occasions a physiotherapist. An assessment video was produced with a feedback session from a driver.

Following collation of results from both groups of warehouse operatives and drivers certain actions were taken regarding workplace design (this consisted of the work layout); use of appropriate manual handling equipment (sack barrow compared with automatic stair climbers), introduction of job rotation (picking stock and fork lift truck driving) and two way team feed back focusing on each individual's manual handling behaviour.

Results comparing the individual total scores in the warehouse scenario were taken a year later and driver assessments following six months as well as retrospective long term absence statistics (comparing 2002, 2003 and 2004 years) and occupational heath referral statistics (comparing management referrals; self referrals and client reviews over two years).

## Results

The MAC was implemented in the warehouse during 2003–2004. The driver assessments took place in 2004. The site had a total of 1820 days lost due to attendance issues in 2002, of which 296 days were reported as musculoskeletal related. By the end of November 2004 the total lost days due to attendance was 710 of which 184 were musculoskeletal related.

### *Action planning in warehouse*

Once results and feedback were collated from the warehouse operatives; an action plan was outlined:

1. *The redesign of the warehouse*
   It was highlighted that operatives were often restricted in their movement due to positioning their trucks too close to the drink product. On questioning, most operatives advised that the aisles were not wide enough for two way traffic. This promoted a change to where the "popular" stock was placed and/or placing of "popular" stock in two sections of the warehouse, this led to a reduction of traffic in the aisles which enabled warehouse operatives greater space to park and manoeuvre when handling stock.
2. *Job Rotation*
   Prior to the MAC the warehouse operatives consisted predominantly of pickers and fork lift truck drivers. The decision was taken to train all operatives on fork lift trucks so rotation between picking stock and driving a fork lift truck was a realistic

option. A fair rotation timetable was produced by the warehouse co-ordinator for all warehouse operatives to work on fork lift trucks as well as picking stock.

3. *Manual handling behaviour*
   Feedback was given to each individual warehouse operative on their manual handling techniques and they were challenged on technique, time keeping, load planning and parking near to stock with enough space to move.

*Action planning with the distribution drivers*

1. *Manual handling behaviour*
   Static postures when handling loads were noted as a primary cause for concern. Drivers were challenged on the repositioning of stock when off loading from the side of a vehicle to allow more movement. During the trial, two physiotherapists (one from general practice and another who specialised in biomechanics) attended the site to give direct feedback concerning posture.
2. *Use of automatic stair climbers*
   During the trial it was noted that only one driver opted to use an automatic stair climber for stair work. This was a concern and most drivers scored highly on the MAC by using a sack barrow. The majority of drivers explained they did not feel comfortable using an automatic stair climber. Following the first phase of the trial drivers were placed in pairs and the stair climber was taken with them to aid competence and comfort.

## Discussion

It has been reported that when advising on MSDs, it is often the case that consultants will indicate changes necessary, such as modifications to equipment, workplace layout and/or working practices and the acceptance of such advice being dependent on the attitudes and beliefs regarding MSDs held by the managers and their employees (Whysall et al., 2004). This study has focused primarily on coaching and/or facilitating the user groups for their input on design and behavioural solutions to combat MSDs.

In conjunction with the trial the site was again given back care and manual handling education and all individuals underwent health assessment consisting of musculoskeletal assessment and measurement of BMI by occupational health. "Behaviour change theory suggests that if change is to take place, the user group need to hold positive attitudes and beliefs relating to the desired behaviour" (Whysall et al., 2004). It was anticipated the back care education and musculoskeletal assessments outlining specific individual weaknesses would motivate each person to want to change their approach to manual handling technique and promote back care for life.

### *Warehouse operatives*

Before the commencement of the trial in 2003, the warehouse operatives consisted of pickers and/or MHE drivers (manual handling equipment drivers on fork lift and reach trucks). Pickers at the time were routinely expected to lift in excess of 10 ton, each case of stock weighing between 8–23 kg; (the majority of product weighing approximately 13 kg). The MAC assessments promoted change: job rotation; warehouse planning and individual advice regarding posture and back care. The results over one year has shown a significant reduction in non – attendance from 650 days in 2002 to 291 days in 2003.

### *Distribution drivers*

Before commencement of the trial (June 2004); the drivers had been provided with automatic stair climbers to assist with stair work in 2000. At the beginning of the trial,

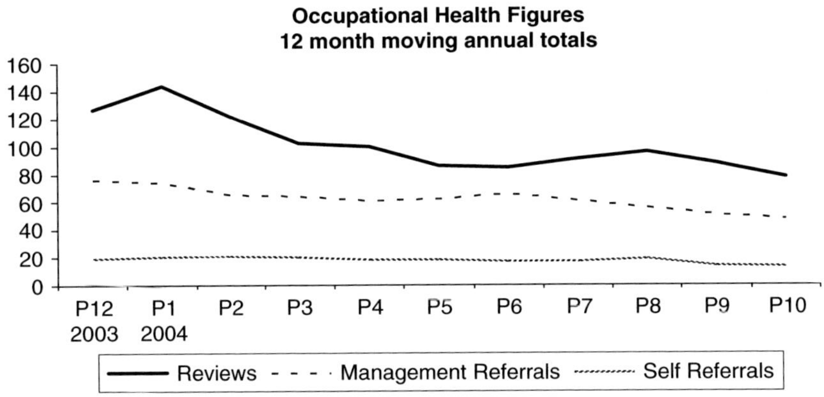


**Figure 1. Occupational health figures for site.**

very few drivers had felt comfortable with the use of an automatic star climber as a manual handling tool, and preferred to use a manual sack barrow for stair work, which they were familiar with. All stair climbers had been adapted following drivers input with design. Following challenges over the automatic stair climbers in comparison with sack barrows for carrying stock up stairs and extra training given to drivers; the results after six months have shown an increase in the use the automatic stair climbers; drivers who know they have stair work on their route take the automatic stair climber to calls. Non-attendance has also reduced significantly.

*Occupational health*

For both warehouse operatives and drivers there has been a gradual decrease in the number of referrals and ongoing reviews to occupational health (Figure 1).

It is well acknowledged that psychological aspects also play a part in work attendance and recovery times of MSDs (Bartys, 2003), although this was not measured, trust and respect were observed between the managers and the employees during the trial which was deemed necessary when warehouse operatives and drivers were challenged and advised on the techniques to gain an overall manual handling behaviour change.

The MAC study is ongoing with the making of a site training video in collaboration with HSE MAC training website.

## Conclusion

The results from the on site MAC assessments highlighted the need to focus on work place design, promoting the correct use of equipment available and individual behaviour, when moving and handling loads. Once actions were taken from the assessments and measures put in place to assist with change; the results proved the time and effort taken was worthwhile.

This study highlights the rewards business's can make by using ergonomic principles combined with effective team inclusion; and making time available for thinking of solutions and action planning. The knowledge of the job role is invaluable to determining realistic solutions which primarily come from the user group. This site has overcome reduced space and high productivity, to have devised effective solutions to manual handling and MSDs. The rest of the business has now seen the benefit of this ergonomic intervention and the plan is for a MAC roll out across the company.

The MAC has worked in a distribution environment; it has proved to be a user friendly tool where not only manual handling experts feel comfortable with its application, but equally as important, the operatives required to work in that environment, provided the appropriate training is given.

## Acknowledgements

Simon Orchard, Shaun Lansfield, Peter Wright, Brian Bass and Steve Molloy for all their time and commitment to the MAC and assisting with solutions and to Carole Anness, Glen Davies and team for the promotion of ergonomics in the workplace.

## References

Health and Safety Executive, 2002. Upper Limb disorders in the workplace. *HSE Books*, Sudbury,UK.

HSE, 1992. Manual handling Operations Regulations 1992. Guidance on Regulations L23 (Second Edition) *HSE books* 1998 ISBN 0 7176 2415 3

Coztug, O. and Buckle, P. W. (2004). The Effects of Musculoskeletal Pain on Work Performance: Which Pain Measurement Tool should be used. *Contemporary Ergonomics*. The Ergonomics Society 2004, 197–201.

Buckle, P.W. and Devereaux, J.J. (2002). The nature of work related neck and upper limb musculoskeletal disorders. *Applied Ergonomics* 33, 207–217.

Quarrie, C. Human Factors unit, Musculoskeletal, Ergonomics and Performance Section. *HSE Health Directorate*, Magdalen House, Bootle, Merseyside L20 3QZ.

Whysall, Z.J., Haslam, R.A. and Haslam, C. Contemporary ergonomics consultancy practices: reducing the risk of MSDs. *Contemporary Ergonomics*. The Ergonomics Society 2004, 207–211.

Bartys, S. Psychological risk factors and absence due to MSDs. *Occupational Health Review.* September/October 2003, 31–34.

# CONFESSIONS OF AN EXPERT WITNESS

**Alan Jackson**

*Health and Safety Executive, Edgar Allen House, 241 Glossop Road, Sheffield S10 2GW*

Many human factors specialists will at some time be called upon to present expert evidence at a coroners inquest or in civil and criminal proceedings. Training courses and guidance are available for experts who are required to present expert witness evidence. Whilst these are good resources for finding out how things are supposed to happen the reality of the courtroom can still be very daunting. The author of this paper is a Human Factors Specialist Inspector with the Health and Safety Executive who has experienced the reality of presenting Human Factors evidence in court. In this paper he will attempt to share some of his practical experiences by discussing issues such as what is supposed to happen vs what actually tends to happen, challenges to expert status, potential pitfalls and some ways to avoid them. The paper will not deal with descriptions of the English legal system, the assumption being that the reader has some knowledge or will seek it elsewhere.

## Introduction

Many human factors specialists will, at some point in their career, find themselves presenting expert evidence in a court of law. Whilst there are many sources of advice and training available on the English legal system and being an expert witness (e.g. Bond et al., 1999) it is still a daunting prospect when you have to do it for real.

The author of this paper has worked as a human factors expert in the Health and Safety Executive for many years. During this time he has presented expert evidence at several coroners inquests and criminal prosecutions. In this paper he will discuss his experiences and some of the practical survival strategies he has adopted and developed in order to survive the ordeal of being an expert witness in court.

As the expert you are not the one on trial. However, it can often feel like you are as the opposition will see you as a target to be attacked. Your basic armour is good solid evidence but you still need to present it undistorted by the attack. Suggestions for survival techniques are presented for Human Factors practitioners to adapt for their own purposes. It is not intended to give topic specific advice, so if you deal with MSD cases or human error the advice given here should be appropriate. The paper will not deal with descriptions of the English legal system, the assumption being that the reader has some knowledge or will seek it elsewhere (see references at end of paper). The paper will present the kind of practical advice not necessarily found in formal law text books, although the above reference does include some useful tips.

## Preparation for court

When you are called to be a witness you should receive information on where and when you are expected to appear. It is important to get the domestic arrangements sorted out fairly early in your preparation as you do not want last minute panics over hotel reservations or transport arrangements. A good place to start is the court service web site (www.courtservice.gov.uk) which will give you contact details and maps along with a whole range of information on what to expect when you go to court. The Home Office is responsible for matters relating to coroners and there is a link on the court service web site. Coroners inquests are not always held at court buildings, with town halls being a common venue. With the domestics out of the way you can concentrate on preparing to present your evidence.

At a coroners inquest the aim should be to present the evidence relating to the circumstances by which the deceased met their death. It is not within the remit of the inquest to apportion blame. As such it is your job to present the technical facts and their interpretation, by way of your expert opinion, to assist the jury in reaching a verdict on the cause of death. In reality, it is likely that there will be council present to represent all interested parties in any subsequent criminal and/or civil action. Experience shows that there will always be some jostling for position on blame and mitigation. It is up to the coroner to keep control of the extent to which this occurs. However, your preparation should be to look at your expert report and consider how you can tell the "story" to the jury, assuming none of them has any technical background. I use the term story to illustrate how you may organise the facts and not to suggest you enter the realms of fiction. All good stories have a beginning, a middle and an end, ensuring the audience can follow what is going on. In the beginning set the scene and introduce the characters, what is it you have been called in to give an expert opinion on? In the middle develop the characters and plot e.g. RSI is caused by … and the effects are … The end is where you sum up and generally give your opinion. In this way you can lead the court through your evidence in a logical sequence which they should find easy to follow.

If you are giving evidence in a defended prosecution then you are caught up in the adversarial legal system. Whether you are called by the prosecution or the defence your job as an expert is still to help the court/jury to understand the technical evidence pertaining to the case being heard. However, your council will want to adopt an approach which will present the evidence in the best way to suite their aims. It is a good idea to find out what this approach will be to help you prepare. It can be very useful to attend a case conference prior to the hearing in order to discuss your teams approach and get a feel for how your evidence fits in with the bigger picture. This is an ideal which is not always possible but you should at least speak with your council. Conversely, the opposition will have their own approach which will likely involve the discrediting of your evidence. It is better to identify potential weaknesses in your evidence at an early stage, and discuss them with your council, than to hope no one spots them. It can be useful to anticipate some cross-examination scenarios and practice your responses.

When preparing your report and statements be careful of the wording. It can often be the case that your evidence is attacked on semantics as opposed to technical content. For example, "it may be considered that this repetition rate is unacceptable" will suggest to the opposition council that it may not be considered unacceptable. If you can't say for sure "this repetition rate is unacceptable" then be prepared for an attack on what is a weak statement.

Finally, make sure you have all the papers you will need, including exhibits where appropriate. You are not expected to memorise every detail and providing they have been exhibited you can refer to reports and photographs, etc. Visual aids, as exhibits, can help a lot in "telling your story" to the court. It is also important to get hold of a copy of the opposition expert's report where possible. It is unlikely that two experts will agree on everything, and highly unlikely when they are on opposite sides of the court!

## Giving evidence

First and foremost the court has to accept you as an expert. There is no room for modesty here so include all academic and professional qualifications along with relevant experience. Do not be tempted to claim expertise where it can not be substantiated. It is not uncommon for adversaries to play the "my expert's bigger than yours" game. You can score points in this game by admitting a lack of knowledge at the start rather than being caught out and discredited later. Learn the phrase "I'm sorry, that is outside my area of expertise" and be ready to use it. It is important to be accepted as an expert by the court as this brings important privileges. The two key ones are sitting through other witnesses evidence and being allowed to give opinions as well as factual evidence. There may be occasions where you will be asked to comment on evidence given by other witnesses to assist the court in it's interpretation. You may be asked for technical advice on evidence presented during the proceedings by your own council. In these situations it is of great help to make notes and written comments as evidence is given.

In a defended case you will give evidence in chief by answering questions posed by your council, be cross examined by the opposition and possibly re-examined by your own council on issues arising from the evidence. The cross examination is likely to be the most challenging as there is usually an attempt to discredit you, your evidence or both. Here is where you need to bring some survival strategies into play. A few tried and tested techniques are presented in Table 1 below.

As discussed above, it should be less confrontational giving evidence at an inquest than in a prosecution. It is good practice though to be prepared for challenges to your evidence and opinion regardless of the setting. Generally a coroner will ask you to "tell your story" and then tease out any further evidence or opinion by asking questions and allowing examination by other legal representatives.

**Table 1. Some useful do's and don'ts.**

| Do's | Don'ts |
|---|---|
| Listen carefully to the question you are being asked. | Ramble on beyond the question you have been asked. |
| Look at the person asking the question then turn away and direct your answer to the jury or bench as appropriate. | Allow yourself to be drawn into intimidating body language and vocal inflections from the opposition. |
| Take a short time to consider how best to present your answers. | Enter into long technical descriptions of your evidence. |
| Assume no technical background on the part of the people you are addressing. | Attempt to show off your expertise with highly technical language. |
| Qualify your answers where when not to do so could mislead the court/jury. | Be afraid to express concern where you suspect you have been misunderstood. |
| Stay within the limits of your expertise as accepted by the court. | Give opinions on items outside the area of your accepted expertise. |
| Take notice of the response of the court/jury to your answers to check they are still following your "story". | Talk too fast for people to record what you say (stenographers, jury, judge, etc). |
| Remain composed, professional and as objective as possible. | Take any attacks personally. |

## Presenting human factors/ergonomics concepts

The first technical point you will be asked to explain is most likely to be "what is human factors/ergonomics?" Don't go for the technically complete and correct definition. Think about what it is you are going to be talking about and put your definition in context. If you are dealing with human error then define human factors as "the study of how people understand and interact with the things around them". If you are dealing with repetitive strain injuries (RSI) then an appropriate definition of ergonomics may be "the study of how the human body responds to the work environment". RSI brings in another issue, technical acronyms. If you must use them, give a clear definition of what you are describing with the acronym e.g. the damage caused to tissue around a joint due to repeated movements is known as repetitive strain injury or RSI for short. You should be prepared to explain all technical terms used or better still, use common language descriptions e.g. instead of referring to anthropometric data, call it tables of human body dimensions. The same goes for anatomical terms. As experts, we will wish to be technically correct but a member of a jury will be much happier with the height to the back of the knee rather than popliteal height.

One of the advantages we have in our field of expertise is that we are usually dealing with people in particular situations. This can help in finding a context which most people can relate to. An example would be to illustrate human error due to inappropriate display information on a process plant control panel by comparison with trying to drive your car at exactly 30 miles per hour when the only display on the dashboard is the engine rev counter. Be careful with units also. Younger people may handle metric units better than older people e.g. 95 kg for a body weight of 15 stones. Some units may be foreign to all but the experts. It may be more technically correct to present force data in Newtons but more people will understand kilograms force, or even equivalent bags of sugar.

Don't think of this process as dumbing down, you do not want to come across as patronising. Consider it more in terms of a beginners guide to the subject you are dealing with. You can get some clues as to how well the jury are following your evidence from their body language. A sea of vacant stares suggests a little more explanation may be required.

## In conclusion

There are a few basic survival techniques which the author has developed through experience which can be summarised as follows (in no particular order of importance):

- Remember, you are not the one on trial.
- Your job is to assist the court in understanding the technical evidence.
- It is not your job to prove guilt or apportion blame.
- Be as careful with the semantics as you are with the facts.
- Listen carefully and answer the question you are asked.
- Learn the phrase "I'm sorry, that is outside the area of my expertise" and use it where appropriate.

As an expert witness you will always be a legitimate target for the opposition in the English courts adversarial system. However, with a little planning and mental preparation it can be a challenging yet rewarding experience.

The comments in this paper are those of the author, based on personal experience, and do not necessarily reflect Health and Safety Executive policy. They are not intended to be exhaustive or prescriptive.

## References

Bond, C., Solon, M., Harper, P. 1999, *The expert witness in court a practical guide*, (Shaw & Sons, UK).
University of Leeds Innovations, 1995, *In my opinion an introductory guide for the expert witness*.
Darbyshire, P. 2001, *English legal system in a nutshell*, Fifth Edition, (Sweet and Maxwell, London).
Partington, M. 2003, *An introduction to the English legal system*, Second Edition, (Oxford University Press).
http://www.courtservice.gov.uk/

# USABILITY EVALUATION METHODS FOR CHILD PARTICIPANTS

**Helen Edwards & Rachel Benedyk**

*University College London Interaction Centre, Remax House, 32 Alfred Place, London WC1E 7DP*

This study assesses three usability evaluation methods (Active Intervention, Peer Tutoring and Cross-Age Tutoring) with children aged 6–8 years old within a school setting, using an interactive, educational multimedia product. Cross-Age Tutoring elicited significantly fewer comments than the other two methods, and "plan" comments were significantly rarer than "action" and "perception and cognition" comments. In terms of the suitability of these evaluation methods for child participants, and context of use in this particular setting, Peer Tutoring appears to have the most potential. Methods need to better reflect children's motivations and the variety of contexts in which they use multimedia products. The implications for HCI practitioners are explored.

## Introduction

In interaction design, children's products must be evaluated by children (Hanna et al., 1997), as this group of users differ from adults in many ways e.g. cognitive development, motor control, language comprehension and reading ability. In addition to objective performance measures, it is important that such usability tests include subjective user data via verbal protocols. This is often done using a technique that is known as "think-aloud".

### *The think-aloud technique*

The origins of think-aloud can be traced back to the work of the experimental psychologist Duncker (Nielsen et al., 2002), though the work of Ericsson and Simon (1984) is often cited as justification (Boren and Ramey, 2000). In usability evaluation, verbal protocols are claimed to uncover the cognitive processes of test participants. This technique has been successfully conducted with participants aged 8 upwards (Branch, 2000).

Despite its popularity, a number of criticisms have been levelled against the technique, such as: the extent to which models developed in cognitive psychology apply to usability testing in HCI; the validity of the data collected; the influence on the tasks performed; and the increase on the subject's cognitive load (Karentsy, 2001). Taking these criticisms into account, traditional think-aloud methods are likely to be unsuitable for younger children as they rely on skills that this age group may not yet have. Verbalising thoughts necessitates firstly having self-awareness, a capacity which younger children are likely to still be developing (Cole and Cole, 1993) and secondly an ability for verbal expression, requiring a mature vocabulary.

### *Techniques used with young children*

Van Kesteren et al. (2003) looked at the ability of children aged 6–7 years to provide verbal comments in user tests, by comparing six techniques; Usability Testing, Think-Aloud, Active Intervention, Retrospection, Co-Discovery and Peer Tutoring. Peer

Tutoring (where a tutor spends some time with a product and then teaches a friend to carry out a number of tasks), has been developed specifically for use with children and has been carried out with younger participants aged 5–9 years (Hoysniemi et al., 2003).

Van Kesteren et al. counted the number of comments made by participants and categorised the comments as "plans", "actions" or "perceptions and cognitions" using Norman's action cycle (Norman, 1986). They found that most verbal comments were elicited from Active Intervention and that participants verbalised fewer plans than the other types of comments. However, the influence of their low subject numbers and the range of individual differences meant they were unable to conclude that differences were due to evaluation methods used.

### *Outline of this study*

This study set out to revisit usability evaluation techniques for young children in order to assess their value to developers of interactive products for children. To achieve this, an educational multimedia product was evaluated with children aged 6–8 years, using three usability evaluation methods: Active Intervention, Peer Tutoring and Cross-Age Tutoring. These techniques do not require participants to provide concurrent verbal protocols and are therefore viable alternatives to traditional Think-Aloud when user testing with younger children. Active Intervention and Peer Tutoring were chosen as van Kesteren et al. suggested these may elicit more verbal comments than other methods. Cross-Age Tutoring (an older tutor teaching a younger child) was included, since such an age gap may facilitate assistive comments (Wellman and Lempers, 1977). In addition, the study was designed to allow statistical comparison between the methods, in terms of the numbers of verbal comments elicited by participants, prompts given by the experimenter and incidents of task repetition. Categorisation of comments used the same coding scheme as van Kesteren et al. to allow comparison. Effects of task on numbers of comments elicited and effects of method on types of comments were scrutinised.

The study took place in a one-form entry primary school in the UK, with participants aged 6–8 years. An interactive history CD ROM, for children aged 6–10 years, was used for evaluation. Participants were unfamiliar with the product but familiar with the content. Six children were allocated to Active Intervention, twelve to Peer Tutoring and six in each of two age groups to Cross-Age Tutoring. The test conditions were balanced for gender and range of ability. The children were provided with six representative tasks to complete, in a set order, using the multimedia product. Following a briefing, a training session and a period of exploration, their understanding was checked. All sessions were recorded using a video camera, which focussed on the participants. From the recordings, all verbal utterances of the experimenter and the participants were transcribed.

During Active Intervention participants were asked questions designed to elicit verbal responses about their "plans", "actions", and "perceptions and cognitions" whilst carrying out each task. In Peer Tutoring and Cross-Age, tutors were told that they would later be asked to teach these tasks to another pupil, and intervention was kept to a minimum so that participants would use their own strategies and vocabulary as much as possible.

## Results

Significantly fewer comments were elicited from Cross-Age Tutoring than the other two methods ($p < 0.05$). Significantly more prompts were provided by the experimenter in Active Intervention than in the other two methods ($p < 0.05$). Significantly fewer plans were elicited than action or perception and cognition comments ($p < 0.05$). Significant differences were found between the number of perception and cognition comments elicited in each evaluation method ($p < 0.05$), but not between other comment types. No significant differences were found on any other tests. There were no fatigue effects nor effects of practice over tasks.

These findings are comparable with those of van Kesteren et al. (2003) who also found that the Active Intervention sessions elicited the most verbal comments and required the most prompts and task repetitions from the experimenter, and that plans accounted for the least number of comments made. Their results also indicate that the greatest number of planning and action comments were made during Peer Tutoring. However, in contrast, the present study found that the most planning comments were made during Active Intervention.

## Discussion

Participants in Active Intervention may have elicited more verbal comments because they were actively probed for responses, and were using the product for the first time. Participants in both types of tutoring sessions may have made fewer verbal comments as they were more experienced with the product and likely to be teaching more efficient ways to complete the tasks.

Significantly fewer plans were elicited compared to actions or perceptions and cognitions. This may be because eliciting plans requires metacognitive skills that participants of this age (6–8 years) are still to develop; reporting actions and perceptions and cognitions does not require such skills.

It may also be that children of this age do not in fact formulate plans when faced with a task to complete. Ellis and Siegler (1997) argue that children fail to plan for many reasons, e.g. planning requires effort and delay or suppression of easier and more automatic behaviours (just jumping into the task), or because it is difficult, time consuming, boring and not guaranteed to pay off.

The differences in the types of comments may be because participants adapted their speech according to the characteristics of the listener. Tutors in Cross-Age Tutoring provided fewer plans and perceptions and cognitions but more comments pertaining to actions. Tutors in these sessions tended to tell younger tutees exactly what to do, instead of providing them with strategies. In contrast, tutors in Peer Tutoring taught friends of roughly the same age and were more likely to suggest a strategy for completing the task. In Active Intervention, participants were communicating with an adult, and produced the most plans, possibly due to the probing of the experimenter. Participants may have commented less on their actions, as they were aware of observation by the experimenter.

All this is consistent with findings in cognitive psychology, which show that children will adapt their speech production and comprehension to account for properties of the speakers, listeners and the settings in which the speech takes place (Flavell et al., 2002).

However, it may be argued that aspects of the procedure in all methods tested may invalidate the data gathered. According to Ericsson and Simon (1984) prompts and comments from the experimenter can interrupt the normal flow of information and invalidate data. Moreover, comments provided by the tutors in these sessions could be seen as instructions and may not be an account of their own cognitive processing.

### *Issues with child participants*

Child participants of the age in this study may require intervention from the experimenter; they expect guidance when asked to use something (a feature of school culture) and when they show signs of losing motivation, they need encouragement. The issue of intervention also raises ethical questions for child participants, such as the maintenance of self-esteem. Hence, it may be necessary to doubt the applicability of the model presented by Ericsson and Simon (1984) to usability testing with children.

The methods do differ in terms of the cognitive demands they place on the participant. The two tutoring methods offer an advantage over the Active Intervention method, as tutors are able to focus their cognitive resources on carrying out the tasks and both participants are able to focus on a single aspect of the task. However, a goal-driven view of children as users may be inappropriate, particularly for educational multimedia products. Druin (1999) argues that children's activities are often exploratory, especially when not directed by adults. Effective multimedia must facilitate interactions that are based on exploratory and generative approaches to learning. The challenge of using tasks that facilitate exploration has implications for the way that verbal data is elicited.

*Context*

Bodker and Madsen (1998) argue that it is important to investigate the situation of use so that realistic test data may be retrieved. A number of key aspects of context must be considered; work materials, artefacts, time, cooperation, location and organisational setting. In terms of the "cooperation" aspect the tutoring methods better match typical use context. Within a classroom, pupils often work without help from the teacher and in pairs when using CD ROMs (Cummings, 1985). It seems that the Peer Tutoring method best matches this aspect of use context intervention.

The aspects of "time" and "location" raise questions with regard to evaluation methods used for evaluating educational multimedia. Time pressure in the classroom, as well as the normal disruptions and background noise, were absent from all the testing methods in this study. These nuances of classroom life may affect the usability of educational multimedia in ways that may only become apparent when products are tested in the field.

## Implications for practice

Although exploratory tasks may reveal more typical behaviour, they may be of less use to practitioners wishing to test a specific aspect of the system, as such tasks may not easily be directed. Thus, it may be necessary to incorporate both exploratory and goal-driven tasks in usability testing sessions. If practitioners were to forge partnerships with schools this might provide valuable insights into the activities in the field that educational products are designed to support. Other usability evaluation methods may be able to better integrate exploratory tasks, such as Co-Discovery where two participants work together to perform the tasks (Kennedy, 1989). However, previous research has reported that child participants find cooperation difficult (van Kesteren et al., 2003). It may be that case studies are an alternative way of gaining further insights. It seems that multiple evaluation methods should be considered when evaluating educational multimedia.

As this study shows, cognitive models developed within psychology such as Ericsson and Simon's cannot be directly applied to user testing when young children are testers. Moreover, there are specific issues surrounding experiments with child participants, such as large individual differences in capability and motivation. This study discusses some of these issues and highlights a need for caution in this area.

## References

Bodker, S., and Madsen, K. 1998, Methods and tools: context: an active choice in usability work, *Interactions*, **5** (4), 17–25

Boren, M., and Ramey, J. 2000, Thinking aloud; reconciling theory and practice. *IEEE Transactions on Professional Communication*, **43** (3), 261–278

Branch, J. 2000, Investigating the information-seeking processes of adolescents: the value of using think alouds and think afters. *Library and Information Science Research*, **22** (4), 371–392

Cole, M., and Cole, S. 1993, *The Development of Children*, Second Edition, (Scientific American Books, USA)

Cummings, R. 1985, Small group discussions and the microcomputer. *Journal of Computer Assisted Learning,* **1**, 149–158. In P. Light and K. Littleton (ed.), *Social Processes in Children's Learning*, (Cambridge University Press, Cambridge)

Druin, A. 1999, Cooperative inquiry: developing new technologies for children with children. *Proceedings of the SIGCHI conference on Human Factors in Computing Systems: the CHI is the Limit*, 592–599

Ellis, S., and Siegler, R. 1997, Planning and strategy choice, or why don't children plan when they should? In S. L. Friedman and E. K. Scholnick (ed.), *Why, How, and When Do We Plan: The Developmental Psychology of Planning*, (Erlbaum, Hillsdale, NJ)

Ericsson, T., and Simon, H. 1984, *Protocol Analysis: Verbal Reports as Data.* (MIT Press, Cambridge, MA)

Flavell, J., Miller, P., and Miller, S. 2002, *Cognitive Development,* 4th Edition, (Prentice Hall, NJ)

Hanna, L., Risden, K., and Alexander, K. 1997, Guidelines for usability testing with children. *Interactions*, **4** (5), 9–14

Hoysniemi, J., Hamalainen, P., and Turkki, L. 2003, Using peer tutoring in evaluating usability. *Interacting with Computers*, **15**, 203–225

Karentsy, L. 2001, Adapting verbal protocol methods to investigate speech systems use. *Applied Ergonomics*, **32** (1), 15–22

Kennedy, S. 1989, Using video in the BNR usability lab. *SIGCHI Bulletin*, 92–95

Nielsen, J., Clemmensen, T., and Yssing, C. 2002, Getting access to what goes on in people's heads? – Reflections on the think-aloud technique. *Proceedings of the second Nordic conference on Human-Computer Interaction*, 101–110

Norman, D. 1986, Cognitive Engineering. In D. Norman and S. Draper (eds.), *User Centred System Design*, (Lawrence Erlbaum Associates, Hillsdale, NJ), 31–61

van Kesteren, I., Bekker, M., Vermeeren, A., and Lloyd, P. 2003, Assessing Usability Evaluation Methods On Their Effectiveness To Elicit Verbal Comments From Children Subjects. *Proceeding of the 2003 conference on Interaction design and children*, 41–49

Wellman, H., and Lempers, J. 1977, The naturalistic communication abilities of two-year olds. *Child Development*, **48**, 1052–1057

# AIR TRAFFIC CONTROL

# HUMAN FACTORS IN AIR TRAFFIC MANAGEMENT: MAKING A DIFFERENCE

**Barry Kirwan[1], Mark Rodgers[2] & Dirk Schaefer[1]**

[1]*Eurocontrol Experimental Centre, Bretigny/Orge France*
[2]*FAA Headquarters Washington, USA*

This paper outlines a set of case studies in Europe and the US concerning successful applications of Human Factors in the Air Traffic Management (ATM) domain. These case studies indicate the state of the art in applied Human Factors in ATM, and show the types of impacts that can be achieved. They also suggest where Human Factors needs to improve, both technically and strategically, to achieve more comprehensive, consistent and sustainable impact in this demanding area.

## Introduction

Human Performance and Human Factors have long been considered to be important to Air Traffic Systems performance. This has led to many air traffic service provision companies and organisations either taking on Human Factors professionals or at least taking active interest in the field and acting on generic research and good practices made available in the public domain. Organisations such the US Federal Aviation Administration (FAA) and Eurocontrol in Europe have invested heavily in this area, and have qualified staff and programmes to continue to develop insight, guidance and solutions related to Human Factors issues in contemporary and future air traffic management (ATM) activities and organisations.

However, in the late '90's, at several conferences in Europe, a serious question was raised by members of the air traffic community. Essentially it was a challenge, and an important one. Whilst it was not disputed that Human Factors was important and even critical, the ability of Human Factors to deliver realistic solutions that would translate into adopted operational practice, was questioned. Where were the published case studies showing Human Factors impacts? Where was the evidence that Human Factors could do more than carry out applied research and write academic papers? Where were the results that would mean something both to controllers at the operational level, and to managers and chief executives in air navigation service provision organisations? These questions made certain people amongst the ATM Human Factors community realise two things. Firstly there was a potential "image problem" for ATM Human Factors. Secondly, and more importantly, it was indeed difficult to find this evidence in the published literature, in readily accessible locations. Although there are many books and articles on Human Factors aspects in ATM, many of these are either written with a peer (Human Factors) audience in mind, or else are dispersed in diverse locations. Often Human Factors case studies do not deal with the ultimate result of how, or even whether, a solution was adopted and had sustained impact in an operational system.

The authors therefore surveyed a number of practitioners in Europe and the US to find "evidence" of effective Human Factors in ATM – success stories. The rest of this paper concerns those successes that were found.

**Table 1. Summary of Human Factors Impacts Found[1].**

| Title | "Driver" for the work | ATM function | HF approach | Impact achieved |
|---|---|---|---|---|
| Development and Implementation of a Position Hand-Over Checklist and Best Practice Process for Air Traffic Controllers | High rate of incidents following position hand-over | TMA | Interviews; observation | *Position hand-over procedure derived and widely implemented* |
| Human Factors issues in airport surface operations | Need to reduce number of incidents in US airports | Airport | Analysis of incidents | *Critical areas of airport operations identified; development of intervention strategies* |
| Human Error in European Air Traffic Management: from Theory to Practice | Need to look "deeper" into a particular incident | All | Human error modeling and derived incident classification scheme | *Enhanced tool for incident investigation developed; more detailed insights for incident reduction gained* |
| FAA Strategies for Reducing Operational Error Causal Factors | Need to reduce contribution of human error to operational errors | All | Incident recreation and operational error classification; investigation decision aid | *Method and tool for incident investigation implemented and validated* |
| Reducing Separation in the Open Flight Information Region: Insights into a Human Factors Safety Case | Desire to reduce separation minima in UK OFIR | Military airspace (OFIR) | Human reliability approaches & simulations | *Vertical separation minima reduction from 5000′ to 3000′* |
| Collaborative Decision Making and Post Operations Evaluation Tool | Desire to provide NAS users with feedback to reduce traffic flow delays and increase capacity | Traffic Flow Mgmt | Application of design guidelines; Analysis of communication techniques in collaborative decision making | *National deployment of Post-Operations Evaluation Tool (POET)* |
| Human Factors Longitudinal Study to Support the Improvement of Air Traffic Controller Training | High ATCO student fail rate | All | Longitudinal study; interviews | *Initial improvement in pass rate by 25%; sustained improvement* |

| | | | | |
|---|---|---|---|---|
| Selection in ATM – post-US-strike recovery | US strike led to sudden need to select large number of controllers | Air traffic control centers (including towers) | Selection tests | *US selected an effective and sustainable controller workforce* |
| Implementation of Critical Incident Stress Management at German Air Navigation Services | Need to support health and performance after critical incidents | Air traffic control centres (including towers) | Critical Incident Stress Management (CISM) | *CISM approach developed, implemented, and used following incidents and accidents* |
| Team Resource Management in Europe | Desire to improve teamwork in ATM operations | All | Team Resource Management (TRM) | *TRM taken up by 20 European countries* |
| Shiftwork in ATM: from research to the workforce | Health and safety considerations related to shiftwork | Air traffic control centers | A range of Human Factors, psychological, and psycho-physiological tests | *Shiftwork effects on performance and well-being better understood; workforce provided with educational material on mitigating fatigue and improving performance* |
| Measuring controller performance in the 21st century | Need to measure user performance to support system design & development | All | Range of measures including workload and situation awareness | *Successful "batteries" of measures available for a range of issues – proven ability to resolve HF issues scientifically* |
| Using modeling techniques to ensure human systems integration | Need to access performance in design & development work | All | Principally computer simulation and modeling approaches | *Use of models to refine airspace design for efficiency* |
| Performance Prediction in Air Traffic Management: Applying Human Error Analysis Approaches to New Concepts | Safety assessment and the desire to protect future ATM from human error | En Route, Approach, Military airspace | Human error identification approaches (TRACER & HAZOP) | *Improvements to interface and safety assurance for live trials, warning of future risk areas* |
| Managing US ATM HF Acquisition Programs | Need to integrate HF into the acquisition process | All | Human Factors study areas tied to Technology Readiness Levels (TRLs); integrated team approaches; HF staffing requirements | *Establishment of a strong set of procedures, competencies and processes for HF integration into projects* |

(*Continued*)

**Table 1. (*Continued*)**

| Title | "Driver" for the work | ATM function | HF approach | Impact achieved |
|---|---|---|---|---|
| The STARS story | Need to upgrade the controller radar terminal | TRACON facilities and control towers | Usability, prototyping and simulation; integrated teams | *Significant improvement of STARS interface design* |
| Transitioning to a new air traffic control centre and system in Malmo, Sweden | Need to upgrade the operations room technology | En route and TMA | HMI design, TRM, HERA, safety climate measures | *Improvement in the interface (e.g. stripless systems) and understanding of human error in ATM incidents* |
| A HF Programme in NATS | Company see HF as a growth area – development of a HF team and program | All | 3-year HF program; integration strategies | *Establishment of a strong HF team in the company – many impacts* |
| Integrating Human Factors Into Company Policy & Working Practice | Integration of Human Factors into company processes during a period of privatization | All | Total systems approach; integration into business activities; risk management approaches | *Consolidated integration of HF into company processes; risk reduction and training improvement* |

[1]For individual author details for each case study please see Kirwan, Rodgers and Schaefer (2005).

## Human factors impacts in ATM – overview of results

The full results are being published in Kirwan, Rodgers and Schaefer, 2005. Nineteen case studies were identified that showed the types of impacts applied Human Factors was having. These case studies varied in terms of what part of ATM they were affecting – e.g. airports, en route air traffic management, terminal manoeuvring areas around the larger airports (TMAs – also called TRACONs in the US), and approaches to airports. The studies also occurred at different times in the system development life cycles, from early design to operations. The studies themselves also differed in terms of why they occurred in the first place, i.e. the "driver" for the application or research. Table 1 summarises the results obtained.

The results show that Human Factors is indeed having real impacts on Air Traffic Management systems. Moreover, these impacts involve the full range of Human Factors application areas, from interface design and workplace layout, to training and procedures development, and from shiftwork and selection to management of the psychological well-being of controllers following an incident involving loss of separation between two aircraft. Also, the case studies demonstrate that Human Factors is being applied in various stages of the system development life cycle, and in the different parts of the ATM system itself, including military, airports, TMA and En Route air traffic control centres.

A major conclusion from the review is that Human Factors needs to focus on developing itself as an effective partner in future ATM system design. This necessitates having effective and relevant programmes of activities that are integrated with other system design and development functions. No-one in ATM denies the importance of the role of the human in ATM now or in the mid-term future. These case studies show that Human Factors can indeed "deliver", and is ready to take up its full role in assisting the development of more efficient and safe air traffic travel.

## Reference

Kirwan, B., Rodgers, M., and Schaefer, D. (2005: in press) *Human Factors Impacts in Air Traffic Management*, (Ashgate, Aldershot).

# DISTRIBUTED COLLABORATION OF A REMOTE SEARCH TASK: COMPARISON OF DATALINK AND AUDIO COMMUNICATIONS

**C. Baber, J. Cross & R.J. Houghton**

*Electronic, Electrical and Computer Engineering, The University of Birmingham, B15 2TT, UK Email: j.cross@iee.org; c.baber@bham.ac.uk*

This paper presents the development and initial user trial of a system of interconnected units, used to collect data and position from the field and relay these data back to a command post where is it visualised. The command centre can then make decisions based on the data collected and relay basic instructions back to the field units. The underlying goal of this research is to investigate the human factors involve when managing multiple field units from a hierarchical command structure with real-time two-way information flow capabilities. In this paper, a study shows that using datalink does not impede performance and significantly reduces non-task communications.

## Introduction

The sharing of data between field operatives and their colleagues has played a significant role in research into wearable computers since their inception. For example, work at Carnegie-Mellon University explored the use of a video link between a field operative and expert-at-base for vehicle maintenance (Siegel et al., 1995) and work at British Telecom explored the use of shared images of trauma victims between paramedics-on-scene and consultants-at-hospital (Garner et al., 1997). In these applications, the system supported two-way audio-visual communications between two people. As soon as more people become involved in the system, then it is likely that video (and to some extent audio) communications would start to overload the system. The overloading might be in terms of both technological constraints, such as bandwidth, and also human factors, such as the need to attend to multiple sources of information or the need to separate messages in a broadcast (i.e., one-to-many) stream. Consequently, in systems which involve multiple agents, there has been a trend away from audio communications to systems based on exchanging other data, e.g., voice communications have been replaced with datalink technologies so that data and commands can be sent without the need for verbal communication.

Research into datalink has tended to focus on communications between Air Traffic Control (ATC) and aeroplane cockpits. The general conclusions from this work would appear to be that, as might be expected, datalink changes the manner in which communication is performed. However, it is not clear whether the impact is always in favour of datalink. For example, Gibson et al. (2001), from a review of aviation near-miss incidents, argued that datalink could significantly reduce the likelihood of communication errors (although it is not clear whether errors would be eliminated or whether they might migrate to other error pathways). The suggestion is that datalink ought to lead to more efficient communications. On the other hand, Harvey et al. (2002) argued that the implementation of datalink would reduce the opportunity for shared awareness amongst flight-crew, e.g., by substituting the "broadcast" display of messages which could be overhead by all crew, with a single presentation of messages via the datalink display that one individual could read. Their study suggested that the introduction of datalink led to an increase in audio communications, with crew discussing the content of the datalink messages (although this was dependent on where the datalink was located in the cockpit). The implication, therefore, is that one might anticipate audio communications to be somewhat less efficient than datalink, but that other aspects of performance arising from shared awareness might be affected. To this end, the main question to be addressed in this work was whether a datalink system impairs performance, in comparison with more conventional audio communications. A relatively novel aspect of the study is the implementation of a datalink system in a context outside of ATC.

## Study

A simple search task was designed for this study. Cork tiles were laid out to form a letter on the floor of a tennis court, and then additional tiles were placed to serve as distracters. The aim of the task was for two people to work together to check the tiles and then decide which letter had been hidden. In this way, each participant would have access to part of the task-relevant information: the person on the tennis court (Field Operator) would be able to see the tiles but not whether they were targets or not; the person in the control room (Controller) would be able to see whether there were targets or not, but not where other tiles might be positioned. By passing information from the tennis court to the control room, the field operator would be able to supply intelligence that could support decision-making, and by passing commands back to the field operator, the controller would be able to manage the search task. The field equipment consisted on an HP Ipaq 5450 Pocket PC which receives data from a Global Positioning System (GPS) unit and a metal detector (see figure 1), and then relays these data to a remote server via a standard Wireless LAN.

In the Audio condition, the controller issues commands to change direction via a handheld radio. The set of commands are {North, South, East, West, and Stop}. In the Datalink condition, the direction commands are issued by clicking (using a mouse) onto a command palette. This causes data to be sent to the Ipaq, which then plays an audio (.wav) file recording to the field operator. Thus, in both conditions the person in the field hears a command to move in a specific direction. When the person in the field moves, the dot on the server screen also moves.

In order to compare a datalink with an audio condition, participants were recruited in pairs. They were paid £11 each to take part in the experiment which (together with

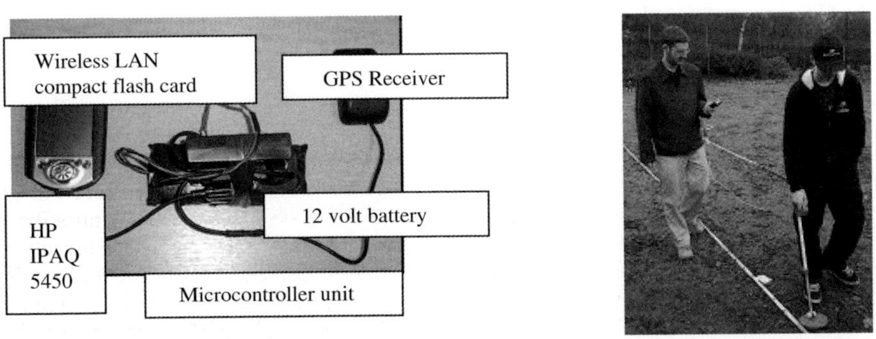


**Figure 1. Hardware used in the study.**

**Figure 2. Control Room Computer and its Display.**

training and debriefing) lasted approximately one hour. In total, sixteen people took part in the study. This gave four groups in the audio condition and four in the datalink condition. The study began with an initial familiarisation session, in which the participants were shown the equipment to be worn in the field and provided with a brief explanation of its function. Then they were shown the control room equipment and given an explanation of how to issue commands. During this training, it was made clear to participants that a clear, consistent and well-paced pattern of issuing commands would be most effective, i.e., it would be little use issuing a command to "Go North" twenty times over a few seconds, but would be preferable to issue commands only when direction of movement needed to be changed or when the person in the field did not appear to be walking. During this phase, the nature of the patterns was also explained (although participants were not told which letter they were looking for). Following the introduction, an initial training phase was undertaken in which one participant donned the kit in the field and the other issued commands from the control room. The aim of this phase was to ensure consistency of performance and to check that all equipment was operating correctly. On completion of this phase, participants engaged in two search tasks. Each search task was performed on separate but adjacent tennis courts (with a short break between each search). Data were recorded from the GPS unit and from the metal detector.

The server consists of a Visual Basic program that is configured to accept incoming connections from remote units as described above, and then displays a visual map of the unit's location and an indication of the data that has been collected at that point. At a remote location, in this case the office, the server displayed the data in real-time (see figure 2). When the remote unit is connected, its sensor data and position are sampled once per second and the visualization tool plots the positions on a map. The position points are represented by green blobs. When the received data meets a certain criteria, the blobs are increased in size and turn red.

## Results

The studies results are divided into two parts. Overall performance data are shown in table one. The average time to confirm target indicates that participants in the Audio condition took approximately 10fi minutes to complete each search task, and participants in the Datalink condition took approximately 8 minutes. The search performance (as defined by the numbers of hits and misses) did not differ appreciably between groups. This suggests that any difference in time was unlikely to relate to the task of searching per se. However, if one looks at the communications data, it is clear that the Audio group made, on average, fewer direction commands but around ten times as many "other" communications, i.e., confirmations, comments, commands to stop etc.

**Table 1. Summary of data.**

| Metrics | Audio | Datalink |
|---|---|---|
| Time to confirm target | 631.3 (234.2) s | 481.4 (127) s |
| Number of hits | 17.13 (7.02) | 17.67 (4.89) |
| Number of misses | 54.75 (22.08) | 50 (18.47) |
| Direction commands | 19.4 (9.18) | 23.5 (13.18) |
| Other communications | 20 (3.35) | 2.0 (1.26) |

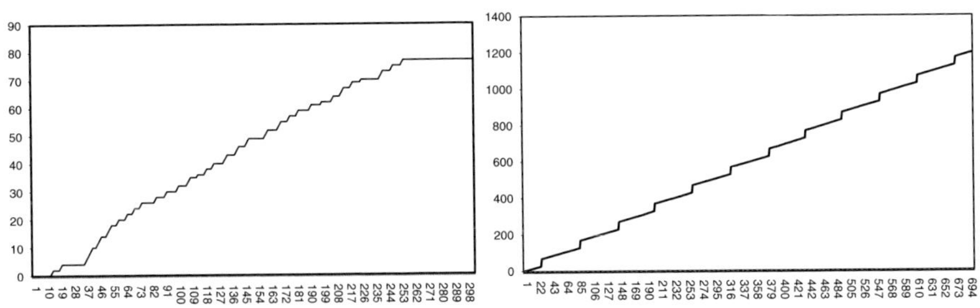


**Figure 3. Cumulative target cquisition for (a) audio and (b) datalink conditions.**

Given that the performance time between conditions appeared to be contingent on patterns of communication, we wondered whether participants in the field would interrupt searching to engage in communications. In other words, would there be an impact on search activity that could be related to the manner in which the search task was performend. Figures 3a and 3b show the cumulative time to acquire targets. Figure 3a (Audio) shows a clear stepping, suggesting that the search task was being interrupted on a regular basis, with the person in the field stopping the search activity in order to engage in communication. On the other hand, figure 3b (Datalink) shows an almost linear increase with time, suggesting far less interruption to the search task, which implies that communications were performed in parallel with moving around and searching.

## Conclusions

In the introduction, it was noted that comparison of datalink with audio communications tended to indicate that the change in technology lead to changes in performance. Typically, datalink is shown to be more efficient, than audio communications (which is one of the primary justifications of datalink), but that the loss of audio communications could lead to a reduction in the opportunity for shared awareness. The results from this study indicate that, in terms of efficiency (as measured by time and amount of communication), datalink appears to be superior to audio communications. There is no obvious impact on performance effectiveness (with both conditions performing similarly). There looks to be relationship between communication and performance time, with the audio condition supporting far more "non-direction" communications. The question is whether these "non-direction" communications would contribute to a sense of shared awareness. Obviously the task used in the study was sufficiently simple to reduce the potential benefits of such sharing of awareness – in effect, the experiment consisted on a "smart" controller and a "dumb" drone moving around outside.

Inspection of the audio transcripts suggests that much of the "non-direction" communication was not aimed at the task per se, i.e., it was not obviously related to sharing of awareness. Rather, it was concerned with managing what might be termed "social protocols" in terms of addressing any perceived imbalance of status between the controller and the drone. This was done either through the controller's commands being offered in the form of requests, e.g., "Now could you go north a bit" or in terms of discussions between the two, e.g., C: "Go east", D: "I can't go any further that way", C: "OK – go north then". The issue to address is whether it is necessary for the system to support communications (in order allow social protocols to be followed) or whether one simply needs to add extra information in order to justify a command, e.g., if the command "Go North" was paired with additional information indicating that the letter being searched for was an "L", this might be sufficient to disambiguate the command and also provide enough information for the person in the field to accept the command.

## Acknowledgements

This work is supported by a grant from EPSRC: GR/S85115 MsSAM (Methods to Support Shared Analysis for Mobile Investigators) and from the Human Factors Integration Defence Technology Centre, part-funded by the Human Sciences Domain of the UK Ministry of Defence Scientific Research Programme.

## References

Garner, P., Collins, M., Webster, S.M. and Rose, D.A., 1997, The application of telepresence in medicine, *BT Technology Journal*, **15**: 181–187

Gibson, H., Megaw, T. and Donohoe, L., 2001, Failures in Pilot-Controller Communications and Their Implications for Datalink , In (ed.) D. Harris, *Engineering Psychology and Cognitive Ergonomics. Volume Five: Aerospace and Transportation Systems*, Edited by. Ashgate Publishing, Aldershot, Hampshire, 325–333

Harvey, C.M., Reynolds, M., Pacley, A.L., Koubek, R.J. and Rehmann, A.J., 2002, Effects of the Controller-to-Pilot Data Link (Datalink) on Crew Communication, *Bridging Fundamentals and New Opportunities. Proceedings of the 46th Annual Meeting of the Human Factors and Ergonomics Society*, Santa Monica, CA: Human Factors and Ergonomics Society, 61–65

Siegel, J., Kraut, R., John, B. and Carley, K., 1995, An empirical study of collaborative wearable computer systems, *CHI'95 Conference Companion*, New York: ACM, 312–313

# RECOVERY FROM EQUIPMENT FAILURES IN AIR TRAFFIC CONTROL

**Branka Subotic[1] & Dr Washington Y. Ochieng[2]**

*Centre for Transport Studies, Department of Civil and Environmental Engineering, Imperial College London, London SW7 2AZ*
*[1]branka.subotic@imperial.ac.uk and [2]w.ochieng@imperial.ac.uk*

This paper gives an overview of equipment failures in Air Traffic Control (ATC). It presents data to show that equipment failures exist in spite of designed technical defences. A preliminary analysis of operational data from two countries has been carried out to determine the average annual number of failures and to highlight the major subsystems affected. The paper concludes with a discussion on the process of recovery from failures and proposes the methods for its improvement.

## Introduction

The Air Traffic Control (ATC) system represents a set of components that act together in order to achieve safe and efficient flow of traffic in any given airspace. The elements of the ATC system are the human operators, equipment, procedures, and environment, as well as all the links and interactions between them. Aviation safety and reliability management has mainly focussed on the prevention of technical (equipment) failures, human failures (also known as human errors), as well as organizational failures.

EUROCONTROL (the body responsible for the safety of air navigation in Europe) defines failures in the ATC domain as "the inability of any element of that system to perform its intended function or to perform it correctly within specified limits" (EUROCONTROL, 2002). This paper focuses solely on failures of one component of the ATC system, namely equipment and its impact on the correct operation of the whole ATC system.

The system architecture of any ATC Centre incorporates technical defences which are effective in mitigating the impact of the majority of the failures. However, there are occasions when these defences are not sufficient to maintain the normal system state and to protect against negative outcomes. The preliminary research presented in this paper is focused only on the failures which penetrate these defences and impact the controller. Thus, only failures requiring controller intervention (i.e. human recovery) will be considered hazardous. The human recovery process concludes with an outcome, ranging from positive (i.e. no effect on safety) to negative (with a different level of severity). A graphical representation of these key stages in failure management chain is given in Figure 1.

This paper gives an overview of the nature and characteristics of equipment failures in ATC. It starts with a brief discussion of the technical defences designed into the architecture of every ATC Centre. This is followed by an analysis of failures in general and those classified as hazards (i.e. failures which penetrate the technical defences and require controller intervention). The available literature has been augmented with the results of a preliminary analysis of operational failure reports. The following section addresses the problem of recovery, both at the technical and human levels. The paper concludes with a summary of the key findings, including the relevant failure statistics and the ATC functions mostly affected.

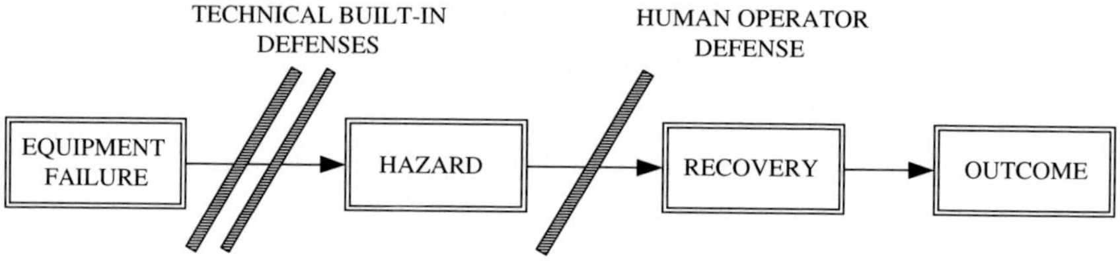


**Figure 1. Failure-hazard-recovery-outcome diagram.**

## Technical built-in defences

Technical built-in defences are design solutions for preventing, mitigating, or alerting the failure. The aim of every design is to assure the delivery of intended functions, as well as to predict possible failures. Experience shows that even the best designed systems fail occasionally. Therefore, it is crucial that every design concept has a means to re-establish system function and provide continuous service. All these design solutions are grouped under the term "technical built-in defences".

From the literature, it has been possible to derive a classification for technical built-in defences based on the function they provide. These categories are as follows:

- *Defences for recovering from failure.* They include safety devices and fault-tolerant system design though redundancy (physically independent spare components) and diversity (prevention of common cause failures).
- *Defences for transmitting information regarding the failure.* They include warning devices. Their function is to alert, inform, guide the initial responses, and to confirm whether the operator's response corrects the problem.

## Equipment failures as hazardous occurrences

An analysis of operational reports from incident databases has been conducted to augment data from available literature. The incident reports used were from two Civil Aviation Authorities (the aviation safety regulators), presented as source A and source B, and for the periods 1999 to 2003 and 2002 to 2003, respectively.

The analysis consisted of several phases: collection of reports, extraction of incidents involving only equipment failures which required the intervention of the controller (considered hazards in this paper), and the classification of the failures according to the ATC function affected. A critical part of the whole process was to prepare the data for a meaningful analysis (i.e. pre-processing). This step in the processing chain dealt with several problems: lack of consistent reporting schemes, ambiguous reports, different taxonomies, and differences in the characteristics of the equipment and the relevant automation level. As a result, the analysis presented in this paper involved a total of 1,089 reports from which 991 were due to equipment failures. The rest were mainly composed of non-ATC equipment, as well as other non-technical types of incidents.

Figure 2 shows the trend of the number of equipment failures over the time periods studied. The average number of failures is 121 and 193, for sources A and B respectively. In other words, on more than one hundred occasions controllers were faced with an unexpected event to which they had to react promptly and efficiently.

Further analysis was conducted to map the equipment failures to the various ATC functions. This showed that the majority of failures were associated with the communication and surveillance functions. These two sub-systems represent 78 percent of all failures reported during the study periods and covering both sources A and B (Figure 3).

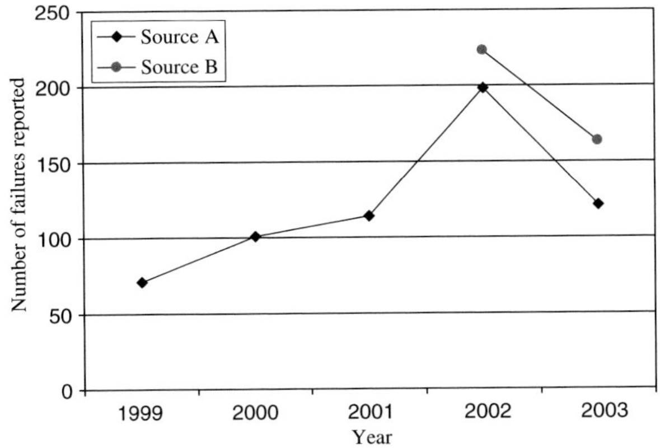


**Figure 2. The number of ATC equipment failures.**

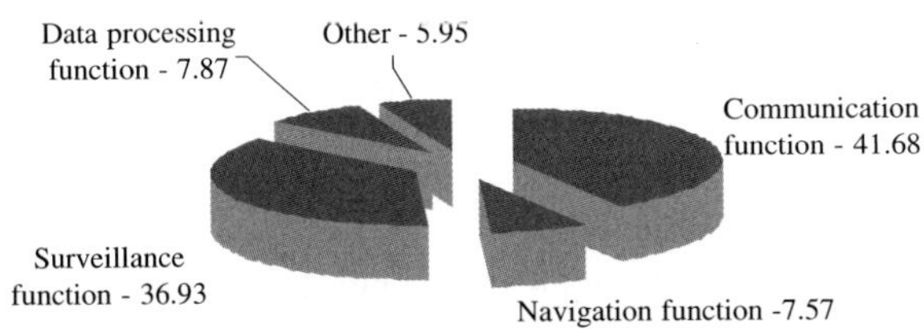


**Figure 3. Distribution of failure reports per ATC functionality.**

However, equipment failures are by default rare events. The higher frequency of communication and surveillance failures mean that controller might acquire experience and skills in handling those occurrences during the professional career. This is the reason why low-probability equipment failures may pose a much bigger threat to aviation safety. With these events, the controllers are unlikely to have any kind of experience, either direct (e.g. the actual occurrence, training), or indirect (i.e. the exchange of past experience with fellow colleagues). In addition, newly implemented or planned ATC systems are introducing higher levels of equipment complexity and thus, increasing the chances for low-probability failures.

## Recovery

At the beginning of this paper it was pointed out that the analysis of failures and recovery mechanisms is equally important and necessary. After the overview of the equipment failures in ATC and their distribution amongst various ATC functionalities, it is important to proceed with the recovery process. There are two separate but complementary types of recovery: technical and human. Technical recovery, as already explained, is achieved through the design solutions that constitute technical built-in defences. Therefore, the following paragraphs will focus on the fundamental principles behind human recovery from failures.

### *The human recovery process*

When all existing defences fail to prevent the occurrence and impact of failure, the controller becomes necessary and crucial for maintaining safe and continuous ATC service. Therefore, a better understanding of human recovery process should provide additional means to improve the system's safety and reliability.

The available literature on human recovery is based around the idea of a sequence of phases that constitutes the process of recovery. A general agreement among researchers is that there are at least three stages in the recovery process (e.g. Zapf and Reason, 1994; Kontogiannis, 1999; Kanse, & van der Schaaf, 2000). These are as follows:

- Detection – awareness that a failure has occurred. Human recovery is a deterministic process which depends highly on detection of failure. In other words, the controller cannot commence the recovery process in the case of missed detection of failures.
- Diagnosis or explanation – knowledge of the source of the failure.
- Correction – consists of the planning of the most appropriate strategy and its implementation.

The failure and recovery process concludes with the outcome phase. The outcome is highly dependant on the contextual issues surrounding the event and time available to react. In general, an outcome can be either positive or negative (Shorrock and Streater, 2003). A positive outcome succeeds a successful recovery which brings the system back to the normal or intermediate (but stable) state. However, if recovery is not successful then the controller will return to either the diagnosis or correction phase and will reassess the problem. These efforts will continue as long as there is time available for recovery. Otherwise, the final outcome may be negative with different levels of severity.

*Improving the human recovery process*

The on-going research presented in this paper further expands the problem of equipment failures in ATC by proposing a generic model of failure and recovery process. This model will aim to identify possible improvements of the recovery process through the solutions in the areas of equipment design, technical defences design, recovery procedures, and training.

## Concluding remarks

The idea of this paper was to present a high-level approach to the assessment of equipment failures in ATC and the processes for recovering from their impact. The data presented in Figure 2 has given an idea of the number of failed functions that occurred during one year in two airspaces. Over 150 failures were registered in both cases, showing that significant failures are a reality and should be given appropriate attention (significant here means that failures influenced controller's performance and were reported to the appropriate Authority). From the data sample studied, it could be seen that communication and surveillance systems contributed the majority of the failures. In spite of the high level of reliability and integrity that is required for all ATC equipment, failures do occur and will continue to occur in future. It is therefore important that recovery processes (both technical and human) are assessed and improved, as suggested in the previous paragraphs.

## References

EUROCONTROL, 2002, *Glossary of Terms and Definitions and List of Acronyms*, from http://www.eurocontrol.int/src/documents/deliverables/srcdoc4e2.pdf.

Kanse, L. and van der Schaaf, T. 2000, Recovery from failures – understanding the positive role of human operators during incidents. In D. de Waard, C. Weikert,

J. Hoonhout and J. Ramaekers (Eds.), *Human System Interaction: Education, Research and Application in the 21st Century*, (Maastricht, Netherlands, Shaker Publishing).

Kontogiannis, T. 1999, User strategies in recovering from system failures in man-machine systems. *Safety science*, **32**(1), 49–68.

Shorrock, S. and Straeter, O. 2003, *Managing System Disturbances in ATM: Background and Contextual Framework,* (EUROCONTROL).

Zapf, D. and Reason, J.T. 1994, Introduction: Human Error and Error Handling, *Applied psychology: An international review*, **43**(4), 412–432.

# VALIDATION OF THE ISA (INSTANTANEOUS SELF ASSESSMENT) SUBJECTIVE WORKLOAD TOOL

**Andrew Leggatt**

*BAE SYSTEMS, Human Factors Department,*
*Advanced Technology Centre, Sowerby Building,*
*Filton, Bristol UK*

This paper provides a summary of applied research undertaken at BAE Systems Advanced Technology Centre to examine the validity and reliability of the subjective ISA (Instantaneous Self Assessment) workload tool. An experimental approach was taken which compared the measure with NASA TLX and the Bedford Rating Scale (BFRS) to establish concurrent validity using an aerospace analogous task MAT-B (Multiple Attribute Task Battery). The results indicated that ISA is a valid tool, with reasonable internal reliability, good retest reliability and no detectable task intrusion. Construct validity was also established as high (.82), which was greater than the other competing workload tools. These combined findings suggest that ISA is a valid subjective workload tool.

## Introduction

This report describes the validity and reliability assessment of the subjective workload measure called Instantaneous Self Assessment (ISA) using data collected at RAF Lyneham and at the BAE SYSTEMS Advanced Technology Centre Human Factors Department, Filton during June, July and August, 2002.

Workload is defined for these purposes as the experience of how hard you are working. Workload measures are used to assess task demand, which is believed to affect perception of workload (Gopher & Donchin, 1986).

There are considered to be 3 types of workload measure:

- Performance measures
- Physiological measures
- Subjective measures

Although performance and physiological measures have some commendable features (e.g. direct measure of task performance is a highly objective measure) there are a number of issues concerning validity and difficulty of implementation associated with these measures. This paper assesses the more practical subjective workload assessments tools.

Subjective measures involve recording user ratings of self-assessed workload and employ rating scales such as ISA, the Modified Cooper-Harper scale, the Bedford Rating Scale (BFRS), and the NASA Task Load Index (TLX) etc. The scales just listed employ user ratings but an alternative is the observer method (OM), where the task is observed and rated by an observer. Whether one opts for user rating scales or the observer method, subjective measures are relatively cheap, easy to apply and simple to understand. However, their validity and reliability vary greatly according to the rating scale employed, (O'Donnell & Eggemeier, 1986), which is why it was important to perform a formal validity and reliability assessment of ISA.

ISA is arguably the simplest subjective workload measurement technique available. It consists of a 5-point workload rating scale represented by 5 buttons usually located on an "ISA box". Each button represents a particular level of demand from very low to very high. ISA is portable and can be used inside a working cockpit, simulator or at a desk etc. Whilst performing the task in question, the user responds to prompts in the form of flashing LEDs at predefined intervals, by pressing one of the five ISA buttons. ISA ratings are instantly recorded and displayed on a monitor out of the user's sight. The electronically collected data are then stored on an associated computer. ISA can also be used by multiple users, where many people perform a task at the same time and give their ratings simultaneously.

ISA was originally conceived at the National Air Traffic Service (NATS) in Bournemouth where it has been used to evaluate Air Traffic Controller workload. On request, the software was given to the Advanced Technology Centre Human Factors Department in 1997 whereupon it was rewritten to allow ISA to be used as a multi-user system.

There is a clear demand for ISA. So far BAE Systems has successfully applied ISA to a variety of programs. These include, among others:

- Nimrod MR2 baseline trials
- C130 Avionics Modernisation Program risk reduction trials
- Nimrod MRA4 Flight Deck Assessment Rig (FDAR) trials
- TRACER Future Scout Combat System (FSCS) reconnaissance assessments
- Brigade HQ Digitisation Stage 1 IT technology assessment

ISA offers several benefits compared to other workload measures. Firstly, ISA provides a simple unambiguous workload measure throughout a task. It is easy to use and simple to understand. ISA is arguably the least intrusive subjective measure available and, therefore, it is possible to gather workload ratings at the same time as achieving other task-specific objectives. Furthermore, whereas other subjective measures rely on tasks being logically segmented to enable the necessary post task assessments, ISA can be easily applied to all types of tasks. Finally, because responses are gathered in real time, ISA ratings are not subject to the memory decay and bias that can occur with measures that rely on post task assessments.

However, despite ISA's apparent ability to cause minimal intrusion upon the task, it still has the *potential* to cause disruption, and it was felt that this should be examined in this study. In addition, the instructions that should be given to the user needed to be looked at, as well as its sensitivity to the task and the correlation between the ISA ratings and actual performance. These issues were also addressed in this study.

Most importantly though, up until this assessment there was no evidence of any formal validity or reliability testing of ISA and this was considered crucial to underpin its use within BAE Systems.

## Method

A mixed experimental design was chosen for this assessment, (i.e. Between Groups and Within Groups variables).

Each participant was exposed to the control (CTRL) condition and two experimental conditions (the observation method and either ISA, TLX or BFRS). The observation method was always carried out at the same time as the other allocated experimental condition and these therefore count as one presentation (A) and the control condition as another presentation (B). The order of presentations was counterbalanced so as to minimise practice and fatigue effects (AB/BA) and therefore half the participants were given the AB order and the other half, the BA order.

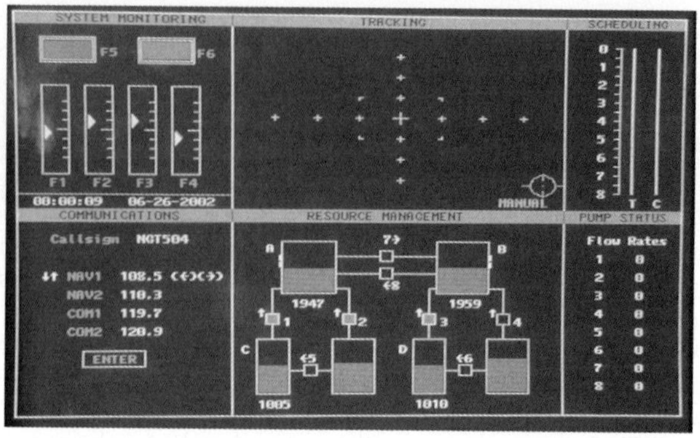


The task (MATB) was also counterbalanced to limit learning effects and this simply involved reversing the original task profile (re-named MATB1) to produce a second task profile (named MATB2). Participants were randomly allocated one of these profiles such that half of the participants were first exposed to MATB1 and half to MATB2.

The independent variables in this assessment were:

| | | |
|---|---|---|
| Workload Measure: | *Within Groups* IV | (CTRL/OM) |
| | *Between Groups* IV | (ISA/BFRS/TLX) |
| ISA Instructions: | *Between Groups* IV | (ISA1/ISA2) |
| Occupation: | *Between Groups* IV | (Pilot/Non-Pilot) |
| Demand Level: | *Within Groups* IV | (Very low 1 to Very high 5) |

The dependent variable (DV) is the variable that is affected in some way by the manipulation of the IV. In this assessment the DVs included:

1. Workload Rating
2. Performance

Thirty-two participants, consisting of 16 pilots and 16 non-pilots completed the experiment. Both groups were opportunity samples.

The participants in the Pilot group comprised RAF C-130 pilots with varied backgrounds (rotary, fast jet etc.) and levels of experience (ranging from a few months to more than a decade). The participants in the Non-Pilot group comprised of RAF Navigators, RAF Engineers, RAF Loadmasters and BAE Systems employees, all with no pilot training.

The experimental task equipment comprised the Multi-Attribute Task Battery (MATB) (see figure above). MATB was developed by the NASA Langley Research Center and was designed to investigate operator workload and performance. MAT-B has been used extensively in a number of laboratory-based experiments (Parasuraman, Molloy, Singh (1993); Comstock & Arnegard (1991)).

A variety of tasks could have been selected for this experiment but MATB was chosen because it incorporates tasks analogous to activities performed by pilots (the "known group"), whilst providing a high degree of experimental control and freedom of use with non-pilot participants.

MAT-B comprises four subtasks (monitoring, tracking, communication, and resource management) that must be performed at the same time for the whole period lasting 30 minutes.

The procedure was standardised for each participant and lasted 2 hours with two 30 minute test sessions during the experiment.

## Results

### *Concurrent validity*

ISA ratings correlated strongly (0.49) with BRFS ratings.
ISA ratings correlated strongly (0.55) with TLX ratings.
The observer method of rating correlated with the participant's ISA ratings at the level of 0.80.

### *Construct validity*

The ISA construct validity was tested by correlating the task demand, as imposed by variations in MATB task, and the workload rating from experimental participants. There was a significant strong correlation of 0.82 between task loading and workload rating. This correlation suggests that the ISA tool has good construct validity. The construct validity for the alternative measures of NASA TLX and BFRS were also tested in this manner. In this evaluation the construct validity of both TLX and BFRS appear to be less strong with correlations of 0.57 and 0.53 respectively.

### *Correlation with performance*

The relationship to performance was tested by correlating workload ratings with MATB task performance using Spearman rank order correlations (a non-parametric test). None of the relationships were statistically significant but as expected, a weak relationship in the anticipated direction was found (0.23–0.42).

### *Intrusiveness*

The intrusiveness of the ISA tool was examined by comparing the task performance with and without making ISA ratings. For all of the comparisons there was no significant main effects difference between the two groups. Likewise there was no significant difference between MATB performance with ISA or any other workload measures.

This is interesting since the task was not suspended for ISA ratings and yet ISA had little effect upon performance.

### *Internal reliability*

The internal reliability of the ISA measure was examined using the Cronbach's Alpha measure. This measure identifies whether the participants are using the ISA measure in a consistent manner by rating the same task loading with a consistent workload rating. The measure was 0.7 which indicates that ISA shows adequate internal reliability.

### *External reliability*

Retest reliability was examined by correlating the workload ratings from one trial with the workload ratings collected on a second trial at an interval of 2 weeks or more. The correlation between first and second workload ratings was 0.84 and is statistically significant. This suggests that ISA has a high level of external reliability.

## Conclusions

The validation experiment was undertaken to discover whether the ISA workload measure was a valid and reliable measure of subjective workload to support its application in the aerospace and defence industry.

The summarised data shown in this paper suggest that ISA has reasonable concurrent validity with both NASA TLX and BFRS. ISA has strong construct validity with a correlation of .82 with the task loading profile, which exceeded that of the alternative workload measures TLX and BFRS. This suggests ISA is a valid measure of workload as it appears to be measuring the same thing as other measures of workload.

The intrusiveness of ISA was assessed by measuring difference in task performance between the control and workload rating conditions. There were no significant intrusions in task performance with ISA. This suggests that ISA measures did not interfere with the task performance.

Internal and external reliability was examined. Internal reliability was assessed for each task level and was found to be satisfactory with a Cronbach's Alpha of 0.7. The retest, or external validity, was found to be excellent over a two week period with a retest reliability of 0.84.

These findings, combined together, suggest that the psychometric properties of the ISA measure provide a valid and reliable measure of workload which can be used as a concurrent measure of subjective workload.

## References

Castle, H. and Leggatt, A. (2002). Instantaneous Self Assessment (ISA) – Validity and Reliability. JS report 14865, Advanced Technology Centre, BAE SYSTEMS report.

Comstock, J.R. and Arnegard, R.J. (1991). The multi-attribute task battery for human operator workload and strategic behavior research. *NASA Technical Memorandum 104174*. Hampton, VA: Langley Research Center.

Gopher, D. and Donchin, E. (1986). Workload – an Examination of the Concept. In: *Handbook of Perception and Human Performance*, **Vol. 2**, *Cognitive Processes and Performance*, Boff, K.R., Kaufman, L. and Thomas, J.P. (Eds.), pp. 41.1–41.49. John Wiley and Sons, Chichester, UK.

Hart, S.G. and Staveland, L.E. (1988). Development of NASA TLX (Task Load indeX): Results of empirical and theoretical research. In P.A. Hancock and N. Meshkati (Eds.), Human *Mental Workload*, (Elseveir: Amsterdam), 139–183.

O'Donnell, R.D. and Eggemeier, F.T. (1986). Workload Assessment Methodology. In: *Handbook of Perception and Human Performance*, **Vol. 2**, *Cognitive Processes and Performance*, Boff, K.R., Kaufman, L. and Thomas, J. P. (Eds.), pp. 42.1–42.49. John Wiley and Sons, Chichester, UK.

Parasuraman, R., Molloy, R. and Singh, I.L. (1993). Performance consequences of automation-induced "complacency". *International Journal of Aviation Psychology*, **Vol. 3**, pp. 1–23.

# AIR TRAFFIC CONFLICT DISPLAYS – FALZON TO TROTSKY

**Hugh David**

*R+D Hastings, 7 High Wickham, Hastings TN35 5PB, UK*

A relative distance display was first suggested by Falzon in 1982. Although the display was not evaluated in its original form, a modified Conflict Risk display was tested in real-time simulation. It was received with enthusiasm by controllers, although its use was not easy. The Short Term Conflict Alert, implemented in real ATC, was more directly usable. Finally, the TROTSKY demonstrator incorporated conflict indications and conflict resolution validation into the controllers' main display. The development reflects the developing awareness of the potential for computer-based systems to reflect human cognitive processes.

## Introduction

The original justification for Air Traffic Control (ATC) was the avoidance of collisions between aircraft in flight. It is therefore surprising to observe that, until recently, although vast quantities of information were displayed to the controller, potential future conflicts were not. In part this may be attributed to the incremental nature of ATC development, which has emphasised continuity with earlier pre-computer systems, and consequently been unaware of the additional services a computer-based data processing system can provide. It may also be explained by the absence of any systematic attempt to design the Air Traffic system, taking into consideration the known strengths and weaknesses of human operators.

## Falzon's relative position display

In 1982, Pierre Falzon was a member of the Orly Group of psychologists and others, carrying out pioneering studies in the field of ATC in France. Some of these studies are reported in Bisseret (1995) which abstracts many of the reports of the Orly Group, most of which have never been published in English language publications.

Falzon (1982) began by pointing out that "The first goal of the work is to maintain safe separations between aircraft. Therefore the object of the controller's activity is neither each individual aircraft, nor the pairs of aircraft, but the future states of these pairs. The diagnosis is a decision on whether the future separations between the two aircraft will be safe."

Referring to earlier work by the Orly Group, (Bisseret, 1995), Falzon noted that the conflict detection process fell into two phases, an initial comparison of alphanumeric information (levels, routes, speeds, etc.) and a subsequent comparison of the relative movements of aircraft on the radar screen. Falzon pointed out that position was displayed graphically on the radar screen, with a history in the form of a trail of dots, providing an indication of speed and heading. Height, on the other hand, was provided only as a three-digit code, with a diagonal arrow pointing up or down where the aircraft

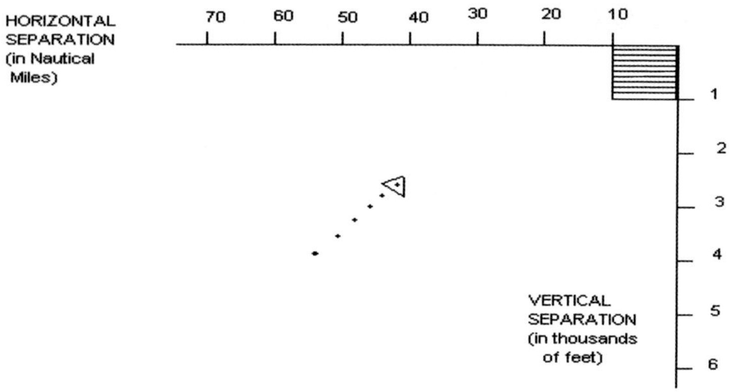


**Figure 1. Falzon's relative position display.**

was climbing or descending. There was no indication of the rate of climb or descent, and controllers in practice assumed that aircraft were not separated vertically.

Falzon rejected three-dimensional displays on the grounds that it is not obvious that a 3-D display would make perceptual or reasoning processes easier, and that operators do not use a 3-D image. Considering one pair of aircraft, he derived an image in which the horizontal and vertical axes corresponded to horizontal and vertical separation, and where a trail of dots provided a history as on a radar display of the period. Figure 1 is re-drawn from the original in Falzon (1982).

Falzon suggested that several pairs of aircraft could be displayed on the same display if properly labelled. He suggested that only those pairs within a certain distance and approaching each other in three dimensions should be displayed, but did not follow up these suggestions.

The initial reaction at the EUROCONTROL Experimental Centre was generally negative. Most aircraft at any time are in level flight at the same level or separated by a multiple of 1000 feet. In practice, the display would show horizontal lines of symbols slowly moving from left to right, and obscuring their trails of previous position dots. For the display to be useful, it should show the future trajectory. Specifically, it should show if the separation would, in the future, infringe the five nautical mile separation minimum.

It would therefore need to take account of the future intentions of the aircraft, and could not rely on simple extrapolation of the existing trajectory.

## Conflict risk display

An alternative display, showing time on the traditional horizontal axis, with horizontal separation on the vertical axis, was developed. The earliest sketches showed hyperboloid curves, extending up to twenty or thirty miles separation. For realistic traffic samples, including traffic in opposite directions at alternate levels on pre-defined routes, most curves became two straight lines. To be intelligible, only those curves where the loss of horizontal separation was accompanied by loss of vertical separation could be shown. Further simplifications were made by showing only the part of the curve for which horizontal separation would be lost, and by showing a horizontal line, at the level corresponding to the minimum separation, in place of that curve. A symbol was placed at the point of minimum separation, to which a number was attached.

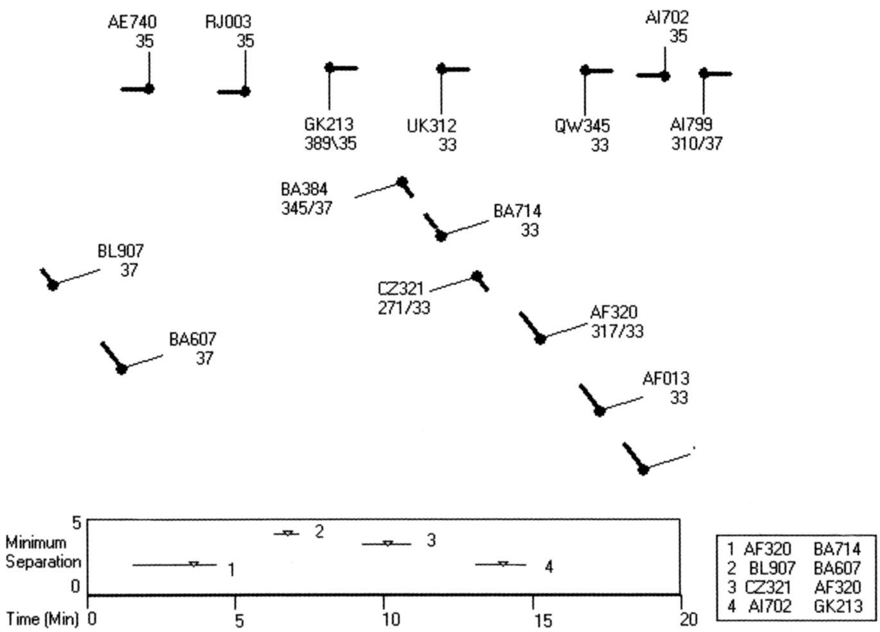


**Figure 2. Conflict risk display.**

A table was provided giving the pairs of call signs corresponding to these points. The controller could then refer to the main display to find the nature of the problem and devise a solution. Figure 2 shows a schematic Conflict Risk Display (CRD), redrawn in black on white.

This display was tested in the ODID III research simulation (Prosser et al., 1991). It was much liked by the controllers, in spite of the indirect trail involved in using it. Significantly, one controller remarked that the display was most useful when it was empty, since it reassured the controller that he had not missed or forgotten a potential future conflict.

Conflict Risk Displays, sometimes extended to show near-conflicts where minimum separation is more than five but less than ten nautical miles, are now widely used in advanced ATC display interfaces.

## Short term conflict alert (STCA)

A conflict display feature was independently developed at Maastricht UAC. Pairs of aircraft that would conflict if they continued on their present headings and height profiles were marked by colour change or flashing markers on their electronic labels. This display feature is rarely triggered in normal operation, since most problems will have been resolved at the planning stage or on sector entry. When it was designed, aircraft positions and speeds were, by modern standards, not well known, and flight plan data could not be reliably integrated into safety-critical applications. Where climbing or descending aircraft were planned to stop at levels which avoided conflicts, this could not always be taken into account, generating irritating false alarms. If an alarm was triggered, the action to be taken was usually obvious, and instructions could be issued in reasonably good time, assuming adequate communications. If several alarms were triggered simultaneously, however, there was a risk of confusion.

## TROTSKY

All these developments, except the STCA, considered separate displays of potential future conflicts. TROTSKY (Dee, 1996), a radical re-design of the ATC system, starting from the controller-system interface, assumed aircraft on direct airway-free routings, using precision satellite navigation and data-link computer-based communications.

TROTSKY defined conflict resolution as the controllers' main task. The four-dimensional trajectories of aircraft were compared with all other aircraft on system entry to detect potential future conflicts (here referred to as conflicts for short). Only aircraft involved in conflicts were displayed. Unresolved conflicts were indicated by linking the symbols of the aircraft with lines of increasing saliency as the urgency of the conflict increased. The most urgent conflict, in terms of time to loss of separation, was selected. Within this conflict, the height and distance profile of one aircraft (No. 1 in Figure 3 above) was shown, and its horizontal trajectory and that of all potentially conflicting aircraft (2, 3) up to their closest point was shown on the main display. The positions of both aircraft at their closest approach were linked by a line showing their relative position. Height differences were indicated by colour coding – Figure 3 shows the original colour-on-black display in black-on-white for ease of printing. The controller could prepare a solution to the problem, seeing the direct and indirect consequences on the main screen as the solution was constructed, keystroke by keystroke. The system verified solutions accepted by the controller, querying those which left conflicts unresolved or did not place the aircraft at the exact space-time exit point required.

In experimental trials (David and Bastien, 2003) 26 university students after one hour of general introduction and 20 short coached trials on TROTSKY, were able to

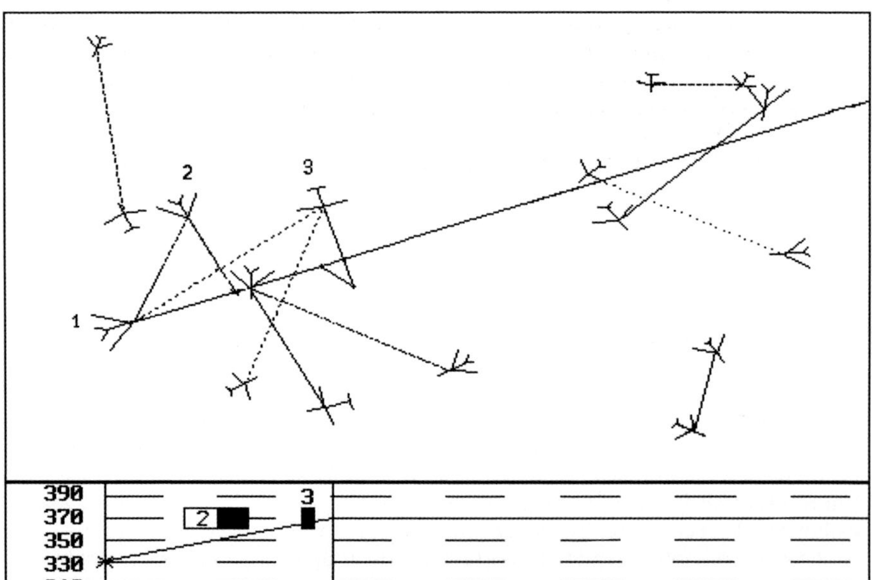


**Figure 3. TRACON conflict display.**

control random traffic at a rate of 200 aircraft per hour, with virtually no errors. Only two out of 1720 conflicts were not resolved correctly, and only 5 out of 10,468 aircraft failed to leave at the exact point, height and time planned. Practically, this implied that there would be little or no need for inter-centre coordination.

## Conclusion

The developments summarised here took place over a period of twenty years and are not yet fully accepted. They have involved a steadily deepening examination of the processes of ATC, with an increasing appreciation of the power and flexibility of modern digital information systems, leading of re-assessment of the roles of operators in the system. Implementing a radical revision of an existing safety-critical system has been described as "re-building a DC-3 as a B747 – *in flight*", and there is always a temptation to avoid difficult decisions in the short-term. Ergonomists may not make these decisions, but they must ensure that those who do are aware of the risks and rewards – of action or inaction.

## References

Bisseret, A., 1995, *Représentation et Décision expert*: Editions Octares (Toulouse, France)

David, H. and Bastien, J-M.C., 2003, Second Evaluation of a Radically Revised En-Route Air Traffic Control System In P.T. McCabe (ed.) *Contemporary Ergonomics 2003*, (Taylor and Francis, London) 305–310

Dee, T.B., 1996, Ergonomic Re-design of Air Traffic Control for increased Capacity and Reduced Stress, in S.A. Robertson (ed.), *Contemporary Ergonomics 1996*, (Taylor and Francis, London) 563–568

Falzon, P., 1982, Display structures: compatibility with the operator's mental representation and reasoning processes. In Johansen G. and Boller, H.E. (eds.) *Second European Annual Conference on Human Decision Making and Manual Control*, (Forschungsinstitut für anthropotechnik, Bonn, Germany) 297–305

Prosser, M., David, H. and Clarke, L., 1991, *ODID III Real-Time Simulation*, EEC Report No. 242. (Eurocontrol Experimental Centre, Bretigny-sur-Orge, France)

# ASSESSING SITUATIONAL AWARENESS IN TEAMS

**Barry McGuinness & Ben Dawson**

*Advanced Technology Centre, BAE Systems, Filton, Bristol BS34 7QW*

This paper describes an experimental evaluation of techniques to assess situational awareness (SA) in teams. Three attributes of SA in teams are assessed: team SA, shared SA and team metacognition. We propose techniques for deriving these measures from a method that combines true/false SA probes with self-ratings of perceived SA and ratings of other team members' SA. To illustrate the techniques, their application to a militarised team-based version of the game of chess is described and the results for team SA, shared SA and team metacognition are presented.

## Introduction

Situational awareness (SA), i.e. a person's real-time awareness and understanding of the dynamic task environment in which he acts, is a central concept in applied human factors research. While the assessment of individual SA is reasonably well developed, there is a need to pay attention to the added complexities of SA in teams. Within teams, some of the SA needed to support effective decision-making and performance will be specific to different individuals, while other aspects of SA must be shared across the group, i.e. the team will need to have a common perception and understanding of certain aspects of the situation. In such cases, team members need to help to avoid or rectify gaps in the team's SA, e.g., by sharing any relevant information with those who may need it.

The aim of the present study is to support the development of measures SA of across teams. Such measures should enable us to investigate differences both within and between teams under different conditions in terms of their SA. The following three attributes of SA in teams are considered fundamental:

- **Team SA**, the proportion of the total SA required that is actually held within a team.
- **Shared SA**, the extent to which common SA requirements are met across the team.
- **Team metacognition**, the extent to which members are aware of each other's SA.

The concepts of team SA and shared SA are illustrated in Figure 1.

Ideally, then, we seek a technique or set of techniques that is able to assess team SA, shared SA and team metacognition.

## The measurement of SA in teams

The approach adopted here is an extension of the QUASA (Quantitative Analysis of Situational Awareness) method developed within BAE Systems. The QUASA method (McGuinness, 2004) uses a combination of true/false SA probes and self-ratings of confidence in each probe response to quantify not only actual SA (through the analysis of probe responses using signal detection theory) but also perceived SA (mean confidence ratings) and, additionally, *SA calibration*, the extent to which perceived SA corresponds with actual SA. The proposed technique for assessing SA in teams involves a

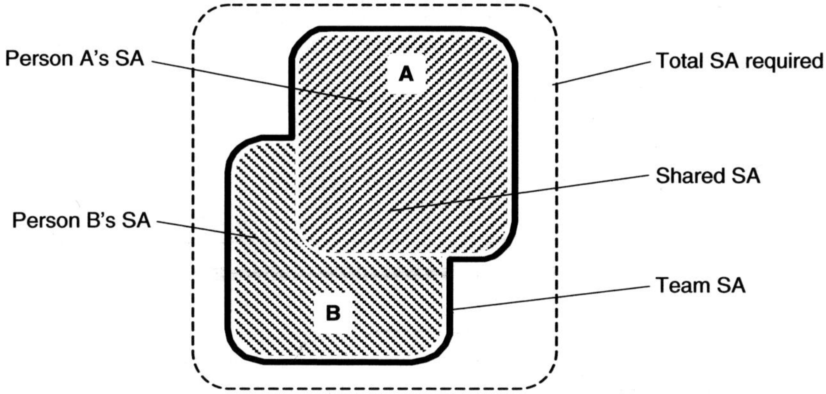


**Figure 1. An illustration of the concepts of team SA and shared SA.**

| SA probe # 2 | **The enemy intends to defend his position** |
|---|---|
| Response: would you say this is true or false? | ☐ **True** ☐ **False** |
| Do you know this to be the case or are you just guessing? | ☐ **I know**<br>☐ **I'm using judgement**<br>☐ **I'm just guessing** |
| Do you think your partner will know this answer or will he/she just guess? | ☐ **He/she will know**<br>☐ **He/she will use judgement**<br>☐ **He/she will just guess** |

**Figure 2. Example of a true/false probe with self-rating scale and partner-rating scale.**

small addition to QUASA: each team member gives ratings of the other team members' SA (termed "partner ratings"); see Figure 2 for an example.

The technique begins with an analysis of the task-relevant SA requirements of the individuals concerned to ensure the appropriate items of SA are being probed. Typically, some SA requirements will be specific to one or another individual (e.g., state of one's own assets), while others will be shared across the team (e.g., state of the overall situation). Having identified the set of SA requirements, probes are then generated by formulating a set of true and false statements about the situation. For example, if an assessment of enemy intent is identified as an SA requirement, then we would formulate a true statement or false (but still plausible) statement describing the enemy's intent, such as "The enemy intends to defend his position". Note that those probes representing shared SA requirements will have to be presented to all team members, while others need only be presented to specific individuals as appropriate.

To analyse *team SA*, we ask for each requirement, "Is the required awareness available within the team as a whole?" If so, we award the team a point. The scoring, however, depends upon whether the SA item in question is specific to an individual or is shared across the team. With regard to team SA, we are only interested in whether *someone* in the team as a whole has the SA. For SA probes that are shared by team members, then, we award a point if at least one member responds correctly, but if all members respond correctly to the same item we still award only one point.

To analyse *shared SA*, we ask "To what extent is this shared item of SA known by *all* relevant team members?" Shared SA assessment thus applies only to shared SA requirements. A "sharedeness" score is awarded for each shared SA probe response that is proportional to the size of the team as well as the number who respond correctly. The following formula is used:

$$\textbf{Sharedness} = \frac{C-1}{N-1}$$

where N = size of the team and C = number of team members answering a shared SA probe item correctly. For example, in a team consisting of N = 5 members, if C = 2 of them answer a particular shared SA item correctly, then the sharedness score awarded would be 1÷4 = 0.25. This formula ensures that when C = N, sharedeness = 1, but when C = 1, sharedeness = 0.

To assess *team metacognition*, we adapt the principle of calibration, i.e. the degree of correspondence between self-perceptions of accuracy and actual accuracy (e.g., Keren, 1991). In this case we analyse correspondence between one person's perceptions of another team member's accuracy, which we assess by means of the partner-rating scale shown in the bottom section of Figure 2.

## Experiment

An experiment was conducted in which eight teams, each with two members, performed a "militarised" version of chess against an "enemy" whose moves were controlled by the experimenter. The area of the chess board used was four times larger than usual (i.e., 16 × 16 squares). Standard chess pieces were replaced with model World War II soldiers and combat vehicles taken from a board game. The conventions of chess piece moves were retained, however. For example, the "commander" piece had the same rules of movement as a King; a "bomber" performed like a Rook and a "fighter" performed like a Bishop.

The two players in a team were both allies with the objective of capturing the enemy commander (King). The enemy maintained a simple strategy of area defence (though this fact was not given to the players). The players could not see the ground truth, but instead had their own maps (16 × 16 boards) on which they could maintain the situational picture as they each perceived it. The players could not see each others' boards, however, as they sat on either side of a screen. Although the players knew their own pieces' starting positions, they were not told of the starting positions of other pieces, neither their partner's nor the enemy's. A special feature of the game was that after each move, a player was presented with information on other pieces that were "visible" to him by virtue of his own pieces' locations. That is, he was informed about the locations of other pieces that lay within the "field of vision" of his own pieces. A piece's "field of vision" was simply the sum of spaces into which it could theoretically move at a given moment. For example a bomber (Rook) could "see" along corridors extending horizontally and vertically from its own position.

During the task, each player was presented with twelve true/false probes. The probes pertained to aspects of the situation defined by the SA requirements analysis and some were tailored in real time to address the state of the game at that point. The players' responses were subsequently compared with the records of ground truth. The teams were divided into two experimental conditions which varied in the amount of communication permitted between the players:

- Condition 1: **High communication** (free discussion allowed)
- Condition 2: **Low communication** (no discussion allowed)

## Results

Team SA was generally high, with means of 89% and 83% in the high and low communication teams respectively. It is clear from Figure 3 that shared SA (based on mean sharedness scores) was more affected by the communications manipulation. Specifically, the teams lacking free discussion had only a moderate shared SA of 50%, in contrast to those in the high communication condition with a mean sharedeness value of 88%. Given that shared SA must be heavily dependent upon the quality and degree of interaction between team members, this result is consistent with expectations.

With regard to team metacognition, the average rating of partners' expected SA accuracy was 1.99 in the high communication teams and 2.45 in the low communication teams (where rating 1 = "will know" and rating 3 = "will just guess"). Thus there were lower expectations of partners' SA when there was no free discussion between team members. When we compare these ratings with actual SA accuracy, we find that

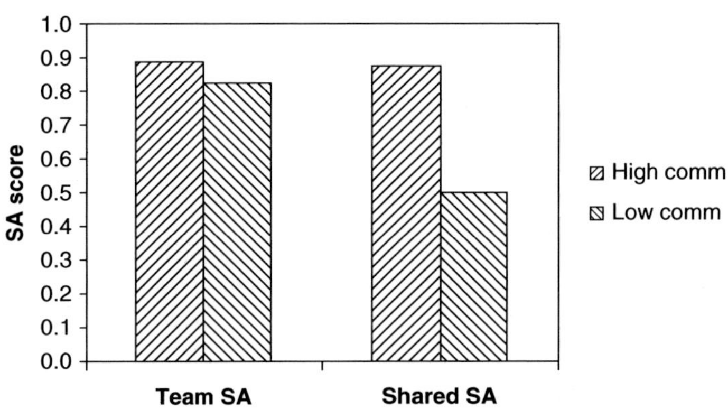


**Figure 3. Team SA and shared SA results.**

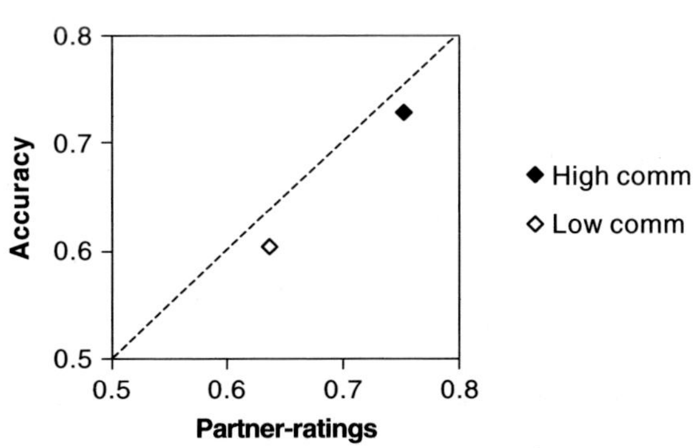


**Figure 4. Calibration of partner-ratings with SA probe accuracy.**

this pattern was not unrealistic. Probe accuracy overall differed between the two conditions: 73% in the high communication teams and 60% in the low communication teams. Normalised partner-ratings are plotted against probe accuracy scores in Figure 4. The diagonal line on the graph indicates perfect calibration between the expected SA accuracy of a partner and that partner's actual SA accuracy: the closer a datapoint is to this line, the better calibrated to partner's actual SA is the rating. Using a simple subtraction formula to obtain a calibration bias score (Yates, 1990), we find that partner-ratings calibration is fairly close to perfect (i.e. zero bias) in both high and low communication conditions (+0.02 and +0.03 respectively). Hence, team metacognition was very good even when free discussion was removed. It seems likely that in the low communication condition, team members effectively compensated by reducing their expectations of their partners' SA accuracy in line with their own reduced SA.

## Conclusion

We have briefly presented a methodology for assessing situational awareness in teams that provides measures of team SA, shared SA and team metacognition based on the results of true/false SA probes and their associated self-ratings and partner-ratings. The results of a small experiment serve to demonstrate the potential utility of the methods. While the technique has been demonstrated in a laboratory task using two-person teams, its validity and value for assessing larger teams in more realistic settings remains a topic for future research.

## References

Keren, G. 1991, Calibration and probability judgments: Conceptual and methodological issues. *Acta Psychologica, 77*, 217–273.

McGuinness, B. 2004, Quantitative analysis of situational awareness (QUASA): Applying signal detection theory to true/false probes and self-ratings. In proceedings of the *2004 Command and Control Research Technology Symposium*, San Diego, 15–17 June, 2004 (online availability: http://www.dodccrp.org)

Yates, J.J. 1990, *Judgment and Decision Making*. Englewood Cliffs, New Jersey: Prentice-Hall.

# HUMAN FACTORS ISSUES IN ATC: THE "USE" OF AN EYE TRACKING METHODOLOGY

**Gemma Cox, Sarah C Sharples[1], Harshada Patel & Alex W Stedmon**

*VIRART(Virtual Reality Applications Research Team), University of Nottingham, Nottingham NG7 2RD, UK*

The Flightdeck and Air Traffic Control Collaboration Evaluation Project (FACE) takes a distributed cognition (DC) perspective to investigate the human factors requirements for the flightdeck of the future in relation to datalink and freeflight. As a theory DC is specifically tailored to understanding interactions among people and technology. Recent literature indicates a general acceptance of datalink technologies, as well as suggesting that the visual information provided by a text based technology may be less disruptive to the aircrew. An experiment using a head mounted Eye Point Of Gaze (EPOG) tracker was employed to investigate how performance on a visuospatial tracking task was affected by speech versus text commands, and the validity and applicability of the method used is discussed and the results will be presented and reported elsewhere.

## Introduction

Datalink (DL) is designed to provide electronic exchange of information between the ATCO and the flight deck. Datalink, in its strongest form, will replace the use of verbal communication with the automatic transfer of visual information. This may potentially lead to less need for clarification of verbal interactions, possible avoidance of error, and lower workload on pilot/controller due to fewer interactions and no need for the pilot to copy down clearances etc.

Recent literature indicates a general acceptance of datalink technologies, as well as suggesting that the visual information provided by a text based technology may be less disruptive to the aircrew. However under high workload, time pressured situations text based datalink may be detrimental to the primary task of flying the aircraft. One of the aims of FACE was to investigate the impact of this change to a more text based, datalink technology.

## Distributed Cognition (DC)

A DC approach differs from the traditional information processing perspective taken to analysing cognitive interaction, rather than just looking at the individual actors and treating their cognition as autonomous to each other, DC looks for a broader class of cognitive events.

Using a DC approach the representation of knowledge is examined as well as the propagation of knowledge through the system. Using this approach, the external representations of knowledge can be used to assess what the internal representations (those

---

[1]Née Nichols

within the cognition of the individual actors) may be. When studying ATCOs, external representations of knowledge constitute artefacts such as flight strips, radars and strip boards. Communication between ATCOs and between ATCO and the flightdeck can be viewed as a system of representations and these verbal communications can only be understood as a function of the system that they exist within (Fields et al., 1998).

The DC approach to human-computer interaction requires real knowledge of the domain. This was achieved within FACE via an extensive literature reviews, familiarisation visits and interviews with subject matter experts. Use of observation techniques and audio and video recording, along with literature reviews identified classes and phenomena that merited further observation and documentation.

Initial cognitive ethnography studies identified the potential effects of the proposed change from speech to text communication. With the introduction of datalink, information from ATCOs may not be available to other members of the flightdeck and vice versa. Loss of intra-crew communication may adversely affect the complex system as the shared mental models and the shared situation awareness of the crew members may change or even breakdown.

Hollans et al. (2000) state that experimental results can contribute to the cognitive ethnography process by providing direct data on performance characteristics of design alternatives. Using an abstract task with novice users does not provide any real ecologically valid data, but due to the nature and set up of ATC it is impossible to test the issue of mode of communication within a real world setting. The experiment instead provides extra information that can be coupled with the cognitive ethnography to provide information on how the current system operates and the potential range of problems that a change to text communication may involve.

## Abstract methodology with novice users

Performance on an ongoing visuo-spatial tracking task was measured in the presence of either speech or text commands.

The primary tracking task tested hand-eye co-ordination, participants tracked the target using a joystick. The secondary task tested the perceived urgency of speech versus text. Participants were required to rate the perceived urgency of a selection of ATC commands on a scale of 1 to 7. An EPOG tracker was used to assess eye movements and measures of tracking task performance were recorded along with the reaction times taken to input the perceived urgency ratings. Subjective measures of workload were taken and a short memory test was employed.

The EPOG video-based tracker utilises a light-weight, head-mounted (HED) tracker which measures the corneal reflection of an infra-red light relative to the location of the pupil centre.

Tracking corneal reflexes on the iris together with the pupil compensates shifts of the camera relative to the head. The use of a HED permits eye-tracking in the presence of large head movements. The gaze image is overlaid on to an optical scene camera image. Video based analysis followed which involved classification of areas of interest (e.g. primary task screen, secondary task screen) and the coding of the eye-movements with respect to these different spatial locations.

The EPOG tracker must be calibrated as this provides a means of linearizing the system and thereby providing a metric for any gaze position within the stimulus field. A 5 point calibration was used in the perpendicular plane that the participant would be working in. This was then checked by asking the participant to fixate on a number of points in the visual scene.

## The visual unit of analysis

The human eye constantly goes through a series of saccades and fixations. Saccades are use to reposition the fovea over different parts of the visual scene, they are fast and visual input is disrupted. Fixations occur when the eye is stationary and the fovea is aligned with a particular pattern in the visual scene. Typically over 90% of viewing time is employed in fixations, with the eye moving about the visual scene less than 10% of the time. Fixation data is therefore important with respect to attentional applications although one cannot infer higher level cognitive processes simply from fixation data. Placement of gaze is not always directed by conscious attention, starring can involve the eyes performing saccadic scanning without any purpose, although passive scanning initiates active looking and it is very difficult to differentiate between the two.

A dwell involves a rest of the eye in a particular location and involves a number of fixations within that parameter. With each fixation the fovea is moved around the area in order to extract information. Dwells have a greater probability than fixations of indicating visual attention, although this does not give any indication of higher cognitive processes. Dwell points indicate prolonged visual perception in a particular area of the visual scene and may be indicative of focussed attention. For this reason dwells were recorded along with their position within the visual scene.

## Method limitations

The iView system is a video-based system that yields a depiction of point of regard projected onto a real time video of the visual scene. The technique provides high external validity in behaviour measurement (i.e. point of gaze at any given time) at the expense of exact statistical analysis (fixation onset, and duration etc).

The use of the head mounted system allows for large head movements, allowing participants this movement provides more ecologically valid data, although during the experiment participants complained that the HED was intrusive and uncomfortable.

As mentioned earlier the EPOG system must be calibrated in order to linearize the system. Calibration times varied greatly between individual participants, making it very difficult to estimate trial length. A five point calibration was performed on the perpendicular plane of the visual scene, this was done by using a laser pointer to mark the points within the participants visual field. It is important to ensure that the distance between the participant and the visual scene is the same during the calibration and the trial. The EPOG tracker can only calculate left to right head movements, if the distance between the participant and the visual scene is altered by a movement of the head (forward or backwards), parallax is introduced. Parallax causes a displacement in the point of gaze due to a change in the position from where it is viewed. This creates inaccuracies in the projected point of gaze but can be overcome by calibrating to the correct distance and using a high-backed chair to encourage the participant to maintain their position.

The accuracy of the calibration is closely related to how well the HED fits the head. Even if an accurate calibration is achieved at the start of the trial, an ill-fitting HED may slip resulting in inaccuracy and loss of data. There was also an occasion where a subject touched the HED and caused it to move which meant a loss of calibration. A compromise between a tight fitting HED and comfort is required. If the HED is too tight it becomes intrusive and can seriously effect performance. A compromise may be frequent re-calibration although this can be time-consuming, and may disrupt the flow of the experiment.

There are a number of factors that influence the calibration process. When calibrating the tracker uses the lightest point in the eye which is the corneal reflection, in combination with the darkest point of the eye which is the pupil. Bright overhead lighting causes lots of reflections within the eye that interfere with the detection of the corneal reflection. Long, dark eye-lashes and mascara form areas that are as dark as the pupil and so interfere with its detection.

As mentioned earlier the tracker is calibrated to the perpendicular plane of the visual field, in the above experiment the tracking of point of gaze yielded accurate results when the participant was looking at either of the computer monitors. When their gaze was directed downwards at the keyboard, the tracker lost the corneal reflex/pupil threshold and this resulted in a loss of data.

There are many limitations with the EPOG system. A lot of these can be remedied by small changes in the set-up and positioning of the participant. In relation to the above experiment, the data collected was adequate to achieve the aim of cataloguing the participants point of gaze with respect to the primary and secondary task screen.

## Method validity

The validity of a method is the extent to which the method measures what it is supposed to measure. If the EPOG tracker is simply used to assess the position of the participants point of regard with respect to the visual scene then the video based output provides an accurate representation. In the present experiment the participants point of gaze was concentrated on the primary task screen for significantly longer in the speech condition than in the text condition and this was coupled with inferior performance on the primary tracking task in the text condition. In order to infer visual attention or higher cognitive processes one must attach some kind of context. The eye-tracking data alone provides no guarantee that the information in the visual scene is actually processed.

The multi-store model of memory stipulates that sensory stores are preattentive, modality-specific and hold information very briefly (e.g. Atkinson & Shiffrin, 1968). Information is held very briefly in the sensory stores, with some of it being attended to and processed further by the short term store. Visual attention is the process that selects what parts of the input will be fed into the mechanisms of higher-level visual cognition.

Verbal data is able to provide information about context, more specifically verbal protocol analysis. Verbal protocol offers one of the few practical techniques for making inferences about cognitive processes and understanding complex behaviour with respect to specific events (Wilson & Corlett, 1995).

Think aloud techniques may be applied in real time, where the user verbalises whilst using the system. Although criticisms have been made claiming that the technique adds extra cognitive load which may interfere with the cognitive requirements of the interaction and hence disturb the task. This calls for the verbal protocols to be carried out retrospectively, whilst the user watches a video of him or herself interacting with a system and can clarify why certain actions and decisions were made. Care need to be taken when using retrospective verbal protocol as thoughts are not held in short term memory (Ericsson & Simon, 1984).

Criticisms of verbal protocol include the lack of correlation or ways of testing the correlation between what the user says they think and what they are actually thinking. However, some studies have been carried out which support the validity of verbal protocol as evidence of thought processes (Bainbridge & Sanderson, 1995).

The working memory model was proposed by Baddeley and Hitch (1974), in this model the unitary short-term store was replaced by a multi component working memory consisting of a central executive, an articulatory loop, and a visuo-spatial sketch pad. The articulatory loop and visuo-spatial sketchpad are slave systems of the central executive, they require a level of perception, although it is the modality free central executive that resembles attention and is needed when dealing with cognitively demanding tasks. The use of eye-tracking in combination with verbal protocol measures information access in addition to measures of current thinking. Verbal protocol information provides data concerning what aspects of the environment were perceived by the visual system and what elements of the visual scene were processed by the central executive.

In tasks where individuals alternate between intentional visual information and verbal communication, the point of gaze may represent an input to the information processing system of the individual (information about the visual scene may be held on the visuo-spatial sketchpad). By combining this gaze data with information from the verbal protocol it may be possible to build up a representation of the individual's cognitive processes.

## Conclusions and future work

Use of an eye-tracking methodology along with verbal protocol analysis offers the opportunity to study information access and perception coupled with attention and higher level cognitive processes. The above experiment was useful in demonstrating the associated head down time connected with communication via text format and its associated detriments in primary task performance. Future work will involve a more ecologically valid exploration of the subject coupled with verbal protocol analysis.

## References

Atkinson, R. C., & Shiffrin, R. M. (1968). Human memory: A proposed system and its control processes. In K. W. Spence & J. T. Spence (Eds), *The psychology of learning and motivation*, Vol. II. London: Academic Press.

Baddeley, A. D., & Hitch, G. (1974). Working memory. In G. H. Bower (Ed.), *The psychology of learning and motivation*, Vol. VIII. London: Academic Press.

Bainbridge, L., & Sanderson, P. (1995). Verbal protocol analysis. In J. R. Wilson & E. N. Corlett (Eds.), *Evaluation of Human Work: A practical ergonomics methodology* (pp. 169–201). London: Taylor & Francis Ltd.

Ericsson, K. A., & Simon, H. A. (1984). *Protocol Analysis.* Cambridge, Mass.: M.I.T. Press.

Fields, R. E., Wright, P., Marti, P., & Palmonari, M. (1998). Air Traffic Control as a Distributed Cognitive System: A Study of External Representations. *Proceedings of the Ninth European Conference on Cognitive Ergonomics*, Edited by T.R.G.

Wilson, J. (1995). A framework and a context for ergonomics methodology. In J. R. Wilson & E. N. Corlett (Eds.), *Evaluation of Human Work: A Practical Ergonomics Methodology* (pp. 1–39). London: Taylor & Francis Ltd.

# COGNITIVE ERGONOMICS

# DISTRIBUTED SITUATIONAL AWARENESS

**Neville Stanton, Chris Baber*, Guy Walker, Paul Salmon & Damian Green**

*Human Factors Integration – Defence Technology Centre, School of Engineering and Design, Brunel University, Uxbridge, Middlesex UB8 3PT, UK*

**Rob Haughton, Richard McMasters & Huw Gibson**

**Human Factors Integration – Defence Technology Centre, Department of Electrical and Electronic Engineering, University of Birmingham, Edgbaston, Birmingham B15 2TT, UK*

The purpose of this paper is to propose foundations for a distributed theory of situation awareness based on the analysis of interactions between agents (i.e., both human and non-human) in subsystems. This approach may help promote a better understanding of technology-mediated interaction in systems, as well as helping in the formulation of hypotheses and predictions concerning distributed situation awareness. It is proposed that agents within a system each hold their own situational awareness which may be very different from (although compatible with) other agents. It is argued that we should not always hope for, or indeed want, sharing of this awareness, as different system agents have different purposes. This view marks situational awareness as a dynamic and collaborative process that binds agents together on tasks on a moment-by-moment basis. Implications of this viewpoint for development of a new theory of, and accompanying methodology for, situational awareness are offered.

*Keywords:* agents, situation awareness, systems, teams, theory.

## Theories of situation awareness

Contemporary theories of situation awareness (SA) tend to focus on the individual actor, for example the embedded-interactive model (Smith & Hancock, 1995; Adams et al., 1995), the cognitive sub-functions model (Bedney & Miester, 1999) and the three-level model (Endsley, 1995). The embedded-interactive model offers an explanation of the dynamic aspects of situational awareness, such as how momentary knowledge is updated and how the search for information from the world is conducted. This view focuses on the interaction of the person with the world. The cognitive sub-functions model proposes how underlying cognitive functions might interact. This view focuses on the information processing activities within the mind. The three-level model describes the types of data that SA activities might produce. This latter model indicates types of SA measures that might be taken (i.e., perception of elements, comprehension, and projection), whereas the embedded-interactive model might argue for data on the status of the world, and the cognitive sub-functions model would require data on the individual. All three models are underpinned by general models of human information processing in individuals.

These three approaches to the study of SA are concerned with describing, explaining and measuring SA in the individual. Whilst this approach may be appropriate for

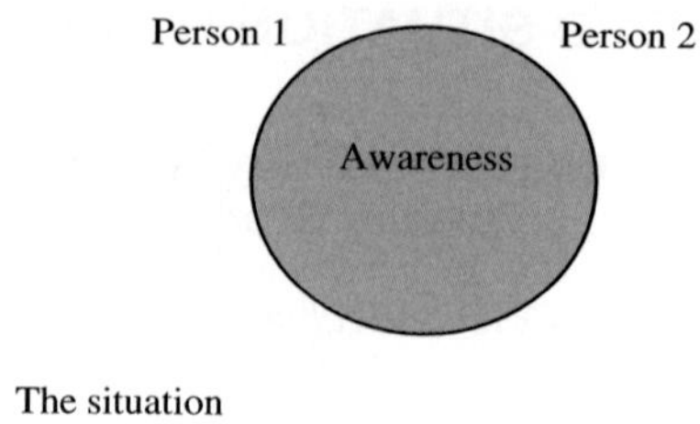


**Figure 1. Shared awareness model.**

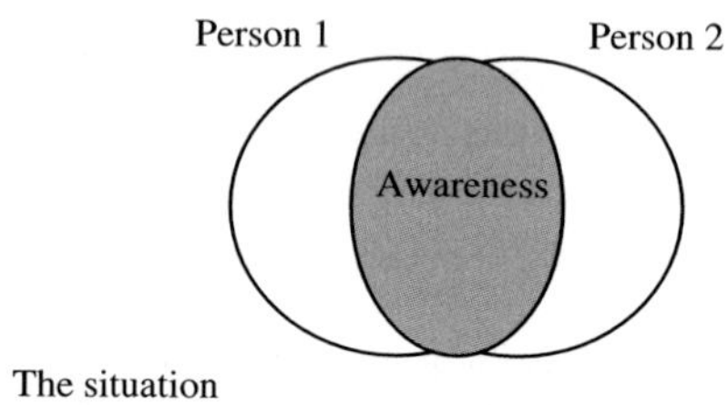


**Figure 2. Overlapping awareness model.**

tasks that are performed by individuals in isolation, it is limited in value when group work is being studied. Few tasks are performed entirely independently of others, especially in complex situations (e.g. landing an aircraft) and when critical decision-making is required – these activities tend to require coordinated activity between several individuals (Cannon-Bowers and Salas, 1990; cited in Salas, Prince, and Baker, 1995). Researchers have begun to address situational awareness in teams. There are three models of team SA: shared, overlapping and distributed. Kaber & Endsley (1998) have argued for research on shared situational awareness, where team members have similar SA requirements, which is illustrated in figure 1.

The shared awareness model assumes that there is a need for identical awareness between person 1 and person 2. To date, much of the team SA research seems to have highlighted the need for a shared understanding (Salas et al., 1995 Jentsch et al. 1999). This is understandable when the research domain addresses people with similar goals, such as pilot and co-pilot, but does not address the SA relationship between the pilot and air traffic controller very well. In the latter example, their may be some common frame-of-reference, as indicated in the overlapping awareness model, shown in figure 2.

The overlapping awareness model assumes that there is an intersection of awareness between person 1 and person 2. Salas et al. (1995) have underlined the importance of team communication for effective situation awareness, and emphasised that degradation in communications can have a negative effect team situation awareness. Endsley (1995) has suggested that team with similar training, experience and goals do not necessarily need to communicate explicitly

By way of an example, a study of air traffic controllers at West Drayton in the UK has shown that they rely upon high levels of verbal communications with aircraft in order to pass instructions about flight levels, air speed, holding stacks, and headings to pilots. There is little time left to communicate with other air traffic controllers regarding the passing of aircraft from one sector of control to another. For this reason, controllers have developed a system of passing aircraft between sectors that does not rely on explicit verbal communication. The system works by putting aircraft into a particular position in the new sector at a pre-agreed height, speed and heading. As soon as the

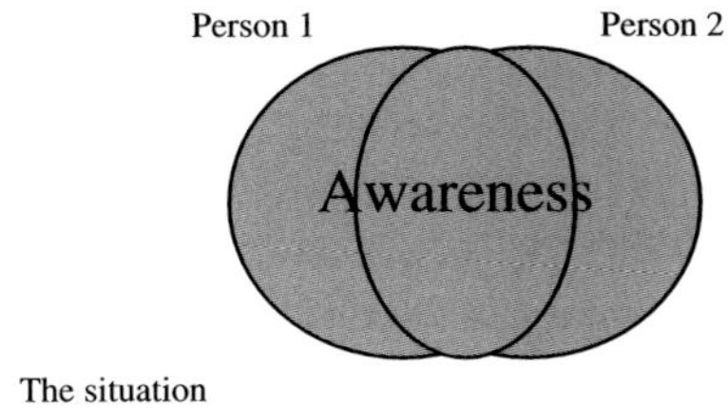


**Figure 3. Distributed awareness model. This assumes that there is compatible awareness between person 1 and person 2.**

controller sees an aircraft in this position they know it is being passed onto them, and they can take control of the aircraft. Thus, the communication between controllers has become implicit. This is a highly evolved form of communication and is constrained by a very precise set of procedures. This example describes distributed situational awareness, where each air traffic controller has awareness of their own sector, but only limited awareness of traffic in other sectors, as shown in figure 3.

This view of SA creates a systems representation view of a situation and a distributed view of awareness. No one agent has a complete picture of any situation, just a facet of the situation at any point in time. As situations are multifaceted and dynamic (in both space and time), tracking any situation (both feedforward and feedback) requires coordination of multiple agents if it is to be performed effectively.

The types of domain that we are exploring can be considered in terms of Klein's (1989) notions of Naturalistic Decision Making, i.e., agents in the field are able to draw upon their experience and expertise to make rapid diagnosis and to perform effective actions in very limited timeframes. In a similar fashion, Smith and Hancock (1995) propose that, "SA is the up-to-the minute comprehension of task relevant information that enables appropriate decision making under stress." As our theory of SA operates at a systems level, it takes a different perspective to the individual and shared SA approaches. We feel that the shared SA approach could misdirect attention to inappropriate aspects of the task, where as there are points in tasks where SA may overlap for brief periods in distributed team working. Overlapping SA requirements is not the same as shared SA however. Overlapping SA directs at point in time where the overlaps are and what information is needed. Shared SA implies shared requirements and purposes.

There are several concepts of "distributed cognition", but they share a common assumption that "cognition" is often performed through the interactions of people and the artefacts they use in the environment. There is a tension in the individual versus social perspective that can be taken on distributed cognition, and in this paper, we resolve this tension through a systems level view proposed by Perry (2003). This is not to deny alternative views, nor to claim priority of one view over others. However, for the purposes of this work, we feel that the systems-level view is useful. The Distributed SA approach appeared only very recently, however it has added a new dimension to our awareness of SA in the few environments in which it has been employed. It also has implications for the design of the working environment, e.g. in terms of team structure, which can impact the flow of information through the system; and HCI, due to the importance of technological artefacts in distributed environments. By viewing SA first as a systems-level phenomenon, it is possible to identify the aspects of a situation about which agents require knowledge. By viewing the active knowledge in each state of an incident, it is possible to determine who knows what at a given time. From this perspective, it becomes possible to indicate how information needs to disseminate through the system for effective performance and to identify possible barriers to effective dissemination. The graphical representation provides a simple but effective means by which system SA can be mapped.

## Propositions for the development of a theory of Distributed SA

On the basis of the evidence presented in this paper, we feel that it is possible to propose a set of tenets that could form the basis of a systemic theory of distributed SA. These propositions are as follows:

a. SA held by human and non-human agents
b. Multiple views on SA of the same scene for different agents
c. Non-overlapping and overlapping SA depends on agent's goals
d. Communication between agents may be non-verbal behaviour, customs and practice (but this may pose problems for non-native system users)
e. SA holds loosely-coupled systems together
f. An agent may compensate for degradation in SA in another agent

The distributed SA approach does not dispute that the individuals in the system will have their own awareness of a situation, or that groups of individuals may share some level of understanding of the situation. It is asserted that complex problem-solving systems will have their own cognitive properties (including SA) that cannot be accounted for by individual cognition and that to study a system at the level of the individual will fail to pick up on these systems level features.

## References


Adams, M. J., Tenney, Y. J. & Pew, R. W. 1995, Situation awareness and the cognitive management of complex systems, *Human Factors* **37**, 85–104

Bedny, G. & Meister, D. 1999, Theory Of Activity And Situation Awareness *International Journal of Cognitive Ergonomics* **3**, 63–72

Endlsey, M. R. 1995, Toward a theory of situation awareness in dynamic systems, *Human Factors*, **37**, 32–64

Jentsch, F., Barnett, J. & Bowers, C. A. 1999, Who is flying this plane anyway? What mishaps tell us about crew member role assignment and air crew situation awareness. *Human Factors* **41**, 1–14

Kaber, D. B. & Endsley, M. R. 1997, Out-of-the-loop performance problems and the use of intermediate levels of automation for improved control system functioning and safety, *Process Safety* **16**, 126–131

Klein, G. 2000, Analysis of situation awareness from critical incident reports, In M.R. Endsley (Ed.) *Situation Awareness Analysis And Measurement*, Mahwah, NJ: LEA.

Perry, M. 2003, Distributed cognition, In J.M. Carroll (ed) *HCI Models, Theories and Frameworks*, San Francisco, CA: Morgan-Kaufmann, 93–224

Salas, E., Prince, C. & Baker, D. P. 1995, Situation awareness in team performance – implications for *measurement* and training, *Human Factors* **37**, 123–136

Smith, K. & Hancock, P. A. 1995, Situation awareness is adaptive, externally directed consciousness, *Human Factors* **37**, 137–148

# HUMAN FACTORS ISSUES IN AIRPORT BAGGAGE SCREENING

**Alastair Gale**

*Applied Vision Research Institute, University of Derby, Kingsway House, Kingsway, Derby DE22 3HL*

Security operators at airports screen passenger baggage by means of X-ray imaging each item and then visually searching the resultant computer display for the presence of objects which pose a potential threat. The performance of screeners on this difficult task is variable. We have theoretically modelled the underlying visual and cognitive reasons for error occurrence in airport baggage screening which demonstrates that threat items can be missed due to failures of visual search, detection or cognitive interpretation of the image information. The key human factors issues underlying variable screener performance are described and a new UK research project introduced which aims to provide key training to improve the identification of possible terrorist items.

## Introduction

In order to secure air passenger safety it is important to examine efficiently all passenger hand luggage and other items of external clothing for the presence of potential threat items; such as knives, guns or improvised explosive devices (IEDs). Almost universally, X-ray imaging of luggage is used which results in a two dimensional image that is examined on a computer monitor. Security operators have various specialised image enhancement algorithms that they can then apply to the image in order to highlight any particular aspects of concern. The image itself is usually viewed dynamically as the bag traverses the X-ray imaging system but it can also be examined statically. Where the operator (screener) is suspicious of some area within the image then the bag can be re-imaged and examined in a different orientation, to help overcome any problems due to overlapping visual image features resulting from the three dimensional object being viewed in only two planes. Any suspicious items can then be hand searched by another security operator if needed. A technique used in airports to maintain high levels of security performance is threat image projection (TIP) where a threat item can be superimposed within the X-ray image of a passenger luggage item and when the operator detects it then s/he receives appropriate feedback.

Recent developments in X-ray imaging has led to the development of 3D imaging of baggage and such systems are now reaching the market place. The advantage of such systems is that it should be easier to distinguish between a possible threat item and something that looks suspicious but is actually simply an overlaying of visual image features on the 2D image (i.e. the approach should produce a reduction in false positive detections). However, currently it is not really known whether the skills required for examining 2D X-ray images transfer readily to the examination of 3D X-ray images or whether the two tasks are identical. What in fact are the relevant skills for examining 2D baggage images?

The airport security screening situation attracts much research but of security necessity most of this is not readily available in the public domain. However, the volume of research in the area is low when compared to similar inspection situations; industrial inspection, radar operators or medical imaging. Additionally, the number of potential threat items in hand baggage is very low and so the task for the screener is similar to many other image inspection tasks where there is a low probability of target occurrence and where the potential target, or targets, appearance is unknown. Sustaining levels of vigilance in baggage screening is therefore important and this is primarily achieved by limiting the amount of time that any one individual spends actually examining images, together with other approaches, such as regular evaluation and training.

The potential for terrorist threat to air passengers is of great current international concern. Recent indicators from the UK, and particularly the USA, demonstrate that it is still all too easy, despite enhanced security measures (e.g. from January 2003 the American Transportation and Security Administration began screening every luggage item at all 429 commercial airports), for air passengers to carry potential threat items on to aircraft.

There is a need to improve airport passenger screening which has been well documented, for instance by the Reason Foundation (Poole, 2002). Additionally, the quality of luggage screening performance has often been questioned, with the US General Accounting Office recently reporting that the average detection rates had dropped from 95% in 1993 to 85% in 1999. In 2001 the Thompson amendment to the US Aviation Security Act was proposed which requires performance standards of these security personnel in an effort to improve potential threat detection levels.

## The luggage inspection task

We have argued previously (Gale et al., 2000) that the security inspection task is similar to the task of examining a medical image for possible abnormality presence. Human visual inspection is an imperfect process and is subject to errors even when the individuals are working to the best of their abilities. In both domains both false positive (identifying a possible threat item or possible medical abnormality) and false negative decisions (failing to spot a threat item or medical abnormality) are made. However, the key interest in both areas is in the nature of any false negative errors and it is important to try to identify the causes of such errors and develop inspection techniques, as well as appropriate training methods, which act to reduce their likelihood. In medical imaging a common experimental method investigating observer imaging performance is to record the visual search process of observers as they inspect a range of images. Such work has given rise to a theoretical model (Nodine and Kundel, 1987; Gale, 1997) which emphasises the role of visual search with such images. Errors are argued to arise due to failures in search, detection or cognitive interpretative processes. Although search is a necessary and important aspect in examining the images most research points to deficiencies in detection and interpretation as underlying the majority of errors (Kundel et al., 1978). The type of error typically made may relate to the type of abnormality (Savage et al., 1994).

In a previous study of ours (Gale et al., 2000) we examined airport security operators in a simulated situation as they inspected a test series of luggage images, some of which contained known IED targets. Three viewing conditions were employed; a 200 ms "flash" presentation (too brief for the observer to change eye fixation location on the image), one second and five seconds viewing. With increasing viewing time the detectability of IEDs increased, possibly not too surprising a finding. However, even in the flash viewing case all of the IEDs were correctly detected by at least one individual, so demonstrating that observers were able to process visual information some distance away from their fixation point on the image. A similar finding has been reported several times in medical imaging studies. In this task we empirically determined the size of the observers' useful field of view and found this to be 2.5° visual angle, a similar figure having previously been found by several researchers to be applicable in the medical imaging domain – pointing to some similarities between the two inspection domains.

Overall, the different lengths of examination times only affected the true positive IED detections and the false negative responses. IED-free images were able to be classed as normal by observers even in very brief viewing times, which gives some insight into the nature of their expertise in examining such images. The false negative errors were examined in the six second viewing condition, which demonstrated that visual search accounted for 14% of errors, detection for 19% and interpretation for 67%. Such a high figure for interpretation errors had not been expected. For instance in medical imaging, Kundel et al. (1978) report circa 45% interpretation errors in a pulmonary detection task. Such a difference may point to both similarities and differences between the baggage inspection situation and the medical imaging domains.

One key difference between the two domains is that in medical imaging, whilst the nature of the abnormal appearance may not be known in advance, the general format of the image type (e.g. MRI scan of the head, mammogram of the breast, ultrasound scan) is known. Consequently the observer has some implicit initial cognitive information before looking at the image, which will aid the subsequent search process. In luggage examination each overall item is very variable in size and shape, and whilst there will be similarities in terms of everyday contents there will also be many differences.

One of the aspects of our previous research has been to try to understand the nature of expertise in the security screening domain. Seamster et al. (1997) used cognitive task analysis to examine the cognitive performance of security screeners and teased out various differences between novice and experienced screeners. However, from our own work we have found that experience in security screening does not necessarily equate with expertise, which is intriguing from both the theoretical and practical standpoints.

In medical imaging, expertise in a somewhat similar screening task (e.g. breast screening) with low rates of abnormal appearances is deemed to be a result of considerable training over years, coupled with regular examination of large number of images which then builds up an appreciation of the wide range of normal appearances possible. In the UK the minimum number of images a year that a radiologist must interpret is 5,000 cases with some examining over 20,000. In the USA the comparable figure has been much lower, circa 200 cases. In security screening it is similarly possible for an individual to build up a high number of examinations of normal luggage items but our work in this domain to date has not been really been able to determine what skills specifically underpin expertise. McCarley et al. (2004) have recently investigated aspects of the development of such domain related expertise using non-security personnel.

In medical imaging, in areas such as breast screening, the radiologists can be helped by computer aided detection (CAD) systems which have had very considerable development in recent years. CAD systems pre-process the image before the radiologist sees it and can highlight potential areas of interest to the observer by means of overlaying prompts on the image. Such systems have had a tendency to produce a high number of false positive prompts of apparent abnormal appearance which can affect the observer's behaviour who may simply choose to ignore such prompts all together as opposed to spending time addressing each prompt location and determining whether it does in fact indicate abnormal appearance. However, due to continual developments in algorithms the number of false positives per case is decreasing and systems are now available in clinical usage (e.g. R2 Image Checker). In baggage screening there exist image enhancement algorithms, which are used by the various equipment manufacturers, but these tend to enhance the image as a whole, rather than prompt the observer for specific image attributes in a particular image location.

## Epaulets

The foregoing addresses the key similarities and differences between the baggage screening scenari o and medical imaging. Building on this research background the EPAULETS project aims to take recent findings from the latter domain and apply these to airport screening. The research utilizes a detailed image database of air passenger baggage items, some of which contain a range of potential threat items of varying conspicuity. These will be presented to screeners in a number of airports across the UK in a series of empirical investigations with the aim of further ascertaining the nature of false negative decisions. Resultant data will then lead to theoretical developments, which will be used to shape new training strategies in this domain. Subsequently, the training strategies themselves will be rigorously evaluated.

## Discussion

Security at airports continues to be a challenge both in practice and for the researcher in trying to suitably apply knowledge from other domains and discipline areas. It is unlikely that there is an easy solution to correctly identifying all potential threat items all of the time. However, the EPAULETS project is targeted at reducing false negative decisions and is planned to make a significant contribution to the area.

## Acknowledgements

This research is supported by the EPSRC "Technologies for Crime Prevention and Detection" programme. The research is performed in collaboration with Tobii Technology and QinetiQ.

## References

Gale A.G. (1997): Human response to visual stimuli. In W. Hendee & P. Wells (Eds.) *Perception of Visual Information – second edition*, (New York, Springer Verlag).

Gale A.G., Mugglestone M., Purdy K.J. and McClumpha A. (2000). Is airport baggage screening just another medical image? In E.A. Krupinski (Ed.) *Medical Imaging 2000: Image perception and performance*, (Bellingham, SPIE), 184–192

Kundel H.L., Nodine C.F., and Carmody D. 1978 Visual scanning, pattern recognition and decision making in pulmonary nodule detection. *Investigative Radiology,* **13**, 175–181

McCarley J.S., Kramer A.F., Wickens C.D., Vidomni E.D. and Boot W.R. 2004 Visual skills in airport-security screening. *Psychological Science*, **15**, 302–306

Nodine C.F. and Kiundel H.L. 1987 Perception and display in diagnostic imaging, *Radiographics,* **7**, 1241–1250

Poole R.W. 2002. Improving Airport Passenger Screening. Policy Study No. 298, The Reason Foundation, http://www.rppi.org/passengerscreening.html

Savage C.J., Gale A.G., Pawley E.F. and Wilson A.R.M. 1994, To err is human, to compute divine? In A.G. Gale, S.M. Astley, D.R. Dance and A.Y. Cairns (Eds.) *Digital mammography* (Elsevier, Amsterdam) 405–414

Seamster T.L., Redding R.E. and Kempf G.L. 1997, *Applied cognitive task analysis in aviation* (Avebury, Aldershot)

# DISTRIBUTED COGNITION IN POLICE OPERATIONS

**R.B. McMaster & C. Baber**

*Electronic, Electrical and Computer Engineering, The University of Birmingham B15 2TT, Email: r.mcmaster@bham.ac.uk; c.baber@bham.ac.uk*

In this paper, we report an investigation into the structure of the command and control network within West Midlands Police, principally focussing upon the role of police control rooms within this network. We present an analysis of police responses to emergency incidents using the systems-level approach of distributed cognition. The study found that the "OASIS" information system plays a key role within the command and control network, enabling communications and the flow of incident information across the various control rooms; from a distributed cognition perspective, the OASIS system is a key artefact, performing a number of roles which support the coordination of police activities.

## Introduction

"Distributed cognition" studies systems-level cognitive processes of groups of individuals and artefacts engaged in the performance of a task (Flor and Hutchins, 1991; Artman and Garbis, 1998). The theory has been successfully applied to a number of domains (Perry, 2003). This research investigates distributed cognition within police command and control, in order to:

1. understand the command and control structures within the police and the function of police control rooms within these structures;
2. investigate how the distributed cognition approach can add value to our understanding of command and control within the emergency services.

Five visits were made to the West Midlands Police Force Communications Centre (FCC) and Bournville Lane Operations Command Unit (OCU). The visits involved interviews with, and observations of, Police Officers and control room support staff and review of emergency incident logs. Analysis of the resultant information provided an overview of the organisation of command and control structures and methods of communications adopted during responses to emergency incidents.

## Command structure

In the United Kingdom, the primary emergency services (Fire, Police and Ambulance) operate three main levels of incident command. These levels are commonly referred to as Bronze, Silver and Gold (HMSO, 1997). Within West Midlands Police, Bronze (*Operational*) Command is performed by one of 21 local OCUs, which handle the majority of emergency incidents. Silver (*Tactical*) Command level is the responsibility of the FCC Duty Inspector, who becomes involved in the response when the severity of the emergency requires the planning, coordination and decision-making activities associated with this higher command level (HM Inspectorate of Constabulary, 1999). Finally, Gold (*Strategic*) Command is only brought into the command structure when the decision

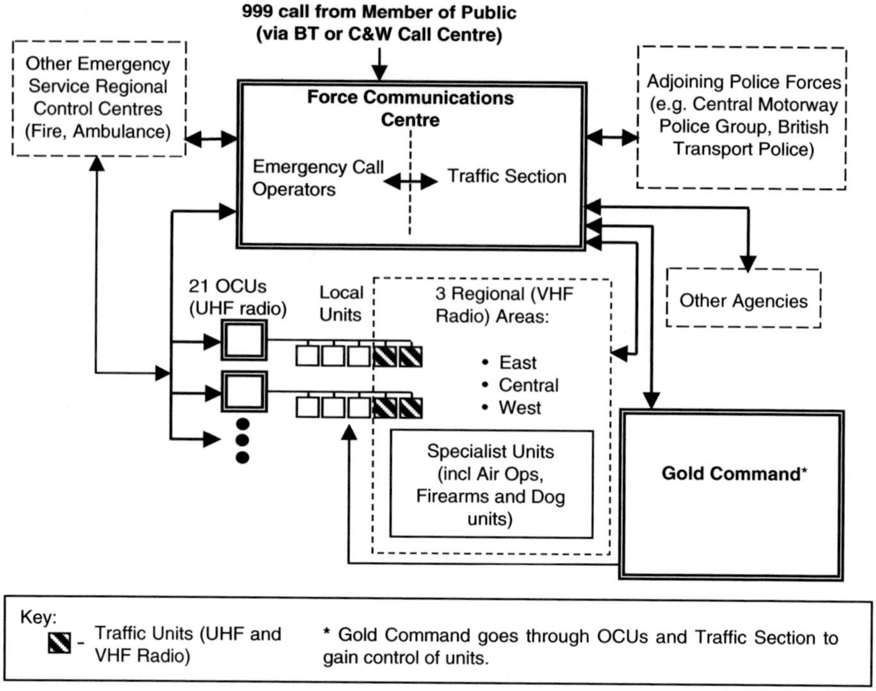


**Figure 1. Primary lines of communication used by West Midlands Police during responses to Emergency Incidents.**

is made to declare the emergency a major incident (Cabinet Office Civil Contingencies Secretariat, 2003).

Figure 1 illustrates the command centres and main lines of communication used during police response to emergency incidents. The FCC studied is situated at Bournville Lane Police Station, Birmingham and is split into two sections: Emergency Call Operators (ECOs) and Traffic Operators. ECOs handle all 999 calls made to the police in the West Midlands area (approximately 2,000 calls per day). Traffic Operators direct the activities of Traffic and Specialist Units and coordinate the police response to incidents with other agencies, such local authorities or neighbouring police forces. An example emergency incident has been summarised above.

## Case study of an emergency incident: car break-in caught on CCTV

The Night Porter of a hotel observes, on his CCTV monitor, three youths attempting to break into cars in the car park. The Porter calls 999 and reports the crime to the ECO, who summarises the information in a new incident log and passes it to the OCU for the incident area. The OCU Operator accepts the log and allocates resources to the incident. The ECO also passes the log to the Traffic Section, who despatch resources to the incident. The ECO remains on the phone to the Night Porter, who is able to provide further details of the offender's descriptions and actions. One of the police units (Unit A) arrives at the incident scene, by which time the offenders have fled by car; Unit A and the Night Porter check the CCTV footage for the offender's vehicle. A second police

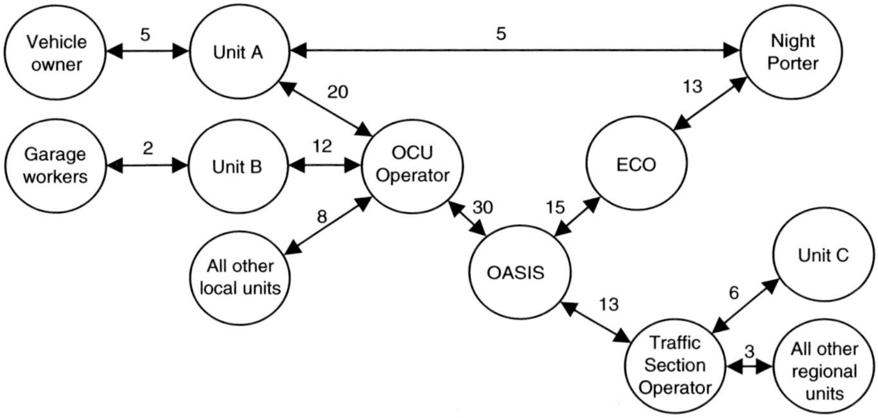


**Figure 2. Social network analysis of Car Break-in Incident.**

unit (Unit B) arrives at the scene and begins a search of the surrounding area. The CCTV footage is found to have captured the offenders, but not their car. The police establish that only one car has been broken into, the owner is located and their ownership of the vehicle verified using the Police National Computer. The owner checks the car and gives a description of the stolen items. The OCU Operator provides a crime reference number, which the Police Officer gives to the owner. The Unit B finishes the search of the area and all police units leave the scene. The OCU Operator notes in the log that this incident was a theft from a motor vehicle and adds the approximate time of the crime. They then close the log. There will be a separate task, initiated after this activity, to complete an investigation of the Crime.

This event was a relatively straightforward police emergency, which lasted less than 40 minutes. In spite of this, a number of actors (in this case 9) cooperated in order to rapidly resolve the incident. Some of the characteristics of the West Midlands Police command and control network are summarised above in order to demonstrate how the network behaves as a distributed cognition system (i.e., through the reliance on and transformations of external representations).

Figure 2 shows the Communications frequency between actors for the emergency incident and highlights the central role of OASIS (Operational and Support Information System).

*Representations and their transformation*

Several artefacts are used to represent information during emergency incidents, including the log held by OASIS, notepads used by Emergency Call and OCU Operators, notebooks used by Officers attending the scene and, in this incident, video footage of the crime taking place. Table 1 summarises the main artefacts that are used to store and transmit information during Emergency Incidents. As key incident information is communicated through the network, a number of transformations are performed upon it. For example, the format in which information is stored and communicated changes a number of times. ECOs summarise the details of the emergency from the caller in concise sentences, rather than typing up the full account from the caller verbatim. This is because 999 callers often repeat themselves, or give irrelevant details first. During this incident, when the Night Porter reports the crime in progress, the 999 Operator listens the Night Porter's account and then summarises the information in a single sentence: *"THERE ARE 3 LADS TRYING TO BREAK INTO CARS I AM WATCHING THEM ON*

**Table 1. A summary of the Main Artefacts in use during West Midlands Police response to emergency incidents.**

| Artefact | Purpose | Users | State | Life | Distribution potential |
|---|---|---|---|---|---|
| ECO's Notepad | A temporary record of call details whilst talking to 999 caller and before entering them into OASIS | ECO | Dynamic | Destroyed at end of shift | Low |
| OASIS Incident Log | 1. The collection and dissemination of emergency call details to relevant C&C agents; 2. A task management tool; 3. A permanent record of the actions taken as a result of the emergency call. | ECO, OCU and Traffic Operators | Both – Static and Dynamic | Permanent | High |
| Police Intranet | 1. Details of Police Officers (including mobile numbers); 2. PNC checks on people and vehicles | Traffic and OCU Operators | Static | Permanent | Medium |
| OCU IT System | 1. Record of the location and availability of police resources; 2. Provide details of incident area (map). | OCU Radio Operators | Dynamic | Permanent | Low |

*CAMERA."* ECOs will also convert call details into standardised Police jargon, for example the phrase *"white lads"* spoken by the Night Porter becomes *"IC1 males"* in the incident log, which is the relevant Police National Computer Ethnicity Classification.

The Communications that take place through the OASIS incident log can be termed implicit, in that they are not specifically directed at a particular person (all OASIS users who have the log in question open will receive the message) and usually do not require any response to be sent. An example of this can be seen when the Night Porter provides further descriptions of the offenders to the ECO, who summarises the information in the log as follows: *"THERE ARE 3 IC1 YOUTHS WEARING DARK CAPS TROUSERS AND JACKETS – ONE HAS WHITE STRIPES DOWN TROUSERS AND WHITE TRAINERS."* This communication does not receive a response, however, it is read and passed on by the Traffic and OCU Operators to the police units proceeding to the scene. As well as these implicit communications, on occasion incident logs are also used for explicit communications. An example of this can be seen when one of the OCU Operators enters the following message into the incident log: *"Location Verify – No amendment made."* This is a request to the ECO to check with the Night

Porter that the address shown in the log is correct, as it has not been altered from the details that were automatically generated from the telephone number the Night Porter called from.

## Discussion

The interviews, observations and analysis of West Midlands Police incident records demonstrate the distributed nature of the activity required in coordinating the response to even straightforward emergency incidents. Actors within the command and control network rely upon a number of artefacts, chief of which is the OASIS system, which takes a central role in the coordination of police activity. The analysis also revealed that OASIS serves multiple roles: a central record (for audit purposes), a cognitive artefact that facilitates collaboration, a shared narrative of the incident (described in a shared language) and a system for allocating responsibility and ownership of incidents in which there is collaboration ("ownership" of the OASIS log is passed between control rooms as the incident unfolds). It is also interesting to speculate as to the further socio-technical impacts OASIS has on police operations; does the terminology used by OASIS (such as its nine classifications for incidents) standardise or perhaps constrain the language of policing itself?

## Acknowledgements

This research from the Human Factors Integration Defence Technology Centre was funded by the Human Sciences Domain of the UK Ministry of Defence Scientific Research Programme.

## References

Artman, H. and Garbis, C. (1998) Situation Awareness as Distributed Cognition. In T.R.G. Green, L. Bannon, C.P. Warren and J. Buckley. (Eds) *ECCE 9: Proceedings of the Ninth European Conference on Cognitive Ergonomics*, 151–156. European Association of Cognitive Ergonomics (EACE), Le Chesnay, France.

Cabinet Office Civil Contingencies Secretariat (2003) Dealing With Disaster – Revised Third Edition. Brodie Publishing Limited, Liverpool.

Flor, N.V. and Hutchins, E.L. (1991). Analyzing Distributed Cognition in Software Teams: A Case Study of Team Programming during Perfective Software Maintenance. In J. Koenemann-Belliveau, T.G. Moher and S.P. Robertson (Eds.) *Empirical Studies of Programmers: Fourth Workshop*. 36–64. Norwood, New Jersey: Ablex Publishing Corporation.

HM Inspectorate of Constabulary (1999) Keeping the Peace: Policing Disorder. HMIC, London. (http://www.homeoffice.gov.uk/docs/poldis.html).

HMSO (1997), *Dealing with Disaster*, London: Her Majesty's Stationery Office

Perry, M. (2003). Distributed Cognition. In J.M. Carroll (Ed.) *HCI Models, Theories and Frameworks. Towards a Multidisciplinary Science*. London: Morgan Kaufmann.

# IMPACTS OF AUTOMATION REVISITED

**Elizabeth Carver**

*Advanced Technology Centre, BAE SYSTEMS, Filton, Bristol BS34 7QW*

This paper revisits the research done in understanding the attributes and impacts of automation on the flightdeck, and considers how these impacts manifest themselves in other complex systems where automation of various types is being introduced. The operator is being removed from real world cues in areas such as the engineering design process and in distributed working environments where teams collaborate with each other electronically. Computer-mediated, rather different cues are provided which are potentially not so rich. The discussion draws on experience from a number of projects looking at human factors aspects of concurrent engineering, knowledge management for design engineers, and e-engineering, in an effort to understand why these processes do not always work as effectively as planned.

## Introduction

Understanding the human impacts of automation has been a key area of human factors research. As early as 1951, Fitts defined the attributes of what machines and men are particularly good at. This was followed by other milestone papers such as Bainbridge's (1983) discussion of the ironies of automation, and more recently the work of researchers such as Sarter and Woods (1995) and Parasuraman (1997), looking at automation on the flightdeck and its impact for pilot performance. Although the research has led to a good understanding of the human impacts of automation, there is a tendency for technology to drive development, not fully taking into account the role of the new technology or its systemic impacts. In the context of the underlying business requirement to build better things, faster and cheaper, automated systems provide significant benefits. However, it is clear from some current design solutions and working environments that automation is not fully understood and that there are still lessons to learn from the research.

## The evolution of automation on the flightdeck

Automation has gradually been introduced onto the flightdeck since the 1920s, with increasing complexity and decreasing direct pilot control. Drivers for development have been technological advancement, as well as accident investigations where automated solutions were implemented to remove human error. Of course we now know that in doing this, one source of error may be removed but further different errors may be caused as a result. Control tasks were the first to be automated, with systems aiding and augmenting human control of flight surfaces. Next to appear were navigation systems requiring speed and position data, with automation gradually increasing to provide the autopilot systems of today. Strategic decision making tasks have been the last to be automated.

Another aspect of cockpit automation is that where as in the early days, information was presented in a multi-sensory manner, with pilots relying on feedback from control surfaces aided by a minimal number of displays; information today is almost entirely presented visually, potentially overloading the pilot. Cockpit displays had been added in a largely random fashion, as the measurement technology improved and the air traffic increased. As more and more systems were automated, the cockpit became very cluttered and pilots had to search for the information that they needed amongst a plethora of different displays. Human factors specialists realized the importance of functionally grouping displays together and designing the displays to aid in understanding and to reduce reading errors. When there was finally no more room for any more displays, multi-function displays were implemented to try to solve this dilemma. The pilot is thus gradually becoming a systems manager rather than a controller. He has a monitoring role, searching for information within the system architecture, and dealing with all information presented via the visual modality. This is in the context of periods of very high workload at take off and landing and quite low workloads during transit. Moreover, if something were to go wrong, the pilot, having been just monitoring the system (a role to which humans are not suited) needs to rapidly gather the information that is needed about the situation in order to solve the problem. Situation awareness by the pilot of the automation status is therefore a crucial aspect.

Reliance on technology is increasing so that aircraft can operate in increasingly poor conditions by supplementing visual real world cues with sensor data. In parallel the cues being removed by the automation processes need to be replaced, and the provision of feedback needs to be managed to counteract the predominance of visual information.

We are now moving towards a situation where information from the real world is no longer raw data, but data that has been sampled and processed and displayed in a synthetic medium – albeit to simulate the real world. Because the data the pilot receives is processed and may conflict with other information from different sources, there is a greater degree of uncertainty attached to it and the pilot needs to have sufficient means to apply checks to maintain confidence that the automation is working correctly. Where once, pilots had a very close, tightly coupled relationship with the control surfaces of the aircraft, there are now many layers of automation between him and the outside world.

## Attributes of automation

Clearly, in an environment where there are multiple automated systems, it is important to implement these so that the pilot can understand what is going on in case he needs to take over control from the automated systems. As a result of studies observing pilots carrying out flight tasks, and also as a result of investigations after critical incidents, there have been good descriptions developed of what "poor" automation looks like (Bainbridge, 1983, Sarter and Woods, 1995).

The major attributes of poor automation can be described as follows:

- It is strong – the system can behave autonomously
- It is silent – the system provides inadequate feedback about activities and intentions
- It can be clumsy – the system interrupts human activity particularly during periods of high workload, or even adds to task loading when workload is already high
- It can be obstructive – it is difficult for the automation to be reconfigured in the desired way and the automation may therefore not be used in the way that the designers intended.

These system attributes may result in less than optimum pilot behaviours such as :

*Automation bias* – where the pilot may use the automation to the exclusion of other systems;

*Automation complacency* – where the pilot may come to rely completely on the automation even if it is faulty or unreliable;

*Automation surprises* – where the pilot may be less than familiar with all the modes of operation of the automation and therefore be surprised by its behaviour asking questions like "what is it doing now?", "why did it do that?" and "what will it do next?" If the system designer understands these attributes, then lessons can be learned.

## Automation in the engineering design process

Having looked at the attributes of poor automation and their likely impacts, it is not a large step to look at the same sorts of things happening in other processes which have more to do with the way we work rather than the product we are building.

### *The drawing office*

The design process has changed dramatically from the pre-1970s when all design activity was done in 2-dimensions on a drawing board. This had several useful features – all the designers were together, and team leaders and experienced designers would be available to novices for advice and would monitor the design activity very closely. It was easy for them to do this because the drawings on the drawing boards were open and available to be seen by all. Any questions could be dealt with in an informal manner, information flow was good, and situation awareness was effective by picking up on the conversations around you along with the visibility of your own work and the work of your colleagues.

The story today is very different. All designers use workstations and design in 3-dimensions. They are very likely to be based at different sites and even in different countries, and will be designing ever more complex systems, using sophisticated analysis and modeling tools and using more and more automated synthesis processes such as those enabled by intelligent computer aided design (ICAD).

### *Automated design processes*

ICAD has proved itself to be exceptionally effective in designing parts that conform to rule sets. There is a high cost at the start with highly skilled designers defining the rules to develop the system, and then variations can be rapidly designed using ICAD. Fewer engineers can achieve more as detailed operations are automated, but the nature of the design role is changed. There are parallels to be drawn with fly-by-wire pilots, as the task becomes one of system operator rather than controller.

Digital Mock-up (DMU) is a method by which designs can be investigated without building full size physical mockups. This saves money, and also means that design reviews can be held in a distributed fashion, with the design geometries being shared by several designers, all potentially based in different geographical locations. It has also lead to improved design for assembly and maintainability. The designer is working removed from the real world and different designers will be working concurrently on different systems within the overall design. Situation awareness of each designer's work is important. The computer mediated solution offers huge benefits, but in some areas provides less rich information. Cues such as scaling factors and qualities of materials are just two such aspects and users need to be aware of the limitations.

Product data management (PDM) systems are another way in which the aerospace industry has been able to manage the huge amounts of product data that are associated with engineering projects as complex as the design of an aircraft. PDM systems usually incorporate a workflow system too, keeping up the pressure to complete tasks on time, to follow the process and to make the workstation the single most important point of contact for engineers. There is a risk that the automated process becomes the equivalent of a conveyor belt working environment for engineers.

As computer environments for design become more sophisticated, it is now a realistic challenge to consider virtual design and certification of an aircraft.

*Automation in distributed collaborative working*

Computer supported co-operative working research has looked at distributed collaboration for some time and there are many tools and environments available to support working in a virtual team – that is, one that is distributed and where individuals are not located in the same place. However, the same process of losing cues and information richness occurs as you move people away from each other and use computer mediated solutions. Moreover, virtual collaborative tools such as video conferencing and even collaborative web-based workspaces, do not replicate the transfer of informal information that forms a very important part of knowledge transfer in organisations. The knowledge management research suggests that one of the most important aspects in growing and sharing the knowledge within organisations, is the informal communication between colleagues for example at photocopiers, by the water fountain or in the corridor. Although these conversations may not immediately appear to be directly work-related, they form the underpinning information flow network in the organisation. This ad hoc serendipitous communication can be termed socialising with purpose. This socialising with purpose is not replicated in current virtual working environments – it is neither supported by the technology nor encouraged in the behaviours of participants. In a business environment where virtual and remote working are becoming more and more necessary, there is a significant challenge to ensure that informal communication networks are facilitated in order to maintain the transfer of tacit knowledge.

Results of experiments carried out looking at a design task in the ENHANCE project (Carver and Leggatt, 2000), indicated that it was more difficult and less successful to carry out a task in a virtual manner. The results showed statistically significant differences between co-located teams and distributed teams. For a distributed team, communicating with e-mail and telephone, the task took longer, communication was less frequent, sequential, and more formal, with some members of the team being isolated. Failure modes constitute potential risks in new ways of working, both in the technology and in their associated behaviours. Failure modes have already been identified in e-mail systems – we could no longer work without e-mail, but many of us now have problems with too many e-mails, difficulties in housekeeping, incomprehensible e-mails, and the potential for inflammatory or unintended comments being sent. The failures are largely due to the fragmented and automated nature of the process: preparation and sending are separated, there is no feedback to confirm understanding and it is easy to broadcast e-mails to many people under the misconception that you are "sharing knowledge".

There needs to be a better assessment of the risks of working remotely, the provision of training to work in this way, and improvements in the technologies available.

## Conclusions

Good human-machine interfaces in terms of the design of the display and the display/control relationship should now be expected. The interaction between the operators as part of the system, with the technology, as well as with the process, is harder to achieve.

The European project HINT (Human Implications of New Technology) concluded that Human Factors (HF) must be integrated as part of the design life-cycle, with HF as part of a multi-disciplinary team. Moreover there needs to be a strategy for the implementation of automation with specific requirements for taking into account the human factors impacts, namely:

| Understanding the role of automation | Appropriate level of functionality |
|---|---|
| Allocation of function and ease of operation<br>Conflict handling | Allocation of responsibility<br>Adequacy of system to human feedback |
| Risk assessment/error tracking – error tolerant design (including recovery from error) | Appropriate levels of workload, distributed between crew/team members members during both normal and abnormal operation |
| Awareness of organisational and cultural differences in dealing with error, attitudes to automation, and in safety culture | Training for complex systems – not just training to configure the system, but training to *use* the system effectively |

Finally, there is still much to be done to ensure that designers understand the impacts of automation, both for flightdeck systems as well as for complex systems which are incorporated in the design *process*. There still remain risks where automation is implemented in more subtle ways which affect the way we work, but that we do not notice until the situation becomes critical.

## References

Fitts, P. M. 1951. *Engineering psychology and equipment design*. In Stevens, S. S. (Ed.), Handbook of Experimental Psychology, New York: Wiley.

Bainbridge, L. 1983. *Increasing levels of automation can increase, rather than decrease, the problems of supporting the human operator*. Automatica, **19**, 775–779.

Sarter, N. and Woods, D.(1995) *"How in the world did we ever get into that mode?" Mode error and awareness in supervisory control.* Human Factors, **37**(1), 5–19, 1995.

Parasuraman, R. 1997. *Humans and Automation: Use, Misuse, Disuse, Abuse*. Human Factors, 1997, **39**(2), 230–253.

Carver, E and Leggatt, A. 2000. *Analysis of Co-located and dispersed communication using the COSITE DMU demonstrator*. ENHANCE project D7.2/3 – V4 Appendix 2.

# DESIGN METHOD FOR DEVELOPING MATHEMATICAL DIAGRAMS

**Kreecha Puphaiboon, Andrée Woodcock & Stephen Scrivener**

*Visual and Information Design Research Centre, Coventry University, Priory Street, Coventry CV1 5FB*

Poorly designed diagrams reduce the effective teaching and learning of mathematics. Typical problems include graphic elements that do not focus on mathematical concepts and their relations. Through the development of series of diagrams to support the teaching of simple geometric concepts, a user centred design method has been developed, which when used in conjunction with guidelines on gestalt principles and critical information cues, helps students to structure and visualise concepts, thereby facilitating learning. Although the diagrams produced using this method did not show a significant quantitative improvement in students' learning, they were shown to encourage reflection on the underlying concepts. Educational diagram designers also believed such a method would enhance current design practice.

*Keywords*: Diagrams, design process, children.

## Introduction

Diagrammatic representations can provide students with a means of understanding and structuring mathematical concepts and problems (Sutherland and Mason, 1993). However, mathematical educationalists (such as Jones and Mooney, 2003) believe that mathematical diagrams are not sufficiently informed by pedagogy i.e., diagrams do not guide students towards an understanding of mathematics or show the meaning of a mathematical process. As a result, the diagrams do not allow students to explore or make conjectures about the concept. Thus it is important to apply pedagogical theory in the design guidelines for designers, so that mathematical diagrams serve the purpose for which they were deserved.

Diagrams can act as cognitive "externalizations" enhancing cognition by mapping problem elements to a display so that solutions become apparent (Sutherland and Mason, 1993). The diagram thus becomes part of a larger cognitive process involving perceptual pattern finding and cognitive symbolic operations. Thus a diagram's effectiveness, to some extent, depends on how well it is designed as an input to the perceptual system. It is important to apply perceptual psychology into the design guideline for designers.

Surveys of professional designers and teachers indicate that typically educational material is not designed in a systematic manner often without evaluation with teachers or students (Cunningham, 1991; Jones, 2000). To improve on this, research suggests that the development should include participation with users. Researchers in educational software (Hughes and Hay, 2001) proposed that individual stakeholders involved in the project development should explicitly describe what they think is important and what they can contribute. Their proposition is helpful in highlighting the need for including key stakeholders but fails to consider the role of the designer. In addition, information theorists suggest that designers must understand users and requirement in order to direct user's attention to key components essential and relevant to the task (Kazmierczak, 2003). However, often designers don't have access to users and clear specification.

Without an understanding of users (e.g., their needs, abilities, interests, and expectations) and clear specification, designers have difficulties in forming design goals, solutions, and schematizations on the users' comprehension (Wai and Siu, 2003).

To resolve the problems of mathematical diagram designs, a design method was developed and tested to support learning of Shape and Space for Year 5. A feature of this research is UCD – involving experts and learners rather than one that prioritises the role of the designer. Thus, this paper will describe the development, evaluation and formalisation of a collaborative, iterative design methodology, which acknowledges user requirements. It will be shown firstly that the use of such a proposed method produces diagrams that engage mathematical learning for the students and secondly that the method was usable by educational diagram designers, and would enhance design practice.

## The development of the diagram design method

The context for the work was the development of diagrams to teach the geometrical concepts of shape and space to middle school children. Given the current decline of performance in the learning and teaching of mathematics, such research was seen as timely.

The resultant design method evolved from work which embraced a study of the background knowledge designers require if they are to develop effective learning aids (eg perceptual psychology, learning theory); observational studies regarding the design and current use of diagrams in educational practice and interviews with teachers, students and designers; and the iterative development and testing of a series of diagrams. This paper outlines the major stages in the development and testing of the model.

## The effectiveness of current diagrams and their use in schools

Observational and interview studies were undertaken with teachers and students at three middle schools to understand the curriculum requirements, the role of diagrams in teaching and problems with current diagrams. The results indicated that: (a) inaccurate diagrams are used in schools; (b) such diagrams confuse children struggling to learn new concepts; (c) the teacher is often unable to guide student learning around inaccurate diagrams.

An evaluation of current diagrams used to teach concepts of shape and space was undertaken with 20 students. The diagrams were presented in a purpose built, CAL environment allowing for pre and post testing and the automatic logging of student keyboard behaviour. The pre and post test evaluation focused on the extent to which the current schoolbook diagrams supported the learning of concepts. A simple paired-samples, *t*-test was carried out showing that the current schoolbook diagrams do not improve students' learning significantly ($t(8) = -0.82$, $p = 0.44 > 0.05$).

Five UK professional educational designers from four companies (with a mean range of 18 years experience), all of whom had designed educational books, were interviewed to determine how diagrams for use in the teaching of mathematics are currently designed. Typically, designers followed the draft diagrams provided by the textbook author and embellished them following recommendations from the marketing department. No discussion occurred between the designer and author during early production; neither was there any evaluation with students or teachers (unless it was major publication) to ascertain the effectiveness of the diagram in supporting learning in different learning contexts. This lack of evaluation could be one of the reasons why erroneous diagrams are still found.

**Table 1. TUTE as applied to diagram design.**

| Category | Detail | Knowledge acquisition |
|---|---|---|
| Task | Learning key concepts<br>Measurable through achievement of learning outcomes | Understanding the curriculum (eg in what has been learnt), how the concept is attained. Accessible through interviews with teachers. |
| User | Cognitive development (in this case mathematical reasoning ability), perception of representations | Cognitive development and reasoning skills (eg Piaget). Perception (especially Gestalt psychology). Accessible through literature review |
| Tool | Visual representations eg diagrams and animations | Principles of diagrammatics and information design |
| Environment | The classroom and way in which diagrams support learning | Observation of lessons. Interviews with teachers. Understanding of educational theory and trends |

## Development of a user centred design method to produce effective diagrams

For the teaching of geometry in middle schools, current diagrams have been found to be erroneous and ineffective in supporting student learning. Although we have not tested whether diagrams in other disciplines are similarly flawed, we could hypothesize that without a rigorous, iterative development process they could well be.

Adopting an action learning approach, we redesigned the existing diagrams, and reflected on this process to create a general diagram design method that would be user centred (in the widest possible sense), iterative and support implicitly the knowledge designers need to produce effective diagrams.

The knowledge required by designers in the production of effective diagrams was broken down into groups relating to **T**ask, **U**ser, **T**ool and **E**nvironment (see Table 1), operating at a macro level. We would argue that effective diagrams can only be produced by the application or understanding of knowledge from each category.

The iterative design and evaluation with users led to the development of six design principles – salience; mapping; causality; visual reinforcement; mathematical generalization and sequential representation – that should be embodied in each representation. These, and other issues were formalised as guidelines to aid in diagram design

## The method for the redesign and generation of diagrams to support learning

The method can be broken down into six stages. Each stage in the development and refinement of diagrams should be underpinned by theoretical guidelines, mediated by the experience of the domain experts (teachers) and feedback from the end users (children)

1. Clarification of the requirements of the diagrams through discussion of the underlying concepts to be taught with teachers or other experts. Such discussion should be supported by existing visual representations where possible.
2. Evaluation of existing diagrams (in the case of redesign) by the teacher, to indicate real design problems and communicate these to the designer.

From these two stages a clear, a written specification is generated which outlines the mathematical concepts to be delivered (eg equal sides and equal angles for an equilateral triangle), the failure of existing solutions and an indication of potential solutions. Additionally mutual understanding is built up – the designer acquires a greater understanding of the way in which the diagrams will be used and the concepts to be delivered. The teacher more fully understands the potential of the diagram.

3. Generation of concept designs, which are discussed with the teacher before detailed design and the development of key frames (for the animation)
4. Detail design and verification, again with the teachers (expert).

The iterative, formative development with the teachers was designed to assess the diagrams in terms of their delivery of key mathematical concepts, their likelihood of confusing the children and whether the design of the diagram's elements, keyframes and motion were appropriate (eg speed of animation, ambiguity and the accompanying text).

5. Formative evaluation with students in which the animations are presented one key frame at a time.

From this evaluation further refinements were made to the designs eg relating to font size and speed of presentation.

6. Summative evaluation with students (representatives of the end user population) to assess the extent to which the revised (or new diagrams) enhance learning.

A comparative study was undertaken with two groups of matched students, a control group using standard material and an experimental one using the redesigned material. No difference in learning attainment was found between the two groups, however the reliability of this finding was suspect due to small sample sizes. Follow up interviews with pairs of students revealed that the diagrams still had some design problems, but more importantly, showed that the new diagrams were engaging and would potentially encourage a greater engagement with the material and promote mathematical thinking.

We believe that the six stages outlined above will lead to the design of effective diagrams. Although the diagrams we produced in the course of the development of the method conveyed simple concepts (such as symmetry and rotation), feedback at each stage was helpful in design refinement, and we would stress the importance of the designer working with domain experts (teachers) and representative end users (children at the mathematical level the diagram addresses).

## Evaluation of the method and diagrams with educational experts

The method was formalized and explained to educational diagram developers. The developers felt that such a method would be of value to their practice, but expressed reservations about the time such a process would take and the need for statistical analysis. Obviously the adoption of a method entailing such a high degree of user involvement would require a major change in the industry. However, if more effective training material is to be produced such a change may be needed.

A complimentary study was undertaken with teachers. Although the method had not been designed primarily for them, they would use the outputs, could be involved with evaluation or use the method to develop their own teaching material. Additionally, dynamic or animated diagrams might provide opportunities for different teaching approaches. They believed that the method could be useful in terms of problem-based design, creation of new mathematical diagrams, and error reduction. Furthermore, the life span of the diagram and textbook could be extended and hence justify for the increased investment. The teachers also highlighted the need for more user centred design in the creation of learning material to take into account individual learning styles and the range of competencies in any one group.

## Conclusion

The study addressed UCD in educational design. This research contributes a method for mathematics diagram design, which can be used by designers and publishers. The method provides guidelines for diagram designers to design effective mathematical diagrams for teaching mathematics by providing a method for developing specifications based on problems and user need and a design guideline grounded in psychological and educational literature. The stakeholder approach giving teachers and students input into the design process leads to alternative diagrams. However, although the method is usable by designers we realize that there are limitations in this research. Further research needs to be carried out with a larger sample size, more diagrams and students and a designer should test the method in a professional context.

## Acknowledgements

Thanks are extended to Washington Karumazondo, John Henry Newman school teachers and students for their willingness to be involved in this research. This work was funded by Kasem Bundit University, Thailand and Coventry University, UK

## References

Cunningham, S. 1991, Visualization Environment for Mathematics Education. In W. Zimmermann and S. Cunningham (ed.) *Visualization in Teaching and Learning Mathematics* 1991, (Mathematical Association of America, Washington), 67–77

Hughes, G. and Hay, D. 2001, Use of Concept Mapping to Integrate the Different Perspectives of Designers and other Stakeholders, *British Journal of Educational Technology* **32**(5): 557–569

Jones, K. and Mooney, C. 2003, Making Space for Geometry in Primary Mathematics. In I. Thompson, *Enhancing Primary Mathematics Teaching and Learning* (Open University Press, London)

Jones, R. 2000, Textbook Trouble, *American School Board Journal* **1**: 1–11

Kazmierczak, E. 2003, Design as Meaning Making: From Making Things to the Design of Thinking, *Design Issues* **19**(2): 45–59

Sutherland, R. and Mason, J. 1993, *Exploiting Mental Imagery with Computers in Mathematics Education*, (Springer Published in Co-operation with NATO Scientific Affairs Division, Germany)

Wai, K. and Siu, M. 2003, Users' Creative Responses and Designers' Roles, *Design Issues* **19**(2): 64–73

# DELIVERING DIFFICULT CONCEPTS USING VISUAL REPRESENTATIONS

**John Burns[1], Elena Gaura[2], Sarah Mount[2], Robert Newman[2] & Andrée Woodcock[1]**

[1]*Visual and Information Design Research Centre, Coventry School of Art and Design,*
[2]*Cogent Computing Group, School of Mathematics and Information Science, Coventry University, Email: A.Woodcock@coventry.ac.uk*

It is difficult to present new, complex ideas in ways that are informative and interesting in situations where a vast amount of information has to be delivered and understood quickly. Given short attention spans, boredom thresholds and limited capacity to assimilate new information, it is essential to use an effective conveyor of the message, which presents the content in a clear and unambiguous manner whilst helping the receiver to remain alert and focused and integrate the information with existing knowledge. This paper introduces a collaboration between computer scientists and artists[1] to develop an effective, animation based information communication tool, in the form of an application scenario, to be used for communicating to students (and wider audiences) dry technical aspects of pervasive computing.

*Key words:* Visual representations, animation, pervasive computing.

## Introduction

An ongoing collaboration between the Cogent Computing Group (CCG) in the School of Mathematics and Information Sciences and the Visual and Information Design Research Centre (VIDE), Coventry School of Art and Design is developing an animation to convey information about a new, relatively complex concept – cogent sensors – in a way that will capture the interest and imagination of audiences, inform them and hopefully instil in them a desire to find out more about the subject.

This paper considers; the role of graphical representations in teaching and training; the development of the animation (the process and the expectations of the group during its development); the usefulness of the visualisation in facilitating debate amongst the CCG and lastly, an evaluation of the material in terms of the added value of the material.

## Animations in practice

We are now at a stage where education is not confined to schools and universities, it occurs throughout life, either within the confines of industry, especially in knowledge-based organisations, or through life-long learning. Audiences and presenters/commissioners of teaching and training material are therefore mature and sophisticated, having

---

[1]Zeshan Ahmed and Alex Jevremovic developed the visual representations and animation.

been exposed to a rich, varied and high quality visual culture. They need to communicate about technical, scientific, conceptual and managerial information within and between large scale companies and institutions: the information may appertain to physical devices and principles such as the operation of machines, chemical compounds or drugs; or address concepts such as safety procedures, aspects of learning and training, the communication of decision making, marketing and research processes.

To meet these demands, companies may hand over responsibility for their external advertising and publicity campaigns to specialist agencies who can produce material to the necessary quality. However, internal training and awareness programmes that are vitally important in terms of making known the activities and aims of the organisation, often do not receive the budget, care or specialist attention they warrant even though the employees, delegates, directors, visitors who attend these sessions are indeed the same persons who are immersed in the televisual culture of the wider community. They tend to judge the informational diet served to them by similar standards to the programmes seen outside. No matter how worthy a subject matter may be, if the method of presentation does not engage and judged as substandard by the viewer, then that subject matter will not be seen to its advantage. This applies equally well to the development of material for teaching within our institutions where material should also be developed bearing in mind the principles of multimedia design for learning (for example, Mayer, 2001).

## The value of animations in education

An increasing amount of research is exploring the potential of different combinations of multimedia to support education. Animations have been found to increase motivation (Rodgers, 1995), aid the quality of learning (Stephenson, 1994), enhance communication of concepts involving time and motion (Hays, 1996) and the visualization of "invisible" dynamic processes that otherwise would need to be understood through abstract concepts (Huk et al., 2003). Researchers have concluded that multimedia presentation enables people to understand things that would be very difficult to grasp from words alone. When both words and pictures support each other, students construct verbal and pictorial mental models and build connections between them. As well as the capacity of animations and illustrations to portray information differently, benefits may also be derived from reductions in cognitive load through the use of a second cognitive channel (Mayer, 2001).

A case in point here is the emerging area of pervasive computing. Pervasiveness is foreseen as the "next big paradigm shift" in computing, making computing available everywhere, anytime. The forecasted social impact is immense and to date, awareness, even amongst computing students is limited. Although the area has so far benefited from enormous research efforts, the tomorrow's "dream applications" are commonly designed (and then presented) using a bottom-up approach- from the level of physical components and computational algorithms, building up to functional systems.

CCG believed that a top-down view, both in the design and the presentation/dissemination of ideas and concepts phases would be more fruitful both technically and in order to ease the communication and exchange of ideas within the technical development team. Rather than look at the minute details of cogent sensor operation, we have developed an animation of a usage scenario, without considering operational details.

## The context – cogent sensors

In the context of Pervasive Computing, CCG have promoted a "dream application" at several international conferences, e.g. Newman and Gaura, 2004. The application is a concept demonstrator for the potential of highly specified and intelligent integrated MEMS (Micro-ElectroMechanical System) based wireless sensing devices. Such a concept is believed feasible using present day technology.

The scenario is for the use of a network formed from a multiplicity of microsensor nodes as a means of planetary exploration (in contrast to using highly functional probes or rovers). Such a scenario would provide the context (and a reason) for the application, and would highlight a number of factors such as the release and deployment of the sensors, the types and quantity of information that might be gathered from the network, and the mechanisms required to transmit that information back to earth. From this scenario, the CCG can outline the design for the node including provision for the required transducers, the signal and data processing circuitry, the data transmission and reception equipment and the actuators required to deploy and maintain the sensor on station.

## The development of the animation

The animation proceeded from initial informal discussions about the requirements of the group and possible scenarios (contexts) between the group and John Burns (illustrator and animator), through to a series of concept sketches – produced by Robert Newman which were elaborated and further refined in discussions between CCG and the co-authors.

These discussions served two purposes: to make the content of the key frames clearer and secondly to provide a vehicle for CCG to discuss how the sensors would work in practice. Examples of the concept sketches and the resultant key frames produced by the research assistants under the direction of John Burns are shown in Figure 1.

At the time of writing we have 15 key frames that can be used as speaker support, and a 5 second animation detailing a meteor hitting out the sensors. The intention is to produce a fully narrated, animated 2.5-minute sequence that will cover the entire concept accompanied by stills that can be used as speaker support to illustrate the presentation.

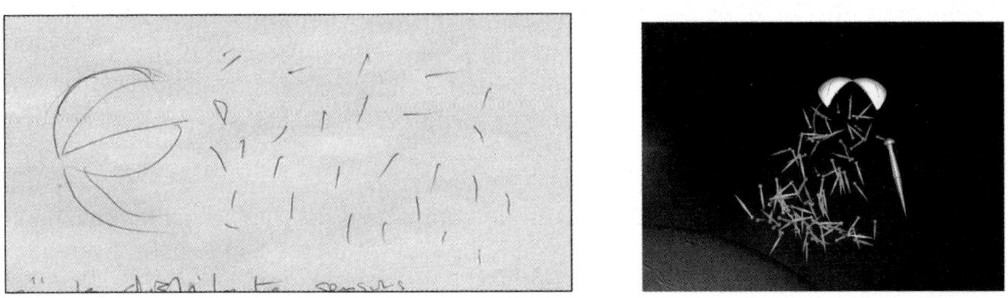

**Figures 1(a) and (b). Concept and detail sketch of the sensors being released on to the Mars landscape.**

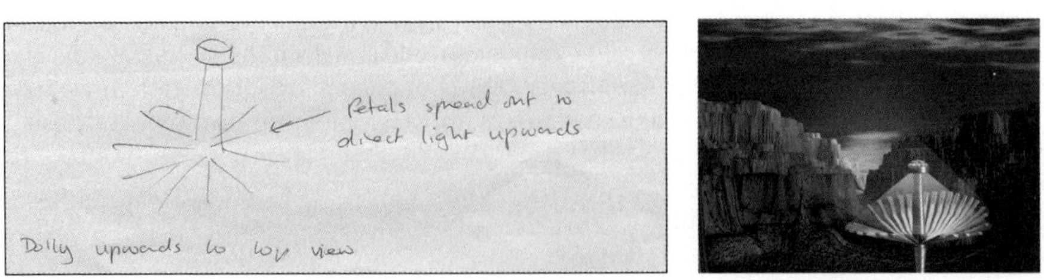


**Figures1 (c) and (d). Concept and detail of the sensor sending out a signal.**

## Stakeholder views of potential deployment and effectiveness of the animation

During the detail design stage a questionnaire was distributed to team members responsible for the development of the animation (ie the coauthors and the animators) to determine whether there were any differences of opinion with regard to the purposes of the animation, process, concepts to be portrayed, benefits, surprises and problems that had arisen. Clearly, such factors address wider issues than can be dealt with here, such as corporate strategy, interdisciplinary working and group dynamics.

Both groups (CCG and VIDE) agreed on the purpose of the animation, the way it would be used, the concepts that were to be portrayed, thereby indicating the success of the initial discussions and the initial concept sketches. Both groups expressed pleasure at working together on a project that furthered the research interests of both groups. Naturally the CCG were more detailed in their outline of the key concepts that could be portrayed. Surprise was expressed on how some apparently clear and unambiguous concept drawings could be (mis)interpreted. In such cases the drawings became a vehicle for more detailed discussion of operational details that had not been previously considered. Additionally some concern was expressed concerning the nature of the scenario and the extent to which it would create an appropriate mindset in the audience.

## The role of the animation in terms of the CCG

Prior to the use of the material for teaching processes, the slides were shown to the wider CCG for comment. This produced a lively debate dealing with issues such as conventions in animations (e.g. the shortening of time, use of light to denote sending of signals), the nature of visual representation, audience characteristics (sophisticated viewers, subject experts or novices) and the impact this would have on their ability to detect/tolerate inaccuracies without losing confidence in the scenario, the main purpose of the animation. At a second level, discrepancies were noted in the modelling of the environment – such as the meteorite impact, the use of light/colour to represent transmission of signals, the density of the network. The third level of debate was about issues the group had not considered, such as clustering of signals and detailed design.

From experience, we know that such debates are common. We have provided a visual representation not a representation of the actual event. Misrepresentations and ambiguities focus the attention of the wider group on issues they may have not addressed thereby enabling opportunities for clarification and new research directions. Providing opportunities for such discussion is clearly a benefit of generating visual representations.

## The animation as a vehicle for delivering information to learners

The material developed so far (the stills and short animated sequence) has been used in the delivery of a lecture on pervasive computing to Masters students. Following the recommendations of researchers such as Mayer (2001) and Roßling and Freisleben (2000) these were integrated into the lecture. The stills were displayed as a PowerPoint, remaining on screen for approximately a minute whilst a particular point was discussed. After the lecture students were given a chance to ask questions on the material before completing questionnaires and engaging in small group discussions about the added value of the animation and screen shots in the lecture. Such a question is believed to be compatible with the argument put forward by Mayer (2001) regarding the affordances of different media in enabling students to build up representations of the material.

All students saw value in the use of graphical material in textbooks and lectures as augmentation, to provide an overview of basic concepts or more detailed explanations. Some regarded the graphical material before the written. The graphics provided "an image in the mind, which will be easy to visualise the whole problem clearly". Comments also confirmed the different affordances of the media and that the graphics had allowed them to focus on problems, and provided additional stimulation. Most students were able to provide detailed answers about cogent system technology and its application. However, most were unable to visualise usage scenarios that did not involve space travel or geographic modelling and were unable to imagine sensors with different appearances.

## Conclusions

Within the computer science communities, most of the concepts involved in the development of even the simplest pervasive applications are difficult to handle. Lack of public understanding of such technologies has resulted in negative publicity, part of this can be attributed to the way in which ideas are communicated. It is the responsibility of researchers to find mechanisms, perhaps visual representations that meet the needs of their audiences. The design teams of complex applications are increasingly multidisciplinary. Communication around visual scenario representations may facilitate interaction in the group. If pervasive computing is to be developed then we have to explain the concepts to students who are not very confident with traditional technologies. Taking them through the logics and visualisation of the application would ease the instructor's task, interest the audience with the animation acting as an anchor for QA sessions on the technical details. A more considered use of visual representations is relevant to communities that need to make results more accessible to the public.

## References

Christel, M.G. 1994 The role of visual fidelity in computer-based instruction. *Human-Computer Interaction*, Vol. 2, No. 9, pp. 183–223

Craig, S.D. et al., 2002 Animated pedagogical agents in multimedia educational environments: Effects of agent properties, picture features, and redundancy. *Journal of Educational Psychology*, Vol. 94, pp. 428–434

Huk, T. Steinke, M. and Floto, C. 2003 Helping teachers developing computer animations for improving learning in science education PROCEEDINGS OF SITE pp. 3022–3025

Mayer, R. 2001 *Multimedia Learning*, Cambridge, UK: CUP

Newman, R.M. and Gaura, E.I. 2004 Design of a generic applications interface for large intelligent sensor networks, Proc. 43rd SICE Annual Conference, pp. 2425–2430, Japan

Roßling, G. and Freisleben, B. 2000 Experiences in using animations in introductory computer lessons in http://:nibbler.tk.informatik.tu darmstadt.de/publications/

Rodger, S. 1995 An interactive lecture approach to teaching Computer Science. *SIGCSE Proceedings*, 278–282

Stephenson, S.D. 1994 The use of small groups in computer based training: A review of the recent literature. *Computers in Human Behaviour*, 10, 243–259

# DEFENCE

# EVALUATION OF THE POSTURE OF TALL MALES IN THE T45 BUNK

**A. Green & R.S. Bridger**

*Institute of Naval Medicine, Crescent Road, Alverstoke PO12 2DL*
*EMail: sohf@inm.mod.uk*

A new bunk has been designed for the Type 45 destroyer. Headroom is greatly increased, to allow occupants to sit up in bed, and personal storage space is included in the bunkspace itself. The purpose of this study was: to quantify the limits imposed by the bunk design in terms of the percentage of users that can sit in the bunk, and to measure the fit and posture of tall people (sitting height greater than the 95th percentile) sitting in the bunk, to establish if it is safe (risk of injury). Eight male subjects all with stature exceeding 95th percentile participated in the trial. The postures of the subjects sitting in the bunk were captured on video and digitized using the Peak Motus® system. The findings indicate that the T45 bunk design and dimensions do not limit the range of users that can recline safely.

## Introduction

A previous assessment (Green et al. 2002) yielded subjective data about two novel bunk designs. This was used to support the selection of one of them for use in the Type 45 destroyer. The Institute of Naval Medicine (INM) reviewed recommendations for bunk dimensions in a 1996 report (Andrew, 1996). The recommended minimum height, width and length of the bunk were 991, 816, and 2037 mm respectively. Cabin size and deck height on the Type 45 impose restrictions on the maximum dimensions of bunks, and the height, width and length of the Type 45 bunk design are below these recommended minima at 920, 810 and 1920 mm respectively. The bunk has a padded, backrest (inclined at 30 degrees to the vertical) for the user to recline against and the clearance above an uncompressed mattress is 790 mm. This bunk is a new departure for the Royal Navy because it provides increased headroom compared to bunks on existing ships. During the prototyping phase of the development of the Type 45 bunk, concern was expressed at Human Factors Subcommittee meetings about encouraging personnel to sit up in bed. In particular, it was suggested that tall males (greater than 95th percentile stature) would have to contort themselves in order to do so, and that the resulting posture would be a hazard. The sitting posture was modelled using a human manikin on a CAD modelling system (the JACK system). The model suggested that a minimum clearance above the uncompressed mattress of 955 mm was needed. However, the clearance above the uncompressed mattress of the new T45 bunk is only 790 mm. Therefore the model indicated that tall people, above 95th percentile stature, and in particular people with sitting heights above 95th percentile (981 mm) (Jenkins et al., 1990), would not fit into the bunk. The purpose of this study was to determine whether there were limits imposed on the range of users that can be accommodated in the T45 bunk, and to evaluate the sitting posture of subjects reclining against the back rest. This involved determining whether the angle of the backrest is adequate to accommodate those with a 95th percentile sitting height (males) and measuring the fit and postural adaptation of tall people sitting up in the bunk.

## Method

### *Subjects*

Eight male subjects participated in the trial (mean age in years = 26, SD = 6, mean stature in mm = 1920, SD = 6.9, mean weight in kg = 91, SD = 10). Their heights all exceeded 95th percentile stature. The mean sitting height was 983 mm (half of the subjects had sitting heights above 95th percentile and all were greater than the 80th percentile).

### *Procedure*

The postures of all the subjects sitting in the bunk were captured on video and digitized using the Peak Motus® system. The following postural angles were measured and the mean was taken: Trunk-thigh angle, Trunk flexion, Hip flexion, Neck flexion, Trunk angle, Knee flexion. Posture markers (small self adhesive spheres) were stuck to the skin overlying the ankle, knee, hip, iliac crest and the shoulder joint. The ear lobe was used as a marker for the ear. Four ultra-sound sensor/emitters (to measure the distance between the sensor/emitters) were placed bilaterally on the skin at the area of the sacro-iliac joints, and between Thoracic vertebra 12 and Lumbar vertebra 1, for recording of lumbar length using the Orthoson Posture Monitor (Friendly Sensors AG).

To calibrate the range of movement (ROM) of the trunk the maximum saggital flexion and extension of each subject was measured from the reference position of the body, which occurs when a standing person looks straight ahead in a relaxed posture. Trunk-thigh angle, hip flexion and trunk flexion were recorded with the subject in standing, full flexion and full extension. After calibration, the subject entered the bunk and reclined against the backrest. When a comfortable resting position was attained, the posture (position of the body markers) was recorded on videotape and the lumbar length was recorded on the posture monitor. All subjects were allowed to use a pillow as a lumbar support and all measurements were repeated three times. The videotapes were analyzed using the Peak Motus system®. The position of the bony landmarks in the saggital plane was digitized and the postural angles mentioned above were measured. To measure mattress compression, the sitting height of 7 subjects (mean weight = 80.5 kg, SD = 16.1 kg), was measured with and without the mattress. The difference in sitting height indicated the thickness of the mattress when compressed by the subject's body weight. The Wilcoxon Signs ranks test was used to compare the data measured in the bunk, with the data recorded when standing.

## Results

### *In bunk measurements*

The following results are based on the difference between the posture when standing and when sitting in the bunk.

*Trunk-thigh angle*: A statistically significant reduction in trunk thigh angle was observed when subjects reclined in the bunk (mean reduction in trunk-thigh angle = 57 degrees, standard deviation (SD) = 6.9, $P < 0.05$). This represented 50% of the maximum range of movement, in flexion, of the trunk and hip joint.

*Trunk flexion*: A statistically significant increase in trunk flexion was observed when subjects reclined in the bunk (mean increase in trunk flexion = 36 degrees, SD = 9.9, $P < 0.05$). This represented 53% of the maximum flexion of the trunk.

*Hip flexion*: A statistically significant increase in hip flexion was observed when subjects reclined in the bunk (mean increase in hip flexion = 24 degrees, SD = 8.6, $P < 0.05$). This represented 47% of full flexion of the hip (from standing).

*Neck Flexion*: A statistically significant increase in neck flexion was observed when subjects reclined in the bunk (mean increase = 22 degrees, SD = 11.8, $P < 0.05$).

*Trunk angle*: A statistically significant reduction in trunk angle was observed when subjects reclined in the bunk (mean angle of repose from vertical was 38 degrees, SD = 7.5, $P < 0.05$).

*Knee Flexion*: A statistically significant increase in knee flexion was observed when subjects reclined in the bunk (mean increase in knee flexion = 5 degrees, SD = 3.8, $P < 0.05$).

### *Orthoson posture monitor data*

Lumbar spine lengths when reclining were significantly different from standing ($P < 0.05$). The mean lumbar length when reclining in the bunk (expressed as a percentage of lumbar length in full flexion using standing as the reference) was 64% (SD 23.9). The lumbar length, measured in the bunk, was within the normal range of movement and indicates that the lumbar spine is not excessively flexed.

### *Mattress compression*

The mean mattress compression expressed as the reduction in thickness of the uncompressed mattress was 46% (giving a mattress thickness of 71 mm).

## Discussion

A reduction in trunk-thigh angle is to be expected when a person sits in a bunk and reclines against a backrest. If most of the postural change is due to trunk flexion, then the spine would be under a worrying amount of stress. However, the data show that this is not the case. When subjects reclined in the bunk, their trunk thigh angles were reduced by a mean of 57 degrees. This was achieved by mean hip flexion of 24 degrees. The remainder of the postural adaptation took place in the lumbar and thoracic spines (mean trunk flexion of 36 degrees) and none of the joints were at the extreme of the range of joint flexion, as the mean trunk flexion was only 53% of total flexion. The amount of neck and trunk flexion measured in the bunk are within the limits for static working postures specified in ISO 11226 (ISO 11226, 2000). The mean trunk angle of 38 degrees from the vertical measured in the bunk results in a backward trunk inclination, but it is fully supported by the backrest which is acceptable within the guidelines in ISO 11226. The mean knee flexion measured in the bunk is negligible (given the backwardly inclined trunk).

Figure 1 compares the mean angles of body posture measured in the bunk (A) and the angles of relaxed posture assumed in weightless conditions (B) (picture adapted from Niebel and Freivalds, 1999). Both the mean neck angles and mean trunk-thigh angles measured in the bunk are very close to those depicted in Figure 1 (B). The postural adaptation from standing that occurs when sitting in the bunk, produces joint angles that are between 24 and 60% of the full flexion of each joint. This indicates that the postural angles found in the bunk are in the midrange for the range of movement of the joints. Good design enables equipment to be used with the joints in the middle of their range of movement (Bridger, 2003). All subjects reported no discomfort when sat in the bunk and all had adequate headroom. The findings indicate that the bunk design and dimensions do not limit the percentage of users that can be accommodated in the bunk.

Contrary to the predictions from the JACK model, this study indicates that males above 95th percentile sitting height can sit safely and comfortably in the Type 45 bunk and recline against the backrest. There are three reasons for this. Firstly, the actual Type 45

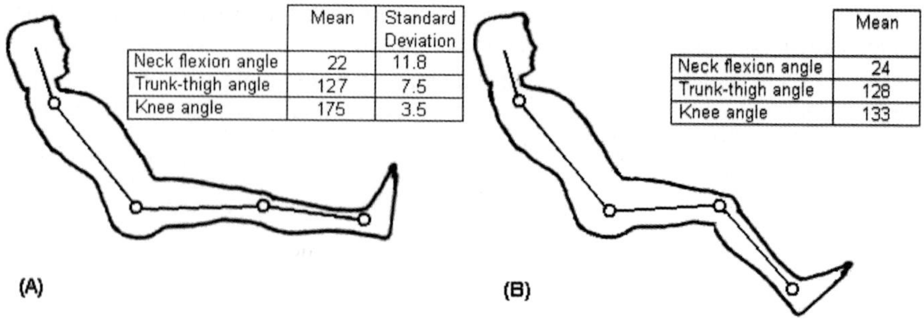


**Figure 1. Mean neck flexion, trunk angle and knee flexion angles in the bunk (A), and angles of a typical relaxed posture assumed by people in weightless conditions (B).**

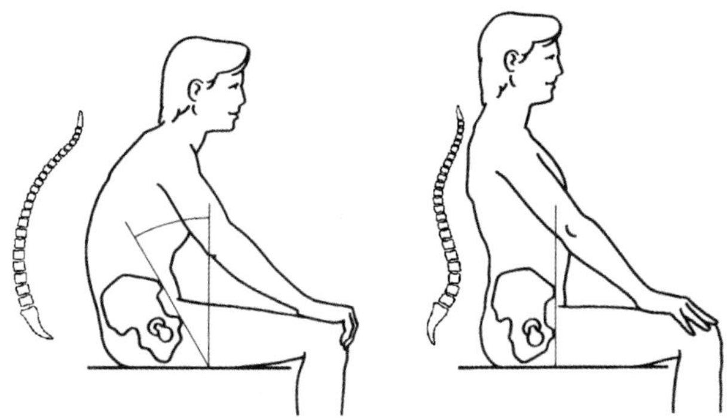

**Figure 2. Reduction of sitting height and the shape of the spine due to backward rotation of the pelvis (picture adapted from Bodyspace, 2nd Edition).**

backrest is at 30 degrees, not 25 degrees (as estimated on the JACK model), which gives (approximately) an extra 50 mm headroom. Secondly, the pelvis may slide forward, away from the base of the backrest, such that the trunk angle is greater than 30 degrees from the vertical. In fact, the current data indicate that this happens anyway to provide clearance for the sacrum against the backrest (the mean trunk angle was 38 degrees, SD 7.5 degrees which is greater than the backrest angle of 30 degrees). Thirdly, the length of the trunk when measured with subjects sitting erect or against a vertical wall (ISO 7250, 1996) appears to change when subjects recline. In other words, sitting height in a reclining posture is **less** than sitting height when sitting erect.

A backward rotation of the pelvis and trunk flexion occurs when sitting (Eklund et al., 1991) at a backward repose angle causing a reduction in lumbar lordosis (Bridger et al., 1992), and a reduction in sitting height (Figure 2). This simply means that the subject does not sit up straight when reclining. Taken together, these mechanisms indicate why people with very large sitting heights may still be safely accommodated in the bunk. It would seem that the JACK simulation does not have sufficient anatomical fidelity to model human postural adaptation when reclining. To illustrate the size of the latter effect, some simple, post hoc calculations were made to re-estimated the Trunk length taking into account the effect of the angular displacement of the trunk when the subjects recline. Hip to head length was 5.3% less. This reduction in the length is undoubtedly caused by the trunk flexion that occurs due to the rotation of the pelvis and the reduction of lumbar curvature that occurs when reclining. Delleman (Delleman et al., 2004) sited in a study of U.S. civilians (HES, 1962) which showed a difference in relaxed sitting height from erect sitting height of 4.1 cm at the mean and similar values at the 5th and 95th percentiles.

Finally, the current dimensions of the new bunk do not appear to impose a restriction on the percentage of users that can be accommodated in the bunk and the backrest angle is adequate to accommodate users with 95th percentile sitting height.

Measurements of sitting height and trunk length made from subjects sitting upright, against a vertical wall, will lead to overestimation of the sitting heights and trunk lengths of subjects reclining against a backrest. Since it does not appear that the CAD manikins used in systems such as JACK have the anatomical sophistication to simulate the postural adaptations that occur when people recline. For these reasons it is recommended that reclining postures, as well as upright postures, be studied in future anthropometric surveys.

## References

Green, A., Bilzon, E., Bridger, R. 2002. *T45 Bunk Assessment Study*. INM's 150/115 dated 30th March 2002

Andrew I. D.C. 1996. Bunk Space – A Review of Previous Work & Recommendations for Future Design. INM Report No. 96045

Jenkins P.B., Pethybridge. R.J., Hooper, H.R. 1990. An Anthropometric Survey of 1333 Royal Naval Personnel 1986–1990. INM Report No. 18/90

ISO 11226:2000. Ergonomics – Evaluation of Static Working Postures

Niebel B, Freivalds A. 1999. Methods, Standards and Work Design (10th Edition), ISBN 0-256-19507-2. (p.122)

Bridger R. S. 2003. *Introduction to Ergonomics*, 2nd Edition. (Taylor and Francis)

ISO 7250:1996. Basic human body measurements for technological design

Eklund J.A.E., Liew M. 1991. Evaluation of seating: The influence of hip and knee angles on spinal posture. *International Journal of Industrial Ergonomics*, 8:67–73

Bridger R.S., Orkin D., Henneberg M. 1992. A quantitative investigation of lumbar and pelvic postures in standing and sitting: Interrelationships with body position and hip muscle length. *International Journal of Industrial Ergonomics*, 9:235–244

Pheasant S. *Bodyspace*, 2nd Edition, 1996. (Taylor & Francis, London)

Delleman N.J., Haslegrave M. and Chaffin D.B. 2004, *Working Postures and Movements*, (CRC Press)

HES (Health Examination Survey), 1962. User Data Set: Physical Measurements, Washington, D.C.: U.S. Department of Health, Education and Welfare

# FEEDING THE FIFTEEN-HUNDRED – DESIGNING THE GALLEYS FOR THE FUTURE AIRCRAFT CARRIER

**Matt Woodward**

*Design Services, BAE Systems, Warwick House,*
*Farnborough Aerospace Centre, Hampshire GU14 6YU*

The galleys on board the Royal Navy's future aircraft carrier will cater for approximately 1500 personnel and will be required to sustain operations around the clock to meet the operational demands of the ship. This paper explains how the application of human factors has been integral to the design process in order to optimise the compartment layout, minimise food safety hazards and determine the number of galley personnel required.

Task analysis was used to assess the workload in the galley and a link analysis was conducted to optimise layout. End-user representatives were consulted throughout to ensure the validity of the studies. The principal output from the work was the implementation of a number of design improvements concerning layout, equipment and storage.

## Introduction

Turn back to the ergonomics textbooks and examples of domestic kitchen design may be found to illustrate the principles and methods of work place layout. Consider now applying these principles to the design of naval galleys required to cater for fifteen hundred, at sea and around the clock. This paper provides a case study demonstrating the integration of human factors with the design of the Future Aircraft Carrier's galleys.

Two carriers are currently being procured for the Navy by the Ministry of Defence. The ships are being designed by a consortium of companies named the Aircraft Carrier Alliance (ACA); at the time of writing the project is in the design phase prior to manufacture. The carriers will be the largest ships in the Royal Navy, around three times the displacement of the existing carriers. Their primary role will be to act as a mobile airfield from which the Joint Strike Fighter (JSF) and other aircraft will be operated. With the air squadrons and supporting operational staff embarked the CVF will accommodate approximately 1,500 personnel. The challenge for the ship's Catering Services sub-department is to sustain this workforce throughout day and night.

## Method

In accordance with best practice a systematic approach was taken to the application of human factors (HF) to the design. The process defined in ISO 13407: Human-centred design processes for interactive systems, was cycled through to support the design.

### *Identify need for human-centred design*

The galleys were chosen as an area for HF involvement as they are compartments where high volumes of materials are manually transferred each day. With such labour

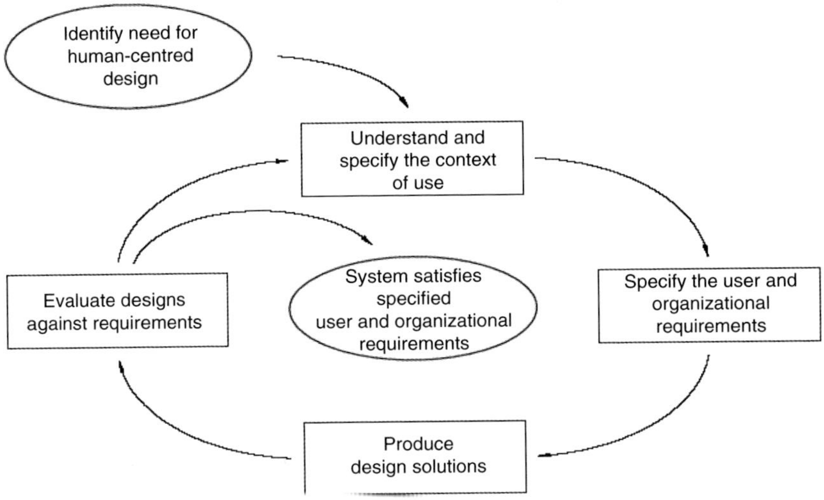


**Figure 1. The interdependence of human-centred design activities.**

intensive processes there is opportunity to improve efficiency by optimising compartment layout. The ACA also needs to determine initial complement numbers in order to scope the sizing of accommodation areas and support systems such as fresh water and waste. Catering is a labour intensive process utilising a relatively large proportion of the complement and was therefore selected for study. In addition there are several recent examples of poor naval galley design where the evidence suggests that ergonomics was not considered. HMS Ocean for example has the meat-slicing machine situated in the vegetable preparation area with associated food safety issues.

*Understand and specify the context of use*

Prior to HF involvement the customer had reached the decision to procure a twin-galley design and two areas had been reserved as placeholders on the General Arrangement (GA). These spaces were sized on the basis of Defence Standard 07-205 Pts 1 & 2: Requirements for Galleys and Associated Spaces. The System Requirements Document (SRD) only provided three high-level galley requirements. In order to better understand the area requirements and the quantity of victuals to be processed an exercise was carried out to map personnel numbers and watch patterns against meal times. For the purposes of the exercise a situation was modelled in which the majority of the ship's company were assumed to be working days (akin to a civilian nine-to-five routine) with the aviation personnel, including embarked squadrons, working one-in-two eight-hour watches.

The output of this activity was a timeline that showed the number of personnel changing watch at any given time and their associated meal requirements. These outputs were used to inform working groups and subsequent analysis. One of the results of early commissariat working groups was the decision to provide a third galley facility on 2-deck to sustain the Aviation personnel. A draft Concept of Operations paper was written to outline throughput requirements and specify a context of use for the galleys.

*Specify the user and organisational requirements*

At this point of the design cycle the focus was upon understanding the tasks and processes employed in the galleys and this was achieved through consultation with the end-users. RN personnel from the Commander-in-Chief Fleet (CinCFleet) organisation, including the Fleet Catering Officer, were engaged to bring knowledge of current operations to the analysis.

A principle activity was to undertake a task analysis using legacy data as a baseline. A comprehensive list of tasks was obtained from HMS Invincible, a current CVS class carrier and then the data organised according to watch and location. Each task was reviewed individually by calculating the task duration for the baseline load (i.e. number of ratings/officers fed for each meal on CVS) and then scaling the task durations according to the throughput data obtained from the watches versus meals exercise. In some cases tasks were added or removed to reflect the draft Concept of Ops and the design intent. By summing the revised task durations a total work effort was reached. Through consultation with the domain experts this was then interpreted in terms of department specific watch routines to arrive at a predicted number of junior ratings for each function. The senior rate management roles were added above this structure, again taking heed of design intent.

*Produce design solutions*

The commissariat lead designer worked up draft compartment layouts based on the requirements, throughputs and project-wide human factors engineering (HFE) policy papers. In practice this activity occurred in parallel with the task analysis, which was beneficial as an understanding of the process was important for galley layout, whilst improvements in equipments and technology impacted the work effort calculations.

The designer produced a main galley layout that enabled a logical flow of materials from the stores lift through the ready-use store onto the preparation, cooking and then serving area. The equipment fit was initially based on defence standards.

*Evaluate designs against requirements*

A link analysis evaluates the importance of relationships between components in a system in order to optimise their layout. This method was selected for assessing the galleys as a significant amount of materials (575 kg of fruit and vegetables for the junior rates for example) are manually handled around the compartment each day. The objectives of the analysis were therefore to:

(1) Optimise the compartment layout to minimise the distance materials and personnel are required to travel;
(2) Identify food safety hazards inherent in the design;
(3) Validate the proposed staff numbers;
(4) Assess storage capacity within the galley.

Before embarking on the full analysis Table 1 above was drawn up and a pilot study run to assess the suitability of each link type. It was important to show the sequence (flow) of materials for the assessment of food safety, and weight/volume was necessary for assessing storage requirements. To optimise compartment layout it would have been preferable to chart the frequency of personnel movement, however the pilot found this to be too ambitious given the number of personnel involved and the range of locations they would cover. Instead the frequency of materials movement was recorded, a measure which correlates reasonably well with personnel movement.

**Table 1. Link analysis scoping.**

| Link type | Weight | Frequency | Sequence |
|---|---|---|---|
| Materials | ✓ | ✓ | ✓ |
| Personnel | NA | ✗ | ✗ |

To identify food safety hazards two principles from the Hazard Analysis Critical Control Point (HACCP) method were incorporated into the analysis. These principles are:

(1) The preparation of flow diagrams to show each stage of the process and the identification of hazards and controls at each stage of the process.
(2) The identification of critical control points. These are stages which if controlled will reduce the risk of that hazard occurring, or eliminate it altogether.

The analysis was conducted at a two-day workshop facilitated by the HF analyst and also involving the lead galley designer and two senior rate end-users from the Navy. Prior to the workshop data such as typical menus and victual consumption had been collated from current carriers. Plan drawings, known as General Arrangements (GA), were reproduced to show the compartment at A0+ size and the sheets fixed to the walls.

Twenty-four hours of galley operations were divided into six evolutions, i.e. groups of discrete activities with a common goal. An example of this is the evening meal preparation, from storage to point of cooking. For each evolution the Fleet Catering Officer outlined the tasks and processes that would be carried out. The movement of materials was marked directly onto the GA using colours to distinguish between raw meat, cooked meat, dairy, vegetables and dry foodstuff. The number of lines between two points represented the frequency of movement or weight (for example one line represents 30 kg) depending on the evolution. Once all lines had been recorded a review of the process was conducted and problems with layout and equipment noted. In addition critical control points were logged.

Once the six evolutions had been analysed the data was reviewed against a 24-hour time-line to clarify those processes that would occur simultaneously. These tasks were then resourced by assigning personnel against them in order to provide a basic validation for the Catering Services complement. At this point potential solutions for the identified hazards and issues were discussed.

## Results

Figure 2 shows a section of the analysis for a single evolution against the original design. It is not intended to convey any detail, but to show the general concept of using links to support layout decisions.

The necessity of reproducing the plan at small scale and in black and white does nothing for clarity (the original plans were sixteen times larger and in colour) but the predicted point of congestion during meal times can be clearly seen.

The results of the analysis were written up and the designers re-worked the compartment to incorporate suggested improvements. The principal issues and implemented solutions are listed below:

(1) **Issue:** Excessive number of journeys made through the single access point in the bulkhead between galley and serving counters. This is predicted to be a bottleneck with approximately 245 journeys occurring within a 90-minute period.
**Solution:** Addition of second door into the bulkhead. This arrangement will allow a one-way door system to be employed.
(2) **Issue:** Transfer of full, hot serving dishes through a standard watertight door with 300 mm lip on lower edge.
**Solution:** Install modern sliding watertight doors with flush threshold. This will improve the efficiency and safety of material transfer.
(3) **Issue:** Large mass of potatoes and vegetables (approximately 575 kg/day) transferred from RU store to the opposite side of the galley for preparation and then returned back across the galley to the prepared vegetable store.

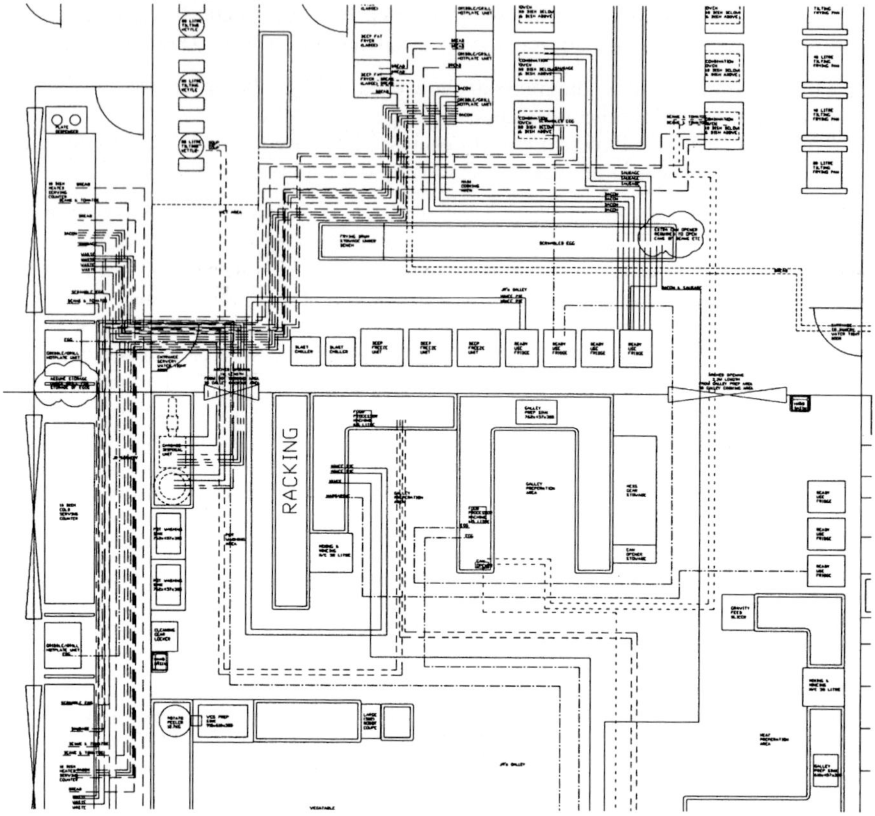


**Figure 2. Link analysis on original layout.**

**Solution:** Re-locate vegetable prep. area adjacent to the vegetable store, minimising distance victuals have to be transported.

(4) **Issue:** Longitudinal cooking bench creates a barrier between the ready-use fridges/freezers and the cooking equipment. Raw meat has to be either carried around the bench or will be placed on bench which creates a food safety hazard.
**Solution:** Divide work surface into two shorter benches to provide access.

(5) **Issue:** All items that are transferred from the lifts to the dining hall have to be taken through the length of the galley and behind the serving counters.
**Solution:** Provide direct access from lift receiving area.

Numerous minor changes such as adjusting equipment fit, storage capacity and storage locations were also made. The link analysis was repeated on the new design to assess the impact of the changes. The results showed a clear benefit in the revised layout with shorter links and a reduced number of crossover points.

## Conclusion and recommendations

There were two key lessons learnt during this work. Firstly it is important to integrate HFI early in the design cycle; in this case the area for the galleys was scoped prior to any analysis, potentially missing an opportunity to optimise the GA footprint. Secondly, the analysis should be scoped at a practical level; with such a complex compartment a pilot study showed that it was preferable to keep the link analysis relatively simple. Although recording personnel movement would have been useful, the amount of data would have impractical to record and difficult to analyse.

A whole-system approach is recommended; in this case it ensured that the design did not develop in isolation from operating policies, personnel numbers and consideration of the end-users. In a constantly evolving project documentation of the analysis and the rationale behind design decisions has proven to be useful for auditing purposes. The development of team working to incorporate a range of skills was particularly important in ensuring a methodical analysis, valid results and implemented improvements.

In summary this course of work has demonstrated that human factors can be effectively integrated with the design process and have an early influence upon design.

## Acknowledgments

WO1 (CS) S Roberts MESH SDA; WO1 M Hickman CINC Fleet; Jason Jones ACA Commissariat Lead Design Engineer.

## References

BS EN ISO 13407:1999, *Human-centred design processes for interactive systems,* (British Standards Institute).

Driver, K. 2003, *HACCP: A Simple Guide*, (Food Technologists Company).

Defence Standard 07-205 Parts 1 & 2: *Requirements for Galleys and Associated Spaces*.

# ONLY A DIGGER: THE HUMAN FACTORS CHALLENGES OF ARMOURED PLANT EQUIPMENT

**Toby Atkinson**

*BAE SYSTEMS CS&S Naval, Design Services,*
*Warwick House, Farnborough Aerospace Centre,*
*Farnborough, Hampshire GU14 6YU*

There are many facets of armoured vehicle design which make it a unique environment for Human Factors practitioners. Not only is there the flexibility to operate the vehicle closed down and with head and shoulders above the roofline of the vehicle, but also there is the limited vision afforded by a vehicle essentially intended to protect the occupants.

This paper discusses some of the Human Factors issues facing the design of armoured plant equipment, and how those issues were addressed through appropriate insertion of technology – particularly with reference to task vision. A key element of the process to assess the efficacy of this technology and the user environment was a full scale mock-up of the crewstation which allowed user interaction with proposed solutions at an early stage in the design lifecycle.

## Introduction

The purpose of this paper is to provide a brief overview of some of the Human Factors Issues facing the design of Armoured Plant Equipment. It is not intended to be a comprehensive exploration of the issues, but instead focuses on a task driven approach to overcoming the problems that an armoured vehicle provides for Plant Equipment operations. This paper summarises work that was conducted during the Product Definition (PD) and Demonstration, Manufacture and Contractor Logistic Support (DMCLS) phase of an armoured plant equipment (vehicle) programme.

The design of armoured plant equipment is driven by the tasks that need to be conducted, and a number of immutable design features. As is so often the case with applied ergonomics, it is the conflict of task requirements with fixed design features that causes problems for the user.

The process by which the problems were identified, defined and resolved is shown in Figure 1.

## Process overview

The process shown in Figure 1 illustrates where the information was acquired at each stage of the process in order to fully understand the scope of the problems.

The starting point of the process was to identify the tasks that were to be conducted with this new vehicle. Some of the capabilities available on this vehicle had not been available previously on armoured plant. To understand those tasks it was therefore necessary to draw parallels with commercial plant equipment.

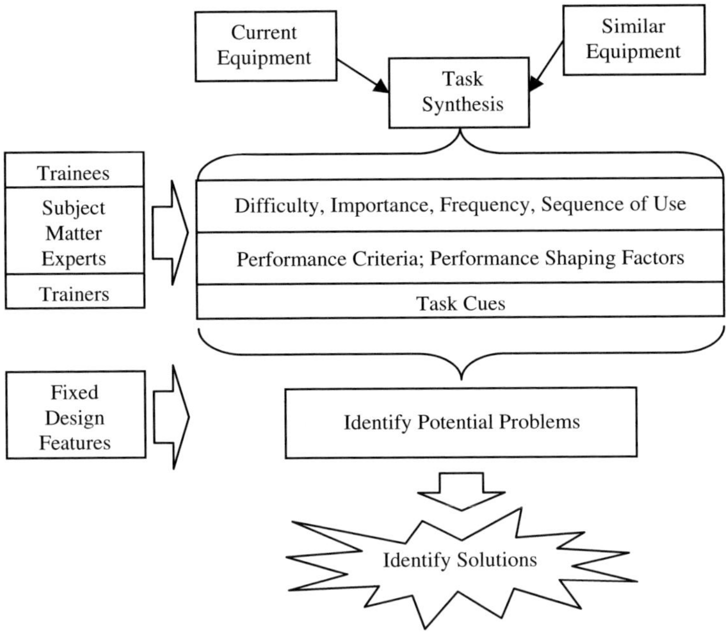


**Figure 1. Process for identifying problems and solutions.**

The tasks identified at this stage were high level based on the functionality provided by the vehicle, rather than specific to any particular equipment. A number of inputs were adopted in order to better understand those tasks. Relying on the qualitative data that could be obtained from focus groups and interviews, structured focus groups were held with military plant operators. These subject matter experts provided information on how frequently tasks were performed, how important those tasks were, how difficult, and more importantly why they were important or difficult.

Through the process of identifying which tasks were difficult and why, the particular performance shaping factors were being identified. A set of general factors were identified which made the task more difficult. In essence these boiled down to "Closed down" and "at night." "Closed down" refers to being enclosed within the vehicle, with only vision blocks to provide a view of the task that is being completed. "At night" refers specifically to the fact that the task area cannot be floodlit and therefore the operator must be able to dig in very low light levels.

The sources of difficulty for these tasks, and the cues that were used to try to make the tasks easier were identified through discussions with the SMEs and Trainers and Trainees. The trainers were selected as they provided an insight from the trainers' perspective of what they perceived trainees having difficulty with. The trainees were selected to provide a more "realistic" view of what may be problematic. As an illustration of this, consider using a clutch when driving a car. The process of using the clutch to engage the gear at just the right point is subconscious for most experienced drivers, but this is certainly not the case for learner drivers, where concentrated effort is required to co-ordinate clutch and accelerator pedal movements with the gear change!

Some of the cues which were identified through this process were:

1. The engine note – if the excavator arm or bucket becomes stuck, the engine note will drop. The user identifies that the equipment is struggling and therefore moves the arm/bucket away from the obstruction.

2. Vehicle movement – if the excavator arm hits a significant obstacle, the pressure that can be applied through the excavator arm means that the plant vehicle moves rather than the obstacle.
3. Lining up the bucket with the edge of a trench is done visually.
4. The depth of a trench is determined visually.
5. The width of a trench is set visually using the bucket width as a guide. If a trench needs to be cut 1.5 bucket widths, then after the first cut, the edge of the bucket will be lined up visually with the centre of the first cut.

As can be seen, and can be derived from a look at commercial plant equipment, there is a great deal of emphasis on the visual aspects of, for example, digging. However, there are inherent design features of an armoured vehicle that constrain this and make the task more difficult. One aspect that has already been mentioned is that armoured vehicles also operate at night, therefore the user needs to be able to have vision of the task under low light viewing conditions.

## Discussion

The role of an armoured vehicle is to protect its occupants whilst conducting whatever task it is that they need to conduct. To this end, a great deal of the bodywork is metal, with little room devoted to glass. This is completely contrary to the requirements of a digging task given the emphasis on the visual aspects of the task. The new vehicle therefore includes an array of cameras which provide vision from a multitude of angles around the vehicle. These supplement the vision provided by the vision blocks. When closed down the user will be able to operate the vehicle from camera views shown on display screens. In Armoured vehicle terms this is a significant step forward in technology. The manipulation of video images in vehicle to achieve tasks is not a widely applied approach.

The introduction of the display screens into the crew compartment and the vision afforded from both the vision blocks and camera systems has been assessed through a series of user trials in a Crew Compartment Mock up. This has been a central feature in ensuring user buy-in throughout the design process.

Given the level of armour on an armoured vehicle, it is not surprising that it is considerably heavier than most commercial plant equipment. This poses two problems in terms of the cues that the operator relies on to conduct their task. Firstly the increased weight of the vehicle means that if the excavator arm hits a significant obstacle, it is unlikely that it would be able to move the vehicle to the extent that commercial plant equipment can. There may be some "rocking" on the suspension but that would be all. Secondly, because of the increased weight of the vehicle, a much greater source of power is required, i.e. a larger engine. This means that the variations in the engine note for the excavator arm working normally and straining against an obstacle are very small in comparison to commercial plant. The user cannot therefore rely on the change of engine note as a cue. Furthermore the crewman is likely to be wearing a communications headset which would also block out some of the noise.

The solutions to the problems have so far focussed on providing support to the visual cues. This has been developed through the use of an interactive crew mock up. This process will continue into a prototype vehicle, which will begin trials in early 2005.

# EMERGENCY EGRESS FROM MULTI-CREW MILITARY AIRCRAFT

**David Harrison & Fiona Sturrock**

*BAE SYSTEMS, Warton Aerodrome W427J, Preston, Lancs PR4 1AX*

There are several factors that can influence a large scale human factors assessment and which need to be carefully considered during the planning stages. One of the main considerations is whether the assessment is to be performed using the real product, a purpose built facility, or software simulation. The paper describes how such issues impacted emergency egress assessments carried out by Human Factors engineers at BAE SYSTEMS on the newly designed Nimrod Maritime Reconnaissance and Attack (MRA 4) aircraft. These assessments were required to prove that the crew could safely egress the aircraft within the time limits set down by design requirements. The paper discusses some of the lessons learned from these assessments and how they may be beneficial for future Human Factors assessments across industry.

## Introduction

Emergency egress from a military aircraft is traditionally conceptualised as the pilot "banging out", i.e. the canopy of a fast jet being explosively removed followed by the pilot being ejected whilst still sitting in the aircraft seat. This dramatic means of emergency egress is not generally an option for the crews of larger military aircraft. The crew must first try to land the aircraft wherever possible, whether that be on the ground or in water, before emergency escape is possible. Once the aircraft has come to a halt, time is of the essence, particularly if the aircraft is on fire or sinking. Egress from the Nimrod Maritime Reconnaissance and Attack (MRA 4) aircraft is achieved either through the aircraft doors using the emergency escape ropes or via the over-wing hatches and down the flaps on the aircraft wings; which exits are used is dependent on the nature of the emergency. Not only must the crew be able to get themselves out of the aircraft they must also ensure that they carry with them emergency equipment such as life jackets, immersion suits and life rafts.

Human Factors engineers at BAE SYSTEMS carried out a series of emergency egress assessments to prove that the full crew complement can evacuate the Nimrod MRA 4 within the time constraints laid down by the design requirements. This paper describes HF issues effecting such assessments and shares some of the lessons learned for future assessments.

## Issues effecting a Human Factors aircraft egress assessment

There are several factors which may influence a large scale assessment and which need to be carefully considered during the planning stages. One of the main considerations is whether the assessment is to be performed using the real aircraft, a purpose built training facility, or software simulation. Each of these options has advantages and dis-advantages, both in terms of cost and assessment logistics; the main issues for each are summarised in Table 1.

**Table 1. Advantages and disadvantages of facilities for egress assessments.**

| Facility | Advantages | Disadvantages |
|---|---|---|
| Real aircraft | • Accurate data<br>• No/few caveats or constraints imposed on the assessment<br>• Potential for highlighting other design issues<br>• Evidence generated for specific aircraft design standards | • Aircraft availability limited<br>• Little opportunity to repeat testing<br>• Less controllable; greater effort required on assessment logistics<br>• Potential for damage to the aircraft and ultimately the project<br>• Difficult to control environmental variables, e.g. lighting levels, smoke generation |
| Purpose built training facility | • Greater availability<br>• Opportunity to repeat tests<br>• Accurate data<br>• Environment is controllable<br>• Greater flexibility; different aircraft configurations can be assessed | • Possible differences between facility and aircraft design which may effect the results<br>• No potential for highlighting other design issues if non-aircraft fittings and equipment are used |
| Simulation | • No requirement for aircraft/ facilities or participants<br>• Infinitely repeatable<br>• Non-intrusive<br>• Flexible; different combinations of procedures and aircraft configurations can be tested | • Time consuming to develop<br>• Physical assessment required (on aircraft or other facility) to validate the simulation<br>• Accuracy of data dependant on fidelity of the physical model and the characteristics attributed to the human participants |

Typically, a large scale industrial assessment does not have the luxury of all three of these options. The decision of which facility should be used generally boils down to what is available within the required timescales. The remainder of this paper focuses on a series of emergency egress assessments carried out by Human Factors engineers at BAE SYSTEMS on the newly designed Nimrod Maritime Reconnaissance and Attack (MRA 4) aircraft. It aims to highlight the impact of the chosen facility on the rest of the assessment logistics.

## Nimrod MRA 4 emergency egress assessments

The Nimrod MRA 4 is the RAF's Maritime Reconnaissance and Attack aircraft operated by a crew of ten aircrew (but has the capacity to carry up to twenty). Human Factors engineers at BAE SYSTEMS carried out a series of emergency egress assessments to prove that the crew could evacuate the Nimrod MRA 4 within the stringent time constraints laid down by the design requirements.

The initial set of assessments were performed on the first development **aircraft**, primarily because this was the only facility available within the required timescale. Carrying out a large scale assessment in an aircraft manufacturing/production environment gave rise to several challenges that would not necessarily have been encountered in the more controllable environment of a training facility or software simulation.

Initially, a great deal of **preparation** was needed to specify the aircraft standard required for the generation of valid evidence. This included the number of seats and workstations, the flooring, trim, lighting, doors and hatches, safety equipment housed within the aircraft: in fact everything that could effect the evacuation of the crew needed to be specified and agreed with the engineers responsible for the aircraft's production. Although this work would also be required if a training facility or simulation were used, the challenge when using the actual aircraft, was to ensure that the specified design was available on the assessment day. With the aircraft in such high demand its configuration can change daily and even seemingly trivial discrepancies in design standard can have a major impact on a human factors assessment. This highlights the importance of **communication** during the planning of a large-scale industrial assessment. Everyone involved must be made aware of the assessment requirements and the potential impact they could have on it. This would be relatively simple if there were only a few people involved but when using the real aircraft there are literally hundreds of people involved, the majority of which have their own priorities and goals.

Not only is this communication vital for preparing the aircraft for the assessment it is also essential for preparing the **participants** for the assessment. From a Human Factors points of view the participants used in the assessment must be representative of the end users of the product. For example, a commercial aircraft assessment should employ novice participants who are not practised in aircraft egress whereas a military aircraft assessment, where the end users are trained in emergency egress, should employ trained participants. Prior to the Nimrod MRA 4 egress assessments the participating crew (who were current and ex Nimrod Mk 2 crew) were given theoretical and practical **training** in the specific emergency egress procedures and techniques to be followed. This required additional time to be built into the aircraft's already busy schedule.

In addition to ensuring that the crew are well trained, the participants used should cover the **anthropometric range** required by the design in order to prove that all users are capable of safely escaping from the aircraft. **Representative clothing** should also be worn by the participants; in case of a military aircraft assessment, such as the Nimrod MRA 4, this included items such as flying suits, gloves, headsets and life vests. Ensuring that the appropriate aircrew equipment was available for the assessment required significant effort.

Another key issue is **assessment fidelity**. Regardless of the representativeness of the chosen assessment facility, or the participating crew, the results will only be accurate if the scenarios/situations that the participants are required to egress from are as realistic as possible. In the case of the Nimrod MRA 4 assessments two scenarios were simulated: a day-time crash landing scenario and a night-time, smoke filled, ditching scenario. The night-time scenario presented the greatest challenge. The interior and exterior of the aircraft had to be darkened to representative levels and the interior of the aircraft had to be filled with representative levels of smoke. With this requirement for fidelity comes the need for increased awareness of the Health and Safety of the participants, i.e. the increased risk of injury whilst egressing from a darkened or smoke filled fuselage. Equally as important, when using the real aircraft, is the potential for damaging the aircraft itself; such damage could halt the entire aircraft programme!

A constraining factor that is often forgotten during the preparation for such an assessment is **availability**. Not just of the aircraft or facility to be used but of the participants, particularly if they are current/serving military aircrew. If concurrent availability cannot be achieved the assessment cannot go ahead. Although this issue sounds relatively straight forward it can be the thorn in the side of such Human Factors assessments. Aircraft development programmes are notoriously volatile and availability of the aircraft is likely to change. Once a period on the aircraft has been agreed there is a great deal of time pressure to get the assessment completed and to return the aircraft to its development schedule. This often means that the assessment is a *one-shot* affair with no time to correct mistakes or re-run the assessment.

A second set of Nimrod MRA 4 egress assessments were carried out on a purpose built RAF **training facility** – the Smoke and Fire Evacuation Trainer (SFET) which became available within the timescales required. This was beneficial not only in terms of the savings (costs and timescales) made on the aircraft programme but also in terms of assessment logistics. The entire assessment process was made easier and was more controllable than the aircraft had been. The SFET design was more stable than the development aircraft and there were fewer people involved which made communication simpler and more efficient. Any constraints on the assessment (such as differences between the SFET and the aircraft) could be identified in advance, the impact assessed and communicated to/agreed with all those involved. The availability of the SFET was also better than the aircraft resulting in reduced time pressure for completing the necessary aircrew training and the actual assessment. This also provided the opportunity to repeat the assessments if necessary. Risks to the SFET and the participating aircrew were also minimised, as the SFET was equipped with environmental controls (such as automatic smoke generators) and safety devices to protect the participants.

*Nimrod MRA 4 results*

Although the results of the MRA 4 assessments are not the focus of the paper the authors thought it noteworthy that the evidence gathered to-date revealed that the crew were capable of evacuating the aircraft (well within the required time constraints) during both day-time crash landing and night-time smoke filled ditch conditions. This evidence has allowed flight clearance to be granted for the first two Nimrod MRA 4 development aircraft. Further testing will be required to clear the Nimrod MRA 4 to carry passengers which would result in a larger crew complement than previously assessed and also involves slightly different egress procedures due to the presence of non military, i.e. novice, passengers. The SFET will be the facility of choice for these assessments given its increased flexibility, controllability and availability compared to the aircraft.

## Lessons learned

Several valuable lessons were learned from these egress asscssments which will be carried forward into future assessments. These lessons are not only applicable to future military aircraft egress assessments as they can, in principle, be applied to any large scale industrial Human Factors assessment. The concluding section of this paper summarises several of the lessons learned:

- **Preparation** – never underestimate the time taken to set up and co-ordinate the logistics of such large scale assessments. They are typically 99% preparation and 1% assessment. Flexibility and **communication** are the keys to a successful assessment;
- **Facilities** – purpose built training facilities are preferred. They offer a greater degree of assessment control and generally have better availability. Requirements for the facility standard should be identified, and agreed, as early as possible;
- **Constraints** – all of the assessment constraints must be briefed to, and agreed with, the participating crew and Customer prior to the assessment. This will ensure that there are no nasty surprises during the assessment and the analyses of the results are not confounded;
- **Participants** – if the assessment is dependent on a specific anthropometric range of participants ensure they are available. Special clothing requirements such as flying suits and headsets must also be identified as early as possible to ensure availability;
- **Training** – all participants must be fully trained prior to the assessment; individual and crew training sessions both theoretical and practical are vital;

- **Single point of contact** – essential to ensure that any changes (to the facility or assessment logistics) are assimilated and disseminated quickly and accurately.

This paper highlights that, as with any large scale Human Factors assessment, there are several issues that can have a bearing on the outcome. When we are taught the principles of Human Factors we are taught to design the perfect assessment in order to create the perfect data set and meaningful results. Unfortunately this is not always possible in applied settings where the perfect assessment is more difficult, but not impossible, to achieve. In industry, large-scale Human Factors assessments tend to be more difficult to control and repeatability for statistical analysis is not the primary aim. Necessary deviations from the perfect academic style assessment are inevitable but they do not have a damaging effect on the results as long as their potential impact is understood and the results are analysed accordingly. Such applied assessments tend to be more realistic than academic style assessments and produce results which have an immediate impact on the product of concern whether that is design change recommendations or clearance for the product to be used. It is only through experience of carrying out such assessments that such valuable lessons are learned. These lessons should be shared throughout the Human Factors community.

# THE EFFECT OF MILITARY LOAD CARRIAGE ON GROUND REACTION FORCES

**Stewart Birrell & Robin Hooper**

*Department Human Sciences, Loughborough University, Loughborough LE11 3TU, UK*
*Email: S.A.Birrell@lboro.ac.uk*

The carriage of heavy loads by military personnel increases injury risk. This study aims to examine the effect of carrying military loads on GRF parameters. Fifteen male participants completed 8 conditions; Boot, rifle, webbing 8 and 16 kg, backpack 16 kg, and a load carriage system of 24, 32 and 40 kg. Results showed that as load was added there was a proportional increase in all major GRF parameters. Rifle carriage anteriorly in the arms increased the impact peak and propulsive forces while decreasing the force minimum. Webbing loaded with 16 kg increased impact peaks compared to the backpack, but the pack increased stance times. As increased GRF are positively linked to overuse injuries the analysis of how carried load affects these are important to understand military injuries.

## Introduction

Military mission requirements often depend on personal mobility. In these situations, personnel carry their own equipment, usually in a Bergen (or backpack) and waist webbing so forming a load carriage system (LCS). A rifle will also be carried. This study aims to investigate what happens to ground reaction forces (GRFs) during conditions that replicate military load carriage. Previous studies have been conducted but these have generally been restricted to loads of 20% and 40% of body mass, usually in a backpack alone. Heavy loads of above 30 kg have not been investigated in depth. This study explored the effects of smaller increments of load rising to 40 kg, differentiating between load added to webbing or the Bergen. The effects of rifle carriage were also studied.

## Methods

Fifteen male participants volunteered for the study (weight 83.2 kg ± 10.0, height 178.8 cm ± 5.4, age 27.8 ± 7.0) and completed 8 conditions after informed consent was gained. Each participant was required to complete 10 successful trials for every condition. The order of the conditions was randomised for each participant. Kinetic data were collected using a portable Kistler force plate, sampling at 400 Hz, embedded centrally and flush in an 8 m walkway. Participants were required to walk at a fixed speed of $1.5\,m \cdot s^{-1}$ (±5%). To ensure this speed was attained two sets of photoelectric cells were used; one set measured the approach speed to the force plate the other measured speed after the force plate. Both speeds had to be within 5% for the trial to be deemed successful, thereby limiting acceleration or deceleration during GRF measurements.

The 8 conditions were as follows:

1. Boot – Wearing non-restrictive clothes and a pair of military boots.
2. Rifle – As (1), but carrying a replica rifle weighing 2.1 kg.

**Table 1. GRF parameters significantly increased ($p < 0.05$) with load.**

| Force | Time | Other |
|---|---|---|
| Impact peak | Time to impact peak | Vertical impulse |
| Thrust maximum | Time to thrust maximum | Braking impulse |
| Force minimum | Time to force minimum | Propulsive impulse |
| Max propulsive Force | Time to max propulsive force | Mediolateral impulse |
| Max braking Force | Time to zero newton's | Loading rate |
| | Stance time | Push-off rate |

**Table 2. Correlations for the main GRF parameters on load (significant, $p \leq 0.001$).**

| | | Regression | | |
|---|---|---|---|---|
| Parameter | Pearson correlation | $R^2$ | Coefficient | Constant ($N \cdot BM^{-1}$) |
| Impact peak | 0.883 | 0.780 | 0.013 | 1.223 |
| Thrust maximum | 0.886 | 0.784 | 0.013 | 1.221 |
| Force minimum | 0.854 | 0.730 | 0.006 | 0.598 |
| Max braking force | −0.567 | 0.322 | −0.003 | −0.289 |
| Max propulsive force | 0.706 | 0 .498 | 0.002 | 0.225 |
| Vertical impulse | 0.922 | 0.851 | 0.013 | 1.084 |
| Loading rate | 0.577 | 0.333 | 0.066 | 9.565 |
| Stance time | 0.419 | 0.176 | 0.001 | 0.666 |

3. Webbing 1 – As (2), with the addition of webbing weighing 8 kg.
4. Webbing 2 – As (3), increasing webbing weight to 16 kg.
5. Backpack – As (2), with the addition of a backpack weighing 16 kg.
6. LCS 1 – As (2), carrying 8 kg webbing and 16 kg backpack.
7. LCS 2 – As (2), carrying 16 kg webbing and 16 kg backpack.
8. LCS 3 – As (7), but with the addition of a final 8 kg.

## Results

To enable direct comparisons between participants all kinetic data were normalised to body mass for the boot condition and system mass (body mass plus rifle) for all other conditions, this is expressed as $N \cdot BM^{-1}$ (or Newton's per body mass). To assess the effect of load a within subjects (repeated measures) ANOVA was adopted, excluding the boot and backpack conditions (1 and 5), with the rifle condition forming the control. The backpack condition was excluded to avoid having two conditions at 16 kg, with the webbing condition being a more realistic method for the military to carry 16 kg.

The addition of load in 8 kg increments has a significant effect ($p < 0.05$) on many GRF parameters measured (table 1). No significant difference was found in mediolateral maximum and minimum forces or in the time for maximum braking force to occur. Further analysis revealed significant, linear regressions between load and the following GRF parameters (table 2). The effect of carrying a rifle was tested in comparison with the boot condition (table 3), as well as 16 kg webbing to the backpack condition (table 4).

**Table 3. Mean forces ($N \cdot BM^{-1}$) in the boot and rifle conditions that differed significantly ($p < 0.05$ using a paired Student t-test).**

| | Impact peak | Force minimum | Max prop force | Mediolateral minimum |
|---|---|---|---|---|
| Boot | 1.203 | 0.622 | 0.215 | −0.064 |
| Rifle | 1.226 | 0.602 | 0.222 | −0.073 |

**Table 4. Mean parameters in the 16 kg webbing and backpack conditions that differed significantly ($p < 0.05$ using a paired Student t-test).**

| | Impact peak ($N \cdot BM^{-1}$) | Loading rate $(N \cdot BM^{-1}).s^{-1}$ | Time to max Prop force (s) | Stance time (s) |
|---|---|---|---|---|
| Web16 | 1.443 | 10.790 | 0.578 | 0.676 |
| Pack | 1.410 | 10.327 | 0.584 | 0.687 |

## Discussion

### *Rifle carriage*

A forward shift in the centre of mass (CoM) generates increased force in the forward and downward direction at heel strike (Hsiang and Chang, 2002). Carrying the replica rifle (at 2.1 kg about half the weight of the SA80) in front of the body has the following effects; significantly increases impact peak, maximum propulsive and the mediolateral minimum forces while decreasing the force minimum in the vertical axis. However, an increase in maximum propulsive force with rifle carriage goes against the inverse pendulum model (Winter, 1980) that suggests load carried in front of the CoM requires less active momentum during push-off. However, other factors like restricted arm movements caused by rifle carriage may be the cause of this increase.

Hinrichs (1990) showed that arm movements tend to reduce the total excursion of the CoM both laterally and horizontally. Eke-Okoro et al. (1997) suggest that arm swing modulates vertical excursions in the CoM and also that they serve to counterbalance the rotation of the body around the vertical axis, concluding that deliberate changes or restrictions to arm swing patterns reduce maximum velocity and stride lengths. The notion that restriction of arm movements influences the basal gait patterns and therefore change the GRF produced is in keeping with this study. A natural arm swing increases the momentum produced by the upper body so aiding forward advancement of the CoM, restricted arm movements may limit this active momentum. Carrying the rifle, so restricting arm movements, may therefore have led to higher maximum propulsive force as increased work is needed to achieve forward momentum.

### *Webbing verses backpack*

The different LCS used here represent a substantial shift in the distribution of the load. The webbing is supported by the shoulders with a yoke, a hip belt connecting at the front. The load is evenly distributed in pouches attached to the hip belt. The webbing distributes load both anteriorly and posteriorly, keeping it closer in vertical distance to the body's neutral CoM compared to the backpack. Higher impact forces observed with the webbing compared to backpack may be due to a larger component of the weight being over the striking foot at the time of initial contact this is supported by the work of Hsiang and Chang (2002). The webbing condition also showed a higher loading rate than the backpack condition, but this could just be a function of the increased impact peak.

Stance time was significantly longer when carrying the backpack compared to webbing. This is similar to the effect of a double pack that distributes load around the anterior and posterior of the trunk (Kinoshita and Bates, 1985; Lloyd and Cooke, 2002). It may be because it takes more time to shift the CoM over the base of support and into the propulsive phase, a possibility supported by the observation that there is a significant increase in time for maximum propulsive force to occur. It could also relate to stability, since the longer the foot is in stance phase the greater the stability of the individual.

### *The effect of load*

Adding load in 8 kg increments increases all vertical and anteroposterior GRF parameters, as previously reported (Harman et al. 2000; Kinoshita and Bates, 1983; Kinoshita, 1985; Lloyd and Cooke, 2000; Tilbury-Davis and Hooper, 1999; Wiese-Bjornstal and Dufek, 1991) with the exception of time for maximum braking force to occur ($p = 0.099$). Loading and push-off rates also increased with load as in Hsiang and Chang (2002). Numerous studies (Harman et al. 2000; Kinoshita and Bates, 1983; Lloyd and Cooke, 2000) have found mediolateral GRF parameters to be un-related to load, reinforced by the current study with the exception of mediolateral impulse that increased with applied load. This increase in mediolateral impulse may be linked to a decrease in stability caused by the increased load or by the continual shift of the body's CoM further away from its neutral position as load is added to the military LCS.

Despite the shifts in LCS CoM, the majority of GRF parameters measured correlate very significantly with added load (table 2), the vertical forces and impulse being very highly correlated with load ($r > 0.85$). These increases in force with applied load seem to be linear, even when heavy loads of 40 kg are used.

High impact forces and frequent exposures, like those experienced during load carriage or running, are a major risk factor for overuse injuries, particularly tibial stress fractures and knee joint problems (Birrell and Hooper, 2004; Keller et al. 1996; Knapik, 2001). For this reason the analysis of impact forces are important as it is likely to relate to injury incidence and prevalence associated with load carriage. It maybe useful to be able to predict the forces produced when heavy loads are carried over known distances and speeds, especially if the maximum stress that can be tolerated by a bone or joint without resulting stress fractures or joint degradation is known. For example, using regression analysis the increase in impact peak with load gave:

$$\text{Impact Peak} = (0.013 \times \text{Load}) + 1.223. \text{ (Load in kg and impact peak as } N \cdot BM^{-1}\text{).}$$

Eventually, combined with observed exposure data in military training, could it even become possible to predict the likelihood of overuse injuries during a forced march or in training for military personnel? At least the correlations show that almost 80% of the change in impact peak results from the addition of load.

## Conclusions

- Vertical and anteroposterior GRFs increase proportionally when load is added.
- Increases in mediolateral impulse occur when load is added that may affect stability.
- Rifle carriage increases the impact peak and decreases the force minimum due to a shift in the body's CoM. Also, greater propulsive forces are needed to advance the body, possibly due to the restriction of arm movements.

- Webbing loaded with 16 kg produces higher impact peaks than a backpack with the same load, but the pack resulted in a longer stance phase.
- Peak impact force may be predicted from load added to a military LCS.

## Acknowledgements

This work was sponsored by the British Ministry of Defence.

## References

Birrell, S. and Hooper, R. 2004, The Biomechanics of Military Load Carriage and Injury Potential: Injury Report. *Loughborough, Loughborough University.*

Eke-Okoro, S.T., Gregoric M. and Larsson, L.E. 1997, Alterations in gait resulting from deliberate changes of arm-swing amplitude and phase, *Clinical Biomechanics* **12,** 516–521.

Harman, E., Han, K.-H., Frykman, P. and Pandorf, C. 2000, The effects of walking speed on the biomechanics of backpack load carriage. *U.S. Army Research Institute of Environmental Medicine*, 1–71.

Hinrichs, R. 1990, Upper extremity function in distance running, *Biomechanics of distance running* (Human Kinetics, Champaign), 107–134.

Hsiang, S. M. and Chang, C. 2002, The effect of gait speed and load carrying on the reliability of ground reaction forces, *Safety Science*, **40**, 639–657.

Keller, T. S., Weisberger A. M., Ray, J. L., Hasan, S. S., Shiavi, R. G. and Spengler, D. M. 1996, Relationship between vertical ground reaction force and speed during walking, slow jogging, and running, *Clinical Biomechanics*, **11**, 253–259.

Kinoshita, H. 1985, Effects of different loads and carrying systems on selected biomechanical parameters describing walking gait, *Ergonomics*, **28**, 1347–1362.

Kinoshita, H. and Bates, B.T. 1983, *Effects of two different load carrying systems on ground reaction forces during walking.* Eighth International Congress of Biomechanics, Nagoya, Japan, Human Kinetics Publishers.

Knapik, J. (2001) Discharges during US Army basic training: Injury rates and risk factors, *Military Medicine*, **166**, 641–647.

Lloyd, R. and Cooke, C.B. 2000, Kinetic changes associated with load carriage using two rucksack designs, *Ergonomics*, **43,** 1331–1341.

Tilbury-Davis, D.C. and Hooper, R.H. 1999, The kinetic and kinematic effects of increasing load carriage upon the lower limb, *Human Movement Science*, **18**, 693–700.

Weise-Bjornstal, D. and J. Dufek 1991, "The effect of weightload and footwear on the kinetic and temporal factors in level grade backpacking." *Journal of Human Movement Science* **21**, 167–181.

Winter, D. 1980, Overall principle of lower limb support during stance phase of gait, *Journal of Biomechanics*, **13**, 923–927.

# GENDER DIFFERENCES IN MILITARY LOAD CARRIAGE

**Renee L. Attwells & Robin H. Hooper**

*Department of Human Sciences, Loughborough University, Loughborough LE11 3TU, UK Email: R.L.Attwells@lboro.ac.uk*

Gait and posture parameters were investigated using a CODA motion analysis system. Twenty individuals (10 male and 10 female) with load carriage experience walked at self-selected speed under 3 load carriage conditions; wearing boots, webbing (7% bodyweight (BW)) and backpack (33% BW), with total load equating to 40% BW. No gender differences were seen in temporal parameters. Gait kinematics indicated females with lower range of motion (ROM) in the knee and greater ROM in the femur, although body size was indicated as the more influencing factor here. Posture measures also showed minimal gender differences; females exhibiting greater ROM of the trunk. As the female group were of lower stature, it is suggested that body size may be the key determinant of load carrying ability, rather than gender per se.

## Introduction

Current restrictions in the UK do not allow females to serve in close combat roles in the military following a review of gender differences (MOD, 2002). A number of factors, including physiological and psychological characteristics, have been cited as reasons for this restriction, one issue at the forefront being the ability to carry load.

Physiological differences between genders during load carriage have been widely reported, the main factors including body weight, $VO_{2max}$, muscle strength and changes in stride length and stride frequency (Haisman, 1988, Knapik et al. 1996). In all cases females are placed at a disadvantage when compared to males. Biomechanical gender differences however have received little attention. Most studies have been conducted on males, and, where females have been studied, males have not been included. Extrapolations of results from one study to another has occurred. One exception to this is the study by Martin and Nelson in 1986, examining the effect of military loads on walking patterns of men and women. Results indicated females have smaller stride length; further decreasing with load; greater double support time and smaller single leg contact time. An earlier study by the same authors (Martin and Nelson, 1982) examined carriage of a 26 kg load at a set treadmill pace. Females once again had shorter stride length and greater stride frequency suggesting a greater number of step cycles was required over a given distance, resulting in higher energy expenditure.

Several other studies have examined gender differences in gait biomechanics without the addition of an external load. Kerrigan et al. (1998) concluded there are more similarities than differences between the genders. Bhambhani and Singh (1985) noted women have higher stride frequency and shorter stride length (presumably due to leg length differences), while Nigg et al. (1994) noted no differences between males and females when walking at a controlled speed (1.25 m/s).

In terms of posture measurements limited work has been completed, particularly in the area of dynamic head posture, both with and without load. Increased forward lean is seen with an increase in load on the back (Filaire et al. 2001, Goh et al. 1998, Harman

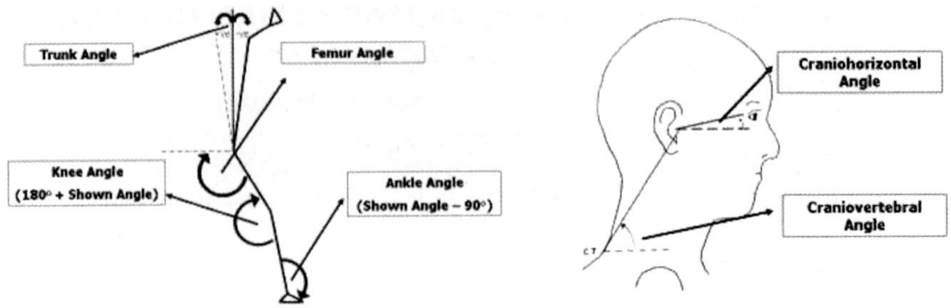


**Figure 1. Angular definitions.**

et al. 2000, Kinoshita, 1985, Martin and Nelson, 1986, Pascoe et al. 1997), and present work on head posture measures are limited to static measures or low loads when measured dynamically (Chansirinukor et al. 2001, Raine and Twomey, 1997). There is a lack of information on dynamic posture, the effect of military loads and differences between the sexes.

This study examined spatiotemporal parameters, and trunk angle when carrying contemporary British military equipment. The previous work cited above is concentrated in America, so a further goal was to consider whether differences in design between American and British equipment warrants further study in this area. The inclusion of specific gait and head posture measures used in the current study also add information to the gender divide.

## Methods

Twenty participants (10 male and 10 female) participated in this study. All participants were experienced in some form of load carriage (University Air Squadron, Army officer training corps, Territorial Army and regular backpackers) but none were members of currently serving military. Participant characteristics were: Male: Age 21.4 ± 1.8 years, height 180.7 ± 7.2 cm and weight 78.6 ± 8.4 kg; Female: Age 27.3 ± 7.2 years, height 170.0 ± 6.7 cm and weight 68.8 ± 6.3 kg. Height and weight for both groups fell within a normal distribution.

In the present study the replication of military load carriage conditions normally experienced by serving members was of importance. For this reason standard issue 90 Pattern (PLCE) Belt webbing and Bergen were used. Participants also wore standard issue boots, socks and carried a replica SA80 rifle in all conditions in order to reproduce the body position imposed on a member of the military. For all conditions participants moved at self selected speed along a 7 m walkway (with embedded Kistler™ force plate), in front of a CODA™ motion analysis system. Markers were placed on key body positions to enable measurement of main gait and posture variables (figure 1).

The 3 loading conditions were as follows (also see figure 2):

1. *Boot:* wearing shorts, t-shirt, military boots, socks and carrying a replica SA80 rifle.
2. *Webbing:* as boot condition with addition of standard issue belt webbing with 7% BW load.
3. *Bergen (backpack):* as webbing condition with addition of standard issue PLCE Bergen with 33% BW load.

In order for comparisons to be made between participants all data were calculated for points of percent stance phase, working from −20% to 120% stance phase, making them independent of time and walking speed. The means for the 7 repeats of each condition were calculated for all participants. Group data for each experimental condition were

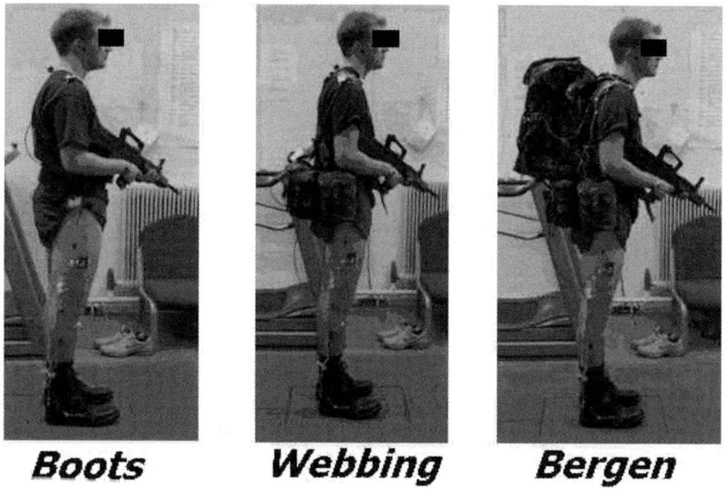


**Figure 2. Experimental conditions completed by all participants.**

**Table 1. Descriptive statistics of gait parameters (mean ± s.d.).**

| | | | |
|---|---|---|---|
| Stride length (m) | $1.50 \pm 0.011$ | $1.50 \pm 0.13$ | $1.48 \pm 0.14^{\#}$ |
| Stride frequency (strides/min) | $52.31 \pm 2.45$ | $51.95 \pm 2.90$ | $51.20 \pm 3.14^{\#*}$ |
| Stance time (seconds) | $0.71 \pm 0.04$ | $0.72 \pm 0.04^{*}$ | $0.45 \pm 0.05^{\#*}$ |
| Speed (m/s) | $1.31 \pm 0.11$ | $1.30 \pm 0.13$ | $1.26 \pm 0.14^{\#*}$ |

* significantly different to boot ($p < 0.05$).
\# significantly different to webbing ($p < 0.05$).

expressed as the mean and 95% confidence intervals for all 20 participants. A 2-way mixed ANOVA was conducted on all spatiotemporal data and maximum, minimum, mean and ROM values for all angles (significance level set at $p = 0.05$).

## Results and discussion

Temporal parameters showed no gender differences, the addition of load resulting in decreases in stride length, stride frequency and speed, therefore increasing stance time and stability (see Table 1).

The nature of these changes is dependent on pace characteristics of locomotion with differing results seen in set paced as opposed to free pace investigations. Results seen in this study are similar to those seen in set paced studies (Charteris, 1998, Harman et al. 1992, Martin and Nelson, 1986), whereas a self-selected pace study by Ghori and Luckwill (1985) showed no changes in stance time for loads up to 50% BW. Response due to gender has also resulted in differing conclusions in the literature; minimal differences at controlled speed (Nigg et al. 1994), women with slower speed and shorter stride length (Cho et al. 2004) and decreased stride length when walking at self selected speed (Kerrigan et al. 1998). The only study examining difference whilst carrying loads however is by Martin and Nelson, (1986), which at a set speed of 1.76 m/s indicated women had decreased stride length and consequential increase in stride frequency. The speed in the current study (see table 1) was significantly slower than that used in the Martin and Nelson study, possibly influencing results. Independent t-test examination

**Table 2. Changes in gait angular data (degrees) with changes in load (mean ± s.d.).**

| | Ankle | | Knee | | Femur | |
|---|---|---|---|---|---|---|
| | Female | Male | Female | Male | Female | Male |
| Boot | 21.00 ± 2.36 | 20.31 ± 2.66 | 38.72 ± 3.29 | 44.03 ± 2.99 | 47.70 ± 3.96 | 43.62 ± 4.03 |
| Webbing | 22.15 ± 2.35 | 21.10 ± 2.01 | 39.60 ± 3.67 | 44.84 ± 2.86 | 49.90 ± 4.23 | 45.76 ± 4.47 |
| Backpack | 22.88 ± 1.90 | 22.01 ± 2.03 | 39.28 ± 5.28 | 45.54 ± 2.83 | 54.28 ± 4.55 | 48.65 ± 4.98 |

of gender differences between boot and backpack conditions indicated women significantly ($p = 0.03$) compensating more in terms of stride frequency as load increased.

Gait kinematics did have some gender specific qualities with females having less range of motion (ROM) in the knee and greater ROM in the femur (average 4.6° for both – these angles linked by definition – table 2). There was also a trend for greater ROM in the ankle for females (Table 2).

Similar results have been established by other authors: Nigg et al. (1994) noting no gender differences in ankle movement, females having greater knee flexion path of motion; Kerrigan et al. (1998) showing trend for increased ankle plantar flexion in females, decreased knee extension and increased peak hip flexion; and Cho et al. (2004) observing no angular ankle or knee differences but women having significantly greater hip flexion than males throughout the gait cycle. When present data was examined relative to the height and weight of individuals, it is possible that these variables were the contributing factors rather than gender in the differences seen.

Posture measures included trunk angle and 2 angles on the head representing the movement of the neck on upper spine (craniovertebral angle) and head on neck (craniohorizontal). These angles have been examined previously in static studies and on children carrying school backpacks (Chansirinukor et al. 2001, Raine and Twomey, 1997). Minimal gender differences were seen in all angles, with trunk ROM being the only variable affected – females exhibiting a greater ROM (average 1.9° over 3 testing conditions). A trend for a more forward head position was also seen. As with previous research the addition of load resulted in greater forward lean (Attwells et al. 2003, Filaire et al. 2001, Goh et al. 1998, Harman et al. 2000, Kinoshita, 1985, Martin and Nelson, 1985, Pascoe et al. 1997) and a more forward head posture (Attwells et al. 2003, Chansirinukor et al. 2001).

As experimental conditions changed females showed differing response to males, being more upright until the addition of the backpack, when they were in a more forwardly inclined position (figure 3). This correlates with females having a trend for a more forward craniovertebral angle in the backpack condition. In contrast to this finding Martin and Nelson (1986) found that without a backpack females had a greater forward lean than males, but when backpacks were added this no longer existed. In terms of static head posture measures without load no differences have been seen in the saggital plane (Raine and Twomey, 1997). Therefore the gender response shown in this investigation may be a result of the load or more likely a result of the dynamic movement taking place. Trunk ROM was not correlated with any particular anthropometric datum, therefore due to a different response to dynamic load carrying by males and females.

## Limitations

The unavailability of military participants may limit the applicability of the present study as training specific adaptations to this type of load carriage most likely occur, some results perhaps being a consequence of the untrained response to military loads. However,

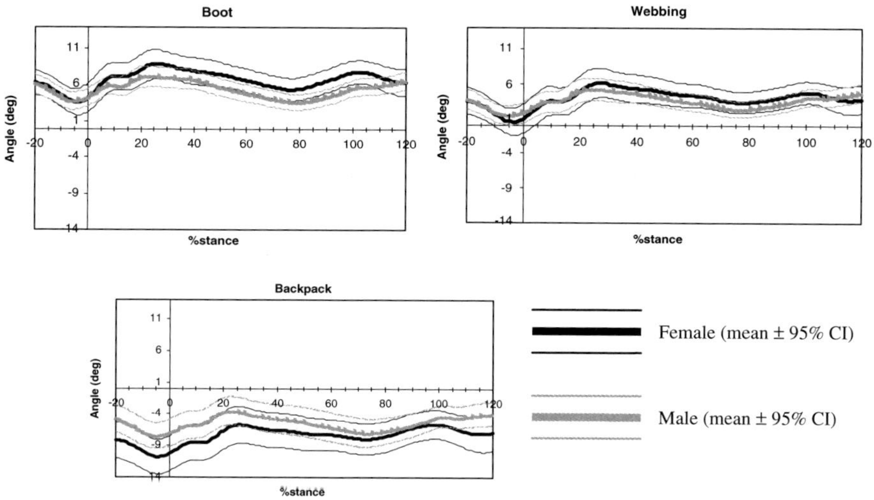


**Figure 3. Gender differences in trunk angle over three loaded conditions.**

participants with some load carriage experience were recruited, minimising this potential impact. Due to equipment restrictions and placing of loads on the body, 2-dimensional analysis is also only reported in the present study. Use of the saggital plane is deemed most reliable (Kadaba et al. 1989) and measures of trunk and head posture measures are only presently defined in this plane. Finally, the duration of load carriage in this study was brief and it is not possible to extrapolate to individuals who carry load for long periods of time, such as during operational activity. This is an area warranting further examination.

## Conclusions

In the current study the gender differences noted may contribute to injury problems due to excess load placed on small muscles of the back, neck and other supporting structures. Increased ROM of any angle also has implications for physiological cost, possibly exacerbated with fatigue. Body size differed between the groups and maybe a crucial determinant. It is possible that a smaller male would display the same response as a larger female. These data raise the question whether members of the military of smaller stature and lower body weight may need different training, or perhaps lighter loads, so as to experience the same relative risk as larger colleagues.

## Acknowledgements

This work was sponsored by the British Ministry of Defence.

## References

Attwells, R.L., Birrell, S.A., Hooper, R.H. and Mansfield, N.J. 2003, Influence of carrying heavy loads on soldiers' dynamic trunk and head posture, *Fifth International Conference on Sport, Leisure and Ergonomics Burton Manor, The Wirral, UK*, 37

Bhambhani, Y. and Singh, M. 1985, Metabolic and cinematographic analysis of walking and running in men and women, *Medicine and Science in Sports and Exercise*, **17**, 131–137

Chansirinukor, W., Wilson, D., Grimmer, K. and Dansie, B. 2001, Effects of backpacks on students: Measurement of cervical and shoulder posture, *Australian Journal of Physiotherapy*, **47**, 110–116

Charteris, J. 1998, Comparison of the effects of backpack loading and of walking speed on foot-floor contact patterns, *Ergonomics*, **41**, 1792–1809

Cho, S.H., Park, J.M. and Kwon, O.Y. 2004, Gender differences in three dimensional gait analysis data from 98 healthy Korean adults, *Clinical Biomechanics*, **19**, 145–152

Filaire, M., Vacheron, J.-J., Vanneuville, G., Pourmarat, G., Garcier, J.-M., Harouna, Y., Guillot, M., Terver, S., Toumi, H. and Thierry, C. 2001, Influence of the mode of load carriage on the static posture of the pelvic girdle and the thoracic and lumbar spine in vivo, *Surgical Radiologic Anatomy*, **23**, 27–31

Ghori, G.M.U. and Luckwill, R.G. 1985, Responses of the lower limb to load carrying in walking man, *European Journal of Applied Physiology*, **54**, 145–150

Goh, J.-H., Thambyah, A. and Bose, K. 1998, Effects of varying backpack loads on peak forces in the lumbosacral spine during walking, *Clinical Biomechanics*, **13**, S23–S31

Haisman, M.F. 1988, Determinants of load carrying ability, *Applied Ergonomics*, **19**, 111–121

Harman, E., Han, K.-H., Frykman, P., Johnson, M., Russell, F. and Rosenstein, M. 1992, The effects of gait timing, kinetics and muscle activity of various loads carried on the back, *Medicine and Science in Sports and Exercise*, **24**, S129

Harman, E., Han, K.-H., Frykman, P. and Pandorf, C. 2000, The effects of backpack weight on the biomechanics of load carriage, *U.S. Army Research Institute of Environmental Medicine Report*, **T00-**17

Kadaba, M.P., Ramakrishnan, H.K., Wootten, M E., Gainey, J., Gorton, G. and Cochran, G.V.B. 1989, Repeatability of kinematic, kinetic and electromyographic data in normal adult gait, *Journal of Orthopaedic Research*, **7**, 849–860

Kerrigan, D.C., Todd, M.K. and Della-Croce, U. 1998, Gender differences in joint biomechanics during walking, *American Journal of Physical Medicine and Rehabilitation*,**77**, 2–7

Kinoshita, H. 1985, Effects of different loads and carrying systems on selected biomechanical paramcters describing walking gait, *Ergonomics*, **28**, 1347–1362

Knapik, J.J., Harman, E.A. and Reynolds, K. 1996, Load carriage using packs: A review of physiological, biomechanical and medical aspects, *Applied Ergonomics*, **27**, 207–216

Martin, P.E. and Nelson, R.C. 1982, Volume III: Effects of gender, load and backpack on the temporal and kinematic characteristics of walking gait, *US Army Natick Research and Development Laboratories Technical Report*, **82/021**

Martin, P.E. and Nelson, R.C. 1985, The effect of carried loads on the combative movement performance of men and women, *Military Medicine*, **150**, 357–362

Martin, P.E. and Nelson, R.C. 1986, The effect of carried loads on the walking patterns of men and women, *Ergonomics*, **29**, 1191–1202

MOD 2002, Women in the armed forces, *http://www.mod.uk/issues/women_af.htm*

Nigg, B.M., Fisher, V. and Ronsky, J.L. 1994 Gait characteristics as a function of age and gender, *Gait and Posture*, **2**, 213–220

Pascoe, D.D., Pascoe, D.E., Wang, Y.T., Shim, D.-M. and Kim, C.K. 1997, Influence of carrying book bags on gait cycle and posture of youths *Ergonomics*, **40**, 631–641

Raine, S. and Twomey, L.T. 1997, Head and shoulder posture variations in 160 asymptomatic women and men, *Archives of Physical Medicine and Rehabilitation*, **78**, 1215–1223

# BIOMECHANICAL EXAMINATION OF BRITISH PROTOTYPE LOAD CARRIAGE SYSTEMS

**Renee L. Attwells & Robin H. Hooper**

*Department Human Sciences, Loughborough University, Loughborough LE11 3TU, UK Email: R.L.Attwells@lboro.ac.uk*

The positioning of load on the body may alter the physiological and biomechanical response of an individual, as well as affecting their comfort whilst carrying the load. Members of the military are required to carry heavy loads for a substantial period of time, hence design of the systems by which these loads are transported is of importance. Twenty individuals carried loads of 40% body weight using three load carriage systems consisting of webbing (carrying all essential items) and the addition of a backpack (Bergen) with additional items. A more upright posture was seen with chest webbing when compared to that worn around the waist, although thermal issues and the negative effect this type of webbing has on soldier profile was not examined. When considering the load carriage system as a whole (i.e. webbing + Bergen) no significant differences in gait and posture variables were seen, however a trend for the new prototype Airmesh system to elicit a more upright posture was apparent. Load carriage duration was brief in this study and extended periods of load carriage with inclusion of subjective data may give better insight into the most beneficial military load carriage system.

## Introduction

Carriage of a load on the person is a critical capability and sustainability issue for the infantryman in carrying out mission essential tasks. Advancements in technology combined with an array of new equipment is placing an increased weight burden on the soldier, as well as reducing the space in which to carry this load. In addition to these space requirements, it is well documented changes in load position result in altered physiological costs, loads placed closest to the centre of mass (COM) resulting in smallest cost (Martin and Nelson, 1986, Obusek et al. 1997, Vacheron et al. 1999). Loads carried in a backpack are expected to result in biomechanical changes due to alteration of position of COM; those carried high on the back as opposed to low resulting in changes to performance of mobility tasks (Holewijn and Lotens, 1992) and affecting activity levels of the back muscles (Bobet and Norman, 1984). Research has been conducted on the use of double packs, thus locating the load closer to the COM. These double packs result in a more upright walking posture (Harman et al. 1994), less discomfort in the lower back region during a 20 km march (Knapik et al. 1997) but increase discomfort in the neck and hip regions and have serious implications for heat illness/distress when compared to single packs (Johnson et al. 1995). Therefore finding the most efficient load carriage system (LCS) requires examination of all of these factors.

Published military research has concentrated on examination of backpack alone rather than the combination of webbing and a backpack/Bergen (i.e. LCS) and including a rifle carried in the arms. Webbing contains all items that are essential for survival, is always

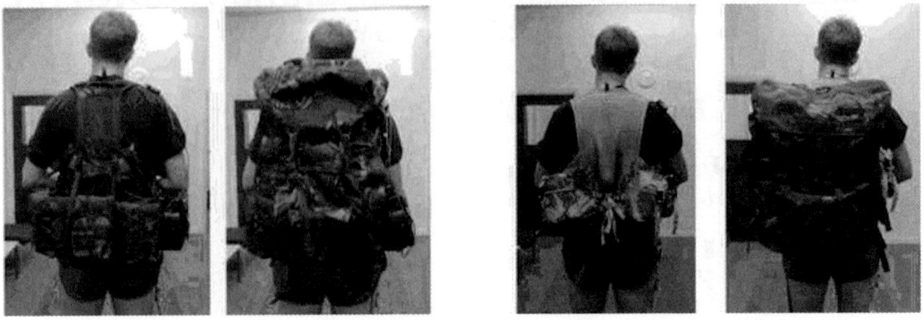

**Figure 1. From left to right; standard issue PLCE webbing, PLCE webbing and Bergen, Chest webbing and chest webbing with Airmesh Bergen.**

worn, and therefore should be considered in any military investigation. Research that has included webbing is currently only published on US equipment (Harman et al. 1999), with differences in design of UK equipment warranting further study, and allowing baseline for comparison of future equipment.

This study examines spatiotemporal, gait and posture variables whilst carrying three different load carriage systems with typical military loads and a replica rifle.

## Methods

Twenty participants (10 male/10 female) were examined under 2 conditions – webbing and webbing + Bergen (see figure 1) – with 3 LCS's, each including a rifle:

(1) Standard (PLCE webbing weighing 7% bodyweight (BW), 90 Pattern Bergen weighing 33% BW);
(2) Airmesh Prototype IV LCS (Chest webbing weighing 7% BW, Airmesh Prototype IV Bergen weighing 33% BW); and
(3) 50/50 Prototype (as Standard with webbing weighing 20% BW and Bergen weighing 20% BW).

All participants carried the standard LCS; half carried Airmesh Prototype IV LCS (5 male/5 female), the other half carrying 50/50 Prototype LCS. Order of completion of LCS conditions was balanced to negate potential order effects. Participants wore standard issue boots, socks and carried a replica SA80 rifle in all conditions in order to reproduce the body position imposed on a member of the military. For all conditions participants moved at self selected speed along a 7 m walkway (with embedded Kistler™ force plate), in front of a CODA™ motion analysis system. Markers were placed on key body positions to enable measurement of main gait and posture variables.

In order for comparisons to be made between participants all data were calculated for points of percent stance phase, working from −20% to 120% stance phase, making them independent of time and walking speed. The means for the 7 repeats of each condition were calculated for all participants. Group data for each experimental condition were expressed as the mean and 95% confidence intervals were calculated for all participants in each load carriage grouping (i.e. Standard + Airmesh OR Standard + 50/50). A repeated measures ANOVA was conducted on all spatiotemporal data and maximum, minimum, mean and ROM values for all angles (significance level set at $p = 0.05$). Comparison of participant characteristics showed no differences between groups when split between prototype load carriage conditions (i.e. Airmesh

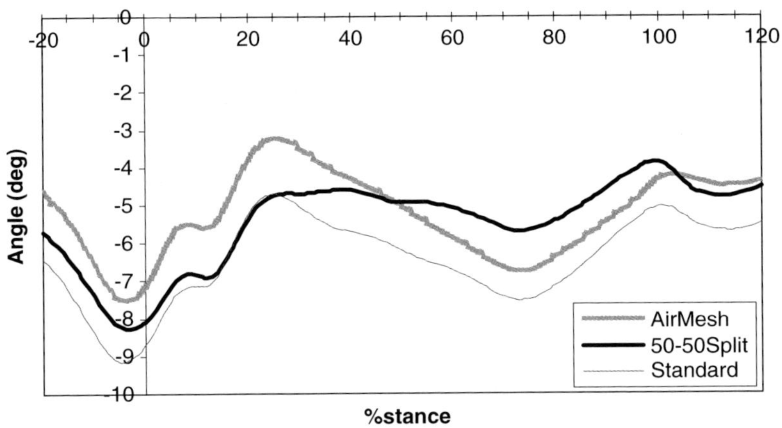


**Figure 2. Trunk angle response from 3 load carriage system conditions.**

and 50/50). Therefore both the webbing and backpack conditions of these systems were compared using independent t-test.

## Results and discussion

The use of chest webbing is warranted both in terms of increasing vision and ease of access to items within pouches, as webbing contains all items necessary for survival, including ammunition. Chest webbing resulted in more upright body posture (5.9°) when compared to the other two webbing possibilities (4.7° standard, 1.8° 50/50) therefore being closer to unloaded locomotion (7.6° – from other connected work with this study). The load is also located closer to the centre of mass, therefore reducing physiological cost (Martin and Nelson, 1986, Obusek et al. 1997, Vacheron et al. 1999). This however may be an unfair comparison as on the 50/50 condition there was a heavier load in the webbing than in the other 2 webbing conditions. Thermal issues have not been considered in either this or other referenced studies above and do play a significant role. Other possible detriments for chest webbing include injury when going to ground and the negative impact on soldier profile.

Considering the LCS as a whole indicates minimal change in any gait or posture variables. It was expected that the Airmesh LCS would result in a more upright posture, with a trend recognised. Significant differences in trunk angle between prototype groups may have masked any other differences between systems. The 50/50 split of load resulted in the least range of motion of the trunk for 80% of stance (figure 2); an important factor in terms of physiological cost to the individual. This should be considered when instructing packing of load, possibly reducing fatigue due to load carriage, therefore improving efficiency and capability of the soldier.

## Conclusions

Future load carriage design must consider the trade off between position of load and the effect this will have on mobility and sustainability. Consideration must also be given to area of body coverage and levels of physiological strain imposed. Further investigation of these LCS's over extended time periods including subjective measures is warranted in order to assess the full effects of changes in design.

## Acknowledgements

This work was sponsored by the British Ministry of Defence.

## References

Bobet, J. and Norman, R.W. 1984, Effects of load placement on back muscle activity in load carriage, *European Journal of Applied Physiology*, **53**, 71–75

Harman, E., Frykman, P., Knapik, J. and Han, K.-H. 1994, Backpack vs. front-back pack: Differential effects of load on walking posture, *Medicine and Science in Sports and Exercise*, **26**, S140

Harman, E.A., Frykman, P.N., Pandorf, C.E., Tharion, W., Mello, R.P., Obusek, J. and Kirk, J. 1999, Physiological, biomechanical and maximal performance comparisons of female soldiers carrying loads using prototype U.S. Marine Corps Modular Lightweight Load-Carrying Equipment (MOLLE) with Interceptor body armour and U.S. Army All-Purpose Lightweight Individual Carrying Equipment (ALICE) with PASGT body armour, *US Army Research Institute of Environmental Medicine Report*, **T99–9**

Holewijn, M. and Lotens, W.A. 1992, The influence of backpack design on physical performance, *Ergonomics*, **35**, 149–157

Johnson, R.F., Knapik, J.J. and Merullo, D.J. 1995, Symptoms during load carrying: Effects of mass and load distribution during a 20 km road march, *Perceptual and Motor Skills*, **81**, 331–338

Knapik, J., Ang, P., Meiselman, H., Johnson, W., Kirk, J., Bensel, C. and Hanlon, W. 1997, Soldier performance and strenuous road marching: Influence of load mass and load distribution, *Military Medicine*, **162**, 62–67

Martin, P.E. and Nelson, R.C. 1986, The effect of carried loads on the walking patterns of men and women, *Ergonomics*, **29**, 1191–1202

Obusek, J.P., Harman, E., Frykman, P., Palmer, C.J. and Bills, R. 1997, The relationship of backpack centre of mass location to the metabolic cost of load carriage, *Medicine and Science in Sports and Exercise*, **29**, S205

Vacheron, J.-J., Pourmarat, G., Chandezon, R. and Vanneuville, G. 1999, The effect of loads carried on the shoulder, *Military Medicine*, **164**, 597–599

# DEFINING ERGONOMICS

# SUCCEEDING AT HUMAN FACTORS INTEGRATION (HFI)

**Monica Sen Gupta**

*Design Services, BAE Systems, Warwick House, Farnborough Aerospace Centre, Hampshire GU14 6YU*

Integrating humans into complex systems involving a wide range of technologies requires considerable skill on the part of the human factors practitioner, and demands consultation with a wide range of stakeholders from end user, to various engineering teams through to the customer who is funding the work. Although structured processes are in place such as the Ministry of Defence (MOD) HFI guides, and ISO 13407 guidance on HFI, that specifies the production of a human factors integration plan (HFIP), systems/equipments are still being designed with HFI issues outstanding. The approach given here promotes the use of an HFI Operability/Usability Log to register and progress human factors design issues that emerges during the course of the design life-cycle, and the notion that the HFI Operability/Usability Log can be used as a design aid to reduce risk of unexpected Human Factors issues emerging at a point in the project when it is too late to correct.

## Introduction

The process referred to in the UK as Human Factors Integration (HFI) originated in eighties in the USA as the MANPRINT programme and was initially developed to assist procurement in the US Army. Re-badged in the nineties using the more politically correct term, "HFI", the human centred design issues that HFI needs to tackle remain essentially the same. Definitions of HFI vary but typically are either referred to as: "a systematic process for identifying, tracking and resolving human related issues ensuring a balanced development of both technological and human aspects of capability" (UK MoD, 2001), and/or "the overall goal of both Human Factors and Human Factors Integration is to design an optimal system consisting of operator, equipment and the environment in which they operate" (UK MoD, 2002).

The driver for recommending that military programmes apply the HFI process was motivated by a number of accidents caused by human error such as the well-known USS Vincennes accident and several fatal military air incidents. HFI is scoped by six interrelated domains, namely; Manpower, Personnel, Human Engineering, Training, Health Hazards and System Safety (see Fig 1) and these domains need to be considered and balanced to produce an operable, cost effective and safe system. HFI for UK Ministry of Defence (MoD) procurements is a mandatory process (UK MoD, 2001), and there are several large military programmes that have successfully integrated HFI into their organisation and project processes. The benefits of such a process being applied outside of military programmes has recently been recognised and the Health and Safety Executive (HSE) has been instrumental in suggesting to oil, gas, nuclear and chemical industries that company's should use HFI as a tool in their overall effort to reduce accidents (HSE Research Report, 2002). Another focus for HFI has been driven by recognition that a number of catastrophic accidents could have been avoided had an HFI process been in place such as the Herald of Free Enterprise and the Paddington rail accident.

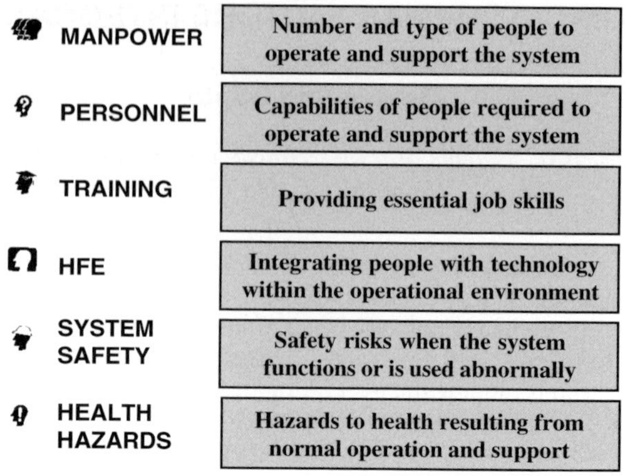


**Figure 1. HFI Domains.**

This paper notes that in spite of the numerous HFI guidance material concerning how to apply HFI that systems/equipments is still not providing adequate support for the people that need to effectively and safely operate them. Good practice recommendations for an aspect of the HFI process that is frequently overlooked that of HFI Operability/Usability log is promoted and examples of its usage and successful outcome in projects cited.

## Human Factors Integration Plans (HFIP)

An integral part of the HFI process is the generation of the Human Factors Integration Plan (HFIP) which provides a plan that scopes, estimates and schedules the human factors activity. The plan can be applied to any project whatever the size at any stage of the design lifecycle. As a minimum the plan should contain the following items:

(1) **HFI Issues and Risks to be addressed in the project:** Identifies the issues that need to be addressed within the 6 domains relevant to that project, e.g. need to minimise numbers of people.
(2) **HFI Constraints:** Refers to elements that cannot be changed, e.g. existing equipment, current legislation and standards that must be met.
(3) **HFI Scope of Work:** This will identify the studies/activities needed to address the issues
(4) **Dependencies between organisations and other project activities,** e.g. safety
(5) **HFI Deliverables, Scheduling and Resourcing**
(6) **Management of the plan:** How all those personnel involved in the project will communicate and progress the human factors issues.

The completed HFIP serves as guidance to the various stakeholders (customer, supplier, project team and end user) concerned with the project on the HFI approach, issues and activities that must be addressed. The HFIP is usually maintained as an evolving document throughout the design lifecycle from concept through to in-service stages. The activities and outputs arising from the HFIP is recorded in an HFI Log, which should contain HFI data and a record of design decisions.

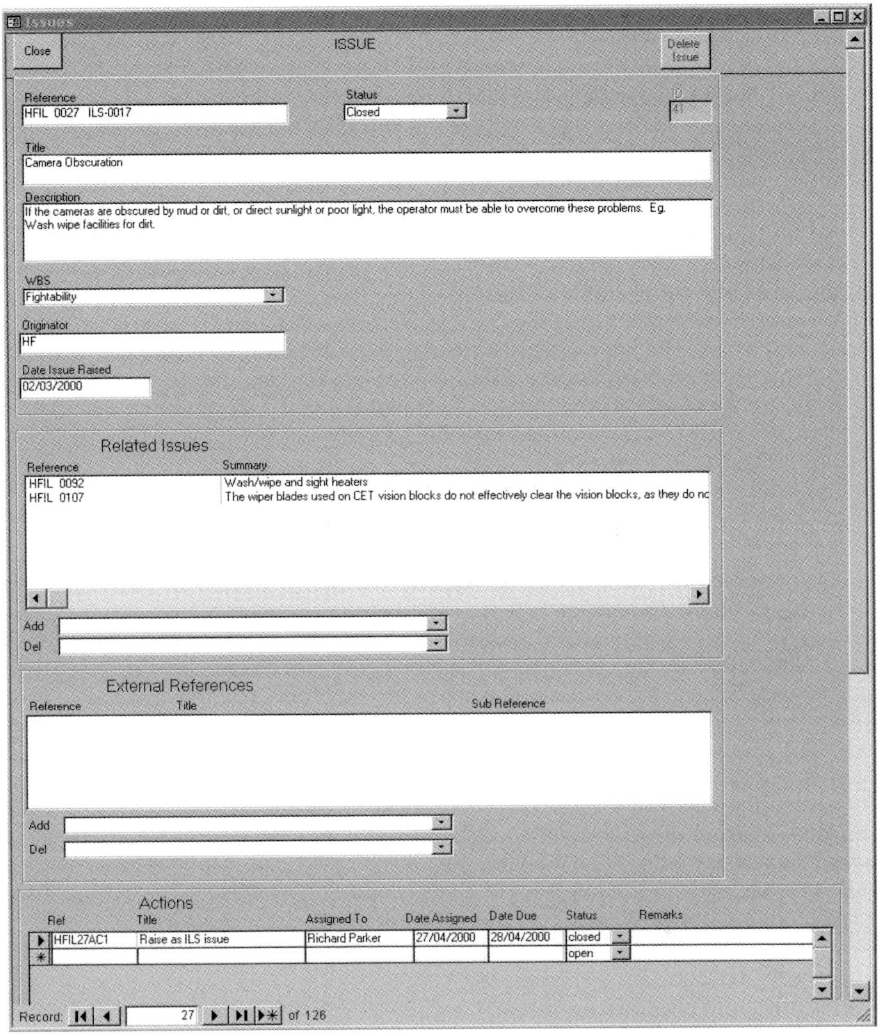


**Figure 2. Example screen shot of HFI operability log.**

## *Limitations with the HFIP*

Production of an HFIP is usually taken as evidence that HFI is being formally adopted by a project yet a dip check of a number of programmes applying HFI has identified several shortcomings:

- Misunderstanding of the purpose of an HFIP
  Too frequently the HFIP has become a Project Manager's deliverable milestone, an item on a long list of project deliverables that needs to be produced and delivered, and once completed and accepted by the customer remains just that. What is needed is for Project Managers within an organisation to have a greater awareness of HFI, which can ensure that the HFIP is fully implemented with full participation from the key players cited in the HFIP.

- Too little too late in the design life cycle
  It is important to integrate HFI early into the design life cycle and frequently an HFIP can be completed too late to exert any real influence on the design. The organisation concerned should ensure that the HFI process is integrated into their company's policies and processes to ensure that this does not happen.
- Inconsistent use of HFI Log
  In the event of an accident an HFI Log (audit trail) is required and frequently on projects this document is often not produced. If it is available, it is wrongly identified with the HFI Issues Log that is generated to register design issues and define the scope of the HFI programme.
- Need for Operability/Usability Issues Log
  A common problem with new equipment procurement is that "lessons learnt" is not always formally and adequately captured into the design. An Operability/Usability Register that is progressed within HFI would ensure that this is mitigated. This should not be confused with another register that is often created during the concept design phase of the HFI Issues Register (UK MoD, yyyy).

## Human factors operability/usability log

The HFI Operability/Usability Log can capture within a project all the HFI issues that could impact on the system/equipment design.

The HFI issues are typically recorded in the form of "lessons learnt" and/or list of user's "sore thumbs". Fig. 2 provides an example of an HFI Operability Log being used on a project. For complex systems this is a particularly powerful tool and can be used to both drive and influence the design and act as an acceptance checklist. It also serves to reduce project risk, and can provide greater confidence that the system will be acceptable to the target audience. The HFI issues need to be actively managed and progressed through Project team working groups to ensure that the issue is resolved and closed with a clear description of the result. Once the project has reached the Acceptance phase the HFI Operability/Usability register can still be used as another means of achieving user acceptance and this is currently underway in a current naval project.

## Conclusion and recommendations

Much has been written on HFI and how to implement it into the project within different stages of the design life cycle, and it is becoming clear that more needs to be achieved for large complex projects to ensure that HFI is successful. An area that is not fully recognised is the capture of "operability/usability issues ("sore thumbs") from users of predecessor systems, and the use of the list as a design agent to influence and drive the design forward. Use of this type of register can be a powerful integration tool and can ensure that the human centric view that HFI takes is fully considered at all levels of the design.

## References

BS EN ISO 13407:1999 *Human-centred Design Processes for Interactive Systems*; (British Standards Institute)

Defence Standard 00–25 Parts 15–20; *Human Factors for Designers of Equipment.*

HSE Research Report 001, 2002; *Human Factors Integration (HFI): Implementation in the Onshore and Offshore Industries*
UK MoD, Issue 1, 2001; *Human Factors Integration (HFI) Practical Guidance for IPTs*
UK MoD, Sea Technology Group, STGP 10, Issue 3, 2002; *HFI Management Guide*
UK MoD, *HFI, Early Human Factors Analysis, User Guide*

# THE ERGONOMIST AS A SKILLED HELPER

**Steven T. Shorrock[1] & David J. Murphy[2]**

[1]*Department of Aviation, The University of New South Wales, Sydney NSW 2052, Australia*
[2]*School of Social Science, Loughborough University, Loughborough, Leicestershire LE113TU, UK*

The role of the ergonomist has, oddly, received little attention from ergonomists. Commonly, the ergonomist is perceived as an "expert", with a solid knowledge-base of theory and experience and a skill-base for application of ergonomics approaches and methodologies. Ergonomics knowledge and technical skill are, of course, prerequisites for ergonomics consulting, and are the basis for most ergonomics research. However, research in allied fields reveals that the processes of "helping" and the factors influencing this process are equally, or perhaps more, important in determining whether an intervention leads to lasting change. This paper therefore explores the role of the consulting ergonomist as a "skilled helper", drawing on research concerning consulting, and integrative and person-centred psychotherapies.

## The ergonomist

Wilson (2000) argues that it is the role – even duty – of a growing discipline such as ergonomics to examine itself in a critical fashion. It seems that ergonomists have indeed reflected on their discipline and its successes over the years, but ergonomists have been rather less reflective of their role and the processes involved in "being an ergonomist". The two considerations are closely linked, of course. Part of the debate on the discipline of ergonomics concerns its status: Wilson recounts debate that suggests ergonomics has elements of science (to explain and predict), engineering (to design for improved performance), and craft (to implement and evaluate). All of these roles are necessary for ergonomics to be successful. In all roles, ergonomists consider human capabilities, limitations, needs, and desires. It is the "craft" role that is of particular interest in this paper. This is probably the most familiar role to consulting ergonomists, and yet is very under-researched. This paper explores some of the processes and personal characteristics that may contribute to effective ergonomics consulting. The ergonomists we envisage may work either externally or internally, full- or part-time, within large or small organisations.

## The consultant

Consulting can be described as a series of phases that progress from a presenting problem to a successful intervention (Wallace and Hall, 1996). According to Wallace and

From The Skilled Helper, A Problem-Mangement Approach to Helping 7th. edition by Egan. © 2002. Reprinted with permission of Wadsworth, a division of Thomson Learning: www.thomsonrights.com. Fax 800 730-2215

Hall (1996), consulting is a complex professional activity in which the consultant provides technical and problem-solving skills, specialised knowledge and work systems and processes to improve the function and outcomes for the consultees and client.

To be effective, the consultant must be clear about his or her objectives, and the skills required to fulfil these objectives. Turner (1982) describes eight such objectives (and skill sets): (1) providing information; (2) solving problems; (3) making a diagnosis; (4) recommending actions; (5) implementing change; (6) building consensus; (7) facilitating learning; (8) improving organisational effectiveness. How might ergonomics consultants rate themselves on these eight objectives? We would argue that ergonomists have been rather more successful in 1–4 than in objectives 5–8.

Glasser (2002) states that one of the most widely used scales of consultant effectiveness is the Consultant Effectiveness Scale (Knoff and Hines, 1995). This comprises 52 questions in four factor groupings, which are shown below, along with the behaviours or characteristics most strongly predictive of these skill groups:

- Interpersonal skills – respect, trustworthiness, approachability, positive attitude.
- Problem solving skills – good facilitator, active listening, good rapport-builder, good at problem solving, astute observation.
- Consultation process and application skill – willingness to get involved, evaluation and focus of ideas, active orientation, follow through, identification of clear goals.
- Professional practice skills – maintains confidentiality, trustworthy, clear identity.

If these factors dominate in determining our effectiveness as consultants, then to what extent do ergonomists learn during training? Ergonomics programmes seem to cover content extremely well, but what of process? Content and process are inextricably linked. But it seems that ergonomics is almost exclusively concerned with knowledge and techniques, which are taught and researched extensively. Process and soft skills are neglected in research and teaching, but are an important part of the "craft" of ergonomics.

Glasser (2002) argues that a key factor in consultant effectiveness is credibility. This acts as a filter for all communications between a consultant and consultee. The most prominent credibility-related factors discussed in the literature are interpersonal skills and abilities. This can be seen above as a major skill group that determines consultant effectiveness, and interpersonal factors permeate the other skill groups. Interestingly, technical competence and credentials are de-emphasised in the research on consultant credibility (Glasser, 2002). Again, this runs contrary to the education and training of ergonomists. The rest of this paper therefore examines this interpersonal role more closely. We draw on research in integrative and person-centred psychotherapies and present a vision of the consulting ergonomist as a "skilled helper".

## The skilled helper

Much of the work of the consulting ergonomist aligns with *helping*, a term often used within the confines of counselling. Egan (2002) elucidates two principal goals of helping:

- GOAL ONE: Help clients manage their problems in living (or working) more effectively and develop unused resources and missed opportunities more fully.
- GOAL TWO: Help clients become better at helping themselves in their everyday lives (or work).

With our added emphasis on work, these goals align well with the consulting objectives and skills. The two goals of helping must be linked. For the discipline to survive for over 50 years, ergonomists must have been quite effective at helping clients to manage their ergonomics-related problems. Less certain is the success that ergonomists have had with helping clients develop unused resources and missed opportunities, and

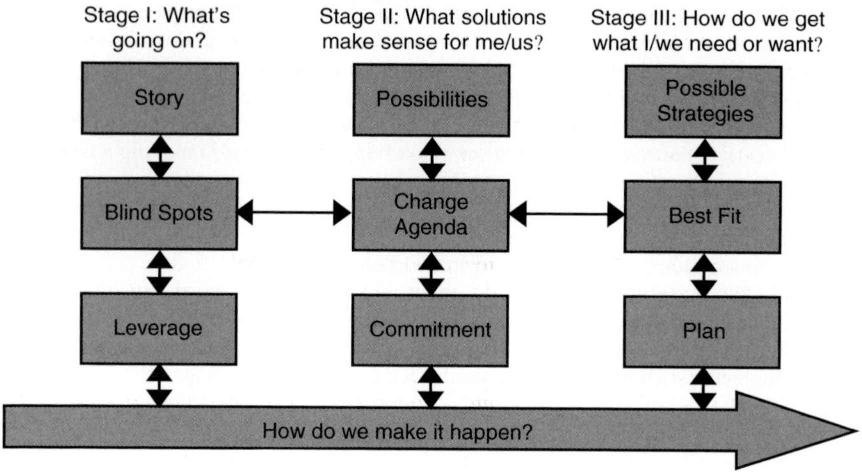


**Figure 1. The skilled helper model (adapted from Egan, 2002).**

with helping clients become better at helping themselves. This may be a somewhat costly oversight. Over the last forty years or so, researchers in the field of psychotherapy have provided strong evidence that a set of common factors is largely responsible for change. Lambert (1992) presented figures showing the percentage improvement in psychotherapy patients as a function of therapeutic factors: Extra-therapeutic change – 40%; Therapeutic relationship – 30%; Expectancy and hope – 15%; Techniques – 15%. While therapy and ergonomics are obviously different, both are quite "technique driven". In light of this, our question is, are ergonomists being effective helpers by concentrating on "hard skills" (techniques) at the expense of factors that may actually instigate change more effectively?

How do skilled helpers work? Egan (2002) presents a problem-management and opportunity-development approach to helping guided by a model shown in Figure 1.

The three "stages" are associated with three "steps" or "tasks" to be addressed at some point in the helping process. These are not sequential; it may be necessary to jump between stages. From an ergonomics viewpoint, it may be useful to see the arrows as representing the application of ergonomics methodologies and techniques, moderating and accommodating common factors in the helping process. The following adapts the model to ergonomics consultancy.

Stage I concerns problem/opportunity clarification and ownership.

*Step I-A: STORIES. Help clients tell their stories.* It is important that the client and ergonomist review the problem situation, missed opportunity or unused resource. Ergonomists may try to find out the various perspectives on the problem.

*Step I-B: BLIND SPOTS. Help clients break through blind spots.* Blind spots prevent clients from seeing themselves, their problem situations and unexplored opportunities as they really are. Blind spots may include unawareness, self-deception, or knowing but not caring. Ergonomists may skilfully challenge mind-sets held by individuals or groups.

*Step I-C: LEVERAGE. Help clients chose the right problems and/or opportunities to work on.* The ergonomist and client will probably start with issues that are seen as crisis issues, or that are causing most trouble, or that the client sees as important and is willing to work on. It may also be wise to begin with a manageable sub-problem, or a problem that will provide a motivating "quick-win" if handled well.

Stage II is about goal setting and commitment.

*Step II-A: POSSIBILITIES. Help clients spell out possibilities for an improved situation.* Ergonomists may help clients and consultees maintain an open mind in generating options to help envisage an improved situation.

*Step II-B: CHOICES. Help clients choose appropriate goals that are real solutions to the key problems and unexplored opportunities identified in Stage I.* Ergonomists may help clients to shape their change agenda to ensure that goals are clear, related to the issue at hand, substantive, realistic, flexible, and time-based.

*Step II-C. COMMITMENT. Help clients find the incentives to help them commit to their change agenda.* Clients should feel ownership of the goals, efficacy in achieving them, and clearly see the benefits of achieving them. Ergonomists may help clients to consider what they are prepared to do, and to deal with any misgivings or competing agendas.

Stage III is about strategies and plans for accomplishing goals.

*Step III-A: STRATEGIES. Help clients see the possible ways of achieving their goal(s).* There may be more than one way of achieving a goal in ergonomics. An organisation could opt for specialised procedures and training, or an automated system; bespoke or commercial-off-the-shelf software system; or one of a number of shift systems.

*Step III-B: BEST-FIT STRATEGIES. Help clients choose best fit strategies.* Only a subset of the strategies envisaged will be suitable for the client. The best-fit strategy must be appropriate in terms of the client's resources, environment, timetable, and goals.

*Step III-C: PLANS. Help clients craft a plan.* The enthusiasm to see change can result in poor planning. Plans map the path from the current situation to the improved situation, thereby breaking down otherwise overwhelming tasks, highlighting required resources, and predicting potential obstacles. Like a map, a balance must be struck between detail and simplicity. Highly detailed, complex and tightly-coupled plans may be self-defeating.

Steps I–III are good examples of working with client variables in the helping process. In line with Lambert's findings (1992), we suggest that frameworks are of little use if they are not implemented skilfully. Egan (2002) outlines four key values for skilled helpers, partly derived from Carl Rogers (1957) "core conditions" – three of the "six necessary and sufficient" conditions required for personality change (Rogers, 1957). According to Rogers, the core conditions of "empathy", "unconditional positive regard" and "congruence" are attitudinal qualities held by the therapist which, when they are present in the therapeutic relationship, create a climate conducive for growth and development.

- *Respect* – a foundation value. This involves doing no harm, being competent and committed, taking the client and consultees seriously, assuming the client's goodwill, and keeping the client's agenda in focus.
- *Empathy* – an orientation value. Whether we realise it or not, in ergonomics consulting we are using empathy to understand the consultee's or client's situation, and we try to do this from their viewpoint, in the context of their organisation and work setting. Empathy is a neglected concept in ergonomics.
- *Genuineness* – a professional value. Rogers proposed that genuineness or "congruence" was the most crucial and the hardest "core condition" to achieve. Congruence is the state in which inner experiences are accurately represented and consciously experienced and communicated to the client. The ergonomist consultant must be congruent within the relationship and the consultee must perceive this.
- *Client empowerment* – an outcome value. The responsibility for running a business remains the job of the organisation. The ergonomist therefore helps clients identify, develop and use resources that will make them more effective change agents, while also understanding when specialised help is needed.

Several communication skills build on these values. These include active listening, paraphrasing, questioning and probing, summarising, and challenging. These vital "soft skills" are generally not taught in ergonomics programs. We believe they should be. They are vital to effective communication, and therefore consultancy.

## Conclusion

The consulting ergonomist can be seen as a "skilled helper". When conceived as such, ergonomics has a lot to learn from allied disciplines. Following Egan's (2002) integrative model of the helping process, we have demonstrated some parallels with the role of the ergonomics consultant. Based on research in consultancy and psychotherapy, we hypothesise that key determinants of effective ergonomics intervention are properties of the client and the relationship between consultant ergonomist, client and consultees. Hence, developing interventions that address client potentialities, and developing a positive alliance between ergonomist and client will engender a significantly more successful and satisfactory outcome for all parties. Surprisingly, "technique" may be less important than ergonomists think. More research is required into the role of the ergonomist and the outcome of ergonomics intervention. We have begun work to study ergonomist and client perceptions of helpful factors within ergonomics interventions.

## References

Egan, G. (2002). *The Skilled Helper: A Problem Management and Opportunity-Development Approach to Helping* (7th Edition). Pacific Grove, CA: Brooks/Cole.

Glaser, J.K. (2002). Factors related to consultant credibility. *Consulting Psychology Journal: Practice and Research*, **54** (1), 28–42.

Lambert, M.J. (1992). Implications of outcome research for psychotherapy integration. In J.C. Norcross and M.R. Goldfried (eds.), *Handbook of Psychotherapy Integration*. New York: Basic Books, pp. 94–129.

Rogers, C.R. (1957). The necessary and sufficient conditions of therapeutic personality change. *Journal of Consulting Psychology*, **21**, 95–103.

Wallace, W.A. and Hall, D.L. (1996). *Psychological Consultation: Perspectives and Applications*. Pacific Grove, CA: Brooks/Cole.

Wilson, J.R. (2000). Fundamentals of ergonomics in theory and practice. *Applied Ergonomics*, **31**, 557–567.

# DESIGN

# PURPOSE AND CONTEXT: SELECTING MULTIPLE METHODS FOR EVALUATING A PILOT PUBLIC INFORMATION TERMINAL

**Suzette Keith, Mark Springett & William Wong**

*Interaction Design Centre, Middlesex University,*
*Trent Park, Bramley Road, London N14 4YZ*

In developing new technological solutions a pilot study provides an important opportunity to evaluate usability and to learn more about the user requirements. We report here on a multi-modal approach adopted to serve the dual requirement to evaluate the current pilot design and to elicit future user requirements. Our evaluation made use of log records, and deployed a number of indirect evaluation methods to analyse usability issues. In addition documentary evidence was used to define the user profiles from which scenarios were developed to examine the specific requirements of the target user group and task demands of a "fit for purpose" evaluation. The methods applied in the case study are examined in relation to the ACquisition of REquirements (ACRE) framework (Maiden and Rugg, 1996) in order to explore issues of purpose and context affecting the selection of methods.

## Introduction

The need to evaluate a pilot project, a public information terminal to provide the socially disadvantaged with social service information (Keith and Wong, 2004), provided an opportunity to consider criteria for the selection of appropriate evaluation methods. In the case study described in this paper there were usability issues such as ease of use, efficiency and learnability to be considered. However the primary concern related to contextual issues such as whether the targeted socially disadvantaged user group would be able to access and interpret the required information. We called this a "fit for purpose" evaluation. It involved an assessment of the goodness of fit between the targeted users' needs, and their ability to access appropriate information from the information terminal. Thus the purpose of the evaluation exercise extended beyond the usual focus of a usability assessment. The evaluation activity thus needed to address three key objectives.

- To describe the purpose of the system including expanding on the characteristics and information needs of the target user group
- To determine whether the terminal was fit for purpose
- To identify gaps and opportunities for further development

We report how these three objectives informed the choice of methods and the decision to adopt a multi-modal approach. We used logged usage data from the pilot study, analytical usability techniques applied to sample tasks and a co-discovery exercise using a partial representation of the system. We also made use of available documentary sources such as client visit reports to provide an analysis of the user characteristics, skills and information needs. The case study evaluation identified a number of simple innovative opportunities to improve accessibility of the information content and has led to an exploratory study of an alternative graphical interface.

A pilot project holds a special place within the design process in which development activity is well advanced but work is still in progress. It offers a dual purpose of assessment and requirements gathering which are best served by a systematic selection of appropriate methods. A pilot study also typically presents some practical constraints, such as incomplete elements of the system and limited access to users, which demand a flexible and innovative approach to method selection. We review our case study in order to explore how the ACquisition of REquirements (ACRE) framework, proposed by Maiden and Rugg (1996) could inform the selection of methods in a pilot programme evaluation.

## The case study

A key objective of the pilot project was to increase opportunities to access information and advice on community issues and health by people who might otherwise be excluded. An important goal for the design was to encourage and empower people to take early action, and thus to reduce pressure on advice agencies caused by late presentation of critical problems.

The information terminal was developed as a web portal with links to over 250 specially selected national websites covering health and advice information, and also a locally held database with links to local clubs and societies. The home page offers a welcome and links to browse information in four categories including the local database, health education, information on medical conditions and an advice service (Figure 1).

A touch screen interface was developed that supported exploration and browsing through a shallow 3 layer hierarchy of up to 30 websites, with the other sites largely accessed as further links from the top level websites. In addition there was a database of local organisations and locally developed health education content.

The selected location of the terminal and the confidential nature of the service favoured the use of indirect and analytic methods of investigation at this stage of the development programme. The evaluation activity progressed through three phases to determine the client profiles, to assess the information delivery activity and to model selected usage scenarios (Keith and Wong, 2004).

### *Client profiles*

The user profiles were derived from documentary evidence including demographic data. Data for the immediate neighbourhood showed that there were 42 different ethnic groups including 14% who do not use English. Indications of low achievement and social deprivation included living in rented accommodation (81%) and having no educational qualifications (30%). Computer ownership in the area was low with 70% not owning a computer. A client profile was derived from summary of data collected for record keeping purposes that preserved client anonymity. This facilitated an analysis of

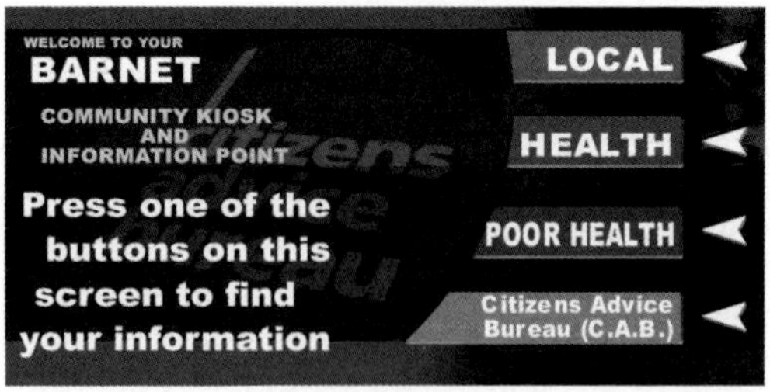


**Figure 1. Home page of the kiosk portal.**

information need including clustering which emphasised some of the complexity of problems faced within the community.

*Information delivery assessment*

A site visit showed "just looking" was not uncommon but no interaction was observed. The analysis of Log data (Figure 2) revealed usage was patchy with occasional high spikes of use, over 1000 pages in the day with a median daily average of 363 pages.

Over 3 months of the pilot trial, 36,000 webpages were visited. An analysis of the log data revealed the favourite sites – these included the community advice site and those on personal relationships and mental health. A co-discovery evaluation invited 2 pairs of young male computer enthusiasts to replicate some of the effect of the peaks of activity within a lab-based setting. Using only a partial copy of the interface, they quickly identified one of the popular sites aimed at teenagers and were also able to quickly find answers to some simple queries. Broad principles of usability were assessed using Nielsen's Heuristics (1994) as a guide, and a Cognitive Walkthrough (Wharton et al., 1994) of a sample interaction was used to consider support for goal formation, choice of links and assessment of progress to goal.

*Modelling*

The client profile data and the usage data were used to construct scenarios of use, characterised by high or low skills in using web search facilities and with information needs of low or high complexity (Table 1). These scenarios were used in discussions with the developers to explore opportunities for developments.

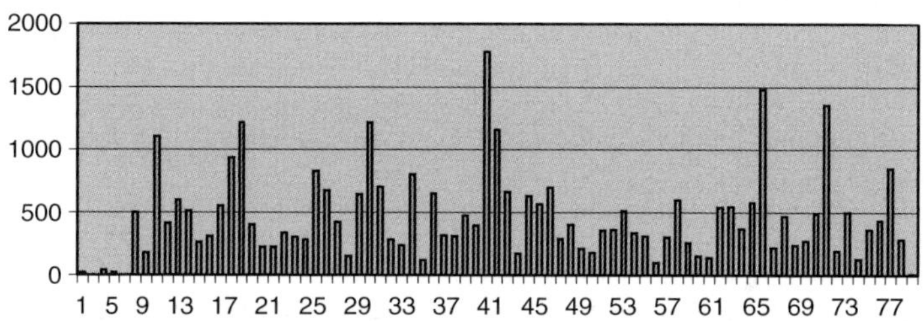


**Figure 2. Log data analysis over 77 days of activity.**

**Table 1. Ability and needs profile.**

| **Ability: literacy and info skills** | **Information needs** Don't know what they need to know | Simple needs | Complex needs |
|---|---|---|---|
| High | e-brochure & kiosk ware (awareness, articulation) | Independent web search e-brochure & kiosk ware (bite-size take-away) | Independent web search Advisor required |
| Low | e-brochure & ware (awareness, articulation) | e-brochure & kiosk ware (bite-size take-away) | Advisor required |

*Results of the case study*

On the positive side it was found that the information terminal provided access in a location that was accessible to a broad range of the community, through a browsable interface to websites that had been selected for relevance and quality. The log data and the co-discovery evaluation suggested the browsing strategy was relatively simple to follow and that some users at least were able to search effectively and finding information of interest.

Minor usability issues were identified, some of which were a consequence of the limitations of a pilot programme. An important refinement to the user requirements was to identify an important class of users who needed much greater encouragement to approach the terminal, perhaps through an e-brochure attractor screen and by providing links to sites in ethnic minority languages from the top level.

## The ACRE framework

This case study to evaluate a pilot project to design an information terminal adopted a multi-modal approach. The ACRE (ACqusition and REquirements) framework provides a theoretical basis for selecting methods that support requirements acquisition. In this section we consider how the framework might offer developers of pilot projects guidance on selecting evaluation methods that support the acquisition of requirements. The faceted framework offers 12 acquisition methods that were divided into 4 broad categories: observation, unstructured and structured interviews and protocol analysis, and describes 6 facets that inform method selection:

- Purpose of the requirements such as to support specification or system selection
- Knowledge types including behaviours and processes
- Internal filtering of knowledge or how the method deals for instance with tacit knowledge
- Observable phenomena such as how objects are used
- Acquisition context including the constraints and limitations imposed on the use of the method
- Method independencies including the selection and ordering of methods

These facets take account of the practical issues of purpose and context that are typical of real world studies. Blandford and Rugg (2002) commented on the real world effect of: "working with inevitable commercial and resource constraints" when considering the place of analytical evaluation methods in requirements acquisition.

We examine here the selection of methods, and the practical demands of purpose and context which led us to use documentary evidence, observation and analytical evaluation methods.

*Selection of methods*

ACRE focuses on methods that directly support knowledge acquisition through observation and interview techniques. We selected other methods that took account of available data resources such as log data and client records, and those which enabled us to model user characteristics and needs in relation to the use of the device.

*Purpose*

A pilot evaluation study offers the developer an opportunity to review progress of the development programme. We consider the purpose of some of the methods we used:

- Log data analysis revealed usage and usage patterns over an extended trial period. From this we were able to identify favourite sites and peaks of activity

- Site observations recorded low usage but without any explanation that would support further development work
- Client profiles using client records and the analysis of the users needs were used to improve understanding of user characteristics and information goals
- Co-discovery was used to explore information content relevance issues
- Modelling explored domain and task based usage, using scenarios that represented different user profiles and needs.

*Context*

The pilot study offered an important context for understanding how users interact with new technology with a fully operational system, over an extended period. Access to the terminal and the users was constrained due to operational considerations in relation to the placement of the terminal and the confidential nature of the information seeking task. At this stage of the development process, it was necessary to select methods that could best utilise the opportunities available for usability analysis and requirements acquisition.

## Conclusion

A pilot study offers a context in which evaluation informs requirements acquisition through an iterative cycle of development. The constraints imposed by the context of this pilot study demanded that we deploy a range of methods to evaluate usability and the task based requirements. While ACRE primarily addresses the selection of methods that support knowledge acquisition through observation and interview, the methods we used in this case study are a useful addition to the repertoire of methods to be considered. Each of the methods supported the purpose of the study to either inform the evaluation or the user requirements or both.

We have reported here primarily on the issue of purpose and context as described within the ACRE framework to support the selection of methods. Further work is ongoing to explore how the ACRE framework can further support developers and practitioners select effective evaluation methods and strategies.

## References

Blandford, A., Rugg, G. 2002, A case study on integrating contextual information with analytical usability evaluation. *International Journal of Human – Computer Studies* **57**, 75–99

Maiden, N.A.M., Rugg, G. 1996, ACRE: selecting methods for requirements acquisition. *Software Engineering Journal*. May 183–192

Nielsen, J., 1994, Heuristic evaluation. In Nielsen, J., Mack, R., *Usability Inspection Methods*. (John Wiley and Sons Inc)

Wharton, C., Rieman, J., Lewis, C., Polson, P. 1994, The cognitive walkthrough method: a practitioner's guide. In Nielsen J, Mack R, *Usability Inspection Methods*. (John Wiley and Sons Inc)

Keith, S., Wong, B.L.W. 2004, Fit for purpose: Accessibility of On-line Public Service Information through the Barnet CAB Electronic Advice Project. Report prepared for the Service Manager, Barnet CAB, Middlesex University Contract No. MUVL-6505 Report Number IDC-001-04, 20 May 2004, Middlesex University, London

# TO DIY FOR: THE CHARACTERISTICS THAT MAKE SELF-ASSEMBLY TASKS DIFFICULT

**Miles Richardson & Gary Jones**

*Centre for Psychological Research in Human Behaviour, University of Derby, Western Road, Mickleover, Derby DE39G*

There has been insufficient research regarding information processing and human cognition in assembly. Richardson et al. (2004) identified a set of task variables that relate to complexity of assembly tasks, but these are based on perceived complexity rather than actual assembly of the products. This study examines the influence of the task variables in the assembly of 8 commercial self-assembly products by 20 participants. The results show that assembly task complexity is increased by: the number of fastenings and available fastenings points to choose between; the number of components available for selection; and the variety and asymmetry of the components, rather than the pure number of components. These results provide further evidence that the task variables proposed by Richardson and colleagues do relate to cognitive workload during assembly.

## Introduction

Self-assembly products and illustration-only instructions have become increasingly common as they offer good value by reducing costs. However, there is anecdotal evidence that people can find self-assembly tasks difficult and it has been noted that there has been insufficient research regarding diagrams, information processing and human cognition in assembly (Novick & Morse, 2000; Prabhu et al., 1995; Shalin et al., 1996). Morrell and Park (1993) conclude that unknown variables play extremely important roles in the performance of procedural assembly tasks with illustration only instructions and that these should be identified.

Much of the assembly research to date has considered the *format* of the assembly instructions. However, we argue that it is the *structure* of the object to be assembled that ultimately impacts on assembly performance. There is little existing knowledge or understanding of how the structure of the assembly object affects the ease with which the assembly can be performed, leaving it almost impossible to build a convincing model of the cognitive processes involved or produce practical guidelines that can help ensure that self-assembly products are easy to assemble.

Previous research has considered certain factors regarding the structure of the assembly object, such as how it affects cognitive load (e.g. Pillay, 1997) and how it impacts on mental representation (e.g. Novick & Morse, 2000). Although it is clear that assembly tasks place a demand on cognitive resources and require the manipulation and transformation of mental representations, there are other factors that also influence assembly complexity.

Recently, Richardson et al. (2004) identified several task variables that relate to complexity during assembly, but these were based on perceived assembly rather than actual assembly of the products. The current study examines the relationship of the task variables to complexity in the construction of scaled-down versions of commercial products (scaled down to avoid simple issues such as the size of components affecting assembly).

## Assembly task variable identification and definition

A Task Analysis of assembly identified the following fundamental assembly operations: select components, orientate components, position components, and fasten components. These sub-operations were mapped on to physical characteristics (task variables) of the assembly that were hypothesised to have an impact on cognition and therefore affect assembly complexity. Two further task variables were related to the general performance of the operations listed above. The task variables identified are listed below with the definition used to operationalise them. This process is detailed fully in Richardson et al. (2004).

Selections – The total number of components available to select from at the start the assembly step or task being evaluated.

Symmetrical Planes – The mean number of symmetrical planes measured in three planes, X, Y and Z per component in the assembly step or task being evaluated.

Fastening Points – The mean number of fastening points available per component in the assembly step or task being evaluated.

Fastenings – The total number of fastenings required in the assembly step or task being evaluated.

Components – The number of components added in the assembly step or task being evaluated (excluding fastening devices such as screws, nuts and bolts).

Novel Assemblies – The number of unique assemblies in the assembly step or task being evaluated.

## Assembly definition

Eight items of commercially available ready-to-assembly furniture were identified. In order to ensure that the task variables could be independently examined, the designs were adjusted so that there were no strong correlations between the task variables. All the assemblies were made at one-third scale using wood and simple screw fastenings and the task variable levels are detailed in table 1. The assembly instructions supplied with the products were used as the basis for the instructions presented to the participants.

## Method

A repeated measures design was used with the task variables as IVs and the DV being thinking time (see coding scheme below). Ten females and ten males, age range 18 to 60 years participated. The assembly tasks were those defined above and were performed on a table. The instructions were presented on a monitor and two video cameras were used to record the assembly. Participants were seated and informed that they would be presented with instructions for 16 separate assemblies in a random order. When the participants were satisfied that each assembly was complete they pressed the spacebar, at which point the completed assembly was collected and the next box of components provided.

## Results

### *Coding*

A coding scheme to remove the variability in the total assembly time due to the time spent on fastening procedures was developed. Time spent on fastening procedures could vary widely for practical reasons, such as an individual's dexterity, and therefore

**Table 1. Task variable levels for each assembly.**

| Assembly | Components | Symmetrical Planes | Novel Assemblies | Fastenings | Fastening points | Selections |
|---|---|---|---|---|---|---|
| 1 Coffee table | 10.0 | 2.1 | 4.0 | 14.0 | 2.8 | 15.0 |
| 2 Bed | 14.0 | 2.1 | 3.0 | 24.0 | 1.7 | 14.0 |
| 3 Chair | 8.0 | 1.3 | 3.0 | 12.0 | 2.8 | 8.0 |
| 4 Corner unit | 9.0 | 1.6 | 5.0 | 20.0 | 2.9 | 9.0 |
| 5 Office chair | 9.0 | 1.7 | 5.0 | 13.0 | 1.4 | 14.0 |
| 6 Picnic table | 14.0 | 1.6 | 5.0 | 25.0 | 1.8 | 14.0 |
| 7 Shelves | 8.0 | 2.5 | 3.0 | 22.0 | 3.3 | 13.0 |
| 8 Side table | 8.0 | 2.9 | 2.0 | 20.0 | 2.5 | 13.0 |
| Mean | 10.0 | 2.0 | 3.8 | 18.8 | 2.4 | 12.5 |

**Table 2. Mean total assembly time and thinking time for the 8 assemblies.**

| Assembly | | 1 | 2 | 3 | 4 | 5 | 6 | 7 | 8 |
|---|---|---|---|---|---|---|---|---|---|
| Thinking time | Mean | 148.5 | 179.2 | 98.8 | 239.7 | 103.8 | 230.4 | 214.0 | 66.6 |
| | sd | 55.9 | 74.1 | 47.7 | 100.2 | 40.2 | 83.4 | 68.1 | 22.2 |
| Total assembly | Mean | 398.2 | 651.2 | 275.5 | 527.7 | 255.9 | 629.2 | 577.1 | 278.9 |
| time | sd | 137.7 | 234.2 | 101.0 | 149.4 | 54.1 | 186.0 | 121.2 | 80.0 |

bear little relationship to complexity and mental effort. A sample of the assemblies were coded by a second researcher. The second researchers timings showed a significant correlation with the main coders ($r(14) = 0.95$, $p < 0.01$). Fastening time was defined as the time spent fastening components together, timed from when the component(s) are ready for assembly or in the final position and the fastening process begins. Timing stops when fastening is complete or fastening actions pause e.g., to check instructions. Thinking time was defined as all time excluding fastening time, such as time spent viewing the instructions, examining the assembly or selecting and manipulating components. Table 2 shows the mean total assembly time and coded thinking time for the 8 assemblies.

*Relationship of task variables to thinking time*

To examine the relationship between thinking time and the assembly task variables, multiple regression analysis was used. Multiple data existed for each participant, meaning that dummy between subject variables (to identify each participant) were entered in the first block in order to control for variability due to individual differences. The main IVs, the six task variables, were then entered as a set into the second block. The between participants variables in model 1 gave R = 0.524 and Adjusted $R^2 = 0.173$. The model including the six task variables gave R = 0.885 and Adjusted $R^2 = 0.783$ ($F(25,131) = 18.94$, $p < 0.01$), with the $R^2$ Change figure suggesting that 50.9% of the variance in thinking time was related to some combination of the task variables. Five of the task variables were found to be significant predictors of thinking time at the $p = 0.05$ level. Standardised regression coefficients for each task variable suggested an increase in thinking time was associated with increases in Novel Assemblies (0.265), the number of Selections (0.320), the number of Fastening Points (0.548), the number of Fastenings (0.517) and a decrease in Symmetrical Planes (−0.457). The number of Components (0.129) was not significant.

## Discussion

The present study was designed to examine how the structure of an object affects how difficult the object is to assemble. A set of task variables that related to the assembly complexity of real world self-assembly products were examined when full assembly takes place. A clear relationship between the task variables and assembly complexity was found with five of the task variables being significant predictors of thinking time: Novel Assemblies, Fastenings, Fastening Points, Symmetrical Planes, and Selections.

It is proposed that Novel Assemblies require additional mental processing because each requires thought, whereas repeat assemblies that are combined in the same orientation and with the same fastenings are likely to become easier to perform as an existing assembly hypothesis exists in working memory. The number of Fastenings involved adds to the amount of information in the task, so an increase in this task variable means the assembly is more difficult. However, more complex explanations may exist. For example, a high number of connections between components leads to a high cognitive load (Marcus et al., 1996). Novel Assemblies and Fastenings were also found to be significant by Richardson et al. (2004) in a study based upon perceived assembly complexity during instruction comprehension.

Fastening Points provide cues and options for positioning components, and the number of alternative points can be related to performance (Lloyd & Jankowski, 1999). With regard to Symmetrical Planes, orientation is critical in component alignment (Shalin et al., 1996) and spatial manipulation is sensitive to the complexity of the information processed (Denis, 1991). For these reasons, component symmetry will reduce the amount of effort required to perform assembly operations. Selections indicate the number of items that are available to select from when performing the assembly. This bears strong relations to component search, and research has already shown that the number of elements to be searched has a dominant effect on search time (e.g. Drury & Clement, 1978).

There was one task variable (Components) that was significant in predicting perceived complexity, but not in the present study, which involved the actual assembly of objects. However, Components was the weakest task variable in the work by Richardson et al. (2004) and the present study reinforces this point. Components is a simple measure that impacts on the number of assembly procedures performed and the amount of information to be processed, but there appears to be more important factors in assembly complexity.

In summary, during assembly the qualitative nature of the information (variety and asymmetry of components) has an affect on assembly complexity and impacts on cognitive load. Furthermore, simple quantitative measures of information and choice (Fastenings, Fastening Points & Selections) also impact on assembly complexity. However, perhaps the simplest measure of assembly complexity, the number of components, was not found to be significant. These results provide further evidence that the approach of using assembly task characteristics to study assembly complexity is reliable and that the task variables proposed by Richardson et al. (2004) relate to cognitive workload and complexity during assembly. It can be concluded that some of the unknown variables that play important roles in the performance of procedural assembly tasks with illustration only instructions suggested by Morrell and Park (1993) have been identified.

Understanding the affect of task variables on complexity allows for greater understanding of cognition during assembly and the methodology adopted also provides a method for the prediction of assembly complexity. The methodology allows the evaluation of the complexity of assemblies and their instructions in applied settings or in order to control and manipulate assembly complexity in a research setting.

The standardised regression coefficients for each task variable indicate what attributes of an assembly task add to encoding time. They suggest that the number of Fastenings, Component Groups' and Novel Assemblies in an assembly task increase encoding time. Encoding time is reduced as the number of Components and Symmetrical Planes is increased, although these are weak affects.

## References

Denis, M. 1991, Imagery and Thinking. In C. Cornoldi & M.A. McDaniel (Eds.), *Imagery and Cognition*. New York: Springer-Verlag.

Drury, C.G., and Clement, M.R. 1978. The effect of area, density, and number of background characters on visual search. *Human Factors*, **20** (5), 597–602.

Lloyd, K.B., and Jankowski, D.J. 1999, A cognitive information processing and information theory approach to diagram clarity: A synthesis and experimental investigation. *Journal of Systems and Software*, **45** (3), 203–214.

Marcus, N., Cooper, M., and Sweller, J. 1996, Understanding instructions. *Journal of Educational Psychology*, **88**, 49–63.

Morrell, R.W. and Park, D.C. 1993, The effects of age, illustrations and task variables on the performance of procedural assembly tasks. *Psychology and Ageing*, **8**, 389–399.

Novick, L.R. and Morse, D.L. 2000, Folding a fish, making a mushroom: The role of diagrams in executing assembly procedures. *Memory and Cognition*, **28**, 1242–1256.

Pillay, H.K. 1997. Cognitive load and assembly tasks: effect of instructional formats on learning assembly procedures. *Educational Psychology*, **17**, 285–299.

Prabhu, G.V., Helander, M.G., and Shalin, V.L. 1995, Effect of product structure on manual assembly performance. *International Journal of Human Factors in Manufacturing*, **5**, 149–161.

Richardson, M., Jones, G. and Torrance, M. 2004, Identifying the task variables that influence perceived object assembly complexity. *Ergonomics*, **47** (9), 945–964.

Shalin, V.L., Prabhu, V. and Helander, M.G. 1996, A cognitive perspective on manual assembly. *Ergonomics*, **39,** 108–127.

# ERGONOMICS ASPECTS OF "IMMORTAL" SYSTEMS


## C.E. Siemieniuch[1] & M.A. Sinclair[2]

[1]*Dept of Systems Engineering, Loughborough University, LE11 3TU*
[2]*Dept of Human Sciences, Loughborough University, LE11 3TU*



"Immortal" systems are expected to last at least through several technology transformations, through several changes of suppliers; through legal and operational environment changes; and beyond the working lifetimes of the designers and operators of the systems. Examples are: the National Power Grid, healthcare systems, and the B-52 and Canberra bombers.

The paper discusses some of the ergonomics issues arising (e.g. how to define user requirements over a century or more), and outlines how these issues might be addressed (such as model-driven architectures).


## Introduction

"Immortal" systems can be used in two senses; firstly, systems that cannot be allowed to fail, and secondly those that are going to exist for some long, period with no foreseeable end to their existence. It is the latter with which this paper is concerned; the former are already considered under the titles, "High Reliability Organisations", or "Safety Critical Systems". Immortal systems are with us now; in the software industry, they are more commonly known as "legacy systems", but their significance is steadily growing with the increased pervasiveness of ITC systems into our global society. Examples are national power networks, or national health systems; in product terms, we might consider the B-52 bomber as a military example.

The main characteristics of these systems is that they exist in changing social worlds, with the following implications:

- other systems with which they interoperate will change, perhaps drastically;
- because of these changes, the goals and/or ways of working of the system are likely to undergo significant alteration
- the people who design, operate, maintain, and manage these systems will have all been replaced within a 40-year period
- the continuity of knowledge and experience that leads to safe operations will have to be renewed, perhaps several times, with the significant possibility of knowledge loss

Added to this are the implications of complexity theory (http://www.santafe.edu/). As was said earlier, one characteristic of immortal systems is that they operate in an increasingly-interconnected world, where they become components in larger, federated systems. This "system of system" complexity compounds the internal complexity of the individual sub systems. Theory predicts that this compounded complexity is likely to result in unpredicted and unpredictable emergent behaviour, both desirable and undesirable. Some characteristics of systems known to cause complex behaviour are:

- Many agents, of different kinds
- Some degree of behavioural autonomy for agents
- Multiple steady states for agents

- Interactions between agents in an environment
- Lots of connections between agents
- Communicating in parallel
- An evolving environment
- Evolving agents
- Interactions between different goals within an agent
- Interactions between agents with different goals
- Language/culture differences

If just some of these are present in a system, then emergent behaviour is to be expected. Furthermore, such systems do not have stable equilibria – desirable states require sustained attention and effort.

## Some of the Ergonomics issues that arise

Several issues arise from these considerations; some immediate ones are:

- *How to define the user requirements for the system that will cover usage over a century or more.* In principle, this could be accomplished by extreme abstraction of the requirements. But this shifts the problem to the designers, who have to interpret a (perhaps conflicting) set of abstract requirements. We are all familiar with this; companies have acquired "best in class" applications guaranteed to be standards-compliant, only to discover that these applications do not interoperate due to slightly different interpretations of the standards.
- *How to define the system interfaces that will emerge as time passes and on-going integration with other systems occurs.* This would be easy if one could predict the future systems with which this system would interface. Even then, there would be the extra problem of predicting the range of potential upgrades required for both the system of interest and the related systems.
- *How to maintain design authority and integrity for the system when the system suppliers and maintainers may have a shorter lifetime than the system itself.* For example, it is predicted that " … by the end of 2020, the average lifetime of a corporation on the S&P will have been shortened to about 10 years." (Foster & Kaplan 2001). Even if the supplier companies still exist, the production processes for the system may no longer exist. Worse still, the storage media for the system documentation may become incompatible with more modern display equipment, meaning that in effect the documentation is permanently locked away. As an example, we can still (albeit with difficulty) read the Domesday Book in the UK, written in the 11th century. The BBC made a similar "Domesday book" in the 1970s for schools, using laser disks. Laser disk equipment is no longer on the market and this information can no longer be accessed.
- *How to maintain system information and knowledge when the people involved are unlikely to be in post for more than a decade.* This is an additional problem to the documentation problem above and is two-fold; the first part is that of training incomers to replace those that are leaving – this becomes an exercise in knowledge management, about which much has been written. It is the second part, which is more insidious. In effect, lots of people at different times will make changes to the system. While all of these people may have the best intentions in mind, it is possible that the accumulation of small changes by individuals, not fully aware of the decisions of others and not fully aware of all possible system interactions (a characteristic of complex systems), may in fact create a disaster scenario which will play out at some time in the future; perhaps after they have left the company.

None of these issues is new; they have been known to the ergonomics community for some time, now. The concept of Immortal Systems brings them together, in sharp focus. The next sections begin the discussion of what to do about them collectively.

## Complexity and the design of "Immortal" systems

There are two kinds of complexity in complex systems; inherent complexity due to the nature of the system itself, and induced complexity that is the result of the design process. Difficulties occur over the design life-cycle of "immortal systems" and some of these are discussed below.

### *Induced complexity*

This arises from the necessary complexity of the design and development process. As complexity theory avers, and Ashby's Law of Requisite Variety infers, the process which delivers a complex system must be as complex as the system itself. In turn, this will give rise to emergent behaviour in the design of the delivered system. For example, there may be additional elements designed post-hoc to make nominally-interoperable subsystems actually interoperable or other special-case patches to obviate unexpected negative behaviours of the system under development. Typically, these problems are caused by different design groups proceeding at different rates towards design maturity.

### *Modular design*

Immortal systems will be subject to upgrades, whether by replacement of old modules or the addition of new modules. Hence, in the case of hardware and software, in addition to the usual "Design-for-X" mantras there will be "Design for replacement" as well. This is likely to show itself as a strongly modular approach, with an equally strong, matching emphasis on standardised interfaces. In the case of liveware (i.e. humans), such as approach is not possible (unless one takes the perspective that they are merely units of work), Nevertheless, the unparalleled capabilities of humans to learn and to adapt (compared to hardware and software) imply that retraining is a valid approach compared to firing and hiring, with the disruptions to knowledge richness and continuity implied by this.

### *"Edge-of-chaos" and "frozen" components*

These terms arise from chaos theory and from network analysis. They refer to the fact that in a complex system it is quite possible that most of the components and modules will not change over the lifetime of the system (except for replacement due to wear and tear). Other components may well be in a near-continuous state of flux, due to system "surprises", changes in external systems, and so on. The implications of this are that, if it is possible to identify which modules will fall into which class, then it is possible to simplify the design management burden and adopt different project management styles for different modules. In this respect, the project classifications of de Meyer, Loch et al. (2002) are useful; variation, foreseen uncertainty, unforeseen uncertainty and chaos, with the project management characteristics necessitated by these classes of project.

### *Module maturity*

Perhaps the most significant aspect in the management of design for complex systems is the need for comparable maturity across all modules of the system. "Maturity" in this context refers to "no surprises"; sufficient is known about the module that its behaviour is predictable. A characteristic of complex systems is the rich interconnection between system components, and when the development of these becomes out of alignment there is scope for the designers of the more developed components to overlook the effects of some of these interconnections, leading to redesign, work-arounds, and patches; especially where the underdeveloped modules are of the "unforeseen uncertainty" or "chaos" classes. This points to the need for frequent design reviews, rapid prototyping approaches, incremental introduction to operations, and, wherever possible, the re-use of proven knowledge.

*Never-ending design, and the maintenance of design authority*

An implication of immortal systems is that their design will never end, because of the upgrades that will be needed, and because of the incremental approaches above. To this one must add the current trends in the design of large systems; the specification of system requirements in terms of effects, not operations, and the move of procurement from platforms to services. This brings into focus four aspects:

- The increased importance of Requirements Engineering and the Product Breakdown Structure firstly in creating the logical link between effect and systems structure, and secondly the division of the system modules into stable and edge of chaos classes, as discussed above
- The importance of user input to design; initially from existing precursor systems, and later input from the operational system, including its changed usage. As Rasmussen and Goodstein (1986) expressed it, the operator should be seen as the designer's representative on-site; i.e. part of the design team extended both in space and time.
- The never-ending aspect creates a problem of design authority. This applies throughout the supply chain; while contracts address the legal issues, there are problems of continuity of authority when some companies may cease to exist, or be taken over by competitors, or dislocate their knowledge base due to over-enthusiastic down-sizing. This is significant for interactions between knowledge sets captured several epochs apart (e.g. Siemieniuch & Sinclair 2002; 2004).
- The "never-ending design" aspect has serious implications for knowledge storage and system design tools. Design information and knowledge must be secure and available for decades, despite organisational upheavals and personnel changes, and must be accessible by new design support tools as they replace current tools. Standardisation and continuity are the abstract principles to be followed; the practice has to be developed on an individual project basis.

## Considerations for the design process

A number of gambits have been suggested above; project management according to De Meyer et al., knowledge re-use, close co-ordination of system development; rapid prototyping, closer, long-term links with system operation, and so on. We now consider some other aspects:

- *Product Line Architectures (PLAs).* The concept implies a common structure and interfaces for a range of products within a given class; for example, Boeing civil airline cockpits are said to be very similar from a pilot's perspective, whichever airplane is considered. Given that the architecture is right, then development and upgrading effort will be reduced. But the architecture is critical; it depends on efficient cross-project analysis and excellent knowledge management; these in turn depend on less tangible aspects such as individual wisdom, experience and trust, and organizational culture and leadership.
- *Model-driven approaches (MDAs).* Stemming from PLAs, it becomes possible to utilize simulations from very early stages of design throughout the lifecycle of the

system. There are obvious advantages from this; the disadvantages are in model validity and reliability, especially when several models are linked together.

- *Shelf engineering.* This refers to advanced engineering, which is undertaken on the basis of engineering projections of the future. The results are then kept "on a shelf" until an occasion for their use arises. While the engineering itself may need some further work, the forward view and knowledge entrainment from this is the real value of this approach.
- *Supply chain partnerships which include the customer.* Partnering in the supply chain is well-established; however, these do not always include the final customer who may cling to the benefits of arms-length competition to get better value for money. In the context of immortal systems, this seems a real mistake; their characteristics demand much closer, long-term links between the players
- *Ontologies for system immortality.* As this paper demonstrates, we do not have powerful tools by which to explore conceptually the issues of systems immortality and complexity. There is a need for research into this issue.

This paper has attempted to highlight some of the critical features of what are now termed "immortal systems" and the difficulties these induce in the specification, design and upgrade of such systems. It is clear that the maintenance of knowledge (formal and tacit) about the design and use of such systems is critical but it is also apparent that a multi-disciplinary approach and toolset will also be required and ergonomics methods and tools have an important role to play here.

## References

deMeyer, A., C. H. Loch, et al. (2002). "Managing project uncertainty: from variation to chaos." Sloan Management Review **43**(2): 60–67.

Foster, R. and S. Kaplan (2001). Creative Destruction: Why Companies That Are Built to Last Underperform the Market – and How to Successfully Transform Them. New York, Currency Books.

Rasmussen, J. and L. P. Goodstein (1986). Decision support in supervisory control. Analysis, design and evaluation of man-machine systems, Varese, Italy, 2nd IFAC/IFIP/IFORS/IEA Conference.

Siemieniuch, C. E. and M. A. Sinclair (2002). The role of the Process Owner in avoiding the "drift to disaster" in processes. 2nd International Conference on Occupational Risk Prevention, Maspalomas, Gran Canaria.

Siemieniuch, C. E. and M. A. Sinclair (2004). Long-Cycle, Distributed Situation Awareness and the Avoidance of Disasters. Human Performance, Situation Awareness and Automation: Current Research and Trends, Daytona Beach, USA, Lawrence Erlbaum Associates. Mahwah, NJ.

# EXPLORING IMPORTANT USABILITY ATTRIBUTES OF CELLULAR PHONES BASED ON CUSTOMER NEEDS: A CASE WITH TURKISH USERS

**Ozalp Vayvay[1], Emine Cobanoglu[2] & Oguzhan Erdinc[3]**

[1]*Marmara Uni. Faculty of Engineering Ind. Eng. Dept., Istanbul, Turkey*
[2]*Marmara Uni. Faculty of Business Administration, Istanbul, Turkey*
[3]*Turkish Air Force Academy, Ind. Eng. Dept., Istanbul, Turkey*

Usability of cellular phones have recently gained utmost importance in consideration of user comfort, user preference, product quality and marketing success. It is emphasized in literature that usability efforts should be based on customer needs. Furthermore, cultural differences between user groups play certain roles in shaping attitudes toward usability and behaviours in cellular phone use. In this study, important cellular phone usability attributes for Turkish users were explored. In-depth interviews were conducted for data gathering. Development of the interview was explained. Results of the in-depth interviews were discussed. This study is expected to contribute valuable findings to literature by empirically demonstrating the usability-related characteristics of Turkish cellular phone users.

## Introduction

Along with the rapid advancement of mobile communication, usability of cellular phones has assumed greater importance in consideration of user comfort and product quality. Usability has been proposed to be an important component of user preference (Han et al, 2001). Volaitis et al, (1998), found that increasing densities of cellular phones (i.e. volume/weight) decreased perceived product quality. Linked with ergonomic concerns, usability has become a differentiation factor for cellular phones in marketing terms.

Through their latest features (e.g. agenda, camera, aesthetic properties), cellular phones have become not only a wide spread communication device, but also a common medium of modern life. Attributes regarding to physical operations on hardware (e.g. size, ease of keystroke, key layout, portability), and mental processes on software (e.g. ease of using menu, understandability of displayed information) converge on the usability concept of cellular phones.

Exploration and assessment of customer needs is a central issue in development and improvement of both consumer and industrial products (Karkkainnen et al, 2001, Khalid and Helander, 2004). A product feature that fails to serve their needs would not be utilized by users and such features would have no significance from usability standpoint. Thus, it is imperative that usability attributes should meet and usability efforts should be based on customer needs.

Cultural differences between user groups substantially influence usability related needs and behaviours in cellular phone usage (Honold, 1999). Honold (1999), elaborated differences between German and Chinese users in learning how to use cell phones. He reported that German users preferred to learn functions of cellular phones broadly, to use manual as primary information source, textual information in manual and to consult to their knowledge or written material in learning how to use cellular phones. On the other hand, Chinese users preferred to learn only basic functions of the cellular phones, to use informal and oral means of information (e.g. asking to friends), pictorial information in manual and more than written procedures, they practiced trial-and-error to learn how to use cell phones.

The implications drawn from studies of the kind can be beneficial for ergonomics researchers, designers and manufacturing companies in building sound product development and marketing strategies for various user groups. From ergonomics view, results of such studies would provide hints to enhance usability performance of products and comfort of users from different cultures.

However, as observed by the authors, no such study was conducted with Turkish users. In this respect, present study aims to explore important usability attributes of cellular phones for Turkish users via in-depth interviews.

## Participants

Participant group was formed by 23 cellular phone users (15 males, 8 females), without disabilities on voluntary basis. Age of participants ranged between 21–55, with a mean age of 28.26 years. Length of cellular phone use experience of participants ranged between 1–8 years, with a mean experience of 5.44 years.

## Method

The study was carried out through in-depth interviews. There is evidence that qualitative research methods such as interviews can be successfully employed in product development and marketing (Karkkainnen et al, 2001). Interviews are widely used to extract important product features to form a basis for subsequent quantitative research on product development or customer satisfaction (Wagner and Hansen, 2003).

Dimensions (i.e. components) that comprise overall usability of various products were described in literature (Han et al, 2001, Sing, 2004). For example, Han et al, (2001), suggesting that concept of usability involves both performance and image of the products, set forth a model which includes performance related (i.e. perception/cognition, learning/memorization, control/action) and image/impression related (i.e. basic sense, description of image, evaluative feeling) usability dimensions for consumer electronic products (e.g. VCR, DVD player). This break-down approach is considered helpful to facilitate a systematic usability research. It was considered that above-mentioned model of Han et al, (2001), can be adopted for cellular phones which bear characteristics of consumer electronic products to a large extent.

In present study, in-depth interview included open-ended questions associated with several usability dimensions and cultural aspects of cellular phone usage. The questions were expanded as necessary during interviews. In-depth interview questions were presented in Table 1.

As introduction part of the interviews, participants were briefly informed about the purpose of the study and value of their opinions. They were encouraged to express themselves freely. First two, "opening" questions addressed to general opinions about usability of cellular phones. Question 3, (Q3), addressed to preference of users toward two interrelated issues; aesthetisc and usability (Han et al, 2000 and 2001, Tractinsky et al, 2000, Khalid and Helander, 2004).

Followingly, respondents were asked to suppose themselves making a new cellular phone purchase decision. Q4 was related to importance of social interactions for Turkish users, which was reported to influence behaviours in cell phone usage (Honold, 1999).

**Table 1. In-depth interview questions.**

1. Explain which features of cellular phones are important for you.
2. What do you understand by "usability" of a cellular phone?
3. Explain which one is more important for you; aesthetics or usability of a cellular phone.

Suppose you are to buy a new cellular phone.

4. Assess the importance of advices given by your circle in your preference.
5. Assess the importance of size for you. How should physical characteristics of the phone be?
6. Assess the importance of menu for you. How should menu be designed?
7. Assess the importance of display for you. How should display be designed?
8. Assess the importance of usability of external equipment (e.g. charger) for you.
9. Assess the importance of clarity and understandability of user manual for you.
10. Assess the importance of ease of demounting the phone for you.
11. Assess the importance of understandability of information and symbols displayed on the phone for you.
12. Explain your preference between phones with keys and with touch screens.

Suppose you bought a new cellular phone.

13. How do you learn to use your new cellular phone?

Suppose you are using your new cellular phone.

14. Assess the importance of economy on key stroke for you.
15. Assess the importance of warnings (e.g. auditory and visual) regarding operations (e.g. sending an SMS message) and status of the phone (e.g. low battery) for you.

Q5, Q6 and Q7 were related to main components of cellular phone usability. Due to the mobility needs and constantly changing contexts of use, usability of external equipment (e.g. charger or head set) and ease of demounting the phone (e.g. changing the battery) were considered to have particular importance for users. These two issues were assessed in Q8 and Q10 respectively. Use of information sources is a culture related usability issue (Honold, 1999) and importance of user manual was assessed in Q9. Q11, addressed to understandability of information displayed on the phone. Q12 aimed to extract preference between traditional cellular phones and new generation cellular phones with touch screen.

In second sequence, respondents were asked to suppose that they bought a new cellular phone. In Q13, strategies in learning how to use a new cellular phones were assessed. The purpose of this question is to identify the characteristic attitudes of Turkish users in learning to use cellular phones.

In third sequence, respondents were asked to suppose that they are using their new cellular phone. Importance of economy on key stroke (i.e. maximum operation with minimum key strokes) which relates to operational efficiency was assessed in Q14. Warnings and signals given by cellular phones help users not to err and to have higher awareness about the status of the phone and completion of operations. In Q15, importance of such warnings was assessed. Additionally, personal information (e.g. age, gender, job, education, brand of cellular phone) were recorded.

## Discussion

The findings discussed below included themes and tendencies indicated by majority of participants and should be assessed in two ways. First, to gain a deeper insight on usability requirements, user perception and needs towards important attributes discussed below should be scrutinized in further quantified studies (e.g. questionnaires). Second, from ergonomic design standpoint, it should be noted that a cellular phone should have these attributes in order to be preferred by Turkish users.

Participants mentioned usability attributes (e.g. ease of SMS operation, portability) rather than features such as brand or price in Q1. Among the usability attributes mentioned was alarm clock, which supported the extensive role of cellular phones in daily life. In Q2, participants mentioned "ease of using menu" from different aspects in response to what they understood from "usability". This emphasizes the coral position of menu design in cellular phone usability. Since cellular phones have foreign origins, users might be facing problems in understanding and using menu. In Q3, almost all users reported preference toward usability rather than aesthetics, which points to priority of usability for Turkish users.

In Q4, participants reported that advice of their circle is important in their cellular phone purchase decision which demonstrated the role of social communication for Turkish people.

Physical attributes (e.g. size, key layout), menu design, display design, understandability of information and symbols displayed on the phone, economy on key stroke and warnings (i.e. particularly use of auditory warnings) were consistently reported to be important by majority of participants. Considerable number of participants reported to prefer *light and small* cellular phones. However, it was observed that users have different measures of weight and size of the cellular phones (Oyama and Shiramatsu, 2002). Preferred weight and size of Turkish users could be a subject of future usability research.

In Q8, usability of external equipment (i.e. particularly charger which is used frequently) was reported to be important. A small, portable charger was of common preference. Hence, research on cellular phone usability should involve design of external equipment. Participants reported that ease of demounting the phone was important. Substantial number of Turkish users have multi SIM cards to benefit economically from marketing campaigns of service provider companies. Due to this trend, ease of changing SIM card has gained particular importance for Turkish users.

In Q12, majority of participants preferred keys rather than touch screens. While some participants were not familiar with touch screens, some others mentioned difficulty of using touch screens. Thus, key-related attributes (e.g. key layout, key distance) are likely to continue to be important in cellular phone usability concept. In order to design preferable computerized phones, usability of touch screens should be improved.

It was considered by the authors that Turkish people would rely on the advises given by a trustable friend more than written information sources. This social characteristic was expected to shape the information gathering strategies of Turkish people. However, in Q9, a great portion of the participants reported that understandability and clarity of user manual was important. In Q13, three competing learning strategies came out to be trial-and-error (i.e particularly for younger users), consulting to user manual and social learning (i.e asking to circle) in rank order.

It was observed that responses of the young participants were more detailed and technically specific. Younger users who are incredibly familiar with computers could use new generation cellular phones more effectively and consciously than older users. The behaviours and tendencies of the young users is expected to determine cellular phone usability requirements for the future. Thus, age should be a main consideration in future usability research on cellular phones.

## References

Chuang M. C., Chang C. C., Hsu S. H. 2001, Perceptual factors underlying user preferences toward product form of mobile phones, *International Journal of Industrial Ergonomics*, **27**, 247–258

Han S. H., Yun M. H., Kim K., Kwahk J. 2000, Evaluation of product usability: development and validation of usability dimensions and design elements based on empirical models, *International Journal of Industrial Ergonomics*, **26**, 477–488
Han S. H., Yun M. H., Kwahk J., Hong S. W. 2001, Usability of consumer electronic products, *International Journal of Industrial Ergonomics*, **28**, 143–151
Honold P. 1999, Learning how to use a cellular phone: Comparison between German and Chinese users, *Technical Communication*, **46**, 196–205
Jordan P. W. 1998, Human factors for pleasure in product use, *Applied Ergonomics*, **29**, 25–33
Karkkainen H., Piippo P., Tuominen M. 2001, Ten tools for customer-driven product development in industrial companies, *International Journal of Production Economics*, **69**, 161–176
Khalid H. M., Helander M. G. 2004, A framework for affective customer needs in product design, *Theoretical Issues in Ergonomics Science*, **5**, 27–42
Oyama H., Shiramatsu N. 2002, Smaller and bigger displays, *Displays*, **23**, 31–39
Sing C. K. 2004, The measurement, analysis and application of the perceived usability of electronic stores, *Singapore Management Review*, **26**, 49–64
Tractinsky N., Katz A. S., Ikar D. 2000, What is beautiful is usable, *Interacting with computers*, **13**, 127–145
Volaitis L., Chou R., Wiklund M. 1998, Consumers' expectations of the densities of portable communication devices, *Proceedings of the Human Factors and Ergonomics Society Annual Meeting*, 589–592
Wagner E. R., Hansen E. N. 2003, A method for identifying and assessing key customer group needs, *Industrial Marketing Management*, Article in Press

# AN INVESTIGATION ON THE ERGONOMIC PROBLEMS OF USING MOBILE PHONES TO SEND SMS

**Vimala Balakrishnan[1], Paul H.P. Yeow[2] & David C.L. Ngo[3]**

*Faculty of Information Science & Technology[1,3]*
*Faculty of Business and Law[2]*
*Multimedia University, Jln Ayer Keroh Lama Melaka 75450, Malaysia*
*Tel: +606 252 3761 Fax: +606 −231 8840*
*vimala.balakrishnan@mmu.edu.my[1], hpyeow@mmu.edu.my[2]*
*david.ngo@mmu.edu.my[3]*

This study identified problems related to Short Message Service (SMS). It was accomplished by conducting structured questionnaire interviews with 30 youths (18–25 years old), focusing on mobile phone designs (shapes and sizes), screen designs, keypad designs, text entry methods and health effects. Major problems found were: (i) poor keypad designs, (ii) mobile phone screen size which is small or poorly illuminated, (iii) dissatisfaction with the auto-spelling software; and (iv) tired eyes, sore thumbs and pain in the upper arms caused by prolonged usage of SMS. Some suggestions given include: (i) keypads that are made from a softer material and keys well spaced between one another, (ii) auto-spelling software that supports abbreviations and emoticons (iii) eyes-free input; and (iv) voice input/audio messaging.

## Introduction

SMS is a service that enables the transmission of alphanumeric messages between mobile phone subscribers. Buckingham (2000) reported that SMS are being exchanged throughout the world at a growth rate that approximately doubles annually. Popularity is due to the wide range of information that can be delivered via SMS, like share prices, sports' scores, weather forecasts and jokes. Moreover, it is a fast medium of communication as a message can be delivered to the recipient within a matter of seconds. In addition, SMS can be stored, retrieved and shared among a group of people.

**Figure 1. Theoretical framework.**

Figure 1 shows the theoretical framework of this study in which one dependent variable and five independent variables were identified. The dependent variable was mobile phone and SMS usage satisfaction, while the independent variables were mobile phone design, screen design, keypad design, text entry methods and health effects.

## Method

Structured questionnaire interviews were conducted with 30 subjects (18 males and 12 females) ranging from 20 to 23 years old, with a mean of 21 years and 1.02 as the standard deviation. Subjects were from the three major ethnic groups in the country, i.e. twelve Chinese, ten Malays and eight Indians. The average years of experience using SMS was 2.27 years with a standard deviation of 1.01. Subjects were categorized based on their expertise level determined by the years of experience and frequency of SMS usage. There were 20 intermediate users, seven experts and three novice users.

## Results

Table 1 shows all 30 subjects significantly agreed that keypad designs affect SMS efficiency. 60% of the subjects (18/30) felt that the keys are placed too close to each other, thus increasing their chances of making mistakes. 30% (9/30) of the subjects mentioned that there are too many alphabets supported on each keypad.

Table 2 shows that there is a significant difference between subjects' expertise level and keypad designs effect on health. Intermediate users were more likely to experience health discomfort due to keypad designs after using SMS for a prolonged period compared with novice and expert users. In addition, 36.7% (11/30) of the subjects stated that they experienced sore thumbs after continuously sending SMS.

Table 3 indicates that the screen size of mobile phone is significantly small. 20% (6/30) of the subjects felt that screen's should have a higher brightness level, as improper lighting makes reading of text messages cumbersome.

Table 4 shows that the text entry method used in composing SMS is significantly slow. All 30 subjects in this study used the same text entry method, which is "multi-press timeout". This technique uses multi-press to select a character to be entered to the screen and a timeout (time interval of 1–1.5 seconds) to enter the next character (Butts and Cockburn, 2001).

Table 5 shows that the mobile phone with auto-spelling software is not significantly better than one without the software. Though this software increases messaging speed,

**Table 1. Subjects' expertise level and their perception of keypad designs effect on SMS efficiency.**

| Expertise level | Keypad designs effect on SMS efficiency | | Z | P |
|---|---|---|---|---|
| | Yes | No | | |
| Novice | 3 | 0 | −5.00 | <0.01 |
| Intermediate | 20 | 0 | | |
| Expert | 7 | 0 | | |

Z – Z value in sign test, P – P value, $\alpha = 0.05$.

it requires users' attention and is not capable of handling abbreviations and slangs commonly used while composing messages. Moreover, 80% (24/30) of the subjects stated that they have to focus on the screen while messaging to ensure the correct word selection is made as the software tends to predict undesired words. The software also doesn't support all languages, thus Chinese and Indian users face difficulty. The subjects also highlighted that it would be desirable to have software that supports emoticons, as this will simplify the process of expressing feelings in their messages.

**Table 2. Subjects' expertise level and keypad designs effect on health.**

| Expertise level | Keypad designs effect on health | | $\chi^2$ | P |
|---|---|---|---|---|
| | Yes | No | | |
| Novice | 2 | 1 | 2.64 | 0.25 |
| Intermediate | 19 | 1 | | |
| Novice | 2 | 1 | 1.27 | 0.50 |
| Expert | 2 | 5 | | |
| Intermediate | 19 | 1 | 13.24 | 0.001 |
| Expert | 2 | 5 | | |

P – P value, $\alpha = 0.05$.

**Table 3. Subjects' opinion on mobile phone screen size.**

| Screen | N | Mean* | S.D. | P* |
|---|---|---|---|---|
| Size | 30 | 1.80 | 0.88 | <0.01 |

N – sample size; S.D. – standard deviation; Mean* – mean based on Likert 5-point scale; $\alpha = 0.05$; P* – P value of one sample *t* test, based on $\mu = 2.5$.

**Table 4. Speed of composing SMS using multi-press technique.**

| Multi-press technique | N | Mean* | S.D. | P* |
|---|---|---|---|---|
| Speed | 30 | 1.07 | 0.78 | <0.01 |

N – sample size; S.D. – standard deviation; Mean* – mean based on Likert 5-point scale; $\alpha = 0.05$; P* – P value of one sample *t* test, based on $\mu = 2.5$.

**Table 5. Subjects' satisfaction with auto-spelling software.**

| Auto-spelling software | N | Mean* | S.D. | P* |
|---|---|---|---|---|
| Performance | 30 | 2.13 | 0.93 | 0.41 |

N – sample size; S.D. – standard deviation; Mean* – mean based on Likert 5-point scale; $\alpha = 0.05$; P* – P value of one sample *t* test, based on $\mu = 2.5$.

**Table 6. Subjects' health discomfort after using SMS for a long duration.**

| Health effects | N | Mean* | S.D. | P* |
|---|---|---|---|---|
| Neck pain | 30 | 2.97 | 0.56 | <0.01 |
| Wrist pain | 30 | 3.33 | 0.80 | <0.01 |
| Tired eyes | 30 | 3.7 | 0.79 | <0.01 |
| Shoulder pain | 30 | 3.00 | 0.45 | <0.01 |
| Pain in upper arm | 30 | 3.33 | 0.92 | <0.01 |
| Sore thumbs | 30 | 3.57 | 0.77 | <0.01 |

N – sample size; S.D. – standard deviation; Mean* – mean based on Likert 5-point scale; $\alpha = 0.05$; P* – P value of one sample $t$ test, based on $\mu = 2.5$.

Table 6 shows that mobile phone users experienced significant pain in the neck, wrist, shoulder and upper arms in addition to tired eyes and sore thumbs after prolonged usage of SMS. Tired eyes were due to the small screen size and poor screen brightness, further aggravated by the small fonts of the text. 40% (12/30) of the subjects mentioned that pain in their upper arms was caused by static posture, as the arm muscles remained contracted for a long time, leading to muscular pain. Repetitively pressing onto keypads that are made of hard material contributed to sore-thumbs.

## Discussion and conclusion

From Table 1, all subjects agreed that the mobile phone keypad design affects the efficiency of sending SMS. Key overloading slows down messaging activities, causing a preference for better keypad design with reasonable amount of space. Levy (2002) redesigned the traditional 12-number keypad to give every letter and number its own key. However, this design has not been implemented yet. It was also found (Table 2) that mobile phone keypad designs caused health discomforts when used for a prolonged period to send SMS. Subjects stated that one of the contributing factors to this effect is the hard material used in designing the keypad. Levy (2002) redesigned the 12-number keypad; however, this design did not include using softer material for the keypads.

On the screen size of mobile phones being significantly small as indicated in Table 3, this finding is consistent with a previous research done to overcome the lack of screen space on mobile computers (Brewster et al., 1998). Their study suggested using structured non-speech sound (such as short piece of music) to represent menu structures in interfaces where visual feedback is not possible. Their method of implementing sound encourages the concept of voice input for SMS.

Table 4 shows that the text entry method used in composing SMS is significantly slow. Butts and Cockburn (2001) evaluated three mobile phone text input methods, i.e. multi-press timeout, multi-press with next button and two-key press. Their study showed that multi-press with next button provided the fastest speed in text entry, followed by multi-press timeout and two-key press. However, they also reported that their participants found all three techniques to be cumbersome. Based on these findings, perhaps multi-press with next button should be incorporated in mobile phones to enhance the efficiency of sending SMS. Linguistic knowledge was added to mobile phone system to predict the most probable word a user is trying to key in. However, this study shows that the mobile phone with auto-spelling software is not significantly better than that without software (Table 5). This leads the subjects to switch off the auto-spelling feature, particularly novice and intermediate users, causing a need for further research to improve the existing software.

Finally, Table 6 shows that mobile phone users have experienced health discomfort after a prolonged usage of SMS. These findings are further supported by Batista's (2001) interview with Andrew Chadwick (Director of British Repetitive Strain Injury [RSI] Association) who coined the term Text Message Injuries. Chadwick stated that text messaging could result in RSI, which occurs when people sit or move in an unnatural position for long periods. This is further aggravated by the small keypad design that forces users to make tiny repeated movements to compose SMS. It is also interesting to note that England's Priory rehab clinic reported that SMS addicts also tend to suffer from sleep deprivation, eyestrain, and increased agitation (The Feature, 2003).

In conclusion, the questionnaire interviews conducted identified some existing problems in using mobile phones to send SMS. The mobile phone keypad design seemed to be the major factor that affects SMS efficiency. This study also revealed that the most common health effects reported were having tired eyes, sore-thumbs and pain in the upper arms, due to messaging for a long time. Mobile phone screen size was also found to be small.

## References

Batista, E., 2001, Sore Thumb? Talk More, Type Less. http://www.wired.com/news/wireless/0,1382,43875,00.html

Brewster, S., Leplâtre, G. and Crease, M., 1998, Using Non-speech Sounds in Mobile Computing Devices, Glasgow, UK

Buckingham, S., 2000, SMS (Short Message Service). http://www.gsmworld.com/technology/sms/

Butts, L. and Cockburn, A., 2001, An Evaluation of Mobile Phone Text Input Methods, Human Computer Interaction Lab, Christchurch, New Zealand

Levy, D., 2002, The Fastap Keypad and Pervasive Computing, *Proceedings of the First International Conference: Pervasive 2002*, Zurich, Switzerland, 58–68

The Feature: It's all about the mobile Internet (2003). http://www.thefeature.com/article?articleid=100119

# ANALYSIS OF SITTING DISCOMFORT, A REVIEW

**Niels C.C.M. Moes**

*Delft University of Technology*
*Fac. Industrial Design Engineering, Section CADE*
*Landbergstraat 15, 2628 CE Delft, the Netherlands*
*C.C.M.Moes@IO.TUDelft.nl*

Sitting discomfort is significantly influenced by the shape of a seat. In order to optimise a shape for reduced comfort, knowledge must be obtained about how to quanty discomfort, and how the aspects of discomfort can be synthesized in an optimisation functional. This paper presents a literature survey of the ways to quantify discomfort, and a causal reasoning scheme to analyse the literature. It is concluded that the control of a shape optimisation procedure using an optimisation functional is not yet possible since much knowledge and techniques are still missing.

## Introduction

Designing the shape of a seat, that is optimal from the ergonomics point of view, requires specific criteria to assess the ergonomics quality. When the ergonomics quality can be quantified (Moes and Horváth, 2002) an objective optimisation functional must be developed that functions as a guide to improve the shape. The criteria should cover subjectively experienced discomfort, objective physiological processes, and efficiency and safety of product use. This paper discusses the research about sitting discomfort and presents a reasoning scheme to interpret measured comfort data for the algorithmisation of an objective optimisation functional. Research on comfort or discomfort typically explores the dependencies of one or more aspects of (dis)comfort on independent variables, which are often called "design parameters". Discomfort assessments are usually based on objective quantifications of subjective judgements. Since the reference level, the sensitivity and the resolution for the assessment of discomfort are subject to personal adaptation and motivation, the quantification of discomfort is essentially difficult and uncertain. Many authors have developed tricks to circumvent this problem and have reached objective shape design criteria for minimal discomfort.

The concepts of comfort and discomfort have been defined variously. (Vink et al., 2004) give in a profound discussion a brief summary, and conclude that some issues crop up regularly when (dis)comfort is discussed: it shows aspects of a subjectively defined personal nature, is affected by physical, physiological and psychological factors, and is a reaction to the environment. According to the *Longman Dictionary of Contemporary English*:

> *Comfort is the absence of pain or suffering and having all one's bodily wants satisfied; discomfort is the lack of comfort.*

The last part of this definition suggests a relationship between comfort and discomfort, which turns out to be an issue of discussion, as will be clarified later.

To build a conceptual framework around the idea of discomfort that can be applied in the assessment of discomfort for shape optimization purposes, we need a model that relates discomfort to the causal stimuli and to the physiological, psychological and emotional reactions of the person. (Vink et al., 2004) introduced a comfort model, that shows some relationships, it has to be further developed to serve as a reasoning model to discuss the reasearch on comfort.

We adopt the reasoning that discomfort does only exist if a person has (i) the perception of pain or suffering, and (ii) the appreciation or non-appreciation of the pain or the suffering. Thus the Longman's definition is adapted and modified: *Sitting discomfort, D, is the subjective appreciation, A, of the perception, P, of the effects, E, of the interaction, I, between a person, P, and an artefact, S, with a specific purpose, Pu, and a specific type of usage, U, during or after use.*

Within the focus on sitting discomfort, the basic entities to consider are the person, the seat, the purpose to sit and usage. The person aspects refer to anatomy, physiology, body region, age, gender, physical or psychological problems, nutrition, emotion, experience, etc. The aspects to be considered for the seat are the number of supporting surfaces, posture, shape, hardness, etc. The purpose of sitting refers to relaxing, support for working, socializing, etc. The usage refers to the aspect of posture (sitting upright, reclining, etc.), the time aspect (duration of sitting, frequency, time of the day), and movements.

The interaction is the immediate, objective and measurable consequence of the fact that a person is using the seat. It includes force transmission, pressure distribution, etc.

The effects of the interaction include the stresses and deformations inside the body and the support material, the change of the temperature, mechanical load of tissue resulting in physiological changes of tissue functioning, and the relaxation of other tissues (one possible purpose for sitting). The load may result in pain, tissue deformation, etc. The effects constitute a set of objective measures, although the actual measurement is usually not too easy.

The effects are percieved as complex functions of the effects: the sensation of being cushioned, of being held in position, of having freedom of movement, or of having a specific posture (Kamijo et al., 1982), but also pain and distortion of tissue can be percieved. The perceptions form a group of objective as well as subjective measures.

Whether a perception is appreciated or not depends on many circumstances, such as the environment conditions or work stress. Also the reasons for sitting may play a role: having rest, viewing a control panel, social control. Discomfort can only exist if a seat is used, and if there is a level of appreciation of the percieved effects.

Discomfort relates therefore to the subjective appreciation of the perception. The appreciation reflects (i) the function or purpose of sitting (to relax, to work, to socialize, etc.), (ii) the posture, (iii) the percieved effects, and (iv) the expectations. It is difficult, if possible at all, to predict the appreciation, since each person judges the effects differently. (Vink, 2002) suggested that the appreciation is determined by the following factors: the past experiences, the present situation, the visual impression, smell, temperature, moisture, posture, and movement. Moreover, when comfort is quantified also the tolorance for comfort must be considered. Thus the above formulation can be condensed: $D = f(A(P(E(I(\text{person, product, usage, purpose})))))$

Generally speaking, research on comfort needs physical or behavioural data, that can objectively be measured, and subjective judgement or appreciation data, that can be obtained with respect to an objective reference. The comparison of these two groups of data may give quantified guidelines on how to modify the objective data in order to improve the appreciation.

## Research on sitting discomfort

In ergonomics research the assessment of discomfort is usually done by rating subjective responses on appreciation of participants on an ordinal scale. (Branton, 1969) argues that measurement of "more comfortable than indifference" does not make sense, but other authors do actually extend the scale from indifferent to "very comfortable". In practice, different types of scales are used, the most simple a two point scale of the presence or absence of discomfort. Neither is differentiated further in such setup. The concept of comfort is also described as *neutral feeling* (McLeod et al., 1980).

More elaborate scales have been developed. (Grandjean et al., 1973) used a three point general comfort scale for sitting (uncomfortable, medium, comfortable). They measured the sitting comfort for several body regions and postures by asking the subjective appreciation. (Shvarts et al., 1980) used a seven point scale (1 = extreme discomfort to 7 = extreme comfort). (Shackel et al., 1969) applied a more detailed scale and used eleven levels (*I feel unbearable pain* to *I feel completely relaxed*). But the question remains if participants are able to express their appreciation in so much detail.

A second method of quantifying comfort uses equivalent comfort contours (Sanders and McCormick, 1993), which can be drawn as a three dimensional graph. This method can be used if the assumption is made that comfort depends on two descriptors, for instance vibrational comfort as a function of frequency and intensity. In case of more variables the equivalent comfort contours can still be defined, but its graphical representation becomes more complicated, because of the increased (⩾4) dimensionality.

Although (dis)comfort is usually defined on a one-dimensional scale it has a multidimensional nature (Zhang et al., 1996). They defined 73 aspects or descriptors and found that comfort and discomfort are not just each others' opposites but are really different entities. It was proved that comfort is associated with aesthetics and "neutral feeling", while discomfort is associated with physiological, biomechanical and time descriptors.

(Sember, 1994) assumed that discomfort depends on the level of the maximum pressure: pressures below 5.0 kPa can be tolerated indefinitely and this level is considered a criterion for sitting comfort. (Kamijo et al., 1982) found that the correct positioning of lumbar support and the existence of pressure in that region are highly appreciated for automobile seats, but that the definition of "correct" varies among countries. (Frusti and Hoffman, 1994) measured comfort of automoble seats by a subjects overall comfort rating assessment of various seats and the personal importance assessment of the different regions of the support. They measured the pressure distribution for the respondents own seats. The investigators converted the measured data into a model for predicted comfort rating: *Predicted Comfort Rating* = 10 − (0:334 × [*DEVP*]) − [*DEVF*], where *DEVP* is the weighted deviations from the target pressure percentages in each area of the seat, and *DEVF* is the deviation from the target ratios of back and seat pressure percentages.

(Bronkhorst and Krause, 2002) argued that designing chairs for ergonomics criteria from scratch is much more difficult than redesigning. They redesigned shapes, based on subjective appreciations, by involving the end users. (Yamazaki, 2002) investigated the relationships between the deformations of standard automobile seats and the perception of a set of "feelings", including the feeling of pressure on the seating bones, on the forepart of the thigh, the feeling of seat hardness, etc. Again within the context of automobile seats, (von Rieck, 1969) tested a set of seats by requesting the subjects to express their appreciation of seven different aspects on a linear scale (Fragebogen), where the ends of the scale were daily life quantifiers (wohl–nicht wohl, härter–weicher, etc). Comparable research findings were reported by (Shen and Galer, 1993), who related chair adjustment and pressure parameters to sitting comfort. Although they found significant correlations, these were not strong enough for practical application. (le Carpentier, 1969) found a significant difference between men and women for "easy chair sitting", based on chair dimensions, viz, "the adult population can only be adequately catered for by two nodels of chairs".

## Reasoning scheme to consider research on sitting comfort

(de Looze et al., 2003) have build a theoretical model to consider comfort and discomfort within the framework of context, seat and human. They considered two separate

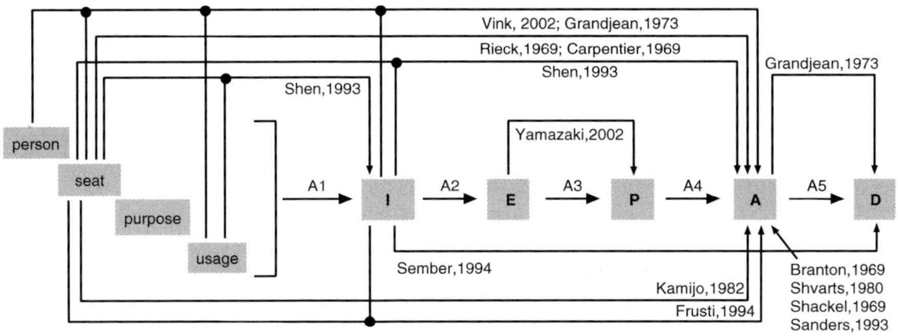


**Figure 1. Overview of research on sitting comfort.**

streams to derive comfort and discomfort, based on the research of (Zhang et al., 1996) and (Helander and Zhang, 1997). Though their scheme shows the main aspects to be considered for the quantification of an optimisation functional, it is still difficult to assess the actual functional relationships in a causal series of events. Moreover, the found research reports that discuss the quantitative assessment of discomfort, can not be given a place in this scheme.

Figure 1 presents our reasoning scheme that (i) is based on a description of causal and functional relationships of discomfort, and (ii) enables to place the research reports within the proper scope of dependency relationships.

If a person uses a seat with a specific purpose, the interaction (I) arises. This interaction consists of, for instance the pressure distribution in the contact area between person and seat. An interaction results in internal body effects (E), such as tissue deformation or the compression of nerves and blood vessels. These effects can be percieved (P) and interpreted, for instance as pain. The next phase is the appreciation of the perception. If it is not appreciatated, it leads to feelings of discomfort (D).

When these steps are expressed as algorithms, it is possible to derive an optimisation functional for shape improvement. The algorithms A1 and A2 have been introduced in (Moes, 2004). The development of the algorithm A3 needs the quantitative application of neuro-physiological knowledge on the perception of physical stimuli. The development of the algorithms A4 and A5 need extended psycho-physiological research on the appreciation or non-appreciation of physical feelings.

## Conclusions

Despite substantial research on sitting comfort it is not yet possible to completely understand and quantitatively formulate the underlying principles of sitting comfort. Therefore the algorithms for the ergonomics optimisation of the shape of a seat, can not yet be defined. Fundamental research on (i) the appreciation of percieved stimuli, and (ii) the assessment of a discomfort scale, are needed before an comprehensive shape optimisation procedure can be developed.

## References

Branton P (1969). Behaviour, Body Mechanics and Discomfort. *Ergonomics*, 12(2):316–327.

Bronkhorst RE, and Krause F (2002). End-Users Help Design Mass Transport Seats. *Human Factors in Seating and Automotive Telematics*, pages 99–104. SAE Techn Papers 2002-01-0780.

de Looze MP, Kuijt-Evers LFM, and van Dieën J (2003). Sitting comfort and discomfort and the relationships with objective measures. *Ergonomics*, 46(10):985–997.

Frusti TM, and Hoffman DJ (1994). Quantifying The Comfortable Seat Developing Measurable Parameters Relating To Subjective Comfort. In *Automotive Body Interior & Safety Systems*, pages 14–18. IBEC'94.

Grandjean E, Hünting W, Wotzka G, and Schärer R (1973). An Ergonomic Investigation of Multipurpose Chairs. *Human Factors*, 15(3):247–255.

Helander G, and Zhang M (1997). Field studies of comfort and discomfort in sitting. *Ergonomics*, 40(9):895–915.

Kamijo K, Tsujimura H, Obara H, and Katsumata M (1982). Evaluation of Seating Comfort. *SAE Transactions*, pages 2615–2620. Society of Automotive Engineers.

le Carpentier EF (1969). Easy Chair Dimensions for Comfort – A Subjective Approach. *Ergonomics*, 12(2):328–337.

McLeod P, Mandel DR, and Malvern F (1980). The effects of seating on human tasks and perceptions. In Poydar HR, editor, *Proceedings of the Symposium Human factors and Industrial Design in Consumer Products*, Tufts University; Medford, Mass. Department of Engineering Design.

Moes CCM (2004). *Advanced human body modelling to support designing products for physical interaction*. Ph.D. Dissertation, Delft University of Technology, ISBN: 90-9018829-0.

Moes CCM, and Horváth I (2002). Optimizing the Product Shape for Ergonomics Goodness Index. Part I: Conceptual Solution. In McCabe Paul T., editor, *Contemporary Ergonomics 2002*, pages 314–318. The Ergonomics society, Taylor & Francis.

Sanders MS, and McCormick EJ (1993). *Human Factors in Engineering and Design*. Psychology Series. McGraw-Hill, Inc., seventh edition.

Sember JA (1994). The biomechanical relationship of seat design to the human anatomy. In Lueder R and Noro K, editors, *Hard facts about Machines: the Ergonomics of seating*, chapter VII-17, pages 221–229. Taylor & Francis.

Shackel B, Chidsey KD, and Shipley P (1969). The assessment of chair comfort. *Ergonomics*, 12(2):269–306.

Shen W, and Galer IAR (1993). Development of a pressure related assessment model of seating discomfort. In *Proceedings of the Human Factors and Ergonomics Society 37th Annual Meeting*, pages 831–835, Seattle, WA.

Shvarts E, Guame JG, Wight RT, Riebold RC, and Glasford EJ (1980). Effect of the circutone seat on hemodynamic, subjective and thermal response during prolonged sitting. In *Proceedings of the Human factors Society 24th Annual Meeting*, pages 639–642.

Vink P (2002). Comfort. Delft University of Technology, Faculty of Industrial Design Engineering. Inaugural Speech.

Vink P, de Looze MP, and Kuijt-Evers LFM (2004). Theory of comfort. In Vink P., editor, *Comfort and Design: Principles and Good Practice*, chapter 2, pages 13–32. CRC Press LLC.

von Rieck A (1969). Über die Messung des Sitzkomforts von Autositzen. *Ergonomics*, 12(2):206–211.

Yamazaki N (2002). Analysis of sitting comfortability of driver's seat by contact shape. *Ergonomics*, 35(5/6):677–692.

Zhang L, Helander MG, and Drury CG (1996). Identifying Factors of Comfort and Discomfort in Sitting. *Human Factors*, 38(3):377–389.

# INFLUENCE OF THE SHAPE OF A SITTING SUPPORT ON THE INTERNAL BODY STRESSES

**Niels C.C.M. Moes**

*Delft University of Technology, Fac. Industrial Design Engineering*
*Section CADE, Landbergstraat 15, 2628 CE Delft, the Netherlands*
*C.C.M.Moes@IO.TUDelft.nl*

The stresses and the strain that occur inside the buttock region during sitting have been computed using a FEM. The computations were done for four different shapes of a seat. The results show that (i) the deformations inside the body depend on the shape of the seat, (ii) the maximum pressure in the contact area decreases with increasing curvature of the seat, (iii) the shear strain and stress are concentrated on the bone surface.

## Introduction

Designing the shape of the sitting surface of a seat includes defining the spatial position, the spatial orientation, and the shape of the seat. Although the overall sitting discomfort is usually related to (i) objective quantities, such as the pressure distribution in the contact area or the activation of muscles in the back, and (ii) subjective measures such as experienced feelings and emotions, such measures are indirect discomfort assessments since the actual initiators of sitting discomfort are stresses and strains inside the body. A designer can not "look inside the body" to know the effects of applied mechanical loads. Therefore we developed a mechanical model of the human body, based on the finite elements technology. This model is shown in the upper part of figure 1-1; a detailed description can be found in (Moes, 2004). It has the following main features. (i) It reflects one specific subject, so that care must be taken in generalising the results. (ii) The geometry was determined for un unloaded human body; its extend was limited to the region around the ischial tuberosities. (iii) The non-linear elasticity is based on the Mooney-Rivlin constitutive model. (iv) The ischial tuberosity is modelled as an undeformable body.

In this paper we present the effects of four different shapes of a seat on (i) the pressure distribution in the contact area, (ii) the deformation of the soft tissue, and (iii) the shear inside the body. We expect that for increasing curvedness an increasing fraction of the sitting force is transmitted to the regions, more peripheral to the ischial tuberosity, so that the maximum contact pressure decreases. Predicting the new location of the maximum pressure point on the bone and on the skin is difficult, since now a large part of the ischial tuberosity is "surrounded" by the support.

## Changing the internal loadings by shape modification

Four different rigid support shapes of varying curvedness were applied, see figure 1A. The first is a flat support (a). The second conforms the shape of the unloaded body (b). The surfaces (c) and (d) were computed by: $(X, Y, Z) = (x, y, \alpha z)$, where $(X, Y, Z)$ represents the new coordinates, and $(x, y, \alpha z)$ the original coordinates. For surface (c) $\alpha = 0.5$; for surface (d) $\alpha = 1.25$.

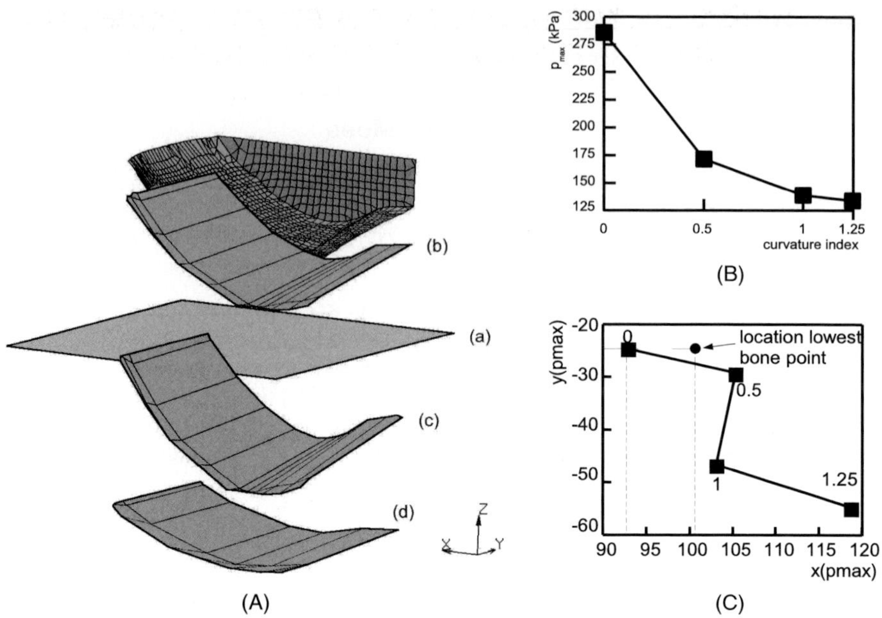


**Figure 1. (A) The FEM (top, dorso-lateral view) and the seat shapes. (B) The computed maximum interface pressure as a function of the curvedness. (C) The shift of the location as projected in a horizontal plane as a function of the curvedness.**

These support surfaces were applied to the deformable body model. The application of the 1 curvedness surface on the deformable body conforms with the experiments of (Brienza and Chung, 1993), who used the unloaded body shape as a starting point for reshaping wheelchair cushions, based on the computed changes of local tissue stiffness.

A detailed overview of the results of the computations is presented in (Moes, 2004). In this paper we discuss (i) value and the location of the maximum pressure in the contact area, (ii) the soft tissue thickness and the vertical tissue relocation during load application, and (iii) the shear stress and shear strain.

## Maximum pressure in the contact area

The highest value, 287 kPa, of the maximum pressure in the contact area was found for the flat surface, which agrees with the computed maximum pressure for a subject with corresponding body characteristics as the virtual subject for which the model was built (Moes, 2000). With the increased curvedness, the maximum pressure decreased until 134 kPa for the surface with the highest curvedness, see figure 1B. This is as expected since the curved surface will transfer a larger part of the sitting force via the peripheral parts.

In figure 1C the shift of the projected location ($x$ and $y$ coordinates) of maximum pressure on a horizontal plane is shown. For the 0.5 curvedness surface the point is still close to the lowest bone point of the ischial tuberosity, but for increasing curvature a tendency of a dorso-lateral movement ($\Delta x > 0$, $\Delta y < 0$) can be observed. The horizontal distance between the location of the maximum pressure point and the location of the lowest bone point was computed to be within 7 mm for the flat surface. For the 0.5 curvedness surface this distance is even smaller, but then it increases with curvedness.

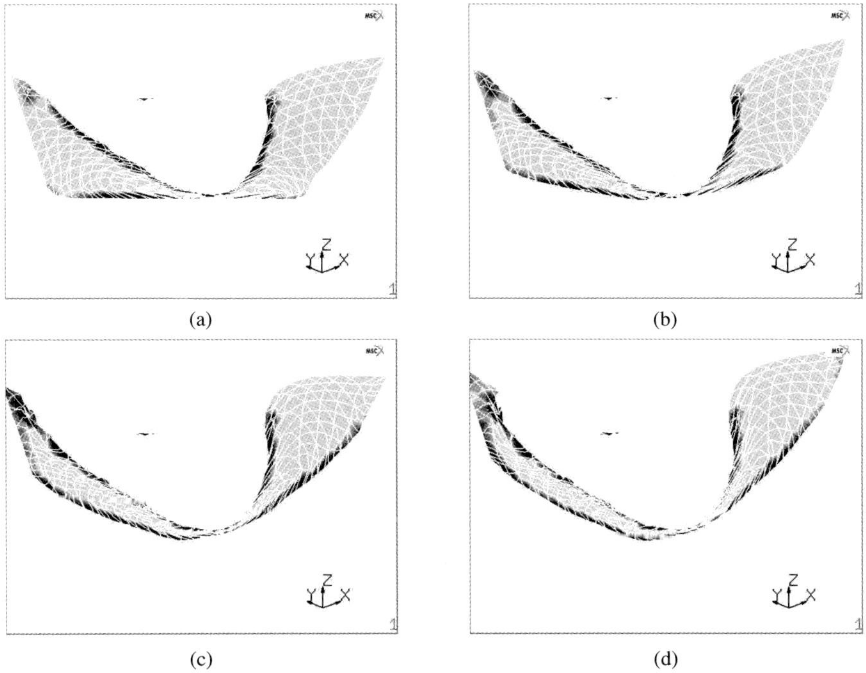


**Figure 2. The deformed body shape for varying curvedness. The left side is the front part of the ischial tuberosity, and the right side the dorsal part. The shading of the grey tones refers to the total shear strain.**

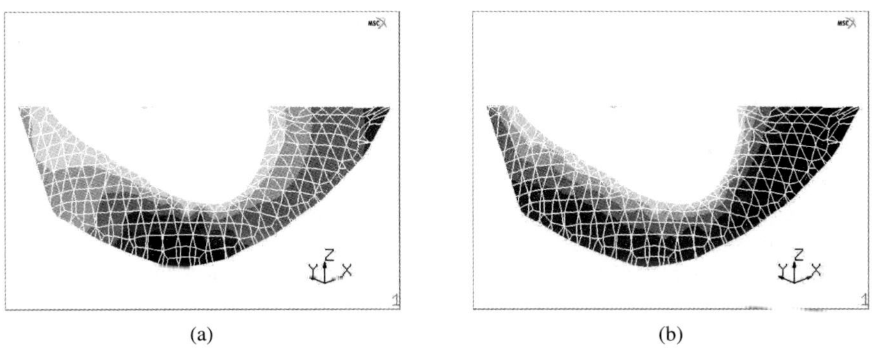


**Figure 3. The z-relocation in the ischial plane for the flat support (a) and the maximum curved support (b).**

Since the elements are extremely flattened, the lateral distance between the nodes increases accordingly, so that the resolution of the stress is reduced.

## Soft tissue thickness and tissue relocation

The figures 2a to d show the shape of the deformed body in a cross section of the pelvis along a plane that (i) runs in the length direction of the ischial tuberosity, and (ii) does

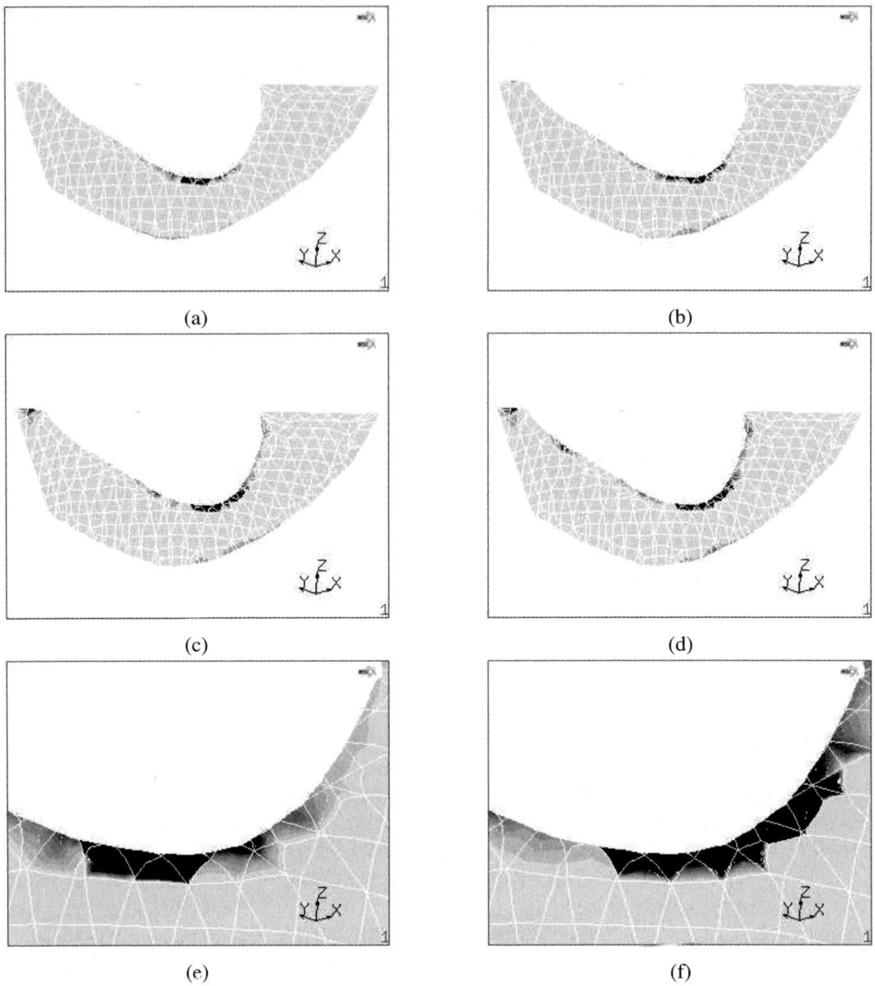


**Figure 4. Figures a to d show the total shear stress for the applied surfaces. The maximum shear stress occurs at the surface of the bones. Figures e and f show a detail the shear stress distribution for the flat and the maximum curved surfaces.**

not touch the walls of the ischial tuberosity. The line of view is dorso-medial: the $y$-axis points to the front and the $x$-axis to the right side of the body. For increasing curvedness the location of the region of minimum soft tissue thickness shifts from the lowest region to the peripheral regions of the ischial tuberosity. A significant change of the minimum thickness for varying curvature was not observed.

The relocation of the soft tissue is computed from the relocation of the nodes of the finite elements. The nodal displacements in the $z$-direction is shown for the ischial plane in figure 3a for the flat surface, and in figure 3b for the maximum curvedness surface. The number of displacement levels was reduced to 5 because of the black and white printing. The scale of the $z$-displacements for the flat support runs from 0 mm until 70 mm (black). For the 1.25 curvedness support the scale is from 0 mm to 50 mm (black). In both pictures the bone was considered a rigid body. In the case of the flat support surface, the compression below the ischial tuberosities is the cause of the main

$z$-relocation. In case of the maximum curvedness surface, the $z$-displacement pattern is different. The increase of the $z$ coordinate of the nodes was more pronounced at the dorsal aspect of the buttock (right part of figure 3b).

## Shear stress and shear strain

The total shear stress, which is considered an important factor for the viability of mechanically loaded soft tissues, is shown in figures 4a to figure 4d (the deformation of the body is not shown). It includes the vectorial sum of the $\sigma_{ij}$ $(i \neq j)$ components of the Cauchy stress tensor, $\sigma$. The low strain regions are rendered light grey. For both support surfaces, the stress appears to be concentrated on the bony surface, where the black regions are subject to shear stresses exceeding 50 kPa. For the flat surface the shear stress is concentrated on the lowest part of the ischial tuberosity, see figure 4a. The curvedness of the seats causes a shift of the location of the maximum shear stress towards the dorsal aspect of the bone.

Although the shear strain is more evenly distributed, the shear stress is concentrated on a relatively small location, see figure 2. Figures 4e and f show an enlarged part of the regions of maximum stress. The different distribution patterns of stress and strain are caused by the non-linearity of the applied constitutive equations (Mooney-Rivlin), and the very large tissue deformation.

The fact that the shear stress is concentrated on the bony surface agrees with the medical evidence that decubitus problems during long term sitting are usually not initiated at the surface of the skin, but at the surface of the ischial tuberosity.

## Conclusions

Sitting on seats of different shape cause corresponding differences of the deformations and stresses inside the body. This is a fundamental observation since it opens up a dimension to design the optimal seat surfaces, considered from an ergonomics and a medical point of view (Moes and Horváth, 2002). Especially in the area of designing body supports for long lasting contacts (wheelchair users or bed-ridden people), the distribution of the internal shear is a major issue for tissue viability and personal comfort. The presented model and the procedures to compute the internal loads can be used by the product designer to assess and control the ergonomics and medical consequences of the different product shapes.

## References

Brienza DM and Chung K (1993). Seat Support Surface Optimization Algorithm Development. In Tarbell JM, editor, *Advances in Bioengineering*, volume BED-Vol. 26, pages 495–498. ASME.

Moes CCM (2000). Pressure Distribution and Ergonomics Shape Conceptualization. In Marjanovic D, editor, *Proceedings of the 6th Int Design Conference DESIGN2000*, pages 233–240, Zagreb. University of Zagreb.

Moes CCM (2004). *Advanced human body modelling to support designing products for physical interaction*. Ph.D. Dissertation, Delft University of Technology, ISBN: 90-9018829-0.

Moes CCM and Horváth I (2002). Optimizing the Product Shape for Ergonomics Goodness Index. Part I: Conceptual Solution. In McCabe Paul T., editor, *Contemporary Ergonomics 2002*, pages 314–318. The Ergonomics society, Taylor & Francis.

# ENVIRONMENTAL ERGONOMICS

# ACCLIMATION AND RESTING CORE-TEMPERATURE

**Martin Schütte[1] & Bernhard Kampmann[2]**

[1] *Institut für Arbeitsphysiologie an der Universität Dortmund*
*Ardeystraße 67, D-44139, Dortmund*
[2] *Bergische Universität Wuppertal, Fachbereich D, Abteilung Sicherheitstechnik*
*Arbeitsbereich Arbeitsphysiologie, Arbeitsmedizin und Infektionsschutz*
*Gaußstraße 20, D-42119, Wuppertal*

It is generally accepted that acclimation results in a lower core temperature during heat exposure. The reduction of core temperature is explained by physiological mechanisms that reduce heat storage via improved heat loss mechanisms.

On the other hand it is hypothesized that acclimation results in a decrease of resting core temperature which might be partly responsible for the attenuation of core temperature during heat exposure. The aim of the present investigation was to examine whether there exists a difference in resting core temperature before and after acclimation.

All in all the data support the hypothesis that acclimation causes a reduction in resting core temperature. The results are discussed with respect to their practical relevance.

## Introduction

It is generally accepted that acclimation results in a lower core temperature during heat exposure. The reduction of core temperature is explained by physiological mechanisms that reduce heat storage via improved heat loss mechanisms (mainly improved peripheral blood flow and increased sweat rates). On the other hand it is hypothesized that acclimation, particularly to humid heat, results in a decrease of resting core temperature (e.g. as measured at the end of a resting period that precedes heat exposure) which might be partly responsible for the lowering of core temperature during heat exposure.

This assumption is supported for example by empirical results (Buono et al., 1998) based on a sample of nine male subjects who completed 7 days of acclimation and performed 2 hours of exercise per day in a hot, humid environment. All participants showed a decrease of resting core temperature ranging from 0.1 to 0.5°C. Other studies show similar results as for example Shido et al., 1999, who found a decrease of core temperature between 0.19 and 0.24°C.

The aim of the present investigation is to examine the hypothesis whether there exists a difference in resting core temperature before and after acclimation on the basis of an independent set of data.

## Experimental conditions

The data of the present study come from two experimental series, where the physiological strain (indicated by core temperature, heart rate, sweat loss etc.) was ascertained in 5 healthy young men (19–21 yrs) during exposures to different air temperatures and water vapour pressures while walking on a treadmill. All experiments were accomplished during the morning starting with the resting period about 8 a.m.

One series, which took place during the cool period of the year, was performed to study the behaviour of the subjects without prior acclimation. Acclimation was avoided by exposing the subjects not more than twice a week to hot climates.

Another series was preceded by a 3 to 4 weeks period where the subjects were exposed to a warm-humid climate (air temperature 33°C–40°C; relative humidity 65%–85%) during five consecutive days a week. The state of acclimation was regarded as sufficient for the following exposures, when rectal temperature did not exceed 38.5°C during a 3 hours exposure. During the following weeks the subjects were again exposed to the warm-humid climate each Monday in order to re-establish the state of acclimation; thereafter, they were exposed to different combinations of air temperatures and water vapour pressures during the following 4 days. The state of acclimation was then considered to correspond to the state of acclimation of persons at hot workplaces.

The analysis performed here bases on rectal temperature measured at the end of the rest periods (minimum duration of 30 minutes) that preceded the heat exposure. Thirty measurements were available for each subject: 15 measurements stem from the rest periods of the non-acclimatized subjects and 15 measurements after the subjects were acclimatized.

## Data analysis

A 3-factorial analysis of variance was accomplished with state (before, after acclimation) and repetition as fixed factors and person as a random factor. Data evaluation is based on a so-called hierarchical model with person (P) and status of acclimation (S) as fully crossed factors since each person is measured before and after acclimation. Because a person cannot be simultaneously acclimatised and non acclimatised, the repeated measurements of core temperature (factor repetition – R) are nested within the factor state (R:S).

Beside the commonly calculated parameters - such as mean squares, F-values – the variance components were additionally estimated for each source of variance using the algorithms for calculating the expected mean squares (cf. table 1).

The estimated variance components show how much persons or status or repetitions differ from one another in their core temperature. Therefore the variance components give information about the deviations from the grand mean. In interpreting the relative magnitude of the variance components usually the sum of the components is taken which reflects the total variance and for each effect the percentage of this sum is determined.

The main proportion of variance – namely 54% – can be attributed to the interaction between the factors "person" and "repetition" (P ∗ R:S, error) which inseparably comprises not only the three way interaction P ∗ S ∗ R but also all other unsystematic or

**Table 1. Estimation of variance components.**

| Source of variation | | Number of levels | Mean squares MSQ | Estimation of variance components $\sigma^2$ |
|---|---|---|---|---|
| Person | P | $n_P$ | $MSQ_P$ | $\sigma^2_P = (MSQ_P/n_S * n_R)$ |
| State | S | $n_S$ | $MSQ_S$ | $\sigma^2_S = (MSQ_S/n_P * n_R) - (\sigma^2_{P*S}/n_P)$ |
| Repetition:S | R | $n_R$ | $MSQ_{R:S}$ | $\sigma^2_{R:S} = (MSQ_{R:S} - \sigma^2_{P*R:S})/n_P$ |
| P ∗ S | | | $MSQ_{P*S}$ | $\sigma^2_{P*S} = MSQ_{P*S}/n_R$ |
| P ∗ R:S,error | | | $MSQ_{P*R:S,error}$ | $\sigma^2_{P*R:S,error} = MSQ_{P*R:S,error}$ |

random errors (cf. table 2). That adds up to more than half of the total variance and represents the so-called residual variance.

The acclimation status (S) explains about 18% of variance and provides a major contribution to the variability of core temperature. The mean core temperature for the period of non acclimation amounts to 36.78°C whereas a value of only 36.62°C is reached when the subjects are acclimatized. The corresponding F-test verifies that the two temperatures differ marginal significantly from each other ($df = 1/4$; $p < 0.06$). The statistical examination has only explorative character because the sample size is small.

But the shift in core temperature due to acclimation varies between subjects as the substantial proportion of variance (14%) explained by the interaction P*S indicates. Even though all individuals concordantly show a reduction in core temperature (cf. figure 1) their responsiveness to acclimation seems to be quite different, as the temperature decrements document, which are ranging between 0.02°C and 0.37°C.

Furthermore 12% of the total variance can be traced back to differences between persons. The mean core temperature of subjects (cf. figure 2) ranges between 36.58°C (subject P2) and 36.78°C (subject P5).

The repeated measurements do not differ substantially as the explained proportion of variance shows a value of less than 1%. The error probability for the calculated F-value considerably exceeds the limit of $p = 0.05$. Since the various measurements produce quite comparable results it is not necessary to have 15 measurements for describing the status of acclimation respectively non-acclimation. One single measurement of each core temperature may be sufficient for a person. Furthermore these results show also that all subjects had reached a stable physiological level under both experimental conditions.

**Table 2. Three-factorial hierarchical analysis of variance.**

| Source of variation | | Error term | df | Mean squares | F value | p value | $\sigma^2$ | % |
|---|---|---|---|---|---|---|---|---|
| Person | P | | 4 | 0.2334 | – | – | 0.0078 | 12.2 |
| Status | S | P ∗ S | 1 | 1.0152 | 7.37 | 0.0532 | 0.0117 | 18.4 |
| Repetition:S | R | P ∗ R:S | 28 | 0.0372 | 1.08 | 0.3796 | 0.0005 | 0.8 |
| P ∗ S | | | 4 | 0.1377 | – | – | 0.0092 | 14.4 |
| P ∗ R:S,error | | | 112 | 0.0345 | – | – | 0.0345 | 54.2 |

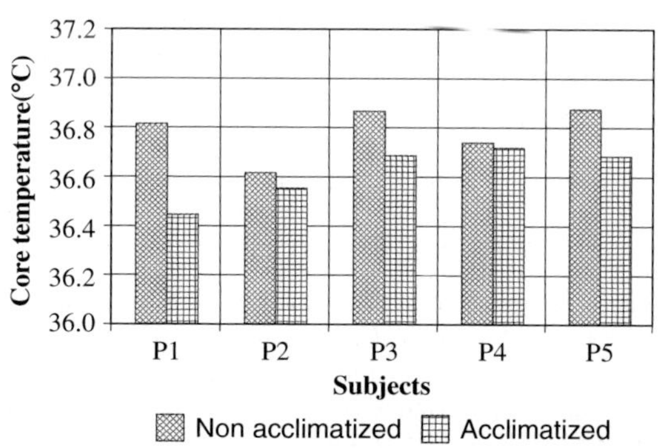


**Figure 1. Interaction between the factors person and status.**

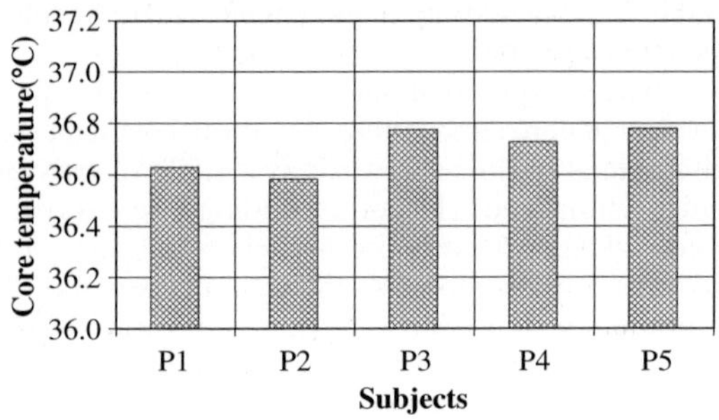


**Figure 2. Mean core temperature (°C) of subjects.**

## Discussion

Concerning all three main effects the factor "status" explaining 18% of the variance, has the strongest influence on rectal temperature. Therefore the results of the present study support the hypothesis that acclimation is accompanied by a reduction in resting core temperature. The difference in rectal temperature between non-acclimated and acclimated resting periods mounts up to 0.16°C and takes a value somewhat lower than the reduction reported in other studies. One possible explanation for this finding is that in the study at hand exposures were broken off at rectal temperatures of 38.5°C whereas Buono et al., 1998 admitted core temperatures well above this value. Additionally in the present investigation a smaller number of consecutive days of exposure were used for acclimation (5 days vs. 7 days). Both aspects could explain that in the present study a lower level of acclimation resulted.

The present data evaluation is characterized by the fact, that the accomplished analysis of variance explicitly takes into consideration the factor "person" and gives information about individual particularities in the responses to climatic conditions. Such indications are important if practical conclusions are derived from the results on hand.

Generally, if resting core temperature is reduced, persons may stay longer in a hot climate until reaching a given threshold value for core temperature. This is of relevance for instance for fire-fighters or mine rescue brigadesmen wearing protective clothing. Taking into consideration that the decrease of core temperature varies considerably between persons as a consequence of acclimation, the calculation of an increase of working time in a given climate seems hardly possible until the heat strain of subjects may be described with high precision by their individual parameters (body height, body mass, fitness etc.). The present experiment supports the assumption that the reduction of core temperature in resting conditions is a benefit of acclimation. It seems to be reasonable that a comprehensive assessment of the acclimation status of individuals takes this effect into account.

## References

Buono, M.J., Heaney, J.H. and Canine, K.M. 1998, Acclimation to humid heat lowers resting core temperature, *American Journal of Physiology*, **274**, R1295–R1299.

Shido, O., Sugimoto, N., Tanabe, M. and Sakurada, S. 1999, Core temperature and sweating onset in humans acclimated to heat given at a fixed daily time, *American Journal of Physiology*, **276**, R1095–R1101.

# PHYSIOLOGICAL AND LABOUR-PRODUCTIVE EFFECTS OF FLUID CONSUMPTION DURING FORESTRY WORK

**Dianne Staal Wästerlund[1] & Janet Chaseling[2]**

[1] *Division of Forest Technology, Department of Silviculture, Swedish University of Agricultural Sciences, 901 83 Umeå, Sweden*
[2] *Faculty of Environmental Sciences, Griffith University, Nathan, Brisbane Queensland 4111, Australia*

The physiological and labour productive effects of two fluid consumption levels were studied on forest workers to compare the effects of a mildly dehydrated physical condition with a fully hydrated condition. Four forest workers each produced manually 2.4 $m^3$ of pulpwood/day on 8 consecutive working days in climatic conditions varying between 8 and 28°C WBGT. They consumed water (0.17 l or 0.6 l) at half hourly intervals with one fluid level assigned to each day. The forest work required a significantly lower percentage of the heart rate reserve at the high fluid consumption level compared with the low level (36.5% versus 39.9%, $p = 0.0024$). The time taken to perform the work was significantly lower with the consumption of 0.6 l each half hour compared to the consumption of 0.17 l (209.0 versus 234.1 min, $p = 0.0003$). Moreover, it was found that the fluid consumption level of the previous day had a significant influence on the time taken.

## Introduction

Forest work in developing countries is mainly performed with manual or motor-manual work methods which according to ISO 8996 may be graded as moderate to very heavy (ISO, 1990). The climate in many developing countries is (sub-) tropical and constitutes an additional and often overlooked burden to the forest worker. As the major way for heat dissipation in a warm environment is through the excretion of water and salts, the risk for heat stress increases if the fluid is not replaced and work is performed while the worker is dehydrated.

Dehydration will also reduce the maximum oxygen uptake capacity (Sawka, 1992). Reductions in physical work capacity between 6 to 48% were reported for dehydration levels between 1 to 8% (Sawka, 1992). This effect is well-appreciated in the sports sector, but in the occupational context it is often assumed that individuals themselves are fully capable of maintaining a good hydration level. In Vietnam however, a reduction in bodyweight of 3.2% was found for forest workers who neither ate nor drank at the work place (Wigaeus Hjelm and Frisk, 1985).

The physiological and performance effects of dehydration have mainly been studied in laboratories under controlled conditions (e.g. Saltin 1964; Hamilton et al., 1991; Below et al., 1995; McConnel et al., 1999). Yet neither the thermal load nor the work load are constant when an individual is performing forest work. Also, the laboratory studies have focussed mainly on the effects during and/or directly after one exercise of relatively short duration (often less than two hours). The purpose of this study was to compare the effects of two fluid consumption levels on physiological factors and labour productivity of forest workers performing a heavy work task in a naturally changing thermal environment. The fluid consumption levels were chosen to compare a mild level of dehydration after a full working day with a level preventing dehydration.

## Materials and methods

Four male forest workers who had at least two years of experience on the job, took part in the study. The test persons ranged from 20 to 40 years of age. Interviews revealed that the workers consumed one to three cups of tea per day, besides the fluids provided during the study.

Each day, each worker was to produce a stack of pulpwood of 2.4 $m^3$ by harvesting trees marked in the stand (i.e. felling 5–7 trees, processing the trees into logs and carrying the logs to the roadside), followed by debarking the pulpwood (i.e. cutting off the bark and stacking the logs). The workers used hand tools only.

The Wet-Bulb-Globe-Temperature (WBGT) was recorded during the work period at intervals of 8 minutes with a Brüel & Kjær WBGT heat stress monitor type 1219. The temperatures ranged from 8 to 28°C WBGT.

During the working day, each individual consumed water at half hourly intervals at the rate of either 0.17 l (low fluid consumption level) or 0.6 l (high fluid consumption level). Before work started each test person was given 0.5 l of water to obtain full hydration at the start of the study day.

A multi-period crossover design (Bodero and Reason, 1986) with unbalanced within block replication was used as study design. Treatment order was assigned at random within each sequence such that replication was balanced across all four participants. As the study included weekends where no work was carried out, the previous day's activities included the two fluid levels and also a condition of unknown (either the first day of study for each person or a Monday).

During the performance of the work, heart rate was recorded each minute with the Polar Sport Tester PE 3000. As a measure of exertion, the percentage of heart rate reserve (*HRres*) was determined according to Rodahl (1989) as the mean heart rate during work minus the resting heart rate, divided by the maximum heart rate minus the resting heart rate. The maximum heart rate was estimated as 220 – age (Rodahl, 1989).

Every hour, the mass of the test person was measured using a Philips HP 5324 scale. A time study was conducted using repetitive timing with a centi-minute stopwatch.

The heart rate reserve (%), body mass change (kg), and time taken to complete the task (min) were analysed using a generalised linear model. All treatment effects were tested against the error term defined by the interactions of individuals with the treatments. To determine whether or not treatment responses varied between individuals, the interaction of treatment and individual was tested and isolated from the error term where necessary.

## Results

### *Physiological effects*

The effect of the fluid consumption levels on the percentage of heart rate reserve was found to vary significantly for the individual test persons during harvesting, debarking and the entire working day. When averaged over all four test persons (table 1), the mean percentage of heart rate reserve was significantly lower for the high compared with the low fluid consumption level for the entire working day ($p = 0.0024$) and during debarking ($p = 0.0050$).

The overall body mass change was consistent for all test persons. When combined across all four persons there was a significantly different mean response between

**Table 1. Least squares means for percentage of heart rate reserve (*HRres*) and body mass change (*BM ch*) for the two fluid consumption levels.**

| Task | Variable | Fluid consumption level (l) | Test persons A | B | C | D | Overall fluid effects |
|---|---|---|---|---|---|---|---|
| Entire | *HRres* | 0.17 | 40.7 | 42.1 | 37.9 | 39.1 | 39.9 |
| working | (%) | 0.6 | 37.2 | 32.4 | 37.4 | 38.8 | 36.5 |
| day | *BM ch* | 0.17 | −0.6 | −0.8 | −0.8 | −0.4 | −0.7 |
| | (kg) | 0.6 | +0.8 | +0.3 | +1.3 | +0.4 | +0.7 |
| Harvesting | *HRres* | 0.17 | 52.7 | 49.7 | 44.8 | 45.7 | 48.2 |
| | (%) | 0.6 | 50.4 | 45.0 | 50.6 | 45.6 | 47.9 |
| Debarking | *HRres* | 0.17 | 35.2 | 38.0 | 34.2 | 35.0 | 35.6 |
| | (%) | 0.6 | 31.6 | 28.6 | 31.9 | 35.7 | 31.9 |

**Table 2. Least squares means for the time taken to complete the tasks for the two fluid consumption levels only and for the fluid consumption levels including the previous day's fluid consumption level.**

| Fluid consumption level (l) Current day | Previous day | Entire working day (min) | Harvesting (min) | Debarking (min) |
|---|---|---|---|---|
| 0.17 | | 234.1 | 103.8 | 130.3 |
| 0.6 | | 209.0 | 91.7 | 117.3 |
| *p*-value | | 0.0003 | 0.0268 | 0.0554 |
| 0.17 | unknown | 241.5 | 121.6 | 132.9 |
| 0.17 | 0.17 | 234.5 | 99.0 | 131.0 |
| 0.17 | 0.6 | 226.3 | 90.7 | 126.9 |
| 0.6 | unknown | 224.2 | 97.8 | 120.1 |
| 0.6 | 0.17 | 202.2 | 85.6 | 119.9 |
| 0.6 | 0.6 | 200.5 | 91.7 | 112.0 |
| *p*-value | | 0.0028 | 0.1007 | 0.0091 |

treatments, with a loss of body mass on the low fluid consumption level and a gain of body mass on the high fluid consumption level ($p = 0.0003$, table 1).

No significant accumulation effect of the previous day's fluid consumption level and the current day's fluid consumption level was found for the percentage of heart rate reserve and the body mass change.

### *Labour-productive effects*

The mean times taken for harvesting and over the entire working day were significantly lower for the high compared with the low fluid consumption level (table 2) while debarking showed a very clear trend in the same direction. This effect was consistent for all test persons.

The combined effect of the previous day fluid consumption level and the current day fluid consumption level showed a significant trend with an appearance of an accumulative effect (table 2). The difference in the mean time taken between the high and low intake was significantly greater following a day on the low intake (32.3) than following a day on the high intake (25.8). Both differences were greater than the difference following an unknown day (17.3) but this difference was not found significant.

## Discussion

As temperatures during the study only occasionally exceeded 25°C WBGT above which according to the WBGT heat stress index (ISO, 1989) risk for heat stress must be considered for this type of work, the effects are discussed only as the results of dehydration.

The statistical differences found between fluid consumption levels for the mean percentage of heart rate reserve were consistent with the findings of other researchers in that there was a higher requirement on the heart rate reserve for the mild dehydration level compared with the full hydration (González-Alonso, 1998; Saltin, 1964). The difference in the present study was found despite the mild level of dehydration (0.7–1.4%) and the higher labour productivity during the high fluid consumption level compared to the low. Significant differences were also found for the body mass change over the entire working day between the fluid consumption levels, although the workers were weighed fully cloathed and with unevaporated sweat still attached to their skin.

The ingestion of 0.6 l of water each half hour reduced the mean time taken for completing the daily task by 25 min or 11% compared with the ingestion of 0.17 l of water each half hour under the described conditions. Similar results were found by Fallowfield et al. (1996), Walsh et al. (1994) and Below et al. (1995), although under much more extreme work load and climate conditions. However, McConell et al. (1999) pointed out that performance might be affected by perceived uncomfortable abdominal bloating. As all test persons gained weight during the high fluid consumption level, uncomfortable abdominal bloating limiting the labour productivity may have occurred.

The results show also that there is a carry-over effect of the fluid consumption level from the previous day on the labour productivity. It appears that the mild dehydration level caused by the low fluid consumption level was insufficiently compensated for by the workers themselves after work, and that this voluntary dehydration level at the start of the next study day was not compensated for by the 0.5 l given before commencing work.

The results of this study showed large differences between the individuals during the work activities. This may indicate that there may be individual differences in how the time reduction is obtained.

As large variations were anticipated due to the practicalities of this field study, it was considered more suitable to study four test persons with four replications for each fluid consumption level rather than more persons with less replication. Further studies are required to determine if the findings of this study are applicable to other forest workers with similar work loads and similar climatic conditions.

## References

Below, P. R., Mora-Rodríguez, R., González-Alonso, J. and Coyle, E. F. 1995, Fluid and carbohydrate ingestion independently improve performance during 1 h of intense exercise, *Medicine and Science in Sports and Exercise*, **27**, 200–210.

Bodero, J. and Reason, G. K. 1986, A multi-period incomplete block cross-over design for research on commercial dairy farms. In: I S Francis, B F Manley and F C Lam (eds). *Proceedings of the Pacific Statistical Congress 1985*, (Amsterdam: North Holland), 175–178.

Fallowfield, J. L., Williams, C., Booth, J., Choo, B. H. and Growns, S. 1996, Effect of water ingestion on endurance capacity during prolonged running, *Journal of Sports Science*, **14**, 497–502.

González-Alonso, J. 1998, Separate and combined influences of dehydration and hyperthermia on cardiovascular responses to exercise, *International Journal of Sports Medicine*, **19 (suppl 2),** 111–114.

Hamilton, M. T., Gonzalez-Alonso, J., Montain, S. J. and Coyle, E. F. 1991, Fluid replacement and glucose infusion during exercise prevent cardiovascular drift, *Journal of Applied Physiology*, **71**, 871–877.

ILO. 1999, *Forestry; wood; pulp and paper*. Sectoral Activities Programme, International Labour Office. http://www.ilo.org/public/english/dialogue/sector/sectors/forest.htm. Update 15 July 1999.

ISO. 1989, *Hot environments – estimation of the heat stress on working man, based on the WBGT-index (wet bulb globe temperature)*. ISO 7243: 1989 (E), (Geneva: International Organization for Standardization).

ISO. 1990, Ergonomics – *Determination of metabolic heat production. ISO 8996: 1990* (E), (Geneva: International Organization of Standardization).

McConell, G. K., Stephens, T. J. and Canny, B. J. 1999, Fluid ingestion does not influence intense 1-h exercise performance in a mild environment, *Medicine and Science in Sports and Exercise*, **31**, 386–392.

Rodahl, K. 1989, The physiology of work, (London: Taylor & Francis).

Saltin, B. 1964, Circulatory response to submaximal and maximal exercise after thermal dehydration, *Journal of Applied Physiology*, **19**, 1125–1132.

Sawka, M. N. 1992, Physiological consequences of dehydration: exercise performance and thermoregulation, *Medicine and Science in Sports and Exercise*, **24**, 657 – 670.

Walsh, R. M., Noakes, T. D., Hawley, J. A. and Dennis, S. C. 1994, Impaired high-intensity cycling performance time at low levels of dehydration, *International Journal of Sports Medicine*, **15**, 392–398.

Wigaeus Hjelm, E. and Frisk, C. 1985, *Physical strain on Vietnamese forest workers*, Undersökningsrapport 1985:5, Arbetarskyddsstyrelsen, Forskningsavdelningen, Solna.

# A MODEL OF THERMAL COMFORT FOR PEOPLE IN RAILWAY CARRIAGES

**Ken Parsons**

*Department of Human Sciences, Loughborough University, Loughborough, Leicestershire LE11 3TU, UK*

Ten participants (7 male and 3 female) sat "side-on" to simulated solar radiation through a plain glass window for 30 minutes and provided, every five minutes, ratings of thermal comfort. Skin temperatures and environmental conditions were also recorded. They wore light trousers and white shirt with their normal underclothes and shoes (0.72 Clo including the seat). Before entering the radiation chamber participants were exposed to a thermal chamber set at 17°C ensuring that they were cooler than neutral. This air temperature was also maintained in the radiation chamber. The intensity of radiation arriving on the subject was 600 $Wm^{-2}$. Results show that the effects of low air temperature combined with solar radiation can be predicted from a Predicted Mean Vote (PMV) model of thermal comfort with a simple modification for intensity of solar radiation.

## Introduction

Parsons (2003) proposed the following thermal comfort index for people exposed to solar radiation.

$$PMV_{solar} = PMV + \frac{RAD}{200} \quad (1)$$

where

$PMV_{solar}$ = Predicted Mean Vote (from +5, extremely hot; through 0, neutral; to −5, extremely cold) taking account of the effects of solar radiation,

PMV = Predicted Mean Vote (ISO 7730; Fanger, 1970) with correction to clothing insulation to take account of the seat, and $t_r$ measured in the shade,

RAD = Direct radiation level received at the position of the person (estimate from 1000 $Wm^{-2}$ = maximum; 600 $Wm^{-2}$ = very high, 200 $Wm^{-2}$ = low; and 0 $Wm^{-2}$ = minimum).

The rationale for the model is that the PMV index is appropriate without direct solar radiation and that Hodder and Parsons (2001) had shown that for every 200 $Wm^{-2}$ of solar radiation, people exposed would on average become "hotter" by one scale point on the ISO thermal sensation scale. For a person sitting in thermal comfort (neutral; PMV = 0) in the shade, if they were exposed to full sun (e.g. 600 $Wm^{-2}$) then they would have a sensation of hot (PMV = +3, i.e. 600 ÷ 200 = 3). The work of Hodder and Parsons (2001) however was derived for people in conditions in the shade in comfort (neutral; PMV = 0). The model had not been tested for people in the shade with sensations away from neutral (e.g. cool; PMV = −2.) For example where the air temperature is low and conditions cool, (i.e. cool in the shade) but made warmer by the sun coming through a window.

The laboratory experiment which is presented in this paper evaluated the effectiveness of the thermal comfort model presented in Equation (1) for people exposed to solar radiation but occupying what would otherwise be cool conditions. The results can be generally applied but are particularly relevant to the thermal comfort of people sitting in railway carriages as the radiation arrived "side-on" to the seated person.

## Method

### *Experimental method*

Ten participants (seven male, three female) were seated in a Fiat Punto front car seat left "side-on" to simulated solar radiation at an intensity of 600 $Wm^{-2}$. The environment was controlled to be thermally "cool" without the direct radiation. Prior to exposure to the radiation, the participants were exposed, for 30 minutes, to cool ($t_a$ = 17 °C) conditions in a separate thermal chamber. The participants wore a white cotton/polyester (65/35%) shirt with sleeves rolled up to the elbows, beige cotton/polyester (65/35%) trousers and their own underwear and shoes. The total estimated insulation including the seat was 0.72 Clo with an estimated metabolic rate of 70 $Wm^{-2}$. The participants were seated "side-on" to a vertical window with the radiation source at 45° to that window (Figure 1). They wore protective eyewear throughout exposure.

The chamber is a purpose built insulated room. There is a 1.0 $m^2$ pane of 3.5 mm clear monolithic glass mounted vertically at one end of the chamber. A bank of four simulated solar radiation lamps (1000 W metal halide CSI lamp manufactured by GE lighting) were fixed to a frame. The intensity was monitored during the investigation using a Skye pyronometer SP1110 to ensure uniformity across the subjects. Air conditioning units were placed inside and on the roof of the chamber to maintain "cool" conditions without draught.

### *Experimental measures*

Environmental conditions were monitored and recorded every ten seconds throughout each experiment with an Eltek/Grant squirrel data logger. Thermistors were positioned around the chamber to record air temperature, $t_a$. Also measured were mean radiant temperature, $t_r$, via a black globe thermometer, relative humidity, rh, and air velocity, v,

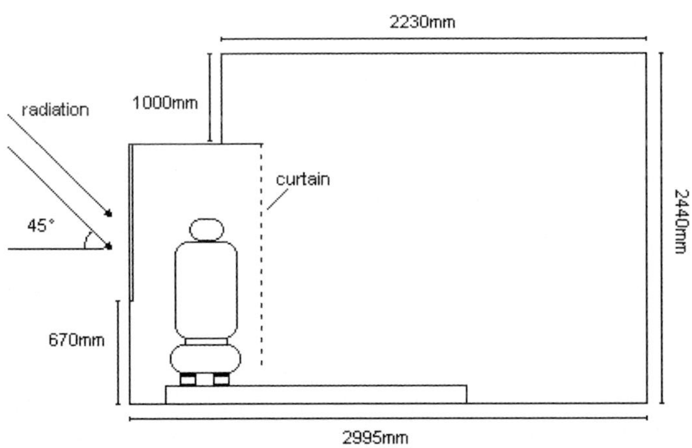


**Figure 1. Schematic of test chamber.**

with a hot wire anemometer. The participants' skin temperature was measured using thermistors positioned at 11 points on the body: right and left head; arm; forearm; thigh; and calf; and one positioned centrally on the chest. These were recorded every ten seconds using another squirrel data logger.

The participants completed a subjective questionnaire every five minutes throughout the experiment, rating their thermal sensation, thermal comfort, stickiness and preference. A 7 point ISO thermal sensation scale was used. All the sensation votes were over 12 individual body parts as well as an overall vote. Preference was measured over a continuous 7 point scale and comfort and stickiness over a 4 point scale.

*Experimental procedure*

The participants sat for 30 minutes in a thermally cool preparation chamber adjacent to the radiation chamber. The thermistors were attached using 3 M transpore tape, and the participants dressed in the standard clothing. They were then taken into the radiation chamber and seated on the car seat protected from the radiation by the curtain. They were asked to complete a questionnaire. Once this was completed the participant was slid into the radiation and the experiment commenced. Questionnaires were administered from this moment, and every five minutes, until the experiment finished 30 minutes later.

## Results

Results are presented in Table 1 for all ten participants and overall thermal sensation. The pre-radiation case is for cool conditions which provides actual thermal sensation of the participant just before exposure to the radiation. The difference between the pre-radiation sensation and that after 30 minutes of exposure provides an evaluation of the model in Equation (1). The model in Equation (1) predicts that the difference will be 3.0 scale units, on average. It can be seen that there is an actual average difference of 2.83 scale units.

## Discussion

Figure 2 shows the results of an experiment by Vaughan et al. (2004) in terms of the mean end vote, after 30 minutes of exposure of eight participants exposed to "side on"

**Table 1. Subjective sensation ratings for 7 male and 3 female participants on a continuous ISO Scale (3 hot; 2 warm; 1 slightly warm; 0 neutral; −1 slightly cool; −2 cool; −3 cold).**

| Participant | Pre-radiation | 0 | 5 | 10 | 15 | 20 | 25 | 30 | Difference |
|---|---|---|---|---|---|---|---|---|---|
| M1 | −1.70 | −0.10 | 0.20 | 0.70 | 1.00 | 1.00 | 2.20 | 1.70 | 3.40 |
| M2 | −1.30 | 1.60 | 2.10 | 2.50 | 2.60 | 2.50 | 2.50 | 2.60 | 3.90 |
| M3 | −1.70 | 1.00 | 1.80 | 1.80 | 2.00 | 2.00 | 1.60 | 1.80 | 3.50 |
| M4 | −0.90 | 0.00 | 1.00 | 0.80 | 1.60 | 1.00 | 1.80 | 1.80 | 2.70 |
| M5 | −0.70 | −0.50 | −0.70 | 0.50 | −0.30 | 0.30 | 0.50 | 1.50 | 2.20 |
| M6 | −2.00 | −1.00 | −0.80 | −1.40 | −0.20 | −0.30 | −0.30 | 0.80 | 2.80 |
| M7 | −1.00 | 0.00 | −1.00 | 0.00 | 0.00 | 0.00 | 0.00 | 0.00 | 1.00 |
| F1 | −1.80 | 0.80 | 1.20 | 1.80 | 1.80 | 1.80 | 1.80 | 1.90 | 3.70 |
| F2 | −2.20 | 1.00 | 1.30 | 2.00 | 2.00 | 2.00 | 2.00 | 2.00 | 4.20 |
| F3 | −1.30 | 0.00 | 0.00 | 0.00 | −0.10 | 0.00 | 0.00 | −0.40 | 0.90 |
| Mean | −1.46 | 0.28 | 0.51 | 0.87 | 1.04 | 1.03 | 1.21 | 1.37 | 2.83 |
| SDev | 0.50 | 0.80 | 1.10 | 1.20 | 1.10 | 1.00 | 1.00 | 0.90 | 1.20 |

radiation at intensities of 0 to 600 $Wm^{-2}$. The participants were in conditions of thermal neutrality in the shade and not cool conditions as were used in the present experiment. This curve shows a distinct and significant increase in the thermal sensation experienced by the subjects corresponding to an increase in radiation intensity. The linear nature of the curve suggests that for every increase in solar radiation level of 200 $Wm^{-2}$, participants are, on average, hotter by 0.8 scale units. So for example for participants otherwise neutral, a solar radiation effect of 600 $Wm^{-2}$ would be expected to increase the thermal sensation from neutral to between warm and hot (average vote of 2.4 on the scale). This corresponds favourably with the value of 2.83 obtained in the present experiment and the observations of Hodder and Parsons (2001) who would have predicted a value of 3.0 on the scale. Hodder and Parsons (2001) however used participants who faced towards the radiation with head shielded.

## Practical application

It is possible to analyse exposure to solar radiation and to relate it to radiant temperatures around the person. The PMV thermal comfort index for example could include this in any prediction and a simple correction for the effects of solar radiation may not be required. It is debatable however whether the PMV index is valid for the specific characteristics of solar radiation having been derived and evaluated mainly for indoor environments. It may also be the case that people react to solar radiation as an "add on" to overall thermal sensation. This is the nature of the model shown in Equation (1). While it would be possible to provide a more complex analysis for thermal comfort assessment involving heat transfer between the environment and the person, the simple model of thermal comfort presented here combines both the established method for the assessment of thermal comfort (PMV, ISO 7730) and a simple additional component for level of solar radiation. The overall model providing a practical assessment method for assessing thermal comfort for people in the sun.

A practical example would be for people who are slightly cool and exposed to full blue sky sun (e.g. 600 $Wm^{-2}$) they will become warm (scale value shift from $-1$ to 2); people who are cold and who are exposed to some sun (e.g. 400 $Wm^{-2}$) will become slightly cool (scale value shift from $-3$ to $-1$); people who are slightly warm and exposed to full sun (e.g. 600 $Wm^{-2}$) will become very hot and so on.

Consider a passenger, wearing a business suit, sitting in a railway carriage. "side-on" to the full sun coming through the carriage window. The train does not generally

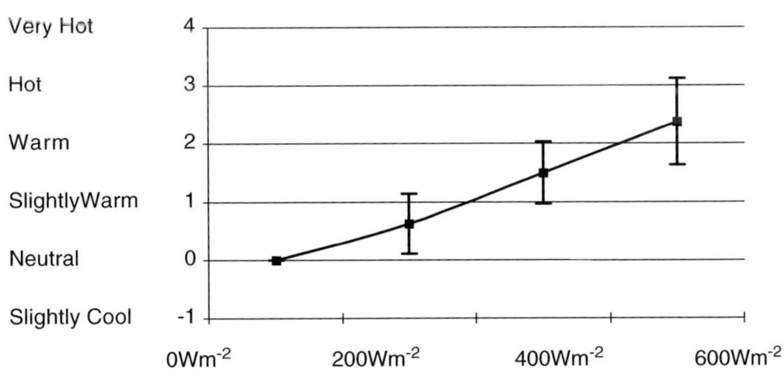


**Figure 2. Mean thermal sensation ratings over 8 participants after 30 minutes of exposure (from Vaughan et al. 2004).**

change direction, so the exposure is continuous. The air temperature around the passenger is 20°C with low air movement and relative humidity around 50%. The Predicted Mean Vote (PMV) value, independently of the effects of the sun, can be calculated from ta = tr = 20°C, v = 0.15 $ms^{-1}$, rh = 50%, Clo = 1.2 (including seat) and M = 65 $Wm^{-2}$.

The value of PMV is −0.4 (neutral to slightly warm) with a corresponding Predicted Percentage of Dissatisfied of 8%. If we estimate the contribution of solar radiation of full sun to be 600 $Wm^{-2}$ then from Equation (1), $PMV_{solar}$ is the PMV (−0.4) plus 3 (i.e. 600 ÷ 200 = 3). That is 2.6 (warm to hot). Further calculation would suggest that if air temperature were reduced to 7°C then the passenger would achieve thermal comfort. It is more likely however that the passenger will take off a jacket, draw the curtains or move away from that seat.

## Conclusion

The relationship between thermal sensation and radiation intensity has been established for people who are otherwise cool, seated "side on" to a radiation source. A simple model for predicting thermal comfort in solar radiation will be effective for conditions where subjects start with a range of thermal sensations and provides support for use of the $PMV_{solar}$ model as a general thermal comfort index.

## References

Fanger, P.O. 1970, *Thermal Comfort,* (Danish Technical Press, Copenhagen)

Hodder S. and Parsons K.C. 2001, The effect of the intensity of simulated solar radiation on human thermal comfort, *Automotive Glazing, Brite Euram Project,* BE97-3020 2.4/001

ISO Standard 7730 1994, Moderate Thermal Environments – Determination of the PMV and PPD indices and specification of the conditions for thermal comfort

Parsons Ken 2003, *Human Thermal Environments*, Second Edition, (Taylor and Francis, London) ISBN 0-415-23793-9

Vaughan Sam, Martensen Jenny and Parsons Ken 2004, The effect of intensity of "side-on" simulated solar radiation on the thermal comfort of seated subjects. *Contemporary Ergonomics 2004* (CRC Press, Florida, USA). ISBN 08493-2342-8

# THE EFFECTS OF COLOUR AND FIT OF CLOTHING ON THERMAL COMFORT IN THE SUN

**Ken Parsons, Steven Naylor & Jamie Wales**

*Department of Human Sciences, Loughborough University, Loughborough, Leicestershire LE11 3TU, UK*

This paper presents two laboratory experiments into the effects of colour and fit of clothing on the thermal comfort of people who are exposed to simulated solar radiation. Participants sat facing solar radiation lamps for 30 minutes when thermal comfort ratings, skin temperatures and environmental conditions were recorded. In the first experiment eight male participants were used to compare the effects of black and white shirts while facing 400 $Wm^{-2}$ of simulated solar radiation. In the second experiment four male participants were exposed to 500 $Wm^{-2}$ of simulated solar radiation and compared black and white shirts as well as loose and tight fit in a 4 × 4 repeated measures Latin Square experimental design. The results show that participants wearing a tight shirt were hotter than when wearing a loose shirt and that wearing a black shirt was hotter than wearing a white shirt.

## Introduction

Thermal radiation from the sun is of shorter wavelength than that from surfaces in indoor environments where surface temperatures are much lower and hence radiation is of longer wavelength. Absorption of radiant heat by a surface will depend upon the wavelength of the radiation. If the radiation is of short wavelength then surface colour will be important. This paper presents two laboratory experiments into the effects of wearing black clothing when compared with wearing white clothing, on the thermal comfort of people exposed to simulated solar radiation. The first experiment hypothesized that people wearing a black shirt would be hotter than people wearing a white shirt. The second experiment also considered the effect of clothing fit. It could be expected that a black shirt would absorb more radiant heat than a white shirt, but also that loose fitting clothing would allow more dissipation of heat by natural (or forced) convection (and evaporation) than tightly fitting clothing.

## Method

### *Experimental method*

Both experiments used identical experimental protocols. In the first experiment, eight healthy male participants carried out two experimental sessions on separate days but at the same time of day. In one session the participants wore a white cotton/polyester (65/35%) shirt and beige cotton/polyester (65/35%) trousers and their own underwear and shoes. In the second session, they wore identical clothing but with an identical shirt that had been dyed black. The shirt sleeves were rolled down and the total estimated clo value including the seat was 0.72 with an estimated metabolic rate of 70 $Wm^{-2}$. Half of the participants wore the white shirt in the first session and half the black shirt. In the second experiment, four male participants carried out an identical protocol (with white and black shirt) to that of experiment one, but with the additional condition of tight (skin – one size too small) and loose (one size too big) fit. A repeated measures 4 × 4 Latin Square experimental design was used to reduce order effects. In experiment one, participants were exposed to 400 $Wm^{-2}$ of simulated solar radiation (at the position of the participant) and in experiment two, 500 $Wm^{-2}$.

Participants were seated in a Fiat Punto front car seat facing the simulated solar radiation. The environment was controlled to be thermally neutral without the direct radiation. The participants were seated facing a sloping window at 45° with the radiation source at 90° to the window (Figure 1). They wore protective eyewear throughout exposure and their head was shielded.

The chamber is a purpose built insulated room that was constructed to provide a thermally neutral environment. There is a pane of 1 m × 1 m × 3.5 mm clear monolithic glass mounted at one end of the chamber. A bank of 4 simulated solar radiation lamps (1000 W metal halide CSI lamp manufactured by GE lighting – Beeson,1 978) were fixed to a frame outside of the chamber. The intensity of the radiation could be altered by varying the distance of the lamps from the participants. This intensity was monitored during the investigation using a Skye pyronometer SP1110 to ensure uniformity across the participants.

*Experimental measures*

Environmental conditions were monitored and recorded every ten seconds throughout each experiment with an Eltek/Grant squirrel data logger. Thermistors were positioned around the chamber to record air temperature, $t_a$. Also measured were mean radiant temperature, $t_r$, via a black globe thermometer, relative humidity, rh, and air velocity, v, with a hot wire anemometer.

The participants' skin temperatures were measured using thermistors positioned at 6 points on the body; forehead, upper arm, lower arm, chest, thigh and calf. These were recorded every ten seconds using another squirrel data logger.

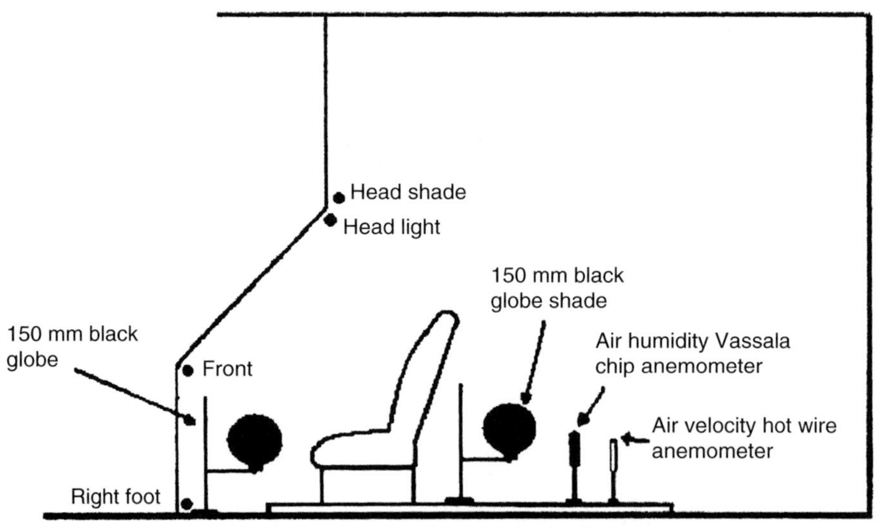


**Figure 1. Schematic of test chamber.**

The participants completed a subjective questionnaire every five minutes throughout the experiment, rating their thermal sensation, thermal comfort, stickiness and preference. The thermal sensation scale was an adapted 7 point ISO scale extended to an 11 point Likert scale, although only the slightly cool negative vote was used at the "cool" side of the scale as it was unlikely the participants would feel anything less than neutral as the chamber temperature was established at a thermally neutral level before radiation exposure. All the sensation votes were over 9 individual body parts as well as an overall vote. Preference was measured over a continuous 7 point scale and comfort and stickiness over a 4 point scale.

*Experimental procedure*

The participants were asked to arrive 30 minutes prior to the experiment and were sat in a thermally neutral preparation room adjacent to the chamber. The thermistors were attached using 3 M transpore tape, and the participants dressed in the appropriate clothing. The participants remained in the preparation room until they felt they were thermally neutral. They were then taken into the chamber and seated on the car seat protected from the radiation by the curtain. They were asked to complete a questionnaire to ensure they were still thermally neutral. Once this was completed the participant was slid into the radiation and the experiment commenced. Questionnaires were administered from this moment, and every five minutes, until the experiment finished 30 minutes later.

## Results

For the present paper, results are presented for overall thermal sensation, after 30 minutes of exposure to the simulated solar radiation. This provides a good indication of the effect of clothing type, as it reflects the accumulation of the effect, it represents a time that should be independent of starting conditions and heat transfer from the body should be in approximate steady state (or dynamic equilibrium). Thermal sensation also provides a direct link with established methods and scales for assessing thermal comfort (ISO 7730; Fanger, 1970; Parsons,2003) (5, extremely hot; 4, very hot; 3, hot; 2, warm; 1, slightly warm; 0, neutral; −1, slightly cool; −2, cool; −3, cold; −4, very cold; −5, extremely cold).

**Table 1. Thermal sensation end votes over 8 participants (30 minutes, experiment 1).**

| | Shirt colour | |
|---|---|---|
| Participant | White | Black |
| A | 1.1 | 1.5 |
| B | 0.7 | 1.3 |
| C | 0.3 | 1.0 |
| D | 2.2 | 1.0 |
| E | 0.2 | 1.0 |
| F | 1.7 | 2.6 |
| G | 2.0 | 2.8 |
| H | 0.4 | 2.3 |
| Mean | 1.25 | 1.68 |
| S.Dev. | 0.78 | 0.75 |

**Table 2. Thermal sensation end votes over 4 participants (30 minutes, experiment 2).**

| Participant | Shirt colour/fit<br>Black/loose | Black/tight | White/loose | White/tight |
|---|---|---|---|---|
| A | 2.8 | 4.1 | 3.2 | 5.0 |
| B | 4.0 | 4.0 | 3.0 | 4.0 |
| C | 2.0 | 2.0 | 1.0 | 1.0 |
| D | 3.0 | 3.0 | 2.0 | 3.0 |
| Mean | 2.95 | 3.275 | 2.3 | 3.25 |
| S.Dev. | 0.82 | 0.98 | 1.01 | 1.71 |

Table 1 shows the results for experiment one. The shirts were of the correct fit for the participants so only colour of shirt is compared. The ratings for each participant are presented. For example for participant A it can be seen that after 30 minutes of exposure in the white shirt a rating of 1.1 (just warmer than slightly warm) was provided. When wearing the black shirt in identical conditions however a rating of 1.5 (half way between slightly warm and warm) was provided. Hence for that participant wearing the black shirt was warmer than wearing the white shirt. It can be seen that participants, on average, were between slightly warm and warm. When wearing the white shirt they were towards slightly warm and when wearing the black shirt, they were towards warm. All participants, with the exception of participant D, were warmer in the black shirt.

The results for experiment two are presented in Table 2. It can be seen that, although only four participants were used, there is a clear indication that participants were hotter when wearing tight shirts and that they were also hotter when wearing a black shirt when compared with wearing a white shirt.

## Discussion

The results from experiment one show that on average, participants were 0.43 scale units warmer when wearing a black shirt than when wearing a white shirt. This implies that , for the levels of radiation investigated ( a "reasonably" sunny day) colour of clothing will have an effect on thermal comfort if people are exposed to solar radiation. Seven of the eight participants were warmer in black shirts and as stated, the difference in the average rating was up to half a scale unit. This margin and level of consistency implies that colour of clothing should be taken into account when assessing thermal comfort in solar radiation.

Table 2 presents the results of only four participants so general conclusions must be made with caution. However there seems to be a trend that indicates that tight fitting clothing is warmer than loose fitting clothing and that black clothing is warmer than white clothing. Surprisingly there is only a very small difference between white and black clothing when tight fitting. Clearly more participants are required to determine whether the results are significant.

## Conclusion

The results of the two experiments show evidence to support the hypotheses that for identical conditions, people will be warmer in the sun when wearing black clothing than when wearing white clothing. They will also be warmer when wearing tight clothing than when wearing loose clothing. The magnitude of the effects are such that they should be taken into account when considering thermal comfort assessment methods for conditions involving solar radiation.

## References

Beeson, E.J.G. 1978, The CSI lamp as a source of radiation for solar simulation, *Lighting Research and Technology*, 10, No 3, 164–166

Fanger, P.O. 1970, Thermal Comfort, (Danish Technical Press, Copenhagen)

ISO Standard 7730 1994, Moderate Thermal Environments – Determination of the PMV and PPD indices and specification of the conditions for thermal comfort

Parsons K.C. 2003, *Human Thermal Environments*, Second Edition (Taylor and Francis, London) ISBN 0-415-23793-9

# DISCOMFORT CAUSED BY SITTING NEXT TO A COLD WINDOW: SIMULATED RAILWAY CARRIAGE AT NIGHT

**Paul Underwood & Ken Parsons**

*Human Thermal Environments Laboratory, Loughborough University, Loughborough, Leicestershire LE11 3TU, UK*

This paper presents the results of an investigation into the thermal comfort of subjects seated "side on" to a vertical cold window, in an otherwise thermally neutral environment. Objective and Subjective measures were taken from eight male subjects whilst they were exposed to a cold window with a surface temperature of 5 ± 1°C The experiment was carried out in a purpose built thermal chamber in which all other environmental conditions were constant. Exposure to the window was found to cause a significant decrease in mean skin temperature, as well as significant radiant asymmetry across participants' bodies. Thermal sensation, preference, discomfort and perception of draught all altered significantly as a result of exposure. Prediction of three models (PMV, Draught Rating and Radiant Asymmetry) were compared with the actual percentage of dissatisfied participants (APD). PMV was found to be the most accurate predictive model for these conditions.

## Introduction

This paper investigates the effects of sitting "side-on" next to a cold window on thermal comfort. The results are applicable to the design of office environments and in particular vehicle environments, particularly at night where outside conditions cause low indoor temperatures. Cold windows can cause discomfort due to a loss of thermal radiation from the body as well as draughts caused by cool falling air. The extent of this discomfort was not known and is established in this paper.

The Predicted Mean Vote (PMV) thermal comfort index (Fanger, 1970; ISO 7730, 1994) considers radiation in the form of mean radiant temperature, but may not be able to incorporate discomfort from the radiant asymmetry caused by a cold window. Olesen (1985; ISO 7730, 1994) investigated radiant asymmetry (RA) caused by vertical cold walls and described a relationship between radiant temperature asymmetry and thermal discomfort. Both PMV and Radiant Asymmetry predict discomfort in the form of a percentage of people dissatisfied with the environment (PPD and PD respectively) and both are used in this investigation and compared with the actual percentage of participants dissatisfied (APD). A draught rating index (DR; ISO 7730) is also compared with the APD and used to investigate the effects of convection currents caused by the cold window.

## Method

### *Experimental method*

Eight healthy male participants were seated in a Fiat Punto car seat "side on" to a cold vertical window (1 m × 1 m × 3.5 mm clear monolithic glass), the internal surface

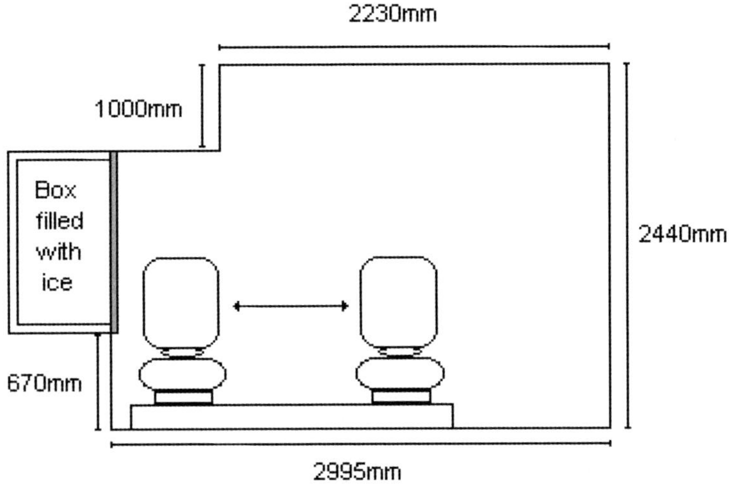


**Figure 1. A schematic of the test chamber.**

temperature of which was kept at a temperature of 5 ± 1°C. The seat, mounted on runners, could be moved next to, and away from the window. The window was cooled by encasing the external surface with an insulated box that was filled with ice. An electric fan was placed inside the box to circulate air (Figure 1). The internal environment of the chamber was controlled to be thermally neutral without exposure to the cold window. A cardboard sheet was used to cover the internal surface of the window, before and after participants were exposed to it. The subjects were dressed in a white cotton/polyester (65%/35%) shirt with the sleeves rolled up to the elbow, beige cotton/polyester (65%/35%) trousers and their own underwear and shoes. The total estimated insulation of the clothing ensemble and seat was 0.72 clo. The estimated metabolic rate of participants was 70 $Wm^{-2}$. Participants were aged between 18 and 23 years.

*Experimental measures*

Air temperature, black globe temperature, relative humidity, air velocity and wall and window temperatures around the participant and the chamber were monitored and recorded throughout the experiment.

The participants' skin temperature was measured using thermistors positioned at eight points around the body, right and left upper chest, right and left upper arm, right and left anterior thigh and right and left shin.

Subjective measurements were taken using a questionnaire that was completed by participants every five minutes throughout the experiment. These questionnaires allowed participants to rate their thermal sensation, thermal comfort, draughtiness and thermal preference (Figure 2).

*Experimental procedure*

The subjects arrived thirty minutes before the experiment and were put in a thermally neutral preparation room to ensure that they were thermally neutral before exposure to the cold window. They were fitted with thermistors, using 3M transpore tape and dressed in the standard clothing ensemble. Once they were thermally neutral, deduced by means of a questionnaire, they were taken to the chamber and sat in the seat. At this

| | | | | | |
|---|---|---|---|---|---|
| **1** | Slightly warm | **3** | Much warmer | **4** | Very uncomfortable/draughty |
| **0** | Neutral | **2** | Warmer | **3** | Uncomfortable/draughty |
| **-1** | Slightly cool | **1** | Slightly warmer | **2** | Slightly uncomfortable/draughty |
| **-2** | Cool | **0** | No change | **1** | Not uncomfortable/draughty |
| **-3** | Cold | **-1** | Slightly cooler | | |
| **-4** | Very cold | **-2** | Cooler | | |
| **-5** | Extremely cold | **-3** | Much cooler | | |

**Figure 2. Example of the modified subjective scales used in the experiment.**

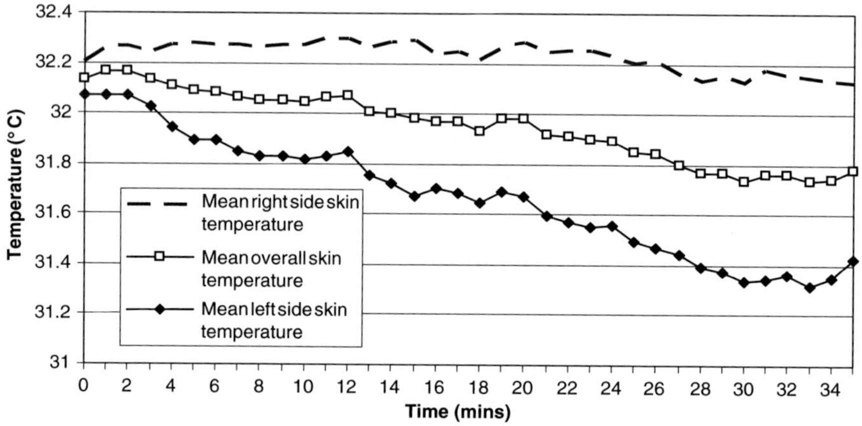


**Figure 3. A graph comparing left side, right side and overall mean skin temperatures for eight participants during the experiment.**

point the seat was positioned away from the window and the cardboard sheet shielded participants from exposure. They were asked to complete another questionnaire to ensure that they were still thermally neutral. The cardboard sheet was then removed and the chair moved next to the window. Questionnaires were completed at this point, and then every five minutes for thirty minutes. The chair was then moved away from the window and the cardboard sheet replaced before a final questionnaire was completed.

## Results

### *Objective results*

Mean skin temperatures were calculated for each participants left and right sides and for their whole body. A Ramanathan (1964) method was used to weight data from the four thermistors on either side, and the mean of these two calculations was used as a measure of whole body mean skin temperature. Figure 3 shows comparisons between left sided, right sided and whole body skin temperatures for the eight participants during the experiment. Using a standard t-test it was found that after thirty minutes exposure to the window, there was a significant difference between left and right sided skin temperatures, $P = 0.01 < 0.05$. There was a temperature gradient of up to 1.5°C across the body which was stated by many participants to be a cause of discomfort. Using another t-test, it was also found that there was a significant decrease in mean

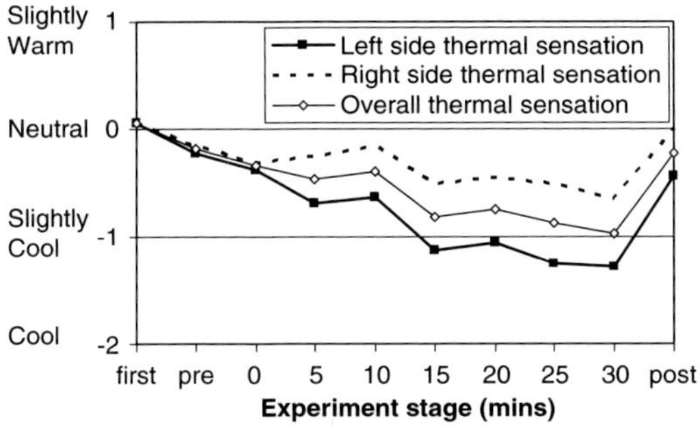


**Figure 4. A graph comparing left side, right side and overall sensation vote.**

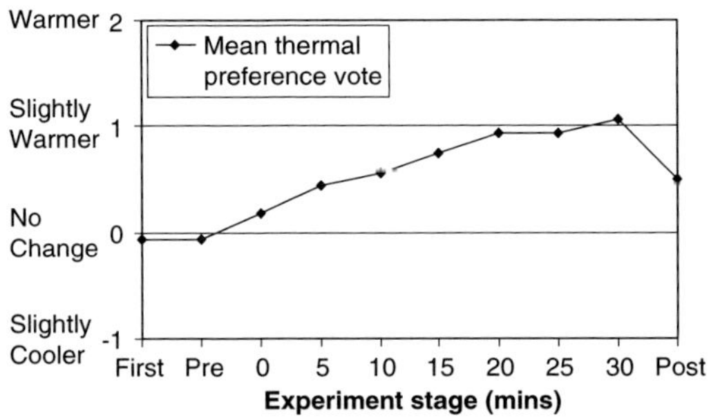


**Figure 5. A graph showing thermal preference vote.**

whole-body skin temperature between the beginning and end points of exposure, $P = 0.001 < 0.05$.

*Subjective results*

Figure 4 shows a comparison of the left side, right side and overall sensation vote of the eight participants. There was a total decrease in overall sensation vote of 0.8 on the scale, from −0.18 (around neutral) to −0.98 (slightly cool). Using a wilcoxon signed rank test there was also found to be a significant difference of 0.62 on the sensation scale between the mean end vote for the left and right sides, $P = 0.018 < 0.05$.

Figure 5 shows the mean thermal preference vote of the eight participants during the experiment. It can be seen that there was a difference of approximately 1 point on the preference scale, from 0 (No Change) to 1 (Slightly Warmer), between participants' mean vote at the start and end points of exposure. Table 1 shows environmental data from the chamber, measured prior to, and at the end point of exposure. Calculated from this information and also shown in Table 1 are PMV, PPD, and percentage dissatisfied given by "draught rating" and "radiant asymmetry". The actual percentage dissatisfied obtained from questionnaires is also given.

## Discussion

The results show that if a person sits next to a cold window then a sensation of up to one scale unit cooler can be expected when compared with conditions without the cold

**Table 1. Environmental data and calculations for predictive models.**

| | "Pre"/control condition | | | At "30 mins" exposure | | |
|---|---|---|---|---|---|---|
| Predicted mean vote | tr = 21.2°C<br>ta = 21.91°C<br>v = 0.15 $ms^{-1}$<br>Relative Humidity = 41.49% | PMV = −0.7<br>PPD = 14% | APD = 0% | tr = 20.15 °C<br>ta = 21.91°C<br>v = 0.31 $ms^{-1}$<br>Relative Humidity = 50.26% | PMV = −1<br>PPD = 25.6% | APD = 25% |
| Draught rating | ta = 21.9°C<br>v = 0.15 $ms^{-1}$<br>Tu = 40% | DR = 15.6% | APD = 0% | ta = 21.9°C<br>v = 0.31 $ms^{-1}$<br>Tu = 40% | DR = 40.6% | APD = 25% |
| Radiant asymmetry | Δtpr < 6°C | PD = 1% | APD = 0% | Δtpr = 15.9 °C | PD = 40% | APD = 25% |

APD = Actual Percentage Dissatisfied, PMV = Predicted Mean Vote, PPD = Predicted Percentage Dissatisfied, PD = Percentage Dissatisfied, DR = Draught Rating (% dissatisfied due to draught), Δtpr = difference between two plane radiant temperatures (the window and the rear wall of the chamber), Tu = local air turbulence (assumed 40%).
*Note* – Any other unmentioned abbreviations have been discussed and defined earlier in this report.

window. An inside window temperature of around 5°C can be regarded as being towards an extreme condition and shielding of the window or increased air temperature may compensate for this effect.

Table 1 shows a comparison of the percentage dissatisfied (PPD, DR and PD) as predicted by three different models, and the actual percentage dissatisfied (APD) obtained from questionnaires. Out of the three models PMV gave the most accurate prediction of the percentage dissatisfied after thirty minutes of exposure, and is therefore the most able to incorporate the effects of the cold window in it's prediction.

## Conclusion

It has been established that side on exposure to a cold window in an otherwise neutral environment causes a significant reduction in skin temperature and significant local and whole body discomfort. It has also been found that Predicted Mean Vote (PMV) is a reasonable predictive model when analysing environments with these conditions.

## References

Fanger, P.O. 1970, *Thermal Comfort.* Copenhagen: Danish Technical Press

ISO 7730, 1994, Moderate thermal environments – Determination of the PMV and PPD indices and specification of the conditions for thermal comfort, Geneva: International Standards Organisation

Olesen, B.W. 1985, Local Thermal Discomfort, in Bruel and Kjaer, *Technical Review No. 1*, Copenhagen

Ramanathan, N.L. 1964, A new weighting system for mean surface temperature of the human body, *Journal of Applied Physiology*, **19**, 531–533

# ERGONOMICS 4 SCHOOLS

# ANALYSING COSTS AND BENEFITS IN THE CONTEXT OF SCHOOL EQUIPMENT AND FURNITURE

**Michael A. Tainsh**

*Centre for Human Sciences, QinetiQ, Farnborough, Hants, UK*

The assessment of costs and benefits for school equipment and furniture in its classroom context, is addressed. This is important as questions are often asked but there are no well-documented approaches, whereas approaches have been developed for large-scale engineering projects. The starting point for this examination is International Standards Organisation 15288 for Systems Engineering. The problem was considered within a set of virtual conferences which generated a set of issues to be taken into account. A high level summary of requirements was generated and a process for moving on to a cost-benefit analysis described.

## Introduction

One important question asked of the "Ergonomics4Schools" Special Interest Group (e4sSIG) has addressed investment costs and benefits of school equipment and furniture. The reply given has generally been weak, and hence it was decided to investigate the subject further as it is clearly a most important issue for many associated with education, and involves substantial sums of central and local governments and private resources.

Typically in the past, ergonomists have addressed requirements, design issues and acceptance, but not taken the additional step of including an analysis of costs with benefits. That is to say, they appear to have concentrated on the technical aspects (including usability, health and safety), while there has been little work in the systematic consideration of costs and benefits. In some areas this is changing as increased emphasis is placed on costs and the handling of acquisition projects is governed by international standards.

The International Standards Organisation (ISO) has produced standard 15288 in the context of large-scale engineering projects and, while intended for general application, has yet to be examined for classroom applications. In a classroom, many of the "elements of the system" are human, rather than equipment, and the goals of the system are likely to be expressed and assessed in terms of individual or small group attributes rather than large scale operations, or equipment characteristics.

Members of the e4sSIG[1] examined the concepts associated with ISO 15288 in the course of an international virtual conference.

## Definitions

It is important to understand the framework described within ISO 15288 and the definitions:

(a) The International Standard concerns those systems that are engineered and may be configured with one or more of the following: hardware, software, Users,

[1]Those involved in the Virtual Conference were Andree Woodcock (UK), Cheryl Bennett (USA), Maria Pineda and Massimo Manzoni (Colombia) and Brenda Buchanan (New Zealand).

processes, procedures, facilities and naturally occurring entities such as water and minerals.

(b) The International Standard applies to the full life cycle of systems including the stages: concept, development, production, utilisation, support and retirement of systems.
(c) The International Standard applies to organisations in their role as both acquirers and suppliers.
(d) The life cycle processes have four components: agreement (contracts), enterprise (management), project management, and technical.

The analysis of costs and benefits is treated here as a technical issue for both acquirers and suppliers. It is most likely to be carried out early in the life cycle i.e. concept or development although it may be developed in subsequent stages. However, the results of the analysis clearly have implications for agreements and enterprise management.

## User and system characteristics

The starting point for any purchase of equipment is to establish a goal for the overall system of elements. The elements include both users and equipment. The goal will drive both User and Technical requirements: the former refer only to issues of use while the latter take into account additional issues associated with, for example, construction.

The consideration of ergonomics/human factors is based on the six HFI domains (Sandom and Harvey, 2004): manpower, personnel, training, human engineering, health hazards, and system safety.

In the case of school furniture, we may consider goals such as: learning outcomes including exam results, health and safety of users, comfort of users, security of users, aesthetic appeal. These goals must be stated explicitly because it will be necessary to compare options against their contribution to achieving the goals and their relative costs and effectiveness.

## The virtual conference and its contribution

Members of the e4s SIG were invited to attend. A "strawman" paper was circulated and participants examined the approach within the context of this application.

The following issues were raised in the course of discussion by the members of the SIG. They are presented in no particular order.

(a) The approach to stating requirements must be inter-disciplinary and involve equipment designers, architects, teaching staff, providers of resources, manufacturers etc.
(b) An approach to requirements based on the six HFI domains appears useful.
(c) It is helpful to consider the use requirements of equipment and furniture.
(d) It is necessary to distinguish between users and stakeholders, and decide how various groups should be categorised.
(e) Schools budgets and the various relevant budget holders need to be identified.
(f) There may be conflicts between the requirements of the users and the stakeholders.
(g) There may be safety implications of the poor equipment design e.g. bad seating may lead to back-pain.
(h) Current purchasing practice may depend solely on current budget available.
(i) The importance of teaching style and learning process was high. (This subject was explored in the context of traditional styles i.e. "seated" children and other styles such as within the "Montessori" approach where more movement may occur.

**Table 1. High level summary of user requirements for classroom, furniture and equipment.**

| Domain-based list of requirements | Ergonomics/HFI contribution | Acceptance of Requirement. |
|---|---|---|
| Classroom manpower – teachers | Teaching styles | To specified roles and standards |
| Classroom manpower – assistants | Teaching styles | To specified roles and standards |
| Student numbers | Teaching styles | To specified roles and standards |
| School manpower – support and maintenance | Maintainability standards | To be assessed through use study |
| Personnel – teachers, assistants, students, maintenance staff | Size characteristics | To include specification of anthropometric standards |
| Personnel – teachers, assistants, students, maintenance staff | Strength characteristics | This may include manual handling requirements |
| Personnel – teachers, assistants, students, maintenance staff | Experience of role | To be assessed through use study |
| Personnel – teachers, assistants, students, maintenance staff | Training in teaching style | To be assessed through use study |
| Training – teachers, assistants, students, maintenance staff | Preparation in use | To be assessed through use study |
| Training – teachers, assistants, students, maintenance staff | On-the-job training/experience | To be assessed through use study |
| Human engineering to match all users and maintenance staff appropriately | Environmental conditions to meet legislative standards | This will include consideration of heat, light and acoustic conditions |
| Human engineering to match all users and maintenance staff appropriately | Security of users, maintainers and their property | To be assessed through use study |
| Human engineering to match all users and maintenance staff appropriately | Size and shape of equipment or Furniture to appropriate standards | This will include consideration of Anthropometric standards |
| Human engineering to match all users and maintenance staff appropriately | To meet standards of comfort | May be based on prior experience or ability of adjust equipment/furniture |
| Health | Environmental conditions | To be met in line with legislation |
| Health | Use of hazardous substances | To be met in line with legislation |
| Safety | Accident and incident control | To be met based on prior reporting |
| Safety | Training to meet legislation and local requirements | To be assessed through use study |

(j) There may be conflicts between the budget holders and the user-based requirements i.e. cost against benefits.
(k) The multi-role nature of classroom activities was emphasised. Children and staff may be carrying out a number of tasks simultaneously with implications for many in the design/purchasing process.
(l) Specific important tasks such as working with computers may need special attention.
(m) The problems of Repetitive Strain Injury (RSI) should be considered.
(n) Costs will need to be justified to purchasing authorities and budget holders, and this approach may help.
(o) The implications of multi-national standards need to be considered by designers.
(p) Knowledge of likely costs may help to save later expenditure.
(q) There is a need to take account of standards, whether formulated by architects, accountants (economists) or others.

These issues above are in line with ISO 15288 but help in understanding its application. (Other important issues were raised in the conference, and can be obtained from the e4sSIG site.)

## Ergonomics/HFI requirements

Following on from a consideration of these issues, it is possible to develop a high-level summary of User requirements for classroom, furniture and equipment as in Table 1.

Table 1 includes in the left hand column a high level reference to the domain and its associated requirements. Then there is a reference to the ergonomics areas, and their potential contribution, and finally a reference to acceptance and the means of meeting the requirement. Table 1 is not intended to be definitive but rather a starting point for further consideration. If a hierarchical breakdown was applied, then one might have a detailed set of tables employing hundreds of statements.

The User requirement will lead to System Requirements which can be considered to be the technical/engineering statement.

## Cost-benefit analysis

There is a need to consider the expression of costs and benefits so that comparisons can be made between options. The approach is shown in Diagram 1 where the design features for each option considered for meeting the requirement, are listed, with an assessment of their benefit.

Obviously, there will be more than one option and there will be many requirements and design features. The Overall Cost for each option is then calculated and this may

| | Design 1 | Assessment of Benefit, against Acceptance Criterion | Cost information |
|---|---|---|---|
| Requirement 1 | Feature 1 | | |
| Requirement 2 | Feature 2 | | |
| Requirement 3 | Feature 3 | | |
| Statement of Overall Costs and Benefits for Option 1 | | | |

**Diagram 1. Representation of design features for an option.**

include Initial Costs (ICs) and Through Life Costs (TLCs) which become particularly important when manpower and personnel domains are involved. The benefits are typically scaled around the acceptance criteria.

## Conclusions

One option is selected dependent on agreed rules. Commonly, the selected option needs to meet all requirements (i.e. to assure benefit) and be the least cost. Other rules may be applied to the selection of options. For example, benefits may be prioritised. It is now common within UK Government acquisitions to place emphasis on Through Life Costs, hence issues associated with manpower and personnel with their high costs have been raised in priority. Similarly, legislation covering health and safety has raised the priority of those domains.

It would appear that ISO 15288 may be applied, but with qualification. The emphasis on the goal of learning, and its implications, is the main difference between the requirements in Table 1, and that which might have emerged from an equipment-centred point of view. The issues raised in the e4sSIG Virtual Conference are important and Table 1 shows that they can be usefully applied.

## References

ISO 15288 *Systems engineering – System life cycle processes.*

Sandom, C. and Harvey, R. 2004 *Human Factors for Engineers*. MPG Books, Published by IEE, London.

# STUDENT RESPONSES TO SCHOOLBAG CARRIAGE: EVIDENCE FOR A RECOMMENDED UPPER SCHOOLBAG WEIGHT

**Hamish Mackie[1] & Stephen Legg[2]**

[1]*School of Sport, Unitec New Zealand, Mt Albert Auckland*
[2]*Centre for Ergonomics, Occupational Safety and Health, Department of Human Resource Management, College of Business, Massey University, Palmerston North New Zealand*

As part of a series of studies addressing schoolbag carriage, two studies were carried out with the objective of providing evidence for a recommended upper schoolbag weight. Sixteen boys were firstly exposed to a simulated school day, while carrying schoolbags of differing loads and shoulder strap adjustments. Student's posture changed proportionately with load. The magnitude of self reported measures of muscular strain and musculoskeletal discomfort (MSD) suggested that 15% BW may be excessive while 10% BW may be acceptable to school students. In the second study, a psychophysical approach demonstrated that student's mean (SD) maximum acceptable schoolbag weight was 10.2(3.8)% BW. The results of these studies support current schoolbag weight recommendations. However, the results of other studies suggest that other schoolbag factors such as design, adjustment, carriage duration and mode also need to be considered when developing schoolbag carriage guidelines.

## Introduction

Despite widespread concern regarding the negative health effects of heavy schoolbags, there is limited guidance for school students, parents and education professionals regarding safe schoolbag carriage. Although guidelines for schoolbag carriage do exist (NBPA 1997) there is limited evidence based information. An exception is a guidance webpage that has been published by the Government of South Australia (Government of South Australia, 2004), which is largely based on peer reviewed research (Grimmer et al. 1999, Grimmer and Williams 2000, Grimmer et al. 2002). Part of the South Australian guidelines recommend that carrying a backpack should not significantly alter young people's posture from the side and front view, backpacks should be worn over two shoulders and that the backpack weight should not exceed 10% of the student's BW. Although there is sufficient evidence to suggest that the mode of carriage can affect student's posture (Maholtra and Sen Gupta 1965, Pascoe et al. 1997) there is not yet any conclusive evidence to support an upper recommended schoolbag weight of 10% BW.

In an attempt to provide objective support to schoolbag carriage recommendations, some studies have examined student's physiological (Hong et al. 2000) postural (Hong and Cheung 2003, Grimmer et al. 2002) and gait (Pascoe et al. 1997) responses to schoolbag carriage. Significant changes in physiological (Hong et al. 2000), postural (Hong and Cheung 2003) and gait (Pascoe et al. 1997) measures have been reported between unloaded and schoolbag carrying conditions of 15–20% BW. In some cases, these findings have been the basis for recommending 10% BW as a schoolbag carriage limit. However, no studies have conclusively demonstrated that 10% BW represents an appropriate upper weight limit for schoolbag carriage.

The first review of studies that contribute to recommended weight limits for schoolbag carriage (Brackley and Stevenson 2004) has recently been published. A schoolbag weight limit of 10–15% BW is recommended by the authors based on the findings of current literature. However, the findings of the physiological and biomechanical studies that contribute to this recommendation support weight limits of 10–20% BW. This range is broad and further work is required to define a more precise schoolbag weight limit.

Five studies were carried out to examine student responses to schoolbag carriage. Specific objectives were to determine the influences of schoolbag design, schoolbag carriage adjustment, the temporal patterns of schoolbag carriage and schoolbag weight on student's responses to schoolbag carriage. The main objective of the final two studies was to provide evidence to support a recommended upper schoolbag weight.

## Study 1. Postural and subjective responses to realistic schoolbag carriage

Never have school student's postural and self-reported responses to different schoolbag weights and adjustments after realistic schoolbag carriage been studied. The objective of this study was to determine school student's postural and self-reported responses to five schoolbag weight and two shoulder strap adjustment conditions, before, during and after a simulated school day. A second objective was to determine an upper weight limit for daily schoolbag carriage.

Sixteen boys [mean (standard deviation) height 161.8(11.5) cm and body weight (BW) 53.1(16.1) kg] aged 13–14 years completed the data collection sessions. They were exposed to unloaded, 5%, 10%, 10% with tight shoulder straps, 12.5% and 15% BW schoolbag carriage conditions, each for the duration of a simulated school day. Postural and self reported measures were gathered at the beginning, middle and end of each day.

Posture, RPE and muscular strain and ability to walk and balance were significantly affected when student's schoolbag load reached 10% BW. Carrying 10% BW was sufficient load to induce a significant ($p < 0.001$) change in horizontal ear-ankle displacement [mean 4.9(Standard deviation 2.9 cm)] from the unloaded condition (figure 1). A significant increase ($p = 0.009$) in RPE from the unloaded condition [7.2(2.1)] was reached [10.5(3.7)] when 10% BW was carried, representing an RPE rating of "fairly light" (figure 2). In addition, a significant increase ($p = 0.004$) in reported muscular strain and ability to walk and balance scores (1–5 scale) from the unloaded condition [1.2(0.3)] was reached [2.0(0.8)] when 10% BW was carried. Despite statistical findings, the magnitude of self reported measures of muscular strain and musculoskeletal discomfort (MSD) suggests that 15% BW may be excessive while 10% BW may be an appropriate upper schoolbag weight limit for a typical school day. Posture, RPE and muscular strain and ability to walk and balance were not significantly affected by the duration of carriage over a school day for loads of up to 15% BW or by shoulder strap adjustment.

## Study 5. A psychophysical approach to determining a maximum acceptable schoolbag weight

The objective of this study was to determine a maximum acceptable schoolbag weight using a psychophysical approach. Snook's (1978) psychophysical methodology was modified to allow for the adjustment and evaluation of schoolbag weight. Sixteen boys [mean(Standard deviation) height 160.8(11.9) cm and weight 52.1(16.5) kg] aged

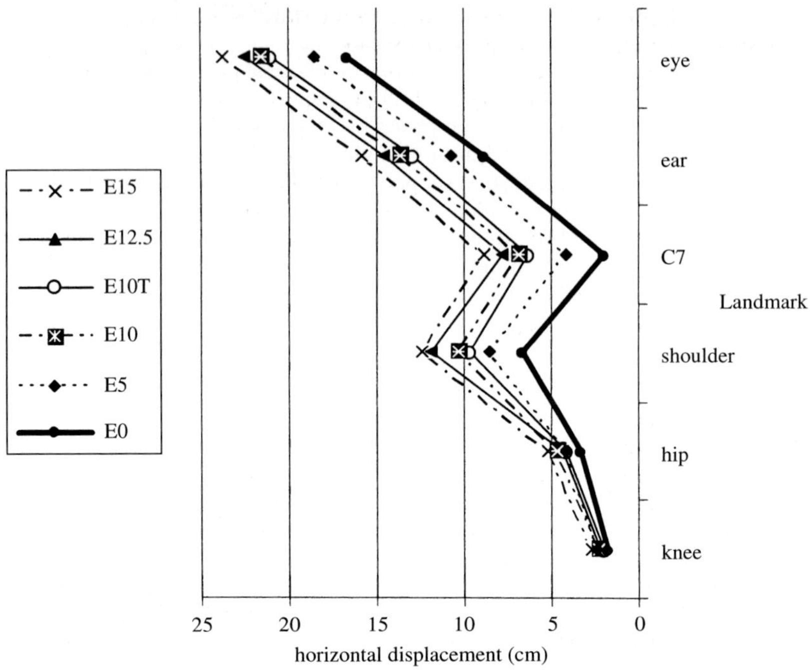


**Figure 1. Relative mean horizontal displacement (cm) of landmarks to the ankle joint (lateral malleolus) for different schoolbag carriage conditions at the end of the simulated school day.**

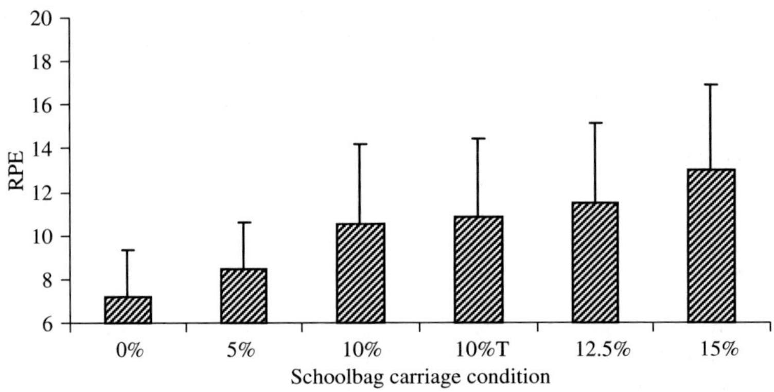


**Figure 2. Mean (Standard deviation) RPE scores for different schoolbag carriage conditions (%BW). 10%T represents 10% BW with tight shoulder straps.**

13–14 years simultaneously completed a 20-minute data collection session whereby they repeatedly adjusted and evaluated their schoolbag weight by adding or removing sand from plastic containers until they reached their "maximum acceptable" schoolbag weight. Participants were divided into two groups of eight. In the first group participant's plastic containers started empty ("start empty"). In the second group participant's plastic

**Table 1. Mean (SD) "start low" and "start high" maximum acceptable schoolbag weight, difference between "start low" and "start high" weights, average of "start low" and "start high" weights and average weight as a percentage of bodyweight.**

| | Start low (kg) | Start high (kg) | Mean absolute difference (kg) | Average | % of bodyweight |
|---|---|---|---|---|---|
| Mean (SD) | 5.0 (1.6) | 5.2 (1.9) | 0.9 (1.2) | 5.1 (1.6) | **10.2 (3.8)** |

containers started filled to the top with approximately 12 kg of sand ("start full"). Prior to data collection, the first group was instructed to begin by adding sand to their containers and the second group was asked to begin by removing sand from their containers. Participants then walked around a set course while wearing their schoolbags. The process was repeated two more times until a final schoolbag weights were achieved. Ten minutes after the end of data collection, the process was repeated, but with the first group starting with full containers and the second group starting with empty containers.

The overall mean maximum acceptable schoolbag weight (as a percentage of BW) was 10.4(3.8) kg (table 1) which is very similar to the commonly recommended schoolbag weight limit of 10% BW and supports objective results the previous study. Based on this psychophysical approach, 10% BW is recommended as a maximum acceptable schoolbag weight, although other schoolbag, school and personal factors also need to be considered.

## Conclusion

Based on postural, self-reported and psychophysical measures, a recommended upper schoolbag weight of 10% BW is supported. However, in addition to a recommended upper schoolbag weight, guidelines for schoolbag carriage should include recommendations on schoolbag design, adjustment and mode of carriage. Schoolbag carriage guidelines may also address the duration of schoolbag carriage, although more research is required in this area.

## References

Brackley, H.M. and Stevenson, J.M. 2004, Are children's backpack weight limits enough? A critical review of the relevant literature, *Spine*, **29**(19), 2184–2190.

Government of South Australia. 2004, Spinal health for South Australian Students, http://www.dete.sa.gov.au/docs/files/communities/docman/1/Spinal_healt_1.pdf

Grimmer, K.A., Dansie, B., Milanese, S., Pirunsan, U. and Trott, P. 2002, Adolescent standing postural response to backpack loads: A randomised controlled experimental study, *BMC Musculoskeletal Disorders*, **3**(10), http://www.biomedcentral.com / 1471-2474/3/10

Grimmer, K.A., Williams, M.T. and Gill, T.K. 1999, The associations between adolescent head-on neck psture, backpack weight, and athropometric features, *Spine*, **24**(21), 2262–2267.

Grimmer, K.A. and Williams, M. 2000, Gender-age environmental associates of adolescent low back pain, *Applied Ergonomics*, **31**, 343–360.

Hong, Y. and Cheung, C.K. 2003, Gait and posture responses to backpack load during level walking in children, *Gait and Posture*, **17**, 28–33.

Hong, Y., li, J.X., Wong, A.S.K. and Robinson, P.D. 2000, Effects of load carriage on heart rate, blod pressure and energy expenditure in children, *Ergonomics*, **43**(6), 717–727.

Malhoutra, M. and Sen Gupta, J. 1965, Carrying of School Bags by Children, *Ergonomics*, **8**, 55–61.

National Back Pain Association. 1997, NBPA school bag survey 97 – findings and recommendations. *Talkback*. Teddington.

Pascoe, D., Pascoe, D.E. and Wang, Y.T. 1997, Influence of carrying book bags on gait cycle and posture of youths, *Ergonomics,* **40**(6), 631–641.

Snook, S.H. 1978, The design of manual handling tasks, *Ergonomics*, **21**, 963–985.

# HOSPITAL ERGONOMICS

# ASSESSMENT OF THE USE OF CARRY CHAIRS BY UK AMBULANCE STAFF

**Jeremy Ferreira Leanne Stanley**

*Health and Safety Laboratory, Harpur Hill, Buxton, Derbyshire SK17 9JN*

The emergency carry chair is an important manual handling aid, used by UK ambulance services as the primary method for transporting patients on stairs. 8 trained ambulance personnel transported a simulated patient up and down a staircase and around a landing using a traditional carry chair and a more recently designed chair that supports the weight of the patient and chair on the stairs. Perceived exertion ratings, biomechanical modelling and interviews were used to assess the tasks. During stair descent, the chair that supported the patient on the stairs provided one promising measure for reducing the physical demands of the task and the risk of injury. During stair ascent, the benefits were found to be smaller and the task remained physically demanding. Chair versatility, particularly in confined spaces, was found to be a primary driver for wide-scale implementation in the field.

## Introduction

The emergency carry chair is an important manual handling aid, used by UK ambulance services as the primary method for transporting patients up and down stairs and into the ambulance (Birtles and Boocock, 2003). With a light, compact design, carry chairs are particularly convenient for transporting patients along narrow stairwells and corridors that are typical of UK residential properties. Lavendre et al. (2000) found that the relative risk of low back disorder was reduced when using carry chairs to transport patients down stairs compared to other patient transport devises such as stretchers and backboards. Nonetheless, the Ambulance Services Working Group of the UK Health Service Advisory Committee still identified the use of carry chairs as one of the three main tasks linked to musculoskeletal injury causation. Within the past few years, new designs in carry chair have emerged that claim to offer significant benefits by removing the need to support the full weight of the patient and the chair (Butt et al., 2002). However, a lack of wide-scale acceptance and implementation within the ambulance services seems to suggest that they incur other unforeseen risks.

The aims of this study were to: (1) determine the musculoskeletal loads that ambulance personnel are exposed to when transporting a seated patient up and down stairs; and (2) investigate how carry chair design and workplace factors impact on the safety of handling procedures performed by ambulance personnel.

## Methods

The experimental methodology was approved by the Health and Safety Executive's (HSE) Research Ethics Committee. Participants had not reported any incidences of back pain within the 6 months prior to the study or previous musculoskeletal injury that they felt would affect their ability to perform the tasks safely.

### *Carry chairs*

To undertake this study, the cooperation of employers and employees of Westcountry Ambulance Service NHS Trust was sought. Two models of carry chair were investigated in the study. Chair Model 1 (13.8 kg) incorporated wheel-based technology, a tension belt and adjustable brake, to support the load of the patient and chair on the stairs. This removed much of the need for lifting. Chair Model 2 (9.5 kg) was considered to represent one of the most common types of chair design in regular use throughout the UK ambulance service. This involved the "traditional" method of lifting and carrying a seated patient up or down the stairs. Participants from this ambulance Trust were previously trained in the operation of both chair models and also possessed experience using these chairs in the field.

### *The task*

Participants transported a 70.6 kg simulated patient mannequin up and down a staircase and around a 180° stair landing. The staircase had a width of 80 cm and consisted of 9 steps, each with a step rise of 15.5 cm and a step going of 28.0 cm. The pitch of the staircase was 29°. A 180° landing, with a width of 80 cm, was marked out on the floor to simulate a narrow staircase landing. Ambulance personnel were requested to perform the tasks within this boundary wherever possible.

Participants transported the patient in teams of two, each acting as the head-end operator (HEO), holding the top handles of the chair close to the patient's head, and as the foot-end operator (FEO), holding the bottom handles of the chair close to the patient's feet. All participants were male, with a mean age of 38.5 years, and represented the upper quartile of the British 18–64 year old population in both stature and weight.

### *Analysis*

A biomechanical modelling approach was used (3D Static Strength Predictor Programme, The University of Michigan, 1993) to estimate the physical loads acting upon the human body during these tasks. The models were scaled using participant anthropometric data. Postures were obtained using two digital video cameras, positioned to provide an orthogonal view of the tasks. To determine the force exerted during lifting and carrying operations, participants lifted and held Chair Model 2 while standing on one of two force platforms (Kistler Instrumente AG). To determine the pushing and pulling force exerted when using Model 2, tri-axial force transducers (MC3A, Advanced Mechanical Technologies Inc) were inserted into the chair frame.

One-way analysis of variance (ANOVA) was performed to examine the effects of chair (Chair Model 1 and Chair Model 2) and operator position (HEO and FEO). Post-hoc analysis of means was carried out using the Tukey HSD test.

## Results

### *Stair descent*

When transporting the patient down stairs with Chair Model 1, the mean force exerted was found to be 16% of the total weight of the load at the HEO position and 15% of the total weight of the load at the FEO position. This was significantly lower than the mean force required to carry the patient seated on Chair Model 2, which was found to be 55% and 45% of the total weight of the load at the HEO and FEO position respectively ($p < 0.05$). Participants reported lower RPE when using Model 1 compared to Model 2.

The biomechanical analysis showed that, on the stairs, L5/S1 compression forces were reduced by 46% at the HEO position when using Chair Model 1 compared to Chair

**Table 1. Mean (and standard deviation) of results for transport down stairs.**

| Chair model | Position | RPE | Mean force exertion (N) | L5/S1 compression force on stairs (N) | Peak L5/S1 compression force (N) |
|---|---|---|---|---|---|
| 1 | HEO | 9.8 (2.4) | 134 (40) | 1122 (482) | 1673 (621) |
| | FEO | 8.4 (2.4) | 127 (36) | 2134 (476) | 4686 (594) |
| 2 | HEO | 11.8 (2.1) | 428 (31) | 2083 (526) | 3049 (475) |
| | FEO | 11.1 (2.1) | 355 (29) | 2035 (321) | 4129 (639) |

**Table 2. Mean (and standard deviation) of results for transport up stairs.**

| Chair model | Position | RPE | Mean force exertion (N) | L5/S1 compression force on stairs (N) | Peak L5/S1 compression force (N) |
|---|---|---|---|---|---|
| 1 | HEO | 11.6 (1.3) | 339 (35) | 1713 (633) | 2567 (934) |
| | FEO | 10.8 (2.2) | 173 (21) | 3291 (470) | 4512 (503) |
| 2 | HEO | 11.8 (2.4) | 428 (31) | 2415 (512) | 3292 (932) |
| | FEO | 11.5 (2.5) | 355 (29) | 1944 (569) | 5818 (1083) |

Model 2 ($p < 0.05$). There was also a 45% reduction in peak L5/S1 compression forces at the HEO position of Chair Model 1 ($p < 0.05$). This appeared to reflect the reduction in force exertion. A similar reduction in L5/S1 compression force was not found at the FEO position. Chair Model 1 required FEO operators to bend forwards to apply a horizontal force at the low foot-end handles and ensure that the tension belt remained in contact with the edge of the step. In this case, L5/S1 compression forces arose primarily from the large mass and moment arm of the operator's trunk.

Where a 70 kg seated patient was carried down stairs with Chair Model 2, the biomechanical analysis predicted that only 80% of males and 20% of females would have sufficient static arm strength to hold the chair handles at waist height and close to the body. Where there is discrepancy between calculated joint moments and population capabilities, an increased risk of musculoskeletal injury is predicted. In comparison, more than 95% of males and females were predicted to have sufficient arm strength to transport the patient down the stairs with Chair Model 1.

### *Stair ascent*

During stair ascent, the benefits of supporting the weight of the patient on the stairs were found to be much smaller. With Chair Model 1, the mean force exerted was found to be 41% of the total weight of the load at the HEO position and 21% of the total weight of the load at the FEO position. This was significantly lower than the static force required to carry the seated patient ($p < 0.05$). However, the operator at the HEO position was found to exert frequent peak forces (averaging 542 N) as the chair, supported on the tension belt, was pulled up one step at a time. This may explain why operators did not report a reduction in perceived exertion when using Chair Model 1 to ascend stairs.

On the stairs, use of Chair Model 1 was not found to reduce the L5/S1 compression forces at the HEO position significantly ($p > 0.05$). At the FEO position, use of Chair Model 1 actually increased L5/S1 compression forces by 70% ($p < 0.05$) on the stairs. By eliminating the lifting component of the task, peak L5/S1 compression forces were reduced ($p < 0.05$). However, the relative importance of this peak compressive force in comparison to the cumulative load on the stairs is unknown.

*Interviews*

Interviews with ambulance personnel revealed that 7 of the 8 participants were found to prefer Model 1 when transporting the patient down a straight set of stairs, such as that provided in the study. With heavy patients or long staircases, a chair that supports the weight of the patient on the stairs would allow operators periods to rest or treat the patient, as the FEO could hold the chair stationary on the stairs with fairly little effort. It was also reported to reduce the unpredictable physical stress that might arise if a patient becomes uncooperative, unbalanced, or in anyway shifts the centre of gravity of the load.

However, participants constantly stressed the importance of equipment versatility and mentioned several risk factors in the field that restricted the widespread use of Model 1. Such risk factors included: spiral staircases, where the step going at the centre of the spiral is too narrow for the chair wheel to rest; stepped staircase landings where there is insufficient space to turn the chair; and narrow landings where there is insufficient space for the FEO to crouch on the landing and assist the chair down the final few steps. In this study, participants at the FEO position were unable to operate Chair Model 1 within the imposed 80 cm landing boundaries.

## Discussion

When transporting patients down stairs, chairs that support the weight of the patient and the chair on the stairwell provide one promising measure for reducing the physical demands of the task and the risk of injury. This may be particularly important where the operator has insufficient arm strength to hold the load at waist height and close to the body. In the event of insufficient arm strength, it is plausible that the operator's elbows could extend and the height of the handles lowered to about hip height. Low handles could obstruct the legs when walking down steps, place increased physical stress on the operator's shoulders and back, compel the operator to adopt a stooped upper body posture and shift the operator's centre of gravity closer to the edge of the stairs. When transporting patients up stairs, the benefits of supporting a patient and chair on the stairs was much smaller and the task remained physically demanding.

The small number of participants limited the extent to which these findings may be generalised, particularly to female ambulance personnel in the UK. However, it appears that neither of the two chairs involved in this study offered a single, optimal solution for reducing the risk of musculoskeletal injury in the full range of circumstances where seated patients may need to be transported. Selection of the most effective piece of equipment for patient transport will depend upon the specific characteristics of the patient, ambulance workers, and the work environment. Ambulance personnel stressed a preference to supply vehicles with both chairs, to complement their training in personal risk assessment and manual handling technique. One ambulance Trust has implemented a successful trial of this solution. However, this measure is unpopular with many ambulance Trusts that report a lack of space within the rear of UK ambulances.

There is an urgent need to improve the designs of chairs that offer mechanical assistance in the transportation of patients up and down stairs. In particular, improving the versatility of more recently developed chairs is thought to be a primary driver for greater implementation across the ambulance sector. Several attempts to attach stair-based handling aid technology to traditional carry chair designs appear to have compromised the primary functional requirements, which made the carry chair appealing to UK ambulance services initially. There now appears to be a need for chairs that can negotiate narrow landings, and chairs that can either support the weight of the patient on the stairs, or in the presence of certain environmental risk factors such as confined spaces or uneven terrain, serve as a light, carrying alternative. To meet this challenge, some manufacturers are working more closely now with some ambulance services on future chair designs. This is driven from a greater awareness that risk management and health and safety in design is a significant factor in purchasing decisions of this nature.

In the meantime, further research is required to support the development, implementation and assessment of safer systems of work that are presently available to ambulance Trusts to reduce the risk of musculoskeletal injury during patient carrying on stairs. Such systems of work may include operational protocols to manage the transport of seated patients up and down stairs, and the deployment of specialised vehicles and equipment to respond to situations that involve particularly high-risk handling.

## Acknowledgements

This work was funded by the Health and Safety Executive (HSE). The authors wish to acknowledge those HSE colleagues and members of Westcountry Ambulance Service NHS Trust who supported the study.

## References

Birtles, M. and Boocock, M.G. 2003, *Evaluation of the manual handling operations of UK ambulance staff, with particular emphasis on the risks associated with the use of carry chairs*. HSL Internal Report ERG/03/18.

Butt, S.E., Fredericks, T.K., Choi, S.D. and Kumar, A.R. 2002, Comparison of commercial stairchairs using data envelopment analysis. *Proceedings of the XVI Annual International Occupational Ergonomics and Safety Conference*, 10–13 June 2002, Toronto, Canada.

Lavender, S.A., Conrad, K.M., Reichelt, P.A. and Johnson, P.W. 2000, Biomechanical analysis of paramedics simulating frequently performed strenuous work tasks. *Applied Ergonomics*, **21**, 167–177.

The University of Michigan, Centre for Ergonomics. 2001, *3D Static Strength Prediction Programme*, Version 4.3. The Regents of the University of Michigan, U.S.A.

# LIMITING THE DISTRACTING EFFECTS OF INTERRUPTIONS IN NURSING CARE

**Kathleen A. Harder[1] & John R. Bloomfield[1,2]**

[1]*University of Minnesota, Minneapolis, Minnesota, USA*
[2]*University of Derby, Derby, UK*

Interruptions are pervasive in medical settings. The objective of this study was to develop recommendations for dealing with them. We focused on interruptions experienced by the nursing staff at a large Midwestern hospital. We observed workflow in two nursing stations with different layouts at various times of day, and conducted focus groups with nursing personnel. We identified several sources of the interruptions including: other people; auditory messages; shortages of equipment; software flaws; inconvenient location of supplies; layout of the nursing stations; work processes; and staffing levels. Our recommendations were aimed at limiting interruptions, reducing distracting effects, improving the working conditions of medical practitioners, and reducing the possibility of error.

## Introduction

The Harvard Medical Practice Study, conducted in the early 1990s, investigated the outcome of more than 30,000 patient cases in 51 randomly selected hospitals in New York State. The results of the study indicated that:

- Adverse events occurred in 3.7 percent of the cases (Brennan, Leape et al., 1991)
- 13.6 percent of the adverse events led to the death of the patient (Leape, Brennan et al.. 1991)

Extrapolation from these results suggests there may be as many as 98,000 deaths each year in the United States that are due to medical error. These findings spurred attempts to reduce medical error and improve patient safety. An important part of this effort has involved applying Human Factors to the investigation of health care systems.

As part of this effort we are conducting a series of Medical Human Factors studies. First, we investigated heparin administration processes at a large Midwestern hospital. In the first quarter following implementation of the recommendations, there was an 11.4% reduction in heparin errors that resulted in increased monitoring or harm to patients on the cardiovascular nursing stations. In the subsequent quarter (4Q2003), there was a 37.8% reduction from pre-implementation baseline data (Harder, Bloomfield et al., in press). Next we investigated procedures used before, during, and after cataract surgery at a second Midwestern medical facility – again there was a reduction in medical errors after our recommendations were implemented (Harder & Bloomfield, 2004).

The study reported here – which is the third in the series – focused on the effect of interruptions on workflow in several nursing stations at a large Midwestern Hospital. Our objective was to develop recommendations aimed at reducing the number distractions and to limit their distracting effects.

## Method

We conducted a Human Factors Review of medical practice at a large Midwestern Hospital (with 622 beds). We observed the workflow at two nursing stations that have different layouts – one circular (Station 35), the other with a pod structure (Station 41). The observations were made at several different times during the day – including busy times when shift changes occur (7:30 am to 9:00 am) and when patients are typically discharged (2:00 pm to 4:00 pm), as well as less busy times during the morning (9:30 am to 11:00 am) and evening (8:00 pm to 9:00 pm). Subsequently, we conducted focus groups with nursing personnel from both nursing stations and with personnel from three other nursing stations (Stations 30, 44, and 53). Our analysis of the information gathered from these activities is presented in the sections that follow.

## Results

Our analysis indicated that the interruptions experienced by the nursing staff have many sources. We categorized these various sources and then developed recommendations for reducing the number of interruptions and/or limiting their distracting effects. Unfortunately, in some categories there do not appear to be ways of mitigating the effects of interruptions. The categorization of the sources of interruptions and recommendations to deal with them, where possible, are discussed below.

*People*: Nurses may be interrupted by the many people, including the following: (1) patients; (2) the family and friends of the patients in the station where the nurses work; (3) the family and friends of the patients staying in other stations; (4) physicians; and (5) other medical staff (including nursing assistants, other nurses, student nurses, new-employee nurses, transport personnel, social workers, dietary support, physical therapy, chaplains, etc.). It is difficult to limit interruptions caused by other people without severely limiting the overall effectiveness of the hospital.

*Auditory messages*: Nurses may be interrupted by: (1) telephones; (2) alarms; (3) announcements made on the public address system. While these sources of interruption cannot be eliminated, the effect of interruptions produced by telephone calls could be reduced by providing more mobile phones for the medical personnel. Nurses who already use mobile phones indicated that they were very convenient. A major advantage for nurses is that they only receive messages intended for them. Because of this dedicated use, other nurses are not distracted by phone calls that are not directed to them. Mobile phones also help to empower the patients. From the point of view of the nurses, this is good when the patient needs urgent medical attention, but may be problematic if the patient has a trivial request. Unfortunately, the mobile phones that are currently used by some nurses are bulky.

We *recommended* that smaller mobile phones be acquired and given to all nurses in the nursing stations. At shift changes, the phones would be transferred from the outgoing nurse to the oncoming nurse.

*Shortage of equipment*: Nurses reported that there was a shortage of the following: (1) pulse oximeters and thermometers; (2) computers; and (3) printers.

Because there are occasions when pulse oximeters and thermometers are being used by others, sometimes nurses must wait before they can use them – this shortage interrupts workflow and causes time delays.

Similarly, because there are too few computers, the nurses may have to wait to access a computer – causing an interruption in their workflow.

Because of the lack of printers, nurses may have to walk a long distance to retrieve print-outs and, sometimes when they get to the printer, their print out may be lost. One nurse told us that she always prints two copies of everything to help to ensure that at least one copy will be at the printer when she arrives to pick it up. In addition, on at least one nursing station (Station 30) lab reports are printed at 4:00 am – the printing takes approximately 45 minutes. This exclusive use of the printer makes it impossible for nurses to print out urgent work-related documents (e.g., heparin protocols) during that time frame.

We made the following *recommendations*: (1) there should be a pulse oximeter and thermometer in each patient's room; (2) more computers should be provided for the nurses; and (3) more printers should be provided – particularly on Station 30.

*Software Programming Flaws*: When the computer-user interface for the administration of heparin is being used, we learned that after the nurse clicks the print button, the information on the monitor disappears. Sometimes, this creates a major problem when the printer is blocked for an extended period of time by another print job – because the nurse can neither access the heparin information on the monitor nor see a print-out.

We *recommended* that: (1) the software used in the heparin administration process should be reprogrammed so that the heparin information remains on the screen after the print job has been ordered; and (2) more printers should be provided.

*Inconvenient location of supplies*: The inconvenient location of supplies (cup dispensers, laundry supplies) can interrupt the workflow of the nurses. We learned that in at least one nursing station the patient rooms do not have cup dispensers. Workflow is disrupted when a nurse must take the time to locate a cup somewhere else in the station when patients request water.

Laundry supplies are not stored in patient rooms. When it is necessary for a nurse to change a patient's linens there would be a smaller disruption to efficient workflow if laundry supplies were stored in patient rooms.

We *recommended* that: (1) cup dispensers should be installed in the patient's bathrooms; and (2) laundry supplies should be stored in each patient's room.

*Layout of the nursing station*: The layout of the nursing stations impacts workflow. The circular layout facilitates communication between the medical personnel. However, it should be recognized that one person's communication can be another's interruption – we observed medical staff talking loudly across the space to each other. Also, at times the circular layout can be chaotic – we observed occasions when the central desk space and terminal area were so crowded with physicians that there was no room for nurses to work.

In contrast, with the pod layout the activity is decentralized activity and less chaotic. However, the pod fingers are so narrow that patient transport is difficult when nurses or physicians are seated at workstations outside the patient rooms. And patient transportation is more difficult if there is equipment parked in the pod finger. With the pod layout, nurses are able to see patients easily, but there are drawbacks. For example, on one nursing station (Station 30) there are no alcoves in which medications can be accessed for administration. Currently, the Pyxis is in a busy hallway. The nurses would like to have a dedicated medication room to prepare medications.

It is difficult to deal with layout issues, and as discussed above there are advantages and disadvantages with both layouts that we observed. We *recommended* that a dedicated medication room be planned for future pod layouts.

*Specific space issue*: During reconstruction of the hospital's parking ramp, one nursing station (Station 30) on the hospital's second floor has become a surrogate reception area for the entire hospital. Routinely, the medical staff on Station 30 are interrupted and asked to give directions to other areas in the hospital.

We *recommended* that, during the reconstruction of the parking lot, a large easily visible sign should be positioned so that visitors to the hospital will be informed that the hospital's reception area is on the first floor. Another alternative – the addition of a receptionist to Station 30 – would compound space issues in an already crowded station. [We include this because it is important to anticipate the unintended consequences of system changes.]

*Work process issues*: We learned that sometimes patients are scheduled to be admitted or discharged during shift changes. Workload is high during the period surrounding shift changes. The scheduling of admissions and discharges during this time can result in delays in patient care. Nurses may have to wait until well past the end of the shift to discharge their patients.

We *recommended* that, if possible, patient admissions and discharges should not be scheduled during shift changes. We also *recommended* that a variation of a procedure currently used on one nursing station (Station 30) could be considered for other nursing stations. On Station 30 reports are tape recorded by the outgoing staff prior to the arrival of the oncoming staff. When each oncoming nurse arrives, he or she listens to the report of the nurse they are replacing. Then, he or she consults with outgoing staff if any clarification is necessary. We recommended that a variation of the tape recording process be adopted: In this variation a separate verbal computer recording would be made by an outgoing nurse for each of his or her patients. Then when oncoming nurses arrive, they would click on the computerized recording for their assigned patients, in their preferred order. As with the tape recorded process, the nurse would consult with outgoing staff if anything required clarification. We believe this will help to facilitate a more efficient transition between outgoing and oncoming staff at shift changes.

*Limited staffing levels*: Interruptions to patient care can occur because of the limited staff available on some nursing stations (e.g., Station 35)

We *recommended* that, if possible, more nurses and/or nursing assistants be assigned to nursing stations with staffing shortages.

## Summary

In this study, we focused on the distracting effects of interruptions. While interruptions are so pervasive that they cannot be eliminated from medical practice, it may be possible to manage their distracting effects. We conducted a Human Factors Review of medical practice at a large Midwestern Hospital. We observed the workflow on two nursing stations with different layouts – one circular, the other with a pod structure. The observations were made at several different times during the day – including the busy times when shift changes occur and when patients are discharged. Subsequently, we conducted focus groups with nursing personnel from these and other nursing stations. We identified several sources of interruptions, including – (1) people (patients and their families and friends, physicians and other health professionals); (2) auditory messages (from phones, alarms, and public address announcements); (3) shortage of equipment (e.g., pulse oximeters, thermometers, computers, and printers); (4) software programming flaws; (5) inconvenient location of supplies; (6) layout of the nursing stations; (7) work processes; and (8) staffing levels. Our recommendations were aimed at limiting interruptions, reducing distracting effects, improving the working conditions of medical practitioners, and reducing the possibility of error.

## References

Brennan, T.A., Leape, L.L., Laird, N.M., Hebert, L., Localio, A.R., Lawthers, A.G., Newhouse, J.P., Weiler, P.C., and Hiatt, H.H. (1991). Incidence of adverse events and negligence in hospitalized patients. Results of the Harvard Medical Practice Study I. *New England Journal of Medicine*, **324**, 377–384.

Harder, K.A., and Bloomfield, J.R. (2004). Medical error and human factors interventions: Two case studies. In: McCabe, P.T. (editor) *Contemporary Ergonomics 2004*. CRC Press: London, U.K., 341–345.

Harder, K.A., Bloomfield, J.R., Sendelbach, S.E., Rush, P.S., Kirschbaum, M., Sinclair, J.S., Shepherd, S.F., and Burns, D. (In press). Improving the Safety of Heparin Administration by Implementing a Human Factors Process Analysis. In: Henriksen, K., Battles, J.B., Marks, E. and Lewin, D.I. (Editors) *Advances in Patient Safety: From Research to Implementation.* Agency for Healthcare Research and Quality.

Leape, L.L., Brennan, T.A., Laird, N.M., Lawthers, A.G., Localio, A.R., Barnes, B.A., Hebert, L., Newhouse, J.P., Weiler, P.C., and Hiatt, H.H. (1991). The nature of adverse events in hospitalized patients. Results of the Harvard Medical Practice Study II. *New England Journal of Medicine*, **324**, 377–384.

# A COMPARATIVE ANALYSIS OF AMBULANCE STRETCHER LOADING SYSTEMS

**Anna Jones & Sue Hignett**

*Hospital Ergonomics and Patient Safety Unit, Dept. of Human Sciences, Loughborough University, LE11 3TU*

This paper will summarise the initial findings of a comparative analysis of stretcher loading systems used by ambulance services in the UK. Over the last decade radical changes have been made to patient transfer systems to reduce the amount of manual handling in ambulance work. As a result new systems have been developed to assist patient loading and unloading tasks for stretchers, carry chairs and independently mobilising patients. There are now three loading systems widely used across the ambulance services: hydraulic ramp and winch (with air suspension), Easiloader stretcher; and Modular tail lift.

## Introduction

The core service provided by Accident and Emergency crews within the ambulance service is the transportation of patients to hospitals. To assist with the safe transportation of patients, attendants often need stretchers and carry chairs. Due to the nature of ambulance work these are often loaded quickly, with due consideration for patient safety, to try to ensure a comfortable journey. The design and functionality of the equipment is therefore very important.

The design of stretcher loading systems is an area in ambulance design which has seen many advances in recent years (Boocock et al., 2000). Vehicle manufacturers and administrators in the ambulance service are placing a high priority on ambulance design and safety to protect patients and staff (Overton, 2001). Since the publication of the BS 1789 standard (British Standards Institution, 2000), new methods of loading/unloading have been considered which has led to the design and purchase of new types of equipment. This has resulted in the gradual phasing out of systems such as easi-loader stretchers and the introduction of new systems such as the ramp and winch and the tail lift (Figure 1). These systems have received a mixed response from front line accident and emergency staff. Although a number of these systems have been analysed in the past (Boocock et al.,

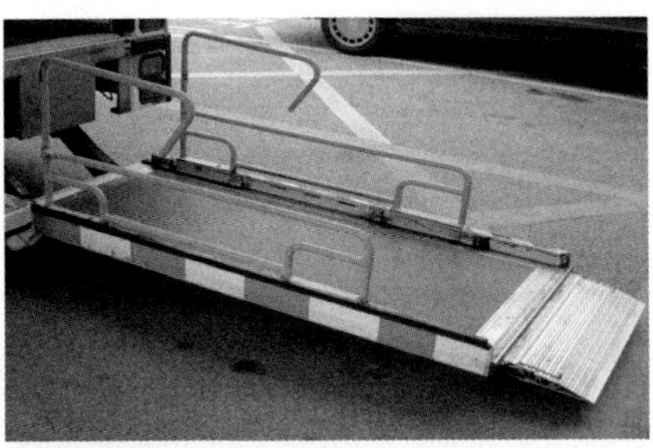

**Figure 1. Ambulance tail lift.**

2002) there has been no comparative evaluation of the three most commonly used systems: Ramp, Tail lift and Easiloader.

This study intends to determine which of the three systems used is the most suitable for use by the ambulance service in the UK. The aim is to generate design guidelines for improved stretcher loading systems to optimise safe systems of work, provide minimum risk of injury, and maximise user comfort.

## Methodology

An initial investigation was carried out to gain an understanding of the ambulance service culture. Background information about the stretcher loading systems and purchasing decisions was collected from each of the three Ambulance NHS trusts collaborating in the study. Interviews were conducted with Fleet Managers, Operation Managers and Ambulance Crews.

Three task analysis techniques were utilized to gain in-depth data for a range of different field environments; Critical Incident Technique (CIT), Link Analysis and Hierarchical Task Analysis (HTA). The CIT used semi-structured interviews to identify safety critical incidents resulting in patient or staff injury. These data were analysed qualitatively using Nvivo (Richards, 1999) and then triangulated with the link analysis and HTA data to produce a taxonomy of design features, key error points and functionality. Link Analysis was used to record and analyse hand movements made between system components whilst the task was being carried out. The task of patient loading and unloading was observed and recorded in a link diagram for each transfer task, system and patient category: e.g. ramp and winch for (1) stretcher, (2) carry chair, and (3) walking patient, resulting in 128 link diagrams. The link diagrams were then summarised by system and patient category to produce one diagram representing the task. The transfer task was also analysed using the Hierarchical Task Analysis (HTA) technique to identify the tasks and sub-tasks for each system. This recorded the number of operations required for each system and assisted with the production of the catalogue of errors (with the CIT data). A final taxonomy is now being developed for each of the stretcher loading systems as part of the detailed comparative analysis.

## Participants

Three NHS Ambulance Trusts were recruited on the study to give a variety of rural and urban environments. Data were collected from nine ambulance stations (three from each Trust). Frontline accident and emergency staff were observed over three shifts at each of the stations (over 300 hours observation). The participants were informed about the project prior to the shift and were given the opportunity to ask questions. Ethical approval was given by MREC (Wales) 04/MRE09/3. Participants were asked to carry out their work in the usual manner to enable the researcher (AJ) to collect the observational data.

## Preliminary results

### *Link analysis*

Link analysis diagrams were drawn for each of the 128 patient loading and unloading tasks observed during phase one of the data collection. Figure 2 shows the results for the task of loading a stretcher using a ramp and winch system. In order to load a stretcher the ramp must be deployed by lowering the air suspension and ramp using the

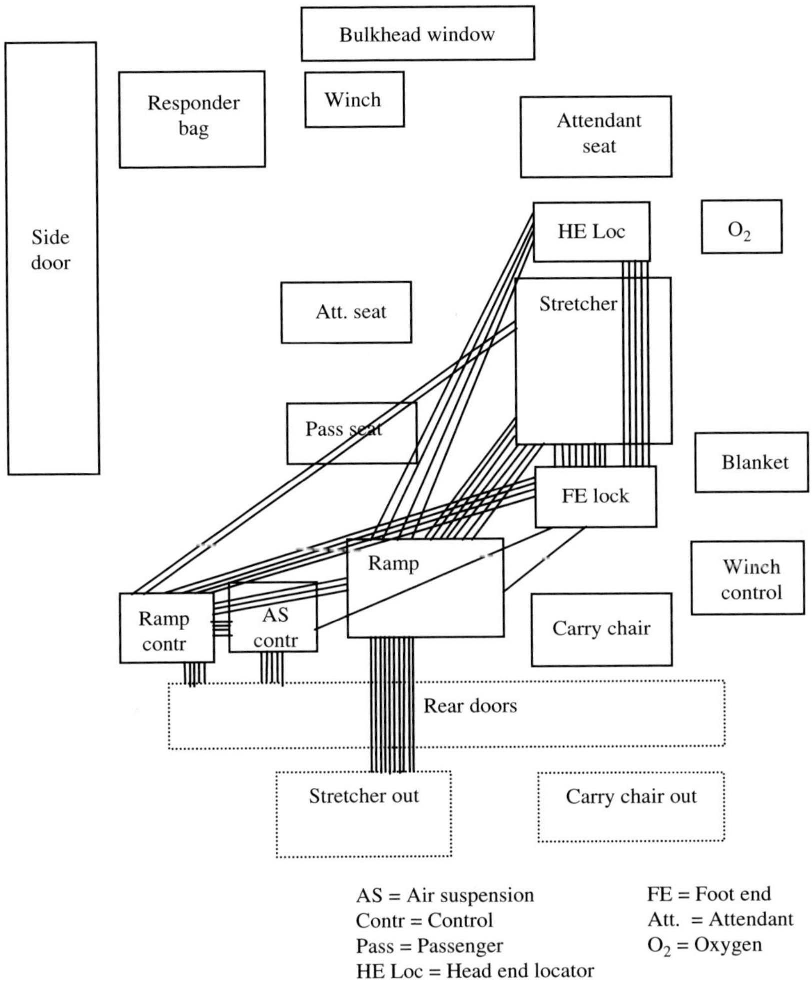


**Figure 2. Link diagram representing stretcher loading using ramp and winch.**

controls inside the patient compartment on the nearside wall. The crew load the stretcher by guiding it up the ramp and into the head end (HE) locator. The trolley is then secured in the foot end (FE) locking mechanism and the ramp is stowed. The links in the diagram show the movement between these components.

From this link analysis is can be seen that the winch was not used during any of the 128 observed tasks. The controls are to the far left of the compartment, the head end locator and foot end lock are in line, to the far right of the vehicle and the ramp is in the centre at the end of the patient compartment. The ramp is operated from the near-side of the rear of the ambulance, resulting in reaching across to the Foot End (FE) lock and stretcher. This might be better located on the off-side wall. The most frequently used components (ramp, controls, foot end lock and head end locator) are all at different angles to each other, requiring the stretcher to be loaded at an inappropriate angle. A better layout might be for the locking mechanism and the ramp to be in line to minimise twisting.

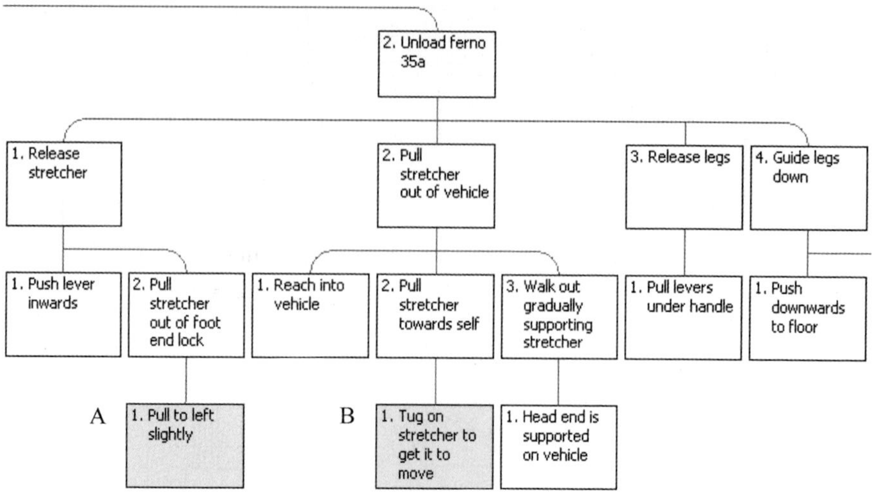


**Figure 3. Hierarchical task analysis of unloading of Easiloader stretcher.**

**Table 1. Triangulation of CIT, HTA and link analysis for the ramp and winch.**

| Node | HTA | CIT | LA | Comments |
|---|---|---|---|---|
| Use of winch | Use force to overcome weight of patient – should use winch | Dislike winch. Too much slack on winch | Winch not used | Observed winch in simulations. Problems of control, timing, and posture |
| Stretcher and control location | Guide stretcher in | Stretcher fixed to side of cabin. Dislike vehicle layout | Operating system at nearside, stretcher offside | Staff work at angle due to position of components |
| Side door access | Open side door | | Use side door more with carry chair and walkers | Carry chair lifted through side door |
| Foot end lock | Kick lever repeatedly, push forcefully on lever | Yellow box in middle of pathway | | Various models differ |

*Hierarchical task analysis*

Figure 3 shows the HTA data for the task of unloading an Easiloader stretcher. To unload, the trolley is released and pulled out at a slight angle away from the locking mechanism and guided out of the patient compartment. The foot end attendant then releases the legs by pulling the levers beneath the handles. At this time the stretcher is supported by the vehicle at the head end and by the attendant at the foot end. The legs are lowered to the ground by the second attendant. When the legs are locked in the stretcher can be wheeled into hospital.

Two of the subtasks (A and B) identified potential risks for the crews. Subtask 1.2.1 (A) "pull slightly to left" requires the attendant to twist together with applying a force, which is likely to increase the postural risk level. A similar issue was identified with subtask 2.2.1 (B) "Tug on stretcher to get it to move".

## Discussion and conclusion

The preliminary analysis triangulated the data from the CIT, link analysis and HTA for the ramp and winch system is shown in table 1.

Other issues included vehicle/equipment used, call criticality, environment and team communication. These issues must be further addressed to enable the development of design guidelines for improved usability.

This paper provided an example of the preliminary data analysis and further analysis is ongoing. The initial results have identified critical scenarios which will be simulated for more detailed data collection, including postural analysis.

## References

Boocock, M.G., Gray, M.I. and Williams, S. 2002, Patient handling in the ambulances services: case study investigations. In McCabe, P., (ed.) *Contemporary Ergonomics 2002*, (Taylor and Francis, London), 33–38.

Boocock, M., Smith, C. and Morris, L. 2000, An ergonomic evaluation of loading operations using the Ferno 35A trolley cot. Health and Safety Laboratory. 1–20.

British Standards Institution. 2000, Medical Vehicles and Their Equipment – Road Ambulances. London: BSI BS EN 1789:1999. ISBN 0 580 35243 9.

Overton, J. 2001, Ambulance Design and Safety, *Journal of Prehospital and Disaster Medicine,* 2001, **16**, (3).

Richards, L. 1999, *Using Nvivo in Qualitative Research.* Melbourne: Qualitative Solutions and Research Pty. ISBN 0-7619-6526-4.

# QUESTIONING THE USE OF NHS ESTATES ERGONOMIC DRAWINGS

**Jun Lu & Sue Hignett**

*Hospital Ergonomics and Patient Safety Unit, Dept. of Human Sciences*
*Loughborough University, Loughborough LEICS, LE11 3TU*

This paper looks at the use of NHS Estates ergonomic drawings, which were developed by the Department of Health in the early 1980s to act as guidance for the design of new hospitals and the adaptation of old buildings. The questions asked include whether the drawings are capable of reflecting changes in clinical practice and medical technology; architects use the ergonomic drawings while designing; and whether there are any problems in the patient room in terms of ergonomic drawings. An overview of the drawings is given in a literature review, pilot study, site visits and expert interviews. The paper also describes the methods used in the project to look at the complex interfaces between the activities of the health care staff and patients (and other stakeholders), the equipment used and the environment in which the care and treatment is provided.

## Introduction

According to the NHS Plan (Department of Health, 2000) creating a safe, efficient, patient-centred environment is the top of the agenda for hospital design team, estates and facilities managers to take steps to satisfy and reflect the requirements of patients and their families in hospitals.

In the early 1980s the Department of Health developed an ergonomic database (Hilliar, 1981; Stanton, 1983) in the form of published Health Building Notes (HBN) to act as guidance for the design of new hospitals and the adaptation of old buildings. The aim of this database was to encourage those involved in hospital design to think in terms of the relationship between a user and a particular component and other components located within a room to produce a more efficient planning of space.

The majority of the drawings were developed over 20 years ago. Although the purpose of producing them could conform to the NHS plan, are they capable of reflecting changes in clinical practice and medical technology, for example day case and keyhole surgery, diagnostic equipment (CAT, NMR), mobile epidural anaesthesia, electric ward beds etc.? Do the people involved in hospital development, e.g. architects, use the drawings while developing programs? Do the drawings work and are there any problems in patient rooms in terms of ergonomic drawings?

The aim of this project is to review and revise ergonomic drawings of single bedrooms, 4-bed bays, and suited-toilet/bathroom in adult acute wards (medicine and surgery) and critical care units in terms of four key clinical activities: manual handling, resuscitation, disability access, and infection control.

## Methodology

Since healthcare is a relatively new area of practice for ergonomics, we propose there are methodologies (specifically task analysis methods) from other safety critical environments

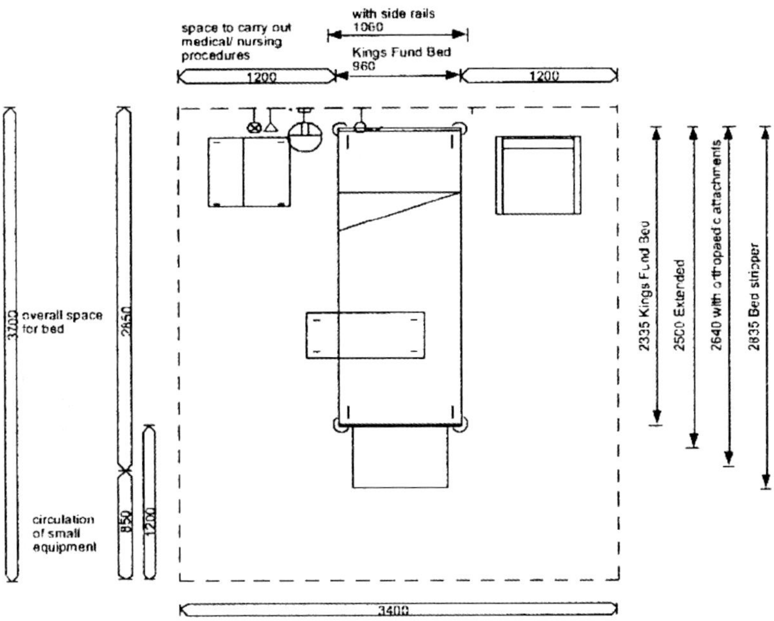


**Figure 1. Core bed space – plan view. (Adapted from HBN 04, NHS Estates, 1997; Reproduced with permission of NHS Estates).**

which have been developed in the last 20 years that could be used for the evaluation of existing ergonomic drawings as well as the development of new ergonomic envelopes. The project phases include:

1. **Literature review** of studies concerned with the development and use of ergonomic drawings in health care in the context of changes in clinical practice over the last 20 years.
2. **Anthropometric benchmarking** to measure the sizes of single bed rooms, 4-bed bays and toilets/bathrooms built in the last 10 years.
3. **Interviews** to elicit the views of architects, clinical staff, project managers, etc about the use of ergonomic drawings and patient rooms.
4. **Observations** in single patient room, 4-bed bay, and toilet/bathroom; and then **simulations** of nursing activities for different layouts to collect data using link analysis, hierarchical task analysis, and critical incident technique.
5. Finally, the **data analysis** will bring together the anthropometric and task analysis data in a comparative review and produce the revised and new ergonomic drawings.

## Findings

The phases of literature review and pilot study are now complete, site visits for measuring the sizes of patient rooms are nearing completion and the interviews with architects, project managers, etc. are on-going.

### *Literature review*

There was found to be little published literature commenting on the configuration of patient's beds and the clinical activities in or around bed space. Most of the literature

**Table 1. Dimensions of bed spaces recommended in different NHS Estates guidance.**

| Publication | Year | Dimension of bed space in single bed room | Dimension of bed space in multi-bed room/bay |
|---|---|---|---|
| HFN 30 – Infection Control in the Built Environment | 2002 | Bed centres at least 3.6 m apart for infection control | |
| HBN 04, Vol. 1 – In-patient Accommodation | 1997 | 3.7 m × 3.4 m | No less than 2.9 m × 2.9 m 3.7 m × 2.7 m (double room – pg. 62) |
| | | 3.7 m × 3.4 m core bed space 3.4 m – minimum acceptable width | |
| HBN 40, Vol. 2 – Treatment Areas | 1995 | 3.7 m – 3.8 m in length 3.1 m – 3.6 m in width | 2.9 m × 2.7 m (called "core bed space") |
| HBN 4 – Adult Acute Wards | 1990 | 3.9 m (3.7 m – restricted minimum) in length 3.3 m (3.1 m – restricted minimum) in width | No less than 2.9 m × 2.5 m |
| HBN 40, Common Activity Spaces Vol. 1 – Example layouts; Common components | 1986 | 3.7 m – 3.8 m in length 2.8 m – 3.6 m in width | 2.9 m × 2.5 m (called "core bed space") |
| HBN 4 – Ward Units | 1961 | 120 – 140 sq. ft (i.e. approx. 3.6 m along the bed axis, 3.3 m in width) | 8 feet – bed centres apart (approx. 2.4 m) |

on hospital space design didn't address the ergonomic factors. The contribution of ergonomics is "both poorly understood and underestimated" (Villeneuve, 2000). From the literature it is difficult to find out whether ergonomic drawings work or if there are problems in patient rooms in terms of using the drawings.

There is little evidence that architects use ergonomic drawings/information well while designing, and have enough knowledge about the process of clinical activities. Held (2003) found that architects made some serious errors in the room layout of the planned ICU owing to lack of detailed knowledge of work procedures in the hospital. On the other hand, HBNs do not explain the "ergonomic" reasons behind the drawings well enough so that architects may ignore or misunderstand them while designing.

In terms of bed space, different documents and publications have different voices. The table one shows the evolution of the recommended dimensions for hospital bed spaces in single bed rooms and multi-bed rooms respectively. It can be seen that NHS Estates have been regularly updating the HBNs. However, very little evidence was found to provide a rationale for the changes in the recommended dimensions.

Additionally, an architect's handbook, *The Metric Handbook* (Adler, 1999) gives the sizes of 3300 mm × 3100 mm for a single room; of 2900 mm × 2500 mm for the bed space in a 6-bed bay, which conflicts with the recommendations in the HBNs and may have led to confusion for architects.

Much of the NHS Estates guidance on building design and planning is dated. For example, HBN 04 Vol.1 – In-patient Accommodation (NHS Estates, 1997) is the most recent guidance on designing hospital in-patient accommodation. However as table one shows the recommended dimensions in HBN 04 do not meet the size of patient bed space recommended in The Health Facilities Note on Infection Control (NHS Estates, 2002). This has already produced controversy with a judicial review into bed space design at University College Hospital (http://www.timesonline.co.uk/article/0,,8122-1362878,00.html).

*Pilot study*

At the beginning of the project, two pilot studies were conducted with experts in manual handling and resuscitation. These provided preliminary results about the dimensions of bed space envelope as 3600 mm in width and 4700 mm along the bed long axis, which is different to the HBN recommendations (Hignett and Keen, 2005).

*Site visits*

Bed spaces have been measured in adult acute wards in five hospitals built or refurbished in the last 10 years. The dimensions of bed spaces in multi-bed bays were found to be between 2600 mm and 2900 mm in length, and 2400 mm and 2660 mm in width. The dimensions of single rooms, excluding the suited sanitary spaces, were found to be between 2900 mm and 4150 mm in length, and 2700 mm and 3680 mm in width. Many of these dimensions are less than the HBN recommendations. These dimensions will be tested during the simulation phase.

## Conclusion

Because of the rapid development in the clinical practice and medical technologies, the hospital building guidance, which once seemed state-of-the-art might become out-of-date very quickly. So there is an urgent need to produce more effective guidance for the design and planning. However, evaluating hospital design should not just involve seeing how well it conforms to the recommendations in the building guidance. The ultimate test should be how well it facilitates clinical staff and patients to perform their activities effectively and safely. This project is empirically reviewing bed spaces and sanitary spaces in recently built hospitals by looking at the clinical activities occurring in them, with the aim of providing up-to-date ergonomic information (drawings) for architects and project managers based on current clinical practice. Preliminary findings indicate a wide variation in the implementation of the HBNs. We suggest that providing robust research evidence to support task-based space allocation will greatly assist with the design of new facilities.

## References

Adler, D., 1999, (ed.) *Metric Handbook. Planning and Design Data,* 2nd Ed. Architecture Press, Oxford.

Department of Health, 2000, *The NHS Plan: a Plan for Investment, a Plan for Reform.* http://www.dh.gov.uk/assetRoot/04/05/57/83/04055783.pdf. Last accessed on 25th Nov. 2004.

Department of Health and the Welsh Office, 1986, *Health Building Note 40, Common Activity Spaces Vol. 1 – Example layouts;* Common components, HMSO, London.
Department of Health and the Welsh Office, 1990, *Health Building Note 4. Adult Acute Wards.* HMSO, London.
Held, J., 2003, Analysis of an Intensive Care Unit. In: *Tagungsband der 49.* Jahrestagung Der Gesellschaft für Arbeitswissenschaft, München, ISBN 3-935089-68-6, S. 673–676.
Hignett, S. and Keen, E. 2005. How much space is needed to operate a mobile and an overhead patient hoist? *Professional Nurse*. (Accepted for publication)
Hilliar, P., 1981, The DHSS Ergonomics Data Bank and the Design of Spaces in Hospitals. *Applied Ergonomics,* 12 (4), 209 – 216.
Ministry of Health, 1961, *Hospital building Note 4. Ward Units.* HMSO, London.
NHS Estates, 1995, *HBN 40 Common Activity Spaces Vol 2: Treatment areas.* The Stationery Office, London.
NHS Estates, 1997, *HBN 04 In-patient Accommodation – Options for Choice.* The Stationery Office, London.
NHS Estates, 2002, *HFN 30 Infection Control in the Built Environment: Design and Planning.* The Stationery Office, London.
Stanton, G., 1983, The Development of Ergonomics Data for Health Building Design Guidance. *Ergonomics,* 26 (8), 375 – 393.
Villeneuve, J., 2000, The Contribution of Ergonomics to the Design of Hospital Architecture. In: *Proceedings of the IEA 2000/HFES 2000 Congress* – a CD-ROM.

# INCLUSIVE DESIGN

# ATTENTION RESPONSIVE TECHNOLOGY AND ERGONOMICS

**Alastair Gale**

*Applied Vision Research Institute, University of Derby, Kingsway House Kingsway, Derby DE22 3HL*

The ART (Attention-Responsive Technology) research project is developing a system to enable mobility-impaired individuals to access technology efficiently. The system monitors both the individual and any ICT devices in his/her environment. It then uses the individual's gaze direction to determine to which ICT device they are potentially attending. This information is relayed to a user-configurable control panel, which then displays only those controls that are appropriate, both to the user and to the particular device in question. The user can then choose to operate the device if s/he wishes. The initial ergonomic challenges in the development of the ART system are described.

## Introduction

Increasingly Information and Communication Technologies (ICT) devices are commonplace, both in the workplace and at home. Whilst these devices offer users many advantages they can pose a problem for some disabled people, who wish either some degree of, or total, autonomy in operating and interacting with them. How best then can individuals who have limited mobility operate such devices?

For some disadvantaged individuals head or breath operated switches can be used. However, various disabilities seriously restrict an individual's mobility but can leave saccadic eye movement control intact. Consequently, this raises the possibility of using such eye movements, which can be voluntarily controlled, to operate both communication or control systems. There are numerous techniques to record saccadic eye movements and various commercial eye movement monitoring systems already exist which have been specifically designed to afford physically impaired individuals the ability to interact with a computer or other devices by using their direction of gaze (e.g. the Eyegaze system). Using such a system the user's gaze direction can be used to operate some selection process, such as "typing" simply by looking at the required keys displayed on an on-screen keyboard, or alternatively the user can select an item which is displayed on-screen in a suitable menu arrangement. More sophisticatedly, Ward and MacKay (2002) describe the Dasher system, based on a language model, which can similarly be operated by an individual's eye movements and where a very reasonable "typing speed" can be attained.

The various issues concerning the development of technologies and standards for such eye controlled communication systems has led recently to the formulation of the COGAIN Network of Excellence (2004) which will run for the next five years.

The major potential drawback to such eye controlled systems is the "always-on" direct linking of the user's gaze direction to some presumed visual attentive selection process (c.f. "what you look at is what you get") and whilst one's direction of gaze is usually associated with the direction of visual attention unfortunately this is not always the case. This is the "Midas touch" problem (Jacob, 1990) where a system which uses the user's point of gaze to activate controls *directly* can be prone to false alarms as the point of gaze is drawn to objects which attract the visual system. In other domains this gives rise to the "look but not see" type of error, for instance in driving or visual inspection scenarios.

Consequently, users generally find that such direct eye movement control systems can be both unreliable and over time somewhat fatiguing to use. A good ergonomic solution to this problem is therefore required.

## Attention responsive technology

The above issues have therefore led to the development of the ART project.

Saccadic eye movements can be both consciously and subconsciously controlled. Therefore the concept behind the current project is to build upon the natural function of the eyes in selecting environmental objects (either accidentally or deliberately), rather than using the eyes actually as a control mechanism to operate such objects.

Land and Hayhoe (2001) well demonstrate that our point of gaze typically precedes action, namely we look at an object before we implement some action concerning it. Consequently, saccadic eye movements can be used as an indicator of the intention of the individual when interacting with his/her environment (Vertegaal, 2002). Limiting the operational function of the eyes in this way offers a distinct advantage. By not using the eyes primarily to control some device should therefore help to overcome the problem of potential false alarms, although this remains to be proved in subsequent experimental trials.

The ART concept is that a user, with very limited physical mobility is located within an environment that contains numerous ICT devices ("objects"). The user may well be able to move about the environment, for instance in a motorized wheelchair. The ICT objects include both domestic and work-related items such as: TV, room lighting, curtains, telephone, air conditioning, door entry system, computer, etc. Whilst some of these objects will be in a fixed spatial location (e.g. window curtains), some will be capable of being moved and repositioned around the room. Such repositioning may be deliberate and occur rarely (e.g. moving the television to a new position) or accidental (e.g. an able-bodied person picking up the telephone and subsequently putting it down in a different location after use) and occur quite often.

Most, if not all, of these objects will need to be controlled by the user and the objective of the project is to accurately determine the user's selection of particular objects for subsequent control. It will do this by monitoring the user's eye gaze vector and if s/he is detected as gazing directly at some object then s/he will be afforded the opportunity to control that object. Therefore control of an object does not necessarily follow immediately and directly by simply looking at it but looking does afford the opportunity for control.

Initial research work is targeting potential end user groups and determining which objects, and object controls are most relevant to them. Additionally this work will ascertain potential user demands from such a system and the outputs from this research phase will then help shape the subsequent technical solutions formulated.

It is necessary then for the ART system to constantly monitor the user's environment to account for potential object, and user, movements. Consequently the environment will be comprehensively digitally imaged and a 3D computer model generated in which each object, its relative size and spatial location, as well as the user location, is accurately located. Each object will be uniquely identified by the imaging system and various techniques to achieve this are being developed and tested.

In order to monitor the user's eye gaze vector two related systems will initially jointly be used. The user's eye movements will be monitored by a head mounted system (ASL 501) which will yield data concerning the rotation of the eye with respect to the user's head-facing direction. Head-in-space position and also head-facing vector information will be separately monitored by using a suitable 3D tracking system. Combining data from both detection systems will therefore give the user's direction of gaze with respect to their physical position within the environment. The computer model will then determine whether the user's gaze falls upon an ICT object in the environment. Eye gaze information will be updated several times a second.

A key usability issue which ART will have to address is the accuracy of using such an eye gaze selection approach. Using ART a user must gaze directly at an object in order to select it – simplistically this would assume that the user is foveally fixating upon the object. However, every eye movement recording technique has associated with it a known theoretical lower limit on its spatial (i.e. how correctly it precisely can indicate the direction of regard of the centre of one's fovea) as well as temporal accuracy. Additionally, the fovea of the eye is of a known discrete size and so it could be assumed that in looking directly at an ICT object then its image could potentially fall anywhere on the fovea. Furthermore the psychological concept of useful field of vision (UFOV – the region of the visual field from which the user can extract information at any one time) is relevant. The size of the UFOV for any user will vary depending upon how crowded the visual field is, as well as various user parameters, such as age.

Consequently the real world accuracy of such an eye gaze based approach has to be determined for each user and this will impact upon the physical positioning of objects and their relationship to each other in the environment. For instance, the system must account for partial or full object occlusion behind other possible ICT objects, which renders a key technical challenge.

## User control

Given an environment crowded with ICT objects then typically a complex menu system of some kind would be needed which would encompass all of the potential objects together with their various levels of control. This could produce either a relatively small display, incorporating a deep menu structure, or else a broad menu which could require a physically large display. Previous work has favoured the use of broad (a menu with fewer levels) rather than a deep menu as being more efficient (Tullis, 1985; Kiger, 1984). With ART an alternative solution to this is possible.

Having selected a potential object then the user requires a suitable means of control. The initial developmental concept is to utilize a GUI on a touch-operated tablet PC with the object selection process triggering the display of only the appropriate controls related to that particular selected object rather than the need for a GUI to present the user with every conceivable ICT device and their associated controls. The user can then decide whether or not to operate the object.

The actual control of an ICT object is not the key research issue in the project and the GUI approach adopted can be replaced with numerous other potential interfaces/controls to suit a particular user. For instance we have previously demonstrated (Wooding et al., 2002) that users with no training can easily employ their eye movements alone to make selections from on-screen alternatives – "eye buttons" (c.f. Sibert and Jacob, 2000). The control interface will be configurable to the individual user, depending upon their physical and cognitive abilities. Where necessary the panel could be replaced by one of several commercially available interaction devices designed for disadvantaged users.

## Current project status

The opinions of potential users have been canvassed and these are being used to direct the project appropriately. A suitable laboratory has been constructed as an environment, which includes a growing range of both domestic and office ICT objects. The digital imaging system and associated computer model are in initial stages of investigation and development as are the GUI control systems. Two eye movement systems, a head mounted and a head free, are under modification for use.

## Conclusions

The ART system arises as a solution to the problem of using gaze based control systems and is designed to overcome the false operation of devices which can come about simply by the user unintentionally looking at the object. The system only provides the user with those controls for the potentially attended-to ICT device whilst overcoming any need for a possibly complex menu control system.

The difference between this system and previous eye movement control systems is that the user's point of gaze does not actually control the object, instead it pre-selects objects for subsequent operation.

## Acknowledgement

This research is supported by the ESRC PACCIT (People At the Centre of Communication and Information Technology) Programme.

## References

COGAIN Communication by Gaze Interaction (2004) EU IST Network of Excellence http://www.cogain.org

Jacob, R.J. 1990, What You Look at is What You Get: Eye Movement-Based Interaction Techniques. In Human Factors in Computing Systems, (CHI '90 Conference Proceedings, ACM Press), 11–18

Kiger, J.I. 1984, The depth/breadth trade off in the design of menu-driven user interfaces. *International Journal of Man-Machine Studies* **20**, 201–213

Land, M.F. and Hayhoe, M.M. 2001, In what ways do eye movements contribute to everyday activities? *Vision Research*, **41**, 3559–3565

Sibert, L.E. and Jacob, R.J.K. 2000, Evaluation of Eye Gaze Interaction. In *Proceedings of ACM CHI 2000 Human Factors in Computing Systems Conference*, (Addison-Wesley/ACM Press) 281–288

Tullis, T.S. 1985, designing a menu-based interface to an operating system. Proceedings of the CHI conference Human factors in Computing Systems (ACM SIGHI, New York) 79–84

Vertegaal, R. 2002, "Designing Attentive Interfaces." In Proceedings of ACM ETRA Symposium on Eye Tracking Research and Applications (ACM Press, New Orleans)

Ward, D.J. and MacKay, D.J.C. Fast hands free writing by gaze direction. *Nature*, August 22 2002

Wooding, D.S., Mugglestone, M.D., Purdy, K.J. and Gale, A.G. 2002, Eye movements of large populations: I. Implementation and performance of an autonomous public eye tracker. *Behavior Research Methods, Instruments, & Computers*, **34**, 509–517

# ABILITYMATCH – BUILDING A DISABILITY MANAGEMENT SYSTEM

**Nadine Geddes & Victoria Haines**

*Ergonomics and Safety Research Institute, Loughborough University, Holywell Building, Holywell Way, Loughborough Leics, LE11 3UZ*

Results from preliminary studies have indicated that locating appropriate information on how to accommodate people with disabilities in the workplace can be a difficult and time consuming task. To help improve this situation research is being conducted to contribute to a comprehensive computer system for managing disabilities in the workplace. This is being done by using an integrated assessment of a person's abilities combined with the requirements of a job to then identify workplace solutions or advice on how to overcome barriers, ultimately working towards the creation of an expert system of reasonable accommodations. This paper outlines the ongoing development of a solutions database for an existing assessment system, AbilityMatch.

## Introduction

Since the employment provisions section of the Disability Discrimination Act (DDA) was introduced in December of 1996, employers have a legal responsibility not to "unreasonably discriminate against existing and potential employees on the basis of their 'disability' ". Additionally employers are charged with a duty to make "reasonable accommodations" in the employment of people with disabilities (AbilityNet Organisation, 2003). This responsibility comes regardless of the extent of an employers training or knowledge in the area of reasonable adjustments and their ignorance can equate legal action. One of the potential issues influencing compliance with this legislation is that employers may not have the education or resources necessary for determining where an individual may require a reasonable accommodation to the workplace or what that accommodation should be. This is an important consideration for any new disability management strategies that are being sought in order to counterbalance the rising proportion of unemployment among the disabled population.

AbilityMatch (formally known as Activity Matching Ability System or AMAS) is a computerised tool designed to assist persons with disabilities enter the labour market by providing information to employees and employers concerning where adaptations may be required to accommodate a person with a disability. This is accomplished by the following three step process:

1. Person Ability Assessment – A list of 78 task related questions that are directed at assessing the abilities and limitations of an individual, including: hearing and communication, vision and perception, posture, general movement, leg mobility, arm mobility, workplace interaction/tolerance, and cognition. All questions are worded in the same style: "If a job involved X (e.g. distinguishing colours, lifting your arms overhead, working in isolation, etc.) could you do that?" For each question that is posed, the individual can answer:
   - "Yes" i.e. they have no difficulty with the task.

- "Yes but…" which allows them to choose from a list of common responses i.e. they can do the task but not for very long, very fast, by themselves etc. Or they can personalize their own answer by selecting "other" and providing an explanation in the notes field.
- "No" i.e. they are completely unable to perform that task.

2. Job Activity Assessment – The questions in this assessment directly correspond to those in the Ability Assessment but are designed to analyse the tasks required for a specific job. Questions are worded "Does the job involve X (e.g. distinguishing colours, lifting your arms overhead, working in isolation, etc.), to which the following answers can be selected:
   - "No requirement"
   - "Some requirement"
   - "Major requirement"
3. Match Report – Selecting "Match Report" produces a document that merges, using an algorithm, the results of the two previous lists, indicating where there may be mismatches between the tasks required to perform the job and the abilities of the individual. The output lists good matches, and where simple or creative resolutions to mismatches are needed.

## Evaluation

Recent field testing of the system found AbilityMatch to be an effective tool for identifying where an individual may encounter a barrier in the workplace (Haines et al., 2003; Brown, 2004). This work has suggested that the system would be enhanced by the development of a solutions database, to assist the individual, their assessor and employer in considering whether a resolution is needed and what that solution might be. This is likely to be job and context specific and have implications for training and reasonable adjustment provision.

Assessment of the feasibility of a solutions database (Geddes, 2003) looked at what the solutions database should contain for it to be most practical for users. A total of 10 participants that worked in the field of disability management (two of whom were also visually impaired) were interviewed, either in person or by telephone, to establish baseline information in three key areas:

1. What are the best resources for obtaining information on adjustments to the workplace?
2. What information should the database contain?
3. What overall improvements could be made to the system of disability management in this country?

Based on many comments from the interviews, it was apparent that even the participants of this study, who were professionals in disability management, found the following difficult:

- Where to start looking for information about a specific impairment.
- How it can affect work tasks or everyday life.
- Where to go for help.
- What can be done to alleviate the barriers.

At present there are some " solutions databases" on the market, however they are mainly based in the United States (SOAR, CAP, ABLEDATA) and therefore contain a large percentage of information (price, availability, funding, support organisations, etc.) that is not relevant to a user in the United Kingdom. The products, strategies and workarounds that are included may be universal but the information on how to go about applying it would need to be adapted specifically for each country. Additionally all the databases that are currently available expect the user to have some basic knowledge of what they are looking for. A typical employer may not have the appropriate background to identify all the potential barriers that an employee with a disability will encounter let alone the ability to make recommendations on how to overcome them.

Services such as Access to Work and AbilityNet assist in the assessment and accommodation of any person with a disability or who has shown need. However little research has been done to determine what percentage of small and medium sized business owners are aware of their existence, how to go about obtaining assistance from them or what assistance they offer. These services aim to have a turn around time of 4–6 weeks which may be too long to be financially feasible for many smaller businesses. Research has shown that 44% of adjustments cost less than £50 (Diversity That Works, 2004) this fact alone could substantiate the reasoning behind building a system that will allow users to quickly identify ways of solving problems that an employee may be experiencing without the need to involve outside organisations.

In the UK there appears to be several excellent resources for people with disabilities however these sources of information or assistance have very little tying them all together. Over the last 10 months several attempts have been made to find various services that can provide advice or equipment for people with disabilities. In most cases the information has not been found, and in some instances it has been found, by accident, long after it was needed. Therefore even though there are several sources available for this type of information it is very difficult to sort through it in order to find what is relevant to the situation. Additionally there are databases that list assistive technology equipment but the user must already have an idea of what they looking for in order to locate the correct information. It is also common to find the information that you want but product information or resource is based on another country.

These interviews identified that there is a strong need to develop a tool to be used in disability management, which will offer alternative assessment and support services for persons with impairments in the workforce. Although the interviewees involved in this study varied greatly in their backgrounds and professions, many of them stated that one of the main problems with keeping or getting persons with disabilities into work is the amount of time that it takes to identify where a reasonable adjustment is needed to accommodate a person with a disability, and once it is identified, how long it takes to institute the adjustment. Some of the participants pointed out that if the integration of persons with disabilities is to extend more fully into the work places of small and medium businesses, who cannot afford to have staff off work or on long training periods, the actual process needs refinement. The process of determining where a person with a disability needs adjustment to their current or potential workplace, along with what the potential adjustment could be, and where or how to go about finding information about that adjustment (including funding, training, access to demonstration equipment, and suppliers) needs to be significantly accelerated. By consolidating all of this information into the proposed solutions database for AbilityMatch, users which could be employers, health professionals, employment counselors, ergonomists, etc. would be able to convey all of this information to most clients within one interview.

One of the major aspects to the construction of the proposed system is the content should be effective for users. Specifically attention should be given to the following details:

- general information about a specific health condition
- how it may affect certain work related tasks
- general advice on the work environment
- specific advice on workplace solutions, where to get them, cost, etc.

- organisations or resources that offer assistance
- accurate and up to date information

## Present research

In order to develop this database further, a number of studies are taking place. These include identifying users of the system (employers, human resource personnel, employment advisors, employees etc), establishing their specific needs (what information they require and how they would use it), identifying the appropriate format for the database (e.g. computer, paper-based, personal service provision), identifying priorities for populating the database (where information is needed most), and ensuring the database is usable (through good interface design, accessibility etc).

The findings from a series of case studies are contributing to this collection of information. To date, a total of 55 assessments using AbilityMatch have been conducted with people who have been unemployed for 6 months or more. Many of these have a disability or health condition and all present some barrier to returning to work. The following table shows the main areas where AbilityMatch has identified a potential barrier to returning to work. Obviously if a job did not involve one of these aspects, the barrier would disappear; however, this highlights the overall scale of limitations that were identified and gives some indication of areas that will warrant attention for workplace solutions.

**Table 1. Common barriers.**

| AbilityMatch Question: If a job involved … could you do that? | Percentage of individuals that indicated a limitation with task |
|---|---|
| driving | 73 |
| working at heights | 56 |
| using a ladder or stepladder | 46 |
| managing other people | 44 |
| seeing objects at a distance | 40 |
| lifting | 40 |
| carrying | 40 |
| working in very hot conditions | 40 |
| working in very cold conditions | 40 |
| using a computer | 40 |
| standing | 36 |
| working around airborne contaminants | 35 |
| using numbers | 34 |
| traveling | 33 |
| leaning over | 31 |
| working in an enclosed space | 31 |
| getting under something low | 29 |
| being exposed to vibration | 29 |
| working with your hands above your head | 27 |
| walking | 27 |
| reaching up high | 25 |
| using a keypad/keyboard | 25 |
| remembering things | 25 |
| doing more than one thing at once | 25 |
| working to deadlines | 25 |

## Future work

The next stage of this research is to develop the user needs assessment and continue to collect data to populate the eventual system. This includes addressing the following issues:

- What are the difficulties encountered by people with a range of disabilities.
- How this can affect their ability to maintain employment.
- Strategies that they currently use to overcome barriers in their day-to-day life, which can be transferred to work situations.
- Various types of assistive technology available to help people with disabilities.
- Organisations that can offer assistance to individuals with disabilities.

The development of this system could eventually be linked to other systems to create a full service disability management system. This would greatly enhance the quality of service provision to employers and people with disabilities, as well as allowing policy makers to be able to track trends in the population. A system such as this could enable front line health care professionals to relay information that is pertinent to an impairment, but that frequently does not reach the person who will be handling the rehabilitation or job adjustments. This system may also include an automatic electronic referral system to reduce the time and effort it takes to book individuals with impairments in for appropriate appointments, as the system could correctly channel it to the best service for the persons needs. This could also decrease the chances that individuals who need assistance will not get access to the services that can help them, because of a lack of knowledge on how to find those services. This should all contribute towards enabling employers to meet their obligations under the DDA as well as ensuring the employment and retention of people with health conditions and disabilities is successful.

## References

Abilitynet Organisation, 12/06/03, 2003-last update, *DDA and implications for employers*. Available: www.abilitynet.org.uk/content/alt_tech/employ/employ.htm (06/08, 2003).

Brown, K., 2004. *Comparative analysis of the activity matching system (AbilityMatch) with traditional ergonomics methodologies via individual case studies.* MSc Thesis, Loughborough University.

Diversity That Works. *Reasonable Adjustments.* Available: http://www.diversity-whatworks.gov.uk/toolkit/2_3.asp (10/11, 2004).

Geddes, N.L., 2003. *A Feasibility Study of a Solutions Database for the Activity Matching Ability System (AMAS).* MSc Thesis, Loughborough University.

Haines, V., Birkin, R., Hitchcock, D., Edwards, N., Duckworth, S. and Hondroudakis, A., 2003. *Matching Ability with Jobs using AMAS.* Final report presented to Government Office South East and European Social Fund.

# WEB ACCESSIBILITY FOR ALL

**Ben Dawson, Bill Wessel, Freyja Lockwood & Caroline Gowing**

*BAE Systems, Human Factors Department, Advanced Technology Centre*
*Sowerby Building, Filton, Bristol BS34 7QW*

This study details the challenges that were encountered in making a large corporate intranet site accessible to disabled users and fully compliant with the Disability Discrimination Act. This report discusses the accessibility problems that were discovered in the original intranet site and also explains how they were eventually solved.

This study also examines the issues of instigating design changes in a very large corporate environment and discusses the challenges that may be encountered when working with disparate multidisciplinary teams.

This study illustrates that with an inclusive and iterative user centred approach it is possible to create a solution that meets the needs of all users within one site, without the need for segregation of disabled users.

## Introduction

The Disability Discrimination Act (DDA) has been in force since 1995. Since then organisations that offer goods and services on the Web already have a legal duty to make their sites accessible. However, according to Bert Massie, Chairman of the Disability Rights Commission (DRC) most websites are inaccessible to many disabled people and fail to satisfy even the most basic standards for accessibility. Despite the obligations created by the DDA, domestic research suggests that compliance, let alone the achievement of best practice on accessibility, has been rare.

Although the act is almost ten years old, it is only recently that attention has focussed on how it applies to websites, intranets and private sites. Part 2 of the DDA states that employers have duties to provide equal access to goods and services and that disabled employees are not treated less favourably as a consequence of their disability.

The UK Disability Rights Commission's policy is to seek improvement through advice and conciliation. However, where the response is inadequate, they have stated that they will 'be vigorous in the use of our enforcement powers; these range from "named-party" Formal Investigations which can lead to sanctions against the owners of inaccessible websites, to the provision of support for test cases being brought by individual disabled people.'

It is clear that these developments could have serious legal implications for almost all British companies. Organisations are now charged with improving the accessibility of their external and internal web sites. So far, formal investigations, such as those by the DRC (2004) have focussed on publicly accessible sites, to which Part 3 of the Disability Discrimination Act 1995 (DDA) applies. It is believed that in the near future the focus will turn to the responsibility of the employer under Part 2 of the act. Beyond this BAE Systems believes that accessibility in the workplace is critically important and that we have a duty of care to our employees. As a result we have carried out a user centred design project to ensure that disabled employees are able to access information and resources as easily as their able bodied colleagues.

The intranet is the main method of information dissemination in many large companies and BAE Systems is no exception in this. Consequently we felt that our intranet was where efforts to improve accessibility would be of greatest benefit to employees.

According to an investigation into web accessibility conducted by the DRC: *"Disabled people must frequently overcome additional obstacles before they can enjoy the full range of information, services, entertainment and social interaction offered by the Web. Nevertheless, the Web has enormous potential for disabled people...with the benefit of assistive technology, potentially tolerant of impairment. Inclusive website design makes it easier to use.... without making a site less attractive to unimpaired users."*

Clearly the accessibility challenges of intranets are linked to how businesses implement and use them. Some businesses use their intranet solely as a means of distributing information from management to their employees, others use it as an information repository and resource. The majority of intranets tend to be a mixture of the two. The BAE Systems intranet certainly falls into this latter camp.

Accessibility and usability are interrelated. If an intranet is not accessible to an individual or group, it clearly cannot be considered usable. As intranets are based on Internet technology they often suffer from the same usability and accessibility issues encountered in the wider Internet world. They also face a few challenges uniquely their own. As the range of disabilities is very wide, so too are the necessary accessibility solutions. However, for a medium so predominately visual, it is those users with visual impairments who often encounter the most difficulties. Visually impaired users can suffer from a whole range of disabilities including: complete blindness, colour blindness, tunnel vision and so on. The most obvious difficulty is not being able to read the text on the page. However issues such as navigation through pages and within the site as a whole are also fundamental.

## Method

### *Expert review*

Firstly an expert review was carried out. This allowed the underlying accessibility issues to be identified. This review served as a "first pass" assessment of the intranet, which also helped to highlight any obvious usability issues that were present. To this end a heuristic review, a review of the site with an automated test and a World Wide Web Consortium (W3C) accessibility audit were conducted.

### *Baseline testing*

Users were asked to complete eight tasks, using only the intranet. They were timed and encouraged to "talk aloud" during the test session so that the evaluators could better understand what they were doing. After each task the users were asked to record whether they had successfully completed the task and whether they were satisfied with the intranet as a means of achieving them. Detailed subjective feedback was collected at the end of the test session.

### *Testing the intranet with disabled users*

This phase of testing involved participants with a range of visual disabilities including macular degeneration, ocular albinism and low vision. Each session took around one hour and comprised three stages, following the same design as for the non-disabled users (Baseline Testing).

### *Iterative prototype testing*

Following the recommendations identified through trials with both disabled and non-disabled users, an iterative prototype redesign of the intranet was developed. Many elements of this design were tested in an ad-hoc fashion on users. Once the designs reached a more mature stage of development a systematic set of user tests was undertaken. Eight participants were used for the prototype design testing, five of whom had normal, or corrected to normal vision and no additional disabilities. Three visually disabled users were also consulted. Participants were asked to complete a reduced set of tasks based on the original baseline testing and give their feedback on the new design.

## Results

### *Expert review*

The expert review identified several accessibility issues. Firstly there was no use of alt text to describe links or pictures. The intranet made use of dynamic drop down menus, which are difficult for screen readers to parse. Html tables were used extensively for layout, which again causes problems for automatic screen readers. These findings were corroborated by an automatic review tool. There were also a number of potential usability problems. The most important involved the dynamic drop down menus. These were designed in such a way that their contents would change dependant upon the Business Unit (*BAE Systems is divided into a number of major Business Groups, each of which is itself divided into a number of Business Units*) the user had currently navigated to. There was also concern about the number of different navigation mechanisms and the quality of the search engine.

### *Baseline testing*

The quantitative questionnaire data revealed areas of concern around consistency, complexity and ease of use, all of which scored poorly on the satisfaction scale. Participants also rated most aspects of site navigation as being below average. This was perhaps evidenced in the variety of different navigation strategies observed during this trial. In terms of qualitative feedback there were a number of important points raised. Users liked the look and feel of the intranet and were keen on its unifying effect over the entire company. They were also impressed at the breadth of information available. However they encountered problems finding information they needed. They often attributed this to an ineffective search engine and to the inconsistencies found in the navigation methods used throughout the intranet.

### *Testing with disabled users*

Compared to the baseline user group, disabled users in general were slightly less satisfied with the intranet. Disabled users were less willing to use the site and they found navigation more difficult than non disabled users. The qualitative data highlighted a number of interesting differences in navigation and usage techniques between the groups. Disabled users tended to make extensive use of memory for coping with navigation and site structure. They also tended to rely far less on visual search than the baseline group, mainly because of the excessive time it takes for a visually impaired person to scan an entire web page.

### *Changes incorporated into iterative design phase*

Based on these 3 stages of review and testing, a number of changes were required, which resulted in a need for the intranet to be redesigned. Changes included; instigating consistent navigation methods, static menus, with far fewer menu items, more extensive use of alt text, higher contrast text, improved support for resizing fonts and an accessible "menu page" for each drop down menu. User feedback on these design changes was positive. An additional number of suggestions were made as a consequence of the changes.

## Discussion

From the results there are a number of accessibility issues exposed that whilst specific to the BAE Systems intranet can also be seen to be relevant to the design of any intranet. In the first instance there are the technical accessibility issues that can be encountered on any web page, be it an intranet or Internet page. Essentially, care must be taken to ensure that the page follows the W3C Accessibility Guidelines or an equivalent set of guidelines. This is a matter of educating the creators of web pages so that they understand the most effective and appropriate ways to employ the guidelines.

The non-technical accessibility challenges are equally as important, but in many ways much harder to control. For example, BAE Systems is a very large company and as such intranet managers and developers are often located on different sites around the country. It is therefore a challenging task to ensure that the importance of accessible and usable design is communicated in a clear and consistent manner to all these individuals.

Another challenge is ensuring that any work involving disabled people is conducted in an ethical and sensitive, but sensible manner. It is, after all, difficult to create an accessible user centred design without talking to disabled users.

It is also important to effectively communicate these accessibility guidelines to intranet developers and to content contributors. When we began this work there were already guidelines in place, but they were rarely used and their impact was limited. This was mainly because the availability of the guidelines had not been widely promoted and as a result their location was known by only a few people. The guidelines themselves were also aimed at an audience with a high level of technical understanding. The majority of contributors to the BAE Systems intranet only do so as a secondary part of their job and they often experienced difficulties understanding the technical language. These difficulties can result in a site design that is inconsistent and varies in quality across the company.

### *Solutions to problems*

The first requirement is for extensive training. Local intranet managers, web site managers and contributors all need to be trained in a consistent manner in web design, accessibility and usability. This can be governed to some extent by providing a well designed and well explained corporate web design template. All intranet contributors should meet regularly, perhaps once or twice a year to be trained and updated on any changes. Once training has been provided, awareness of the availability of design resources and support materials should be promoted. When creating these resources care should be taken to ensure that they fit the needs of the user. This means writing them in plain English, while avoiding, if possible, extensive use of technical jargon. Worked examples and illustrations are especially helpful, as are templates.

It can help to develop relationships with disabled users and their line managers locally. This helps to ensure that their needs are met in future design iterations and often means that participants are happy to take part in future work. This ensures iterative designs keep meeting key users' needs.

It is especially important to reassure local web managers that they will not lose control of content or local ability to instigate changes. They should be helped to understand the need for a generic template and the reasons to follow accessible design principles.

However they should be allowed a degree of flexibility in how pages can be displayed and the type of content that can be included in templates and intranet design iterations. Their support is vital to ensuring that an accessible design is not only accepted, but also maintained over time.

## Conclusion

### *Continuing accessibility challenges*

It is possible to make significant improvements to a large corporate intranet based on expert reviews, audits and heuristic evaluations. These will help meet the basic requirements of W3C guidelines. However, in order to ensure that the intranet is accessible and meets people's needs a user centred iterative design approach is essential. It allows you to understand how people use the intranet, what strategies disabled users have developed to overcome obstacles and how you can best support their use of the site.

*"It is … clear that compliance with the technical guidelines and the use of automated tests are only the first steps towards accessibility: there can be no substitute for involving disabled people themselves in design and testing, and for ensuring that disabled users have the best advice and information available about how to use assistive technology, as well as the access features provided by Web browsers and computer operating systems." Bert Massie Chairman of the Disability Rights Commission*

To ensure that the benefits of a user-centred approach are not diminished over time it is important that training and communication of the design is continued for the life of the intranet.

## References

Disability Discrimination Act, 1995, HMSO

The Web Access and Inclusion for Disabled People. 2004, A Formal Investigation conducted by the Disability Rights Commission. London: TSO

White J., Chisholm W., Caldwell B., Vanderheiden G. 2004, Web Content Accessibility Guidelines 2.0, www.w3.org

# USER REACTIONS TO ANIMATED SIGN LANGUAGE

**Mary Sheard & Jo Kewley**

*RNID, 19–23 Featherstone Street, London EC1Y 8SL*

British sign language (BSL) is the preferred language of 50–70,000 people in the UK. Developments in technology have lead to the possibility of animated or video presentations of BSL being available online. The eSIGN project has developed virtual sign language technology. Animations of signing played through an avatar are potentially an effective and economical way in which organisations can provide signing on their web sites and other screen based displays. User reactions to avatar signing have been studied, in order to gauge the comprehensibility and acceptability of this novel technology. Reactions were generally positive, although users' reports highlighted a number of areas where improvements to the animation and presentation are necessary.

## Introduction

British Sign Language is used by 50–70,000 people in the UK, and for many of these people English is very much their second language. Sign languages have no written form, and BSL has a very different grammar and structure to English. Because of this, many Deaf people find reading and writing English difficult. Media that we might assume are accessible to Deaf people, such as the Internet and subtitled television, depend on English text and can therefore be challenging to sign language users. Until recently BSL could only be used in face-to-face situations. Now, recording and communications technologies using video have now enabled the storage and transmission of sign language, but these technologies are far from ideal for some applications.

The Internet is a vast and dynamic information resource containing information which is, on the whole, available to all. One of the strengths of the Internet is the power that site owners have to update and change information easily and quickly at relatively little cost. If a site was to use videos to offer sign language translation of its content, this would mean the loss of that particular strength. If one detail of a video sequence changes it must be rerecorded, which requires a sign language interpreter (maybe the same individual who made other clips to ensure site uniformity), a studio (again, the same as was used for other clips), video recording equipment, and the means to upload the new clip to the server. The time and expense involved in this process may well result in the signed sections of a site being out of date compared to the text which can be easily edited. In addition, the good quality video files needed for BSL information are large; taking up a lot of server space and being slow to download. Synthetic sign language, produced by animating an avatar, overcomes these problems to a great extent.

As with synthetic speech, user acceptance, naturalness and comprehension are very important issues in synthetic sign language. Signed languages are very different from spoken languages in the way that they are produced, but they are complete, natural languages. Subtle changes in features of a sign or phrase can change the meaning, just as in spoken language, so the accuracy of synthetic signing is key to producing language that can be understood.

In the most eSIGN project, two different computer systems including synthetic signing were developed in the UK. The first of these was a web site run by a Deaf organisation. In this context an icon linking to the signing was placed adjacent to the relevant pieces of English text. The second system was a transaction support tool designed for use in council information centres to enable hearing information clerks to communicate more successfully with Deaf enquirers. The clerk could select phrases from a store in the system and send signed translations to the client. Where no suitable BSL phrase existed, English text could be sent. The Deaf client could then reply using any communication tactic of their choosing, or by typing English messages into the system and sending them to the clerk's computer.

This paper discusses some of the key findings from a series of six user evaluations of virtual signing technologies carried out in the UK as part of the EU funded eSIGN project, and also refers to work carried out by other partners in the consortium in Germany and the Netherlands.

## Comprehension

Comprehension was formally measured for single signs and for phrases presented without context. Here, participants were asked to repeat the sign or phrase back to the experimenter, who was a native BSL user. In addition, some material in context (e.g. in a web site with accompanying text and graphics) was presented.

Comprehension of a range of single BSL signs was tested, including one and two handed, symmetrical and asymmetrical, words, numbers and letters. Seventy percent of the signs tested with nine participants were correctly understood. There were large individual differences with scores ranging from 45% to 91%. After indicating what they thought the sign was and being told the correct answer, participants were asked to comment on each sign, especially if they had not recognised it. Their comments provide insight into how the technology can be improved to maximise comprehension.

**Figure 1. Virtual guido signing the letter T.**

Participants in another study were presented with BSL phrases, and asked to repeat them back to the experimenter. Phrase recognition is a more complex task than recognising single signs, with more scope for errors. Phrases were scored as correct or incorrect, with 104 of the 225 phrase presentations (46.2%) administered in the trials resulting in a "correct" score. Again there were large individual differences with scores ranging from 1/15 to 13/15. There were big differences in recognition rates for individual phrases too, ranging from one phrase that was only understood by one participant, to another that was understood by all.

The comprehension rate was higher for single signs than for phrases; this pattern was also found in similar evaluations of synthetic Dutch sign language. Often when someone could not understand a phrase, it was because they had not understood or had misinterpreted a single sign. This highlighted that a small loss of information could in some cases render the phrase incomprehensible or nonsensical. This is the most likely cause of the relatively lower phrase comprehension scores. This was borne out in the participants' comments, where it can be seen that in phrases that were not correctly understood, often one of the key signs was missed but all or most of the other signs were recognised correctly.

Qualitative analysis and review of the user comments provided valuable insight into why people did not understand some of the signs and phrases. This information is now being used in developments of the technology and working methods used in the production of synthetic signing. These data indicated that in most cases the manual features of the language are conveyed well, but that facial expressions and other non manual features (such as body movements, shoulder shrugs etc.) are less clear. Non-manual features are essential for BSL, so it is important that the avatar and tools are developed more in this area.

In some cases, misunderstandings were due to particular sign choices or to the style of the interpretation. Like any language, BSL can express one concept in a range of different ways. It also has the additional complication of regional differences, whereby users in different parts of the UK will use different signs for the same things, for example, number signs vary between regions. In this work qualified relay interpreters were used to make the translations from English text to BSL, but some users did not like the translations for some phrases. They stated that it was not how they would have signed it, making it harder to understand. These findings have highlighted the importance of the translation process, in particular of selecting the most appropriate signs and phrases possible bearing in mind the target user group for the application.

When signed phrases were presented in context, participants reported using the additional information to help them understand the signing. Comprehension was not formally measured, so direct comparisons with the above data are not possible. However, where the text for most of the signed phrases was directly available on screen, most users reported needing to use the text as well as the animation to gain a full understanding of the meaning. Users were asked whether they would like to see both signing and text on screen, and nearly all said that they would. They appreciated the value of having access through both languages, and of being able to use one to back up and verify their understanding of the other.

## User acceptance

Aside from comprehension, it is also important with any synthetic language to consider how acceptable the technology is to the users. While it is important that the users can understand the language, it is also relevant to ensure that it is pleasant to watch, and that it is a means of receiving information that does not annoy, irritate or otherwise upset the users. The style and appearance of the avatar was found to be key to user acceptance; this is discussed in the section below.

*The avatar*

The avatar, Virtual Guido, is the most recent in a series of signing avatars developed by consortia including RNID. He is the most advanced yet, with high quality skin textures and animation. User evaluations of signing produced by earlier avatars lead us to use a male character. Earlier female avatars had "Barbie doll" complexions which made it very difficult to see facial expressions. Virtual Guido is a more mature, male avatar with more textured skin, which makes it much easier for the users to see the facial expressions. In the past users had also informed the development team that the avatar should look friendly but professional. Human BSL interpreters have a smart dress code and this was also expected of the virtual human.

The acceptability of Virtual Guido and other signing avatars will impact on the quality of the user experience, which is key to the success of the product. Deaf sign language users in Germany, the Netherlands and the UK have evaluated Virtual Guido. In general he proved popular, with those respondents who were familiar with earlier avatars reporting that he is an improvement. In spite of the fact that the signing was hard to understand much of the time, respondents generally found the avatar signing pleasant enough to watch. Most features of Guido, such as his face, hair, clothes and general appearance received positive reports from most respondents.

The naturalness of the hand and arm movements of the avatar were rated overall as being somewhat unnatural, as were the facial expressions. In the evaluations, people used the avatar for a relatively short time. This unnaturalness has the potential to cause annoyance in situations where the signing is used for longer time periods. However, for short sequences of signing, this is much less likely to be a problem.

Some specific areas for improvement came up repeatedly, and these will be taken into account in future developments. The mouthing and non-manual features were often thought to be unclear; as important features of BSL, this would have a strong impact on comprehension and as such improvements in these areas are critical to the future success of synthetic signing. Another criticism that may have impacted on comprehension concerned the thumbs, which were reported to occasionally make unnatural movements. Finally, it was reported that it was hard to perceive depth on the screen, and for many signs it is important to know how far the hands are from the body. With videos of signing this problem would be harder to overcome, however it is possible for the end-user to turn the avatar and view from a different angle. Users who tried this found that they could see depth more easily when viewing from a slight angle.

The temperament of the avatar is important to the user experience, although it may not be as critical for comprehension as other characteristics. It was reported that he looked a bit too serious, and some users went as far as to say he looked unkind or unpleasant. In particular, his eyes were thought to give him a cold look.

## User suggestions for future developments

At the end of the evaluation sessions, many users made comments about the system and gave suggestions for how it could be improved in the future. Many wanted to be able to change the appearance of the avatar themselves. In some cases this was for fun; to make different costumes, backgrounds, races and to interchange males and females. In other cases there were more serious reasons; for example changing colours to create better contrast for people with dual sensory impairment.

Some users expressed a desire for a choice of translation levels. It was suggested that some texts could have a direct translation, as well as being expanded upon and explained in more detail in a second BSL sequence. This would extend the group of users that would benefit from the technology by making it more flexible and accommodating a wider range of needs. In general, users preferred that the text and signing was in short "chunks", to maximise understanding by a wide range of people. Where longer texts were translated, users wanted text currently being signed to be highlighted, maybe using a yellow block.

The way in which signing is incorporated into web sites was reviewed. At present users click on an icon, adjacent to the relevant text, to link to the signing which appears in a pop up window. The idea of having a pop up was liked, as it means that people who do not want to use the signing do not need to have an avatar on their screen unnecessarily. On the UK site, the main and avatar windows automatically resized to make best use of screen space without overlapping; this was also popular as it prevented the common problem where pop ups can obscure the main page underneath.

Another practical suggestion was to include a slide bar to allow the user to see how far through the signed sequence they were, and to allow them to move to different parts; this was implemented in the Netherlands and was well received. Linked to this is the facility to speed up and slow down the avatar. The UK trials demonstrated that comprehension rates were higher when the signing was viewed at 75% of full speed; giving users the options to adjust the speed is likely to impact positively on comprehension.

Finally, users wanted the option to change the avatar window size, and to zoom in on particular parts of the avatar such as the face. It is not known whether the size of the avatar or zooming on key features would aid comprehension, so further research is needed in this area.

## Conclusions

Animated sign language is a relatively new area. As it is developed and released to market it is hoped that it can dramatically increase access to sign language for Deaf people. At present there is a shortage of qualified sign language interpreters, and these individuals may be booked up weeks or months in advance. This technology is not intended to replace human interpreters, but rather to increase the range of situations where sign language is available. Services such as providing signed assistance for filling in official forms on the Internet could enable Deaf people to complete the task without seeking additional help. Both of the applications tested in the UK – a web site and a transaction support tool – were well received, and are applications that could be further developed and used in the future.

## Acknowledgements

The eSIGN project was funded under the Information Societies Technologies programme by the European Union's Fifth Framework under the eContent Programme.

# SYNFACE: PROVIDING LIP-READABLE SUPPORT TO HARD OF HEARING PEOPLE

**Jo Kewley & Mary Sheard**

*RNID, 19–23 Featherstone Street, London EC1Y 8SL*

Many hard of hearing people rely on lip-reading as well as their residual hearing to follow conversations. However, this additional visual support is not available during telephone conversations. Synface aims to alleviate this problem by providing an animated talking head that recreates the lip movements of the caller on a screen for the hard of hearing person to lip-read. This paper discusses the findings of user trials of the Synface system carried out in the UK, their implications for the development of the system, and of other lip readable systems in the future.

## Introduction

In the UK, 9 million people have a measurable degree of hearing loss, a prevalence which is similar throughout Europe. The majority of deaf and hard of hearing people have an age related hearing loss, however the noise pollution of daily life and exposure to loud music are also causing a growing number of people to be affected (RNID Factsheet, 2004).

Although the number of people that rely on lip reading is unknown, everyone uses lipreading to some extent especially in noisy situations. For hard of hearing people as their hearing loss gets worse, more information is obtained through lipreading (RNID Factsheet, 2004). However during telephone conversations this information is missing and people are left to rely only on what they can hear.

At present only two alternatives to telephone communication are available that allow real time conversation when amplification of the speech signal on a telephone is not enough. A textphone allows users to communicate using typed text, which is fed character by character in real time to another textphone user or via a relay operator to a hearing person using a standard voice telephone. A videophone allows a real-time image of each user to be fed from videophone to videophone using a wideband video connection. However these alternatives have drawbacks, with both users needing expensive equipment or for intermediaries to be involved.

Research has demonstrated that synthesized facial animation can be used to derive visual speech information from auditory input to assist individuals with a hearing loss (Beskow et al, 1997; Agelfors et al, 1998). The Synface project aims to increase the accessibility of voice telephones by providing an artificial face where the lip movements are derived from the speech signal using automatic speech recognition. A prototype has been developed for English, Swedish and Dutch, providing animation on a synthetic face with the proper lip, jaw and tongue movements to match the corresponding sounds. This visual speech synthesis is controlled by a speech recognition system that has been designed to work with a very small delay (200 ms). The speech signal is delayed by this amount and presented in sync with animation. For each language the recognisers were developed using an extensive database of speech samples, allowing the system to be used with a variety of voices.

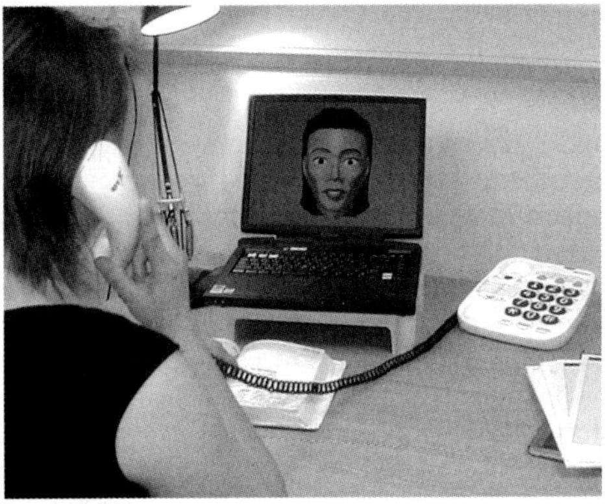

**Figure 1. The Synface prototype in use.**

## Evaluations

This paper discusses the results of lab-based evaluations of Synface that have been carried out in the UK. The aim of these evaluations was to investigate attitudes and opinions towards lipreadable support during telephone conversations, and to evaluate the perceived value, effectiveness and usability of the English prototype.

Participants were recruited to be representative of the hard of hearing population in the UK. Participants varied in age, residual hearing levels and ability to communicate using a voice telephone. As expected the majority of participants stated that they used lipreading, with lipreading becoming more important when trying to understand strangers, rather than friends and family.

Data was collected using questionnaires and a short interview at the end of the evaluation. To introduce participants to the concept of lipreadable support during a telephone call a short video was produced that showed a hard of hearing person using Synface. To allow participants to experience lipreading within optimum conditions participants were then shown a short story about the North Wind produced using hand labelled speech.

To gain a measure of comprehension when using the prototype and to allow the same animation and audio to be shown to all participants 32 pre-recorded sentences were produced. The sentences used were all of a similar length and complexity and were taken from 2 lists of BKB Sentences (Bench and Bamford, 1979). These sentences were originally designed to test speech perceptual ability in children, and have since been extensively used in speech audiometry. The sentences were pre-recorded and produced using an offline version of the recogniser with the same processing time as applies in the prototype.

Participants were then given the opportunity to use the prototype for real calls. To facilitate these calls four scenarios were created. Two scenarios were designed to resemble informal conversation with a friend or family member and the other two scenarios were designed to resemble enquiry type calls, where participants had to find out specific or detailed information. With the order of presentation counterbalanced to prevent order effects. Each scenario was designed to give a theme to the conversation, and although possible questions were suggested, participants were given the freedom to ask any questions they liked and were not assessed on success completing each scenario.

## Findings

### *Attitudes and opinions towards lipreadable support*

Hard of hearing people's need for support during telephone calls and that this need is not met by existing technologies was illustrated by participants' reactions to the concept of lipreadable support during telephone calls. After the concept had been explained to participants, 70% stated that they felt Synface could help them to use a telephone more successfully. After using the prototype only 9% stated that they would prefer to use a voice telephone without Synface. Participants also stated a preference to use Synface over other existing technologies, with 91% of participant preferring Synface over textphone to textphone communication, 88% preferring Synface over a textphone and relay service and 70% preferring Synface over videophone to videophone. Illustrating.

### *Perceived value*

Participants' general impressions of the prototype were positive with 21% stating that Synface had helped a lot and 70% finding it had helped a little, when having conversations using the prototype. Participants' perceptions of the helpfulness of Synface were not found to relate to hearing level, however those participants that found that Synface "helped a lot" found Synface significantly more enjoyable than those who found Synface only "helped a little".

Participants' perceptions of usefulness revealed Synface to be more useful during conversations than when listening to the sentences. Sentences were perceived as being harder to understand because there was no context available, whereas during the conversations context was known allowing guesswork to be used if needed.

The physical presence of the synthetic face was mentioned as providing participants with confidence and reinforcing that someone else was actually there. Lip movements were described as reinforcing what participants thought they were hearing and allowing them to pick up more information, especially details such as names and numbers that are normally difficult to understand. With the extra support available from Synface some participants also stated that they were more willing to ask the other caller to repeat information.

### *Effectiveness*

Comprehension scores were calculated from the number of correct key words participants identified in each of the BKB sentences. Scores varied ranging between 9% and 99% with a mean of 73%. Despite the non significant relationship between hearing loss and comprehension scores, mean comprehension scores were found to decrease as the level of loss increased. How often participants used a voice telephone was found to significantly effect comprehension, with those who use a telephone at least once a week identifying more key words than those who no longer use the telephone. Participants that find communication using a voice telephone very or somewhat effective also identified significantly more key words than those who find telephone communication not at all effective.

Participant comments highlighted the importance of experience and being in a relaxed environment to benefit from Synface. Learning effects were not found to occur over the short duration of the trial. However, many participants commented that they felt it would take time to learn how to use Synface, learn how to lipread the face and remember the face is there to provide support. The concept of learning and extended use is currently being investigated.

*Usability*

Participants agreed that the prototype was very easy to use and make a call, with very few visible controls and working in the same way as a standard telephone. Other findings highlighted issues that need to be resolved before many participants would consider buying Synface.

When questioned about their attitudes and opinions towards the face, participants seemed to feel that the face was not particularly aesthetic or nice to look at, preferring to look at a more realistic human face. However, when questioned further many were unsure how or whether this would improve their ability to lipread. Feelings that lip movements were missing, did not relate to the audio, were the wrong lip movements for a sound or where there were lip movements when there was no sound was noticed by 60% of participants. This lead to feelings of upset, annoyance and confusion for some participants. Only 36% of participants regarded the movements of the face as being natural. When asked how the face could be improved, participants suggested larger more exaggerated lips, better articulation and clear, accurate lip movements. Several participants also highlighted that no information was available from the face or eyes, and that facial expression is an important part of lipreading.

Participants were also asked if they felt the speech and lip movements were in time with each other only 15% felt they were all the time, and 45% felt they were the majority of the time. Only 9% felt that such timing differences caused a lot of upset, annoyance or confusion. When asked about the timing participants seemed to be lacking in confidence as to whether there were timing differences and how frequently these were occurring. With suggestions that observed timing difference may have been wrong lip movements, the lips continuing to move when the audio had stopped or at the start of a sentence the animation not starting at the same time as the audio. These comments highlight that timing differences both effect lipreading and usability.

The delay of the incoming audio in the prototype, necessary for the processing of the signal and the animation of the synthetic lip speaking, causes a potential problem for the usability of the system. While the length of the delay is not long enough to severely effect the flow of conversation it can potentially mean that the users voice would be repeated back into their earpiece 200 ms after it was spoken. In order to prevent this happening the system was designed to be a half-duplex, silencing the audio and suppressing the articulatory movements of the face when a speech signal from the local handset is detected. Not all participants noticed this switching mechanism, 36% noticing it occasionally and 9% frequently and only 3% feeling it was not acceptable, whereas 36% felt it was acceptable but not ideal. However as participants' commented that they still want to hear all of a conversation and that this mechanism occasionally caused participants to miss parts of the conversation, suggests that it would be beneficial to resolve this issue. This mechanism also meant that there was no side tone (repetition of your own voice played back through the earpiece of a telephone), this created difficulties for people using the inductive coil in their hearing aids to regulate the volume of their voice.

During the first few lab based evaluations participants mentioned that they could not imagine speaking to a male voice through a female face. This lead to a both a female and male experimenter being used to carry out the scenario based calls. No significant effects were found between the gender of voice and ratings of naturalness and realisticness. However a significant effect was found between ratings of helpfulness, with those participants listening to a male voice through a female face rating Synface as slightly more helpful than those that had listened to a female voice through a female face. This may have been caused by features of the individual voices.

## Implications

These findings support the need for a method to help hard of hearing people to communicate using a voice telephone. The current prototype was gauged as being beneficial to some participants, however the improvements mentioned above need to be made to increase the market potential of the technology. This is supported by lab trials (Siciliano et al, 2004) that have demonstrated that the automatic speech recognition technology developed within the Synface project can provide important visual speech information, however further refinements are needed before intelligibility is at the same level as hand labelled speech.

## Acknowledgement

Synface is a collaborative project, funded under the Information Societies Technologies programme by the EU Fifth Framework.

## References

Agelfors, A., Beskow, J., Dahlquist, M., Granström, B., Lundeberg, M., Spens, K-E., and Öhman, T. (1998). "Synthetic talking faces as a lipreading support." *Proceedings of the International Conference on Spoken Language Processing*, Sydney, Australia.

Bench, J. and Bamford, J. (eds.) (1979). Speech-hearing Tests and the Spoken Language of Hearing-impaired Children. (Academic, London).

Beskow, J., Dahlquist, M., Granström, B., Lundeberg, M., Spens, K-E. and Öhman, T. (1997). "The Teleface project: Multimodal speech communication for the hearing-impaired." *Proceedings of Eurospeech*, Rhodes, Greece.

RNID Factsheet (2004). "Fact and figures on Deafness and Tinnitus". www.rnid.org.uk, London.

Siciliano, C., Williams, G., Faulkner, A. and Salvi, G. (2004) Intelligibility of an ASR-controlled synthetic talking face. J. Acoust. Soc. Am. 115, 2428.

# MODELS FOR INCLUSION
# EVIDENCE FOR CHOICE AND INNOVATION

**John Mitchell[1], Robert Chesters[2] & Dr John Middleton[3]**

[1]*Director, Ergonova,* [2]*Inclusive Design manager for Medilink West Midlands,*
[3]*Director of Public Health, Rowley, Regis and Tipton PCT*

Thirty disabled people in Sheffield were asked about their aspiration, barriers and their ideas for living more independently at home (Mitchell, 2004). The study exposed barriers against people with different types of impairment in their homes and neighborhoods, in their domestic utilities, systems and products, in their housing services and in the marketplace for homes, contents and services.

These barriers are addressed in three papers in this edition of Contemporary Ergonomics. The first, (Mitchell, 2005a) presents models for gathering evidence about the barriers that restrict disabled home users. The third (Mitchell, 2005b) presents models for using this evidence to enable consumers and providers to make informed choices and, in so doing, help to establish an informed marketplace for inclusion. This paper proposes a model for eradicating barriers during home refurbishment in the West Midlands the "Smart, Inclusive Homes" initiative under the leadership of the Rowley, Regis and Tipton Primary Care Trust and Medilink West Midlands Ltd.

The Model for Inclusive Refurbishment has the following stages

(a) Gathering systematic local user evidence
(b) Achieving an inclusive home carcass
(c) Achieving inclusive home contents
(d) Achieving inclusive smart technological backup
(e) Achieving inclusive home services

## Background and barriers

Despite recent advances in knowledge and technology and the potential benefits of inclusion, new homes and contents still contain historic barriers against users with impairments. During the 19th and 20th centuries, large numbers of "terraced", "council" and "high-rise" homes were built during times of rapid industrial and social change. Although they met their objectives with considerable success, time has exposed flaws in their designs which reflect the prevailing levels of knowledge and technology. If their builders had had access to modern technology, if they had understood barriers and how they were caused and if they had had to make good the costs of excluding people who were disabled by their homes they might well have built much more inclusively. Similarly, if the need for inclusion had been recognised when gas, water, electricity and telephones were introduced into homes, it is quite possible that they might have been easier to find, reach, adapt and use as and when required. There may now be an important opportunity for revisiting the designs of these homes and finding out if they could become fully inclusive during refurbishment.

The government's Decent Homes 2000 program is intended to encourage Local Authorities to renovate their existing housing stock and bring it up to modern constructional standards. The Home Market Renewal Pathfinder Program, on the other hand, is aimed at revitalising areas of underused and derelict housing in order to re-establish its social and economic viability. The Joseph Rowntree Foundation has developed its "Lifetime Homes" guidance for achieving a higher degree of inclusion in new build and refurbished homes. Its aim is to produce homes that have few barriers and that are easy to adapt when required. However, these guidelines are focused mainly on people with physical rather than sensory or cognitive impairments.

Disabled peoples' aspirations, barriers and ideas for greater inclusion at home have emerged from a study of thirty people with a range of cognitive, physical and sensory impairments in Sheffield (Mitchell, 2004). Participants said that barriers restricted their lives and activities in their neighbourhoods, their homes and gardens, their domestic utilities, systems and products, their housing services and in the markets in which they were seeking to choose good quality homes, contents and services.

The Rowley, Regis and Tipton Primary Care Trust is preparing to bid to develop "Smart, Inclusive Homes" in Sandwell to act as a model for inclusive home refurbishment. The team of disabled home users, builders, suppliers, and providers intend to acquire a stock of "terraced", "council" or "high-rise" homes, strip them of their barriers and fittings and refurbish them to the highest available standards of inclusion.

## The challenge

The challenge for the "Smart, Inclusive Homes" project will be to use evidence-based choice and innovation (Bennington et al. 2000) to eliminate barriers during home refurbishment.

## Model for inclusive home refurbishment

The purpose of this five part model is to develop methods for achieving inclusive home refurbishment to the benefit of consumers, providers and commerce.

### *Gathering systematic local user evidence*

Inclusion cannot be achieved without good quality, authentic evidence from disabled people about the barriers they face, about their ability to use one or another product or system and whether new approaches help them achieve inclusion. The role of user and provider evidence is outlined in Table 1 over the page.

**Table 1. User and provider evident for inclusion at home.**

The team will:

- Establish partnerships and feedback networks with local users and providers
- Audit and expose the barriers in local homes
- Develop performance specifications that define inclusive practice and establish the penalties that these barriers impose
- Calibrate local findings with those from Sheffield
- Assess the scope for replacing existing barriers with "inclusive" alternatives
- Compare inclusivity of existing and prototype designs for homes and systems

### *Achieving an inclusive home carcass*

It is important to ensure that the carcasses of homes are barrier-free because the shells of buildings tend to be solid, durable and resistant to adaptations. Factors that can limit the flexibility and adaptability of homes are shown in Table 2 below.

Home adaptations are often carried out on a "one-off" basis as previously non-disabled people find they can no longer cope with the barriers in their homes. Home adaptations have tended to stigmatise their users and to be removed as quickly as possible. Once users have set out the barriers they face in finding, reaching and using their homes the team will consider how these are caused and how they might be overcome by adapting the home carcasses. The team will work with users, providers and the building industry to develop inclusive, systematic and mutually advantageous ways of enabling users to find and reach every part of their homes and gardens. Table 3 sets out possible approaches to overcoming these barriers.

### *Achieving inclusive home contents*

A barrier-free home carcass will eliminate many of the most intractable barriers to inclusive living at home but it cannot prevent barriers from being "imported" in the shape of non-inclusive mainstream fixtures, fittings and contents. Table 4 provides examples of some of the barriers that may be imported in this way.

Attempting to design out the barriers in these examples would be fruitless and impractical during any short term project. However, it is quite practical to "choose out"

**Table 2. Home features that can limit inclusion.**

It can be difficult to remove barriers to inclusion if homes have:

- Different surface levels inside and outside homes
- Steep, narrow stairways
- Restricted circulation spaces in rooms and passages
- Internal walls that are difficult to relocate and cannot bear the loads of subsequent adaptations

**Table 3. Systematic approaches to inclusive home carcasses.**

Systematic, modular approaches may be required for:

- Equalising surface levels inside and outside the home
- Tactile way finding
- Step free access
- The ready re-arrangement of doorways, room spaces and passageways
- The reinforcement of fragile room walls

**Table 4. Barriers in non-inclusive home contents.**

Barriers to finding, reaching, understanding, relating to and controlling any of these items will disable some potential users:

- Doors, windows and double-blazing fittings
- Electrical, gas, water and central heating outlets and controls
- Fixtures and fittings such as bathrooms and storage facilities
- Domestic products such as cookers, washing machines and refrigerators
- Leisure and communication systems such as telephones, TV and sound systems, security and alarm systems

barriers by using user feedback to find out which items contain the fewest barriers for users with different impairments. This feedback could also reveal areas in which innovation could prove most valuable in the longer term.

*Achieving inclusive smart technological backup*

Evidence-based choice and innovation can be seen as ways of "Designing-out" and "Choosing-out" barriers in the carcasses and contents of homes. If these are used successfully, they will greatly reduce the numbers of barriers that restrict disabled peoples' home lives and activities. Inevitably however, some systems will continue to include barriers which reflect existing technical limits or the failure to exploit technical resources fully to achieve inclusion. "Smart technology" can be used to provide the monitoring, communicational, sequencing, co-ordination and control functions that disabled people require when they attempt to use non-inclusive products and systems. For example, disabled home users with a range of impairments (Mitchell, 2004) described the following barriers that prevented them from using their front doors as fully as they wished.

These barriers could be overcome by developing an "inclusive front door". They could also be overcome by assessing the barriers that particular users experience and by choosing and installing packages of smart technology that would make good the deficiencies in their existing front doors as suggested in Table 6 below.

Essentially, control of each of these items involves receiving information in a usable format, choosing a response or setting and activating the desired actions. Each of these functions is actually or potentially within the reach of smart technology and it is not difficult to support functions similar to those listed in Table 7 below.

**Table 5. The "non-inclusive" front door.**

People with a range of impairments reported that their front doors are difficult to:

- Find (non-inclusive signage)
- Reach (non-inclusive access and turning space for wheelchair users, door controls located too high or too close to door framework)
- Understand (non-inclusive means of identifying and communicating with visitors)
- Control (handles and keys difficult to operate)

**Table 6. Barriers to controlling home environments and systems.**

People with impaired sight, hearing, use of their hands, mobility, understanding and health said that they had difficulty in finding, reaching, understanding, relating to or controlling their:

- Central heating systems
- Security and monitoring systems
- Visually and/or audibly mediated communication systems
- Kitchen and laundry facilities

**Table 7. Inclusive home control through smart technology.**

Individual control at home could be achieved through a smart, portable device that could:

- Receive emergency and status signals from all domestic sources
- Present these in whichever format individual users require
- Provide appropriate sequencing and guidance as required
- Transmit the desired commands or settings to the controlled element

*Achieving inclusive home services*

All home users require a range of services, for example, from local and housing authorities, from specialist care and adaptation agencies and from those who provide the choice and acquisition of homes and their contents. User feedback about the quality and appropriateness of these services will be gathered and used in the same way as described under 3 achieving inclusive homes, above.

## References

Bennington et al., 2000, *Autonomy for Disabled People* (King's Fund)
Joseph Rowntree Foundation, *Lifetime Homes*, (York)
Mitchell et al., 2005a, *Evidence for Inclusion at home and elsewhere* (Contemporary Ergonomics), in press
Mitchell et al., 2005b, *Evidence for an Inclusive Home Marketplace* (Contemporary Ergonomics), in press
Mitchell et al., 2004, Why *Can't We Have a Home with nothing wrong in it?* (Inclusive Living Sheffield)

# MODELS FOR INCLUSION
# EVIDENCE FOR AN INCLUSIVE HOME MARKETPLACE

**John Mitchell[1], Robert Chesters[2], Dr John Middleton[3] & Chris Ramsden[4]**

*[1]Director, Ergonova [2]Inclusive Design manager for Medilink West Midlands*
*[3]Director of Public Health, Rowley, Regis and Tipton PCT*
*[4]Advantage West Midlands*

In "Designing-out the Cell of Exclusion", Mitchell and Chesters (2004a) suggested that stakeholders are failing to remove barriers that exclude disabled people because there is little evidence to show that, if they did so, they would increase their markets or decrease their support costs. The problems of this missing evidence and its applications are addressed in three papers in this edition of Contemporary Ergonomics. The first, (Mitchell, 2005a) presents a model for gathering evidence about the barriers that restrict disabled people in their homes. The second (Chesters, 2005) proposes a model for eradicating barriers during a proposed home refurbishment initiative "Smart, Inclusive Homes" in the West Midlands. If this project is successful, it will show how inclusive available homes, contents and services are currently. It may also produce innovations and approaches that can eliminate barriers more effectively than can be done at present. These initiatives can only benefit consumers, providers and commerce if they reach and succeed in the marketplace. This paper therefore presents a model for using evidence about inclusive and non-inclusive products and systems to enable users and providers to make informed choices and, in so doing, help to establish an evidence-based marketplace for inclusion.

## Background and barriers

The marketplace is an essential element in enabling disabled people to move from exclusion into mainstream life. The process will be accelerated as inclusive goods and services become more widely available in the marketplace, as customers begin to purchase a higher proportion of inclusive items and as assistive products are more precisely chosen to match the aspirational and impairment profiles of their users (Bennington, 2000).

However, as a recent study reveals (Mitchell, 2004), disabled people are still held back by barriers in the marketplace for homes, products and for assistive services. These consisted not only of difficulties in obtaining suitable products and systems but also in the "disability premium" which resulted in their having to pay higher prices and charges than other people as shown in Table 1 over the page.

These difficulties are mirrored by the barriers that innovators face in bringing inclusive products to the marketplace. If the "Smart, Inclusive Homes" project is successful, it will collect extensive user feedback about the inclusivity of existing homes, systems and services. At the same time, it may develop inclusive products that could restore autonomy to their users and also reduce the costs of dependence. These initiatives can only benefit users, providers and manufacturers who were not involved in the project through the marketplace and its intelligence systems as suggested in Table 2 below.

**Table 1. The non-inclusive home marketplace.**

People with a variety of impairments said they were prevented from making full use of the market for homes, contents and services because of the following barriers:

- Non-inclusive ways of presenting information
- Poor quality and incomplete information about available homes and products
- The lack of user evidence to guide the choice of homes or systems
- The lack of systems for including disabled peoples' expertise in the choice and innovation of better homes, systems, products and services
- The fragmentation of disabled peoples' and providers' purchasing power
- The "disability premium"

**Table 2. Inclusion and the marketplace.**

The removal of barriers to full inclusion requires that:

- Inclusive mainstream and assistive products are developed, offered in the marketplace and compete successfully with their non-inclusive equivalents
- Purchasers begin to choose inclusive and assistive products effectively and turn away from products that contain barriers

**Table 3. Market intelligence and decision-making.**

Stakeholders require good quality market intelligence to help them to decide:

- Which mainstream and assistive technology to choose
- Whether to innovate inclusive mainstream and assistive products
- Whether to bring new products to the market

Bennington (2000) has discussed the need for better, more systematic evidence for matching product performance and users' profiles of aspirations and barriers. This paper is concerned with the need for good quality market intelligence about products and about market opportunities. The quality of market intelligence is particularly important for inclusive products because they are likely to be sold more on the basis of their quality and performance than on their purchase prices.

## The challenges

The challenges for the market are to develop intelligence processes that enable inclusive products to become established within mainstream markets. This requires widespread commercial and purchaser awareness of:

- The potential size of the market for inclusion
- The effects of barriers in reducing market size
- The support costs that accompany exclusion
- The benefits of choosing inclusive products
- Criteria for distinguishing inclusive from non-inclusive products

This awareness requires routine, systematic methods for incorporating this data into standard market intelligence systems.

## Model for inclusive markets

The model has six parts which correspond to the challenges above and the need to embed information about inclusion in standard intelligence systems.

*Model 1 Defining the potential size of the market for inclusion*

The size and value of this market could be an important motivator for commercial organisations that might be considering the benefits of developing and selling inclusive products. Much of the data is already available in existing sources such as censuses and data on market trends. However, this needs to be reviewed to expose the size and value of the market to assist those who might wish to enter it.

*Model 2 The effects of barriers in reducing market size*

There is growing awareness that, by excluding disabled users, barriers exclude potential customers and users. Once again, some of the data is already available but this would need to be analysed and augmented to compare, for example, the number of disabled people in a locality with the rates of usage of local facilities such as shops, buses or learning facilities and their purchases of products such as mobile phones or PCs.

*Model 3 The support costs that accompany exclusion*

When disabled people are prevented from carrying out activities for themselves they must either forgo them or rely on help from someone else who might be paid or unpaid. As well as reducing disabled peoples' capacities and productivity, barriers also divert others from their own activities to provide support. It is particularly important for service providers to recognise and calculate the costs of dependence because, if they provide homes or other facilities that disable some of their users, they may then have to provide them with support services. At present, there are no routine, systematic methods for collecting this evidence from the wide range of sources involved or of analysing it to expose the human and economic costs of dependence.

*Model 4 The benefits of choosing inclusive product*

It is not surprising that the sale of homes and other products are heavily influenced by their purchase price because this is relatively easy to find out and to compare with others. It is much harder to assess the longer term costs that are imposed when barriers prevent users from carrying out their usual functions and activities. It would be useful and instructive to analyse and compare the effects of inclusive and non-inclusive settings on users' autonomy, dependence and support costs. For example, if the "Smart, Inclusive Homes" refurbishment project (Chesters et al., 2005) is adopted, disabled home users will be asked to compare their lives and activities before and after the barriers in their homes were removed.

*Model 5 Criteria for distinguishing inclusive and non inclusive products*

Once users and purchasers have recognised that they could benefit from inclusion, they then need to be able to find out which homes and products are inclusive and which are not. Currently, there are no established ways of making these distinctions. Some standards have been developed for defining physical accessibility, such as the Lifetime Home Standards, but these cannot provide guidance for achieving full inclusion because they do not cover barriers against the full range of impairments. Recently, however, attention has shifted towards a more "pan-impairment" approach. For example, Yelding (2003) has listed the barriers that are commonly experienced by users with

**Table 4. Mainstreaming information on inclusion.**

To reach all its potential users, Evidence on inclusion would need to be fed into a variety of systems, such as:

- Retail brochures
- Internet purchasing systems
- Specialised professional databases
- Consumer databases and newsletters

a range of impairments when they try to use mainstream domestic products and systems. An analysis of the barriers that people with a range of different impairments experience in their homes was undertaken by Mitchell (2004). This produced the "Sheffield Formula for Inclusion" (Mitchell, 2005). The formula breaks down the essential processes of use into the functions of finding, reaching, and understanding, relating to and controlling whichever products or systems are being used.

The formula could provide a basis for exposing and integrating the various barriers that exclude people with impaired sight, hearing, use of their hands, movement, understanding or health. The findings could then be used to develop "Inclusive Performance Specifications" for these products and systems which would define the type and location of the barriers in their present formats. They could also define the design changes that would be necessary to achieve full inclusion. The Formula and specifications could then be used to rate different products and systems for their inclusivity and for the barriers against people with particular impairments that they currently contain.

*Model 6 Mainstreaming information on inclusion*

The market for inclusive products and systems is not confined to people who consider themselves to be disabled. At times, most people are likely to have difficulties in finding, understanding or controlling the things they wish to use and many people might welcome a system which highlighted items that were easy to use. Data on inclusion would have to be fed through a series of different systems to reach all those who might be interested in receiving it, as shown in Table 4 over the page.

## References

Chesters et al., 2005, *Evidence for Choice and Innovation*, (Contemporary Ergonomics), in press

Bennington et al., 2000, *Autonomy for Disabled People*, (King's Fund)

Commission on Architecture in the Built Environment, 2004

Joseph Rowntree Foundation, *Lifetime Homes*, (York)

Mitchell and Chesters, 2004, *Designing out the Cell of Exclusion*, (Contemporary Ergonomics) pp 445–449, ISBN 0 8493 2342 8

Mitchell et al., 2005, *Evidence for Inclusion*, (Contemporary Ergonomics), in press

Mitchell et al., 2004, Why *Can't We Have a Home with nothing wrong in it?* (Inclusive Living Sheffield)

Yelding D, 2003, *Power to the People*, pp 104–117, in Clarkson et al. Inclusive Design – Design for the Whole Population, (Springer-Verlag, London)

# MODELS FOR INCLUSION
# EVIDENCE FOR INCLUSION AT HOME AND ELSEWHERE

**John Mitchell[1] & Robert Chesters[2]**

[1]*Director, Ergonova*
[2]*Inclusive Design manager for Medilink West Midlands*

An analysis of disabled peoples' aspirations for living at home and the barriers that impede them (Mitchell et al, 2004), has revealed barriers in their homes, their contents, their surroundings, their housing services and in the marketplaces for homes, contents and services. This paper is the first of three in this issue of Contemporary Ergonomics that address different aspects of these barriers. It presents a model for gathering and using evidence to promote the development of inclusive homes and other systems that seek to meet the requirements of all their users. The second (Chesters et al, 2005) presents a model for using evidence-based choice and innovation to achieve inclusive home refurbishment while the third (Mitchell et al, 2005) sets out a model for achieving an inclusive marketplace.

Homes and housing development organisations are seen as systems that require feedback from their users to detect and correct any aberrations in their design and function. It is suggested that systematic methods for obtaining and using this feedback are required to motivate and guide progress towards fully inclusive homes. The model for evidence for Inclusion has the following sections:

- Exposing barriers against people with all types of impairments
- Integrating barriers across the full range of impairments
- Specifying inclusion
- Using evidence-based choice and innovation to achieve inclusion

## Background and barriers

Barriers against people with impairments prevent them from making full use of the facilities, products, services and systems that they wish (Mitchell et al, 2004). The authors suggest that developers continue to produce these barriers because they do not yet recognise the importance, the benefits and the feasibility of designing inclusively for the full spectrum of users. There is growing awareness of the penalties that exclusion places on disabled people and on providers (Commission on Architecture in the Built Environment, 2004) which is leading to an increasing demand for the eradication of barriers and the full inclusion of disabled people.

The levers for achieving this are seen to be the twin benefits of an increased market size for mainstream products and reductions In the support costs that accompany non-inclusive designs. Despite the undoubted power of these levers and the flexibility that is available through modern technology there are few conclusive signs of a concerted move towards full inclusion. Mitchell et al (2004) suggested that, although the levers for inclusion are powerful, the evidence that would allow them to be successfully deployed is either missing or fragmented. They also recommend that, since disabled people experience these barriers at first hand, they should take the lead in exposing and eradicating them.

This paper uses the systems approach to propose a model for achieving inclusion through systematic choice and innovation. The model has been developed from analyses of disabled peoples' aspirations, barriers and ideas for improvement in several areas of application. These include choosing and using homes (Mitchell, 2004), learning (provision Mitchell, 2004), employment provision (Mitchell, 2003), support services (Mitchell, 2002, wheelchairs (Mitchell 1998) and healthcare beds (Mitchell, 1998).

## A systems approach to inclusion

Homes, wheelchairs, learning facilities and other designed outputs can each be regarded as systems in their own right. For example, homes can be seen as "systems for domestic living". An "inclusive" goal for these systems would be To enable all their users to carry out all the domestic functions and activities that they require. Similarly, the organisations that develop and supply homes and other products can be seen as "development and supply systems". An inclusive goal for these systems would be To ensure that homes and their development and supply systems enable all their users to meet their domestic requirements within the prevailing human, technical and economic settings. All development systems requires feedback about their performance to monitor and correct any aberrations from their purposes. Systems without built-in feedback loops cannot detect deviations from their goals. They are likely to perform sub-optimally and, should this occur, they will be unaware of and unable to correct these aberrations.

These problems may well apply in the housing and other development systems in which User feedback is not used systematically to expose barriers against people with the full range of impairments. As well as impeding home users, this lack of feedback also impedes those developers who wish to improve the human, technical and economic performance of their systems.

## Including users with all impairments "in the loop"

Users naturally wish to use their homes and their contents without restriction.

If these are "inclusive", all users will be able to do so without meeting barriers that restrict or exclude them. However, if people with impaired sight, hearing, use of their hands, movement, memory, understanding or mental or physical health cannot use the system, this is likely to indicate either a malfunction or the presence of barriers against users. Both are system aberrations which should be fed back to the appropriate development and supply system as suggested in Table 1 over the page.

### *The challenges of exposing and eradicating barriers*

The main challenges for those who are seeking to achieve full inclusion are to develop systematic methods for:

- Exposing barriers against people with all types of impairments
- Integrating barriers across the range of impairments
- Taking co-ordinated action to remove all barriers

**Table 1. A systematic feedback loop for exposing and removing barriers.**

(1) Routine use by the full spectrum of users
(2) Barriers reported by some users
(3) Reports of barriers are used to expose their:
- Effects on people with different impairments
- Location within the product or system
- Nature and causes
- Possible solutions

(4) Evidence about barriers is then fed back to:
- The appropriate development System
- Prospective purchasers and users
- Innovators

## Model for achieving evidence-based inclusion

This model has four parts which are based on the challenges above.

### *Part 1: Exposing barriers against people with all types of impairments*

While barriers against people with physical impairments are already reasonably well recognised, the same cannot be said of those which face people with impaired sensation, understanding or health. The exposure of these barriers should be based on authentic reports from people with a wide range of different roles and impairments. Some of the potential difficulties of exposing these barriers are illustrated in Table 2.

These difficulties can be addressed according to the model presented in Table 3.

This approach was used to expose the barriers in the homes of 30 disabled people with a wide range of impairments and backgrounds in Sheffield (Mitchell, 2004). The barriers they reported are summarised in Table 4.

### *Part 2: Integrating barriers across the full range of impairments*

Some barriers, such as the lack of high-quality information for choosing homes, contents and services, affected all participants, regardless of their impairment. Others had a more focused effect on people with specific impairments. For example, speech presentations without visible backup mainly exclude people with impaired hearing while printed text without audible or tactile backup mainly excludes people with impaired sight. Complex procedures mainly exclude people with impaired understanding and hostile or unwelcoming settings can have particular significance for people with impaired mental health. It would be advantageous to try to understand and eliminate the full range of barriers together rather than tackling them one by one. Analysis of these barriers showed that they are, to some extent sequential and that people with different impairments experience them at different stages in the process of use. This sequence was used to develop the Sheffield Formula for Inclusion as a means of guiding and testing inclusion.

The Sheffield Formula for Inclusion states that nothing is fully inclusive unless you can, find it, reach it, understand how to work it, relate to it without being intimidated and control it. The formula can be used by people with any type of impairment to expose the precise point where their use of a product or system is blocked by a barrier, such as the lack of simple, "plain language" guidance or instructions or a control arrangement that demand a high degree of manual dexterity from their users.

**Table 2. Difficulties of exposing the full range of barriers.**

People with different impairments may:

- Be difficult to contact and recruit
- Face unrecognised barriers
- Have relinquished their autonomy in the face of their barriers
- Have become "desensitised" to familiar barriers
- Be unaccustomed to discussing their barriers
- Suffer from "research participant fatigue"
- Be reluctant to explain their barriers to those who are not familiar with them

**Table 3. Promoting the disclosure of barriers.**

- Ensure that disabled people control the research
- Use disabled researchers
- Involve people across the full spectrum of impairments
- Structure discussions around shared impairments and barriers
- Collect data in participants' own familiar settings
- Researchers disclose their own impairments, aspirations and barriers, invite participants to follow suit
- Encourage participants to explore their shared experiences
- Encourage participants to explore any differences in experiences

**Table 4. Barriers to inclusive living at home.**

Participants reported that barriers made it difficult or impossible for them to use what they wanted in their:

- Neighbourhoods
- Homes and gardens
- Domestic Utilities, Systems and products
- Local service
- Markets for homes, contents and services

*Part 3: Specifying inclusion*

Once barriers have been exposed in a particular product or system, guidance is needed to ensure that these barriers are "designed-out" of future versions. Performance specifications can provide this guidance in a flexible and testable format that defines the qualities that users require without specifying a particular type of design or construction. Inclusive Performance Specifications could therefore be developed and used to define the performance qualities that would enable all users to carry out the functions and activities they require easily, safely, effectively and comfortably. These specifications could provide the systems goal for an inclusive home, an inclusive housing development system as suggested above under "A Systems approach to Inclusion" or for an inclusive refurbishment system such as the "Smart, Inclusive Homes" initiative outlined by Chesters (2005). They could also be used to define the required performance qualities for domestic utilities, products or systems that could be fully used by all their potential users.

*Part 4: Using evidence-based choice and innovation to achieve inclusion*

The final stage of the model is concerned with achieving inclusion by both "designing-out" and "choosing-out" any barriers that have been reported by users. The processes of systematic, evidence based choice and innovation, as described by Mitchell et al

(1998), rely on routine user feedback as the primary means of alerting users, providers and commerce to the availability, benefits and opportunities of inclusive products and systems in comparison with their non-inclusive counterparts as shown in Table 1.

## References

Bennington et al, 2000, *Autonomy for Disabled People*, (King's Fund)

Chesters et al, 2005, *Evidence for Choice and Innovation*, (Contemporary Ergonomics) (in press)

Commission on Architecture in the Built Environment, 2004.

Mitchell et al, 2004, *"Designing out the Cell of Exclusion"*, (Contemporary Ergonomics), pp 445–449, ISBN 0 8493 2342 8

Mitchell et al, 2004, *"Why can't we have a home with nothing wrong in it?"* (ILS)

Mitchell et al, 2000, *Systematic Choice and Innovation*, (Contemporary Ergonomics)

Mitchell, J, McClenahan, J and Bennington, J, 2000, *"Autonomy for Disabled People: the need for systematic choice and innovation"*, (Contemporary Ergonomics), pp 375–380, Eds McCabe, PT, Hanson, E and Robertson, SA, (Taylor and Francis and the Ergonomics Society), London

Mitchell et al, 2004, *"Disabled Peoples'* Aspirations and Barriers in Learning for Life and Learning for Work", (Ergonova), Sheffield

Mitchell, J, 2002, *"Finding, Getting, Keeping and Progressing in Work: disabled peoples perspectives"*, (Ergonova), Sheffield

Mitchell, J, 2002, *"Disabled Peoples' Perspectives on Autonomy, Dependence, Support and Evidence"*, (Ergonova), Sheffield

Mitchell, J and Bennington, J, 1998, *"Mapping the Mobility and Functional Needs of Consumers in Wheelchairs"*, Health Research Institute, Sheffield Hallam University

# METHODS AND TOOLS

# CAN ERGONOMICS (AND INVESTMENTS IN HEALTH AND SAFETY) BE USEFUL IN SUSTAINING A COMPETITIVE ADVANTAGE?

**Sylvie Nadeau**

*Department of Mechanical Engineering,*
*École de technologie supérieure 1100, Notre-Dame St. West,*
*Montreal, Quebec, H3C 1K3, Canada*

This work evaluates the impact of ergonomics (and investments in health and safety) on corporate performance. Porter's national competitive advantage framework is used to show conditions already in place encourage Quebec firms to invest in ergonomics, health and safety. Then Porter's value chain model, the systems approach and the value-added activity concept are employed to show how an organization can improve profitability and competitive position through ergonomics, health and safety innovations and improvements.

*Keywords:* Ergonomics, Health, Safety, Competitive advantage, Value added activity, Porter.

## Introduction

Ergonomics is the discipline that seeks to promote to the greatest extent possible mutual adaptation between human beings and their work. Its intent is to improve efficiency and productivity among workers while contributing to their well-being.

In practice, many businesses consider health, safety and ergonomics as cost centres (Rouse and Boff 1997). This conclusion implies less about the inherent value of human factors than it does about the usual organizational role of human factors. Moreover, Smallman et al. (2001) consider ergonomics, health and safety a specialist's field that needs auditing and intervention from consulting firms. Finally, investments in health, safety and ergonomics are often perceived as too demanding (Nadeau 2001).

In this paper, we apply the systems approach, the value-added activity concept, Porter's national competitive advantage framework and value chain to demonstrate that ergonomics (and investments in health and safety) can be useful in sustaining a competitive advantage, especially for world scale competitors.

## National competitive advantage, ergonomics, health and safety

Porter models the origins of a national competitive advantage as a diamond (figure 1).

Four national attributes contribute to a national advantage:

1. The nature of national demand for products and services in an industry;
2. The presence of assisting and closely related industries;
3. The strategy, structure and rivals: firm creation and rivalry are ruled by national conditions;
4. The inputs necessary to produce in an industry.

Does the Province of Quebec offer a national advantage in ergonomics, health and safety?

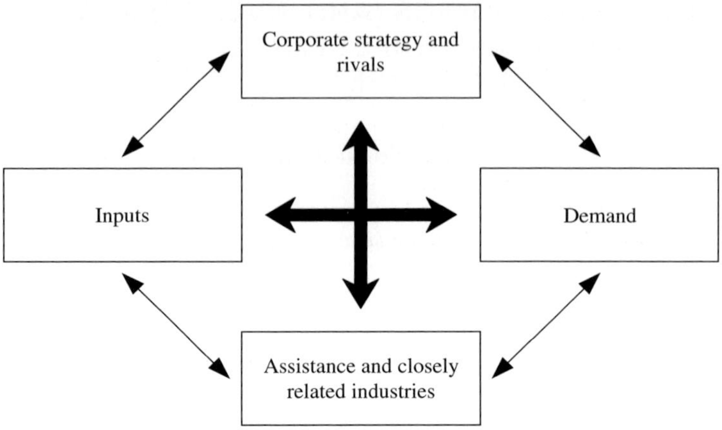


**Figure 1. The origins of national competitive advantage (Porter 1999).**

*Inputs necessary to produce in an industry*

Are inputs needed to solve ergonomics, health and safety problems sufficient to sustain a national advantage?

Firstly, relevant information needs to be collected and compiled on ergonomics, health and safety at work. This task is currently performed by the Quebec Health and Safety Commission (CSST), the physician responsible within the organization, hospitals, the community health department (CLSC) and employers.

Secondly, specialized training programs must exist. These are already available in Quebec. There are many graduate, doctoral and certificate programs at Quebec's University in Montreal (UQÀM), Quebec's University in Chicoutimi (UQAC), Quebec's University in Rimouski (UQAR), Quebec's University in Trois-Rivières (UQTR), Superior Technological School (ÉTS), Laval University, McGill University, Montreal University, Sherbrooke University, Polytechnic School of Montreal, etc. Only drawback is that most programs are in engineering and health sciences courses. Most students in management have scarce knowledge in health and safety (Smallman et al. 2001). At the undergraduate, professional and technical level, health and safety courses are rare.

Thirdly, local university research activities must be encouraged. Again, this is already the case throughout the Province of Quebec. Chairs have been established and many health and safety laboratories exist. A health and safety research network (RSSTQ) brings together approximately 130 researchers in the field throughout the province.

Finally, a strong specialized scientific institute is needed to encourage continuous improvement and innovation (Porter 1999). In Quebec, the Institute of Research in Health and Safety (IRSST) achieves this.

All indicates resources to solve ergonomics, health and safety problems already existing within the province are adequate to sustain a competitive advantage.

*Nature of national demand for products and services in the industry*

Is national demand for products and services in health, safety and ergonomics sufficient to sustain a national advantage?

Innovative, demanding customers play a key role in sustaining a national competitive advantage. Today's customers (intra or extra firm) want useful, usable, safer products. Design for ergonomics in products, in assembly and in field service, seeks to eliminate hazards, inefficiencies in production, and human error at the design phase.

Many Quebec firms already have taken the first step in this direction by proposing knowledge based products.

Rising cost of health and safety compensation, aging population, increased musculoskeletal injuries and mental health problems, automation and difficulties inherent to the introduction of new work organizations, all constitute challenges. To these must be added recent changes in criminal law (Bill C-45) making employers criminally liable for the death or injury of an employee.

Therefore, national demand for ergonomics, health and safety products and services would appear adequate to sustain a national advantage.

*Strategy, structure and rivals: New entrants and rivalry depend on national conditions*

Do national competitive conditions favour investments in ergonomics, health and safety?

In more and more firms, health, safety and ergonomics are considered threatening to the firm's reputation and profits. They are therefore integrated in global initiatives (for example, integral quality management or six sigma) to increase productivity, competitiveness and profitability (Smallman et al. 2001). Furthermore, ergonomics, through just-in-time initiatives, lends itself to production environments.

Conversely, costs and consequences arising from poor management of ergonomics, health and safety would seem to make it good business to invest in these areas.

*Assisting and closely related industries*

Are assisting and closely related industries competitive?

Suppliers of material and services in health and safety (consultants and suppliers of equipment) and many associations (unions, employers' associations, sector-based associations, occupational associations, security groups) are numerous. Advantage springs from efficient communication, rapid and continuous movement of information. Firms and suppliers normally accelerate the innovative rhythm (Kleiner and Drury 1999) but in Quebec a great deal remains to be done.

## Ergonomics, health and safety as value-added activities and tools to sustain a competitive advantage

Are ergonomics, health and safety valued added activities?

Accidents and work related illnesses can be considered a source of waste owing to inefficient use or lower productivity among personnel (Boisvert 1992).This may be considered an opportunity cost for the organization. Furthermore, ergonomic interventions can identify delay and bottleneck problems.

Sound management of health and safety is a value added activity by Porter's definition (Porter 1999). It aims to add value for men at work and for the organization as a system by improving core and support activities as well as improving the product delivered to the customer. Many authors support this position (MacLeod 1995, Gross 1996).

Organizations are analyzed classically using the systems approach (figure 2).

This value chain disaggregates a firm into its strategically relevant activities. To gain a competitive advantage, an organization must create value for its customers through its different strategic activities. Where are ergonomics, health and safety in this systems viewpoint?

The vast majority of our organizations regard all three as an assistance activity linking skills in human resources, accounting and sometimes engineering. In reality, however, both core and assistance activities are supported. Ergonomics, health and safety use the

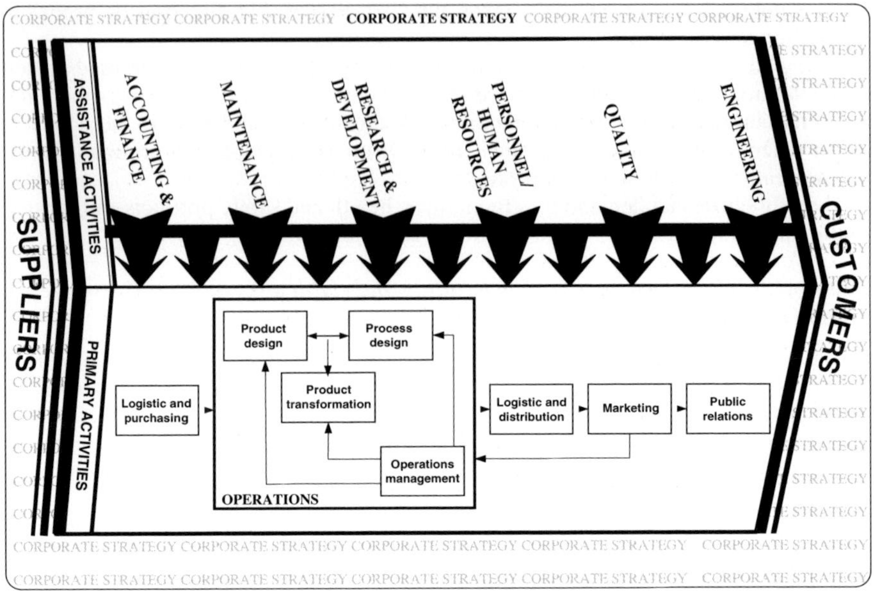


**Figure 2. The organization system. Adapted by Pasin et al. (1998) and Porter (1999).**

systems approach to solve adaptation problems confronting individuals and groups in the workplace. (Karwowski 1996). Problems analyzed often reveal failures in product design, process design, equipment design or operations management. Employing ergonomics, health and safety tools in an organization's strategic activities can be considered an innovative way to pursue new or different approaches to strategic management.

To be efficient, however, ergonomics, health and safety all must be integrated in the value chain and considered as important as commercial and operational activities (Simard 2001, Smallman et al. 2001). Parcelling responsibilities among different intervening parties may diminish the efficiency of health and safety risk control.

## Conclusion

Based on Porter's national competitive advantage framework, we are satisfied that the ergonomics, health and safety industry in Quebec is well structured.

We have also demonstrated that ergonomics (and sound management of health and safety) ought to play an integral part in the best practices followed by an organization seeking ways to improve profitability, especially in the context of a worldwide economy.

Finally, ergonomics, health and safety initiatives can play an important role in delivering the competitive advantage needed for sustained profitability and growth in any organization.

## Bibliography

Boisvert, C.Z., 1992. *Gestion de la santé et de la sécurité au travail.* (Canada: Gaëtan Morin).

Gross, C.M., 1996. *The right fit: the power of ergonomics as a competitive strategy.* (USA: Productivity Press).

Karwowski, W., 1996. *The complexity-compatibility paradigm in the context of organizational design of human-machine systems.* In Brown, O. and Hendrick, H.W. Human factors in organizational design and management. (Holland: Elsevier).

Kleiner, B.M. and Drury, C.G., 1999. *Large-scale regional economic development: macro-ergonomics in theory and practice.* Human Factors and Ergonomics in Manufacturing, 9(2), 151–163.

MacLeod, D., 1995. *The ergonomics edge: improving safety, quality and productivity.* (USA: Van Nostrand Reinhold).

Nadeau, S., 2001. *Outil d'analyse multifactorielle pour la prévention des maux de dos.* Doctoral thesis, École Polytechnique de Montréal, Département de Mathématiques et de Génie Industriel, Canada.

Pasin, F., Jobin, M.H., Handfield, R. et al., 1998. *Introduction à la gestion des opérations.* (Canada: École des Hautes Études Commerciales).

Porter, M., 1999. *La concurrence selon Porter.* (France: Village Mondial).

Rouse, W.B. and Boff, K.R., 1997. *Assessing cost/benefits of human factors*, in Salvendy, G., Handbook of Human Factors and Ergonomics. (USA: Wiley-Interscience).

Simard, M., 2001. *La prise de risques dans le travail*, phénomène organisationel, in Goguelin, P. and Cuny, X. La prise de risque dans le travail, Actes des journées d'étude et de réflexion organisées par les Chaires de psychologie du travail et sécurité du travail (France: Octares).

Smallman, C., Robinson, A. and John, G., 2001. *Des attitudes qui altèrent les performances.* In Financial Times: L'art de la gestion des risques. (USA: Prentice-Hall).

# USE OF LOGIC TREES TO EVALUATE CORPORATE PERFORMANCE IN ERGONOMICS, HEALTH AND SAFETY

**B. Julien[1], S. Nadeau, N. Voyer[2] & C. Roy[3]**

*[1]École de technologie supérieure, Mechanical engineering department,*
*1100 Notre-Dame West, Montreal (Quebec) H3C 1K3 Canada*
*[2]Sous-traitance Industrielle Québec,*
*1080 Beaver Hall, Suite 900, Montreal (Quebec) H2Z 1S8 Canada*
*[3]Mouvement Desjardins, Complexe Desjardins C.P.7, Montreal H5B 1B2*

Poor management of investments in ergonomics, health and safety can have a considerable impact on corporate performance. Working from this hypothesis, a review of the literature and fault tree analysis, we designed a logic tree by which to evaluate performance of an organization in ergonomics, health and safety. Results indicate firms can improve profitability and competitive significantly by conducting adequate audits and by adopting approaches that lead to continuous improvement in ergonomics, health and safety.

*Keywords:* Ergonomics, Health, Safety, Performance, Profitability, Competitive advantage.

## Introduction

Ergonomics is the discipline that seeks to promote, to the greatest extent possible, mutual adaptation between human beings and their work. Intent is to improve efficiency and productivity among workers while contributing to their well-being. Poor management of investments in ergonomics, health and safety has considerable impact on corporate performance. From this hypothesis, a review of the literature and use of fault tree analysis, we have designed a logic tree by which to evaluate performance of an organization in ergonomics, health and safety.

## Logic tree

We relied on an organization and information tool to develop an audit logic tree based on a review of the literature addressing performance audits in health and safety management, and an exhaustive description of those elements which ought to be considered in a health and safety system.

The tool favoured in the study was to construct scenarios using fault tree analysis. Developed in 1961 by Watson of Bell Laboratories (Watson 1962), this tool has been used widely in safety engineering. The rigour of its logic, verification and authentication, makes this an appropriate tool in our context. It needs to be stated, however, that iterative analysis may fail to identify all the risk factors associated with good performance in health and safety. Accordingly, changes to the model may be required in the future. Finally, analysis of interrelationships among these factors has been undertaken using results from a review of the literature and a deductive process similar to fault tree analysis. Our study identified links with other corporate functions: purchasing, engineering, personnel/human resources, transformation/conversion processes, accounting and maintenance. Figure 1 represents the logic tree developed.

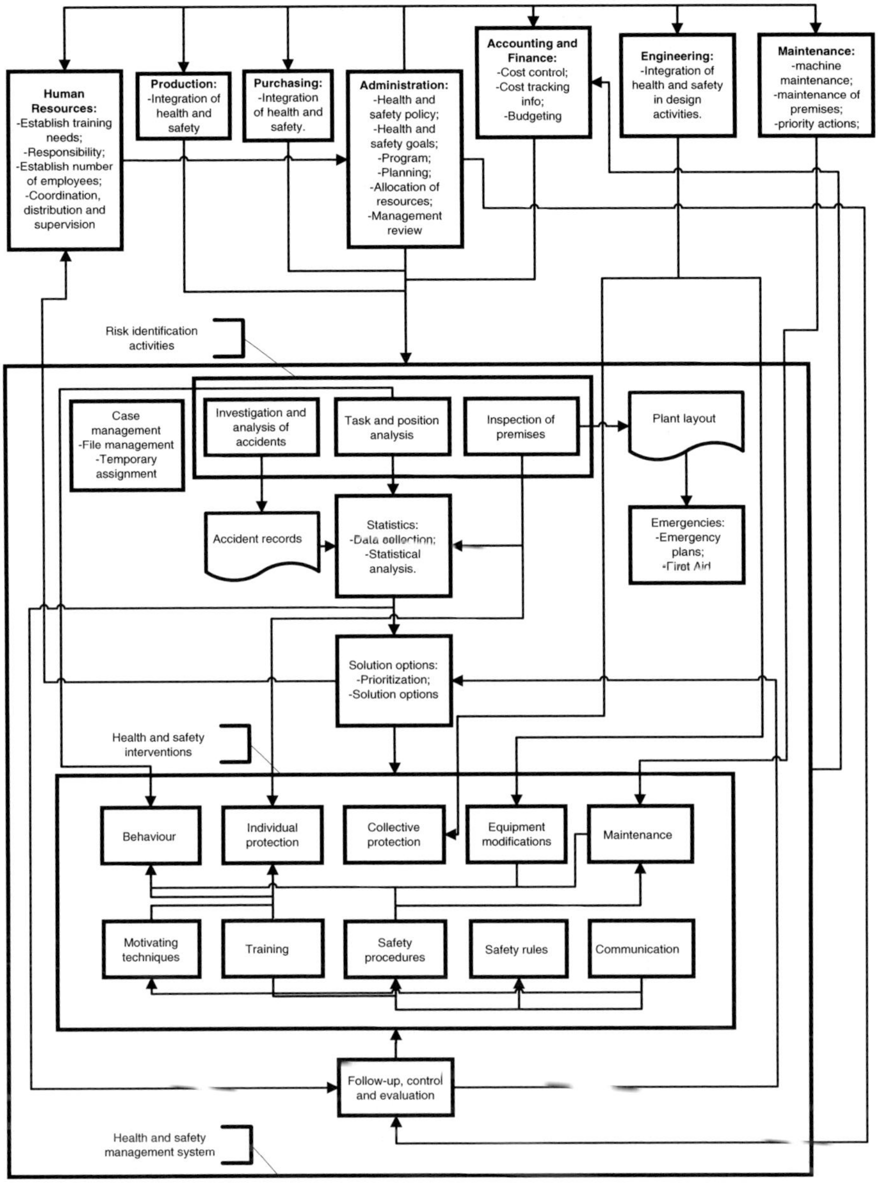


**Figure 1. Health and safety logic tree.**

## Conclusion

We believe an adequate audit coupled with continuous improvement approaches can improve the profitability and competitive position of firms. Many consider an audit a fundamental total quality tool (Périgord 1992, Mitra 1998).

Sous-traitance Industrielle Québec (STIQ) and Mouvement Desjardins, in partnership with the research team in production integrated technologies (LITP) of the École de technologie supérieure (ÉTS) have developed a tool to audit manufacturing firms that covers every aspect of a corporate system: maintenance, personnel/human resources, quality, accounting, engineering, distribution, product design, process design, transformation/conversion processes, operations management, purchasing, marketing and public relations. Results of this research project are in the evaluation phase, through the STIQ DIAGNOSTIC project.

This research project has been sponsored by NSERC (National Sciences and Engineering Research Council), STIQ (Sous-traitance Industrielle Québec) and Mouvement Desjardins.

## Bibliography

Akbar-Khanzadeh, E. and Wagner, D.D. 2001, "*Safety and health program assessment in relation to the number and type of safety and health violations*," AIHAJ, vol. 62, pp. 605–610.

Canada, D.D.R.H. 2002, "Formation et santé et sécurité au travail," *Travail et santé*, vol. 18, pp. 28–29.

Colvin, R.J. 1991, *The guidebook to successful safety programming*: Lewis Publishers Inc.

CSST (2003) "Communiqués de presse," in *http://www.csst.qc.ca/publications/fil_de_presse/2004.html*:CSST.

Daoust, A. 2002, *"Sécurité des machine: faut-il toujours cadenasser les équipements?,"* *Travail et santé*, vol. 18, pp. 27–29.

Donoghue, A.M. 2001, "*A risk-based system to penalize and reward line management for occupational safety and health performance*," Society of occupational medecine, vol. 51, pp. 354–356.

Flin, R. 2003, "*Danger-Men at work: management influence on safety*," Human factors and ergonomics in manufacturing, vol. 13, pp. 261–268.

Genaidy, A., Karwowski, W and Christensen, D. 1999, "*Principles of work system performance optmisation: a business ergonomics approach*," Human factors and ergonomics in manufacturing, vol. 9, pp. 105–192.

Gey, J.-M. and Courdeau, D. 2002, *Pratiquer le management de la santé et de la sécurité au travail*: AFNOR.

Herzberg, F. 1987, "*One more time: how do you motivate employees,*" Professional safety, vol. 65, pp. 109–120.

Jones, S.E., Hosein, H.R., Swalm, G.R. and Yablonsky, J.F. 1990, *Occupationnal hygiene management guide*. Chelsea: Lewis Publishers Inc.

Krause, T.R., Langston, G.W. and Hidley, J.H. 1994, "*Behavior-based task analysis prompt early response to ergonomic response*," Occupational health and safety, vol. 63, pp. 55–59.

Krause, T.R. 1997, "*Trends and developments in behavior-based safety,*" Professional safety, vol. 42, pp. 20–25.

Krause, T.R. 2002, "*Cross-functional improvement: behavior-based safety as a tool for organizational success*," Professional safety, vol. 47, pp. 27–33.

Krause, T.R. 2000, "*Motivating employees for safety success*," Professional safety, vol. 45, pp. 22–25.

"*Loi sur la santé et la sécurité*," 2002, pp. 69.

"*Loi sur les accidents du travail et les maladies professionnelles*," 1991.

Maslow, A.H. 1943, "*A theory for motivation*," Psychological Reviews, pp. 370–396.

Mitra, A. 1998, *Fundamentals of quality control and improvement*, Prentice Hall Inc.

Mouton, J.P. and Chaboud, J. 2003, *La sécurité en entreprise*. Paris: Dunod, 2003.

Périgord, M. 1992, *Réussir la qualité totale*, Les éditions d'Organisation.
Pérusse, M. 1995, *Le coffre à outils de la prévention des accidents en milieu de travail*, ed. Napierville: Le groupe de communication Sansectra Inc.
Pérusse, M. 2001, "*La gestion environnement-santé-sécurité des entrepreneurs*," Travail et santé, vol. 14, pp. 13–17.
Redinger, C.F. and Levine, S.P. 1998, "*Development and evaluation of the Michigan occupational health and safety management system assessment instrument: a universal OHSMS performance measurement tool*," American Industrial Hygiene Association Journal, vol. 59, pp. 572–581.
"*Règlement sur la santé et la sécurité du travail*," 2001.

# EVENT ANALYSIS OF SYSTEMIC TEAMWORK (EAST); A NOVEL INTEGRATION OF ERGONOMICS METHODS TO ANALYSE C4i ACTIVITY

**Dr Guy H Walker[1], Dr Huw Gibson[2], Prof. Neville A. Stanton[1], Dr Chris Baber[2], Paul Salmon[1] & Dr Damian Green[1]**

*Defence Technology Centre for Human Factors Integration (DTC HFI)*
*[1]Brunel University, School of Engineering and Design, Uxbridge UB8 3PH, [UK]*
*[2]University of Birmingham, School of Electronic, Electrical & Computing Engineering, Edgbaston, Birmingham B15 2TT, [UK]*

The EAST methodology has been developed by the DTC HFI as a means to describe and understand C4i scenarios using a multiple method approach. As a demonstration of the capability of EAST it has been applied to four live scenarios involving collaborative activity amongst distributed teams. These scenarios were drawn from the rail industry, and this paper briefly describes the process-based application and outcomes of the analysis.

## Introduction

Command, control, communications, computers and intelligence (C4i) is the management infrastructure for any large, complex and dynamic resource system, usually comprised of human and non-human elements (Harris & White, 1987). A number of existing methodologies relate to the study of C4i, deriving, amongst others from cybernetics, systems engineering and cognitive psychology. Regardless of subject domain, the key descriptive constructs in C4i can be broadly distilled down to:

1. Who (are the agents participating in a C4i scenario)
2. When (do the agents perform tasks and communicate within the C4i scenario)
3. Where (are the tasks being enacted)
4. How (are agents coordinating and communicating to achieve scenario aims)

In order to access these key constructs we propose a novel approach based on methods integration. The approach is called EAST (Event Analysis for Systemic Teamwork) and it aims to provide a unified descriptive overview of C4i scenarios.

## Methods

EAST is based upon the collection of live data using a combination of observation and interviews. The observation involves activity sampling, with a view to constructing a Hierarchical Task Analysis (HTA) of the scenario (Annet, 2003), performing a Coordination Demand Analysis (CDA) (Burke, in press), applying the Critical Decision Method (CDM) (Klein, Calderwood & MacGregor, 1989), deriving propositional networks (described in the accompanying paper by Gibson et al.), producing Communications Usage Diagrams (CUD) and Social Network Analyses (SNA) (Watts &

Monk, 2000; Driskell & Mullen, in press). This approach offers a novel form of triangulation, whereby complementary methods offer multiple perspectives on the same data. The final method is an enhanced form of Operation Sequence Diagram (OSD), used as a means to rationalise these multiple perspectives into one summary representation. Process charts like OSD offer a systematic approach to describing tasks (representing both human and system elements), in a manner that is relatively easy to follow and understand (Kirwan & Ainsworth, 1992). A further advantage in using the OSD representation stems from its potential compatibility with systems analytic approaches (such as Unified Modelling Language). In this manner, we propose that the process-based analysis derived from EAST provides us with a convenient route into Human Factors Integration.

Examples of large, complex and dynamic resource systems are taken from the UK rail industry. The scenarios serve as a source of live data to demonstrate the capability of the method. Data was elicited from recorded transcripts wherein four maintenance activities were identified, all involving coordination between distributed teams and individuals. The scenarios are illustrated in the task networks presented in figure one.

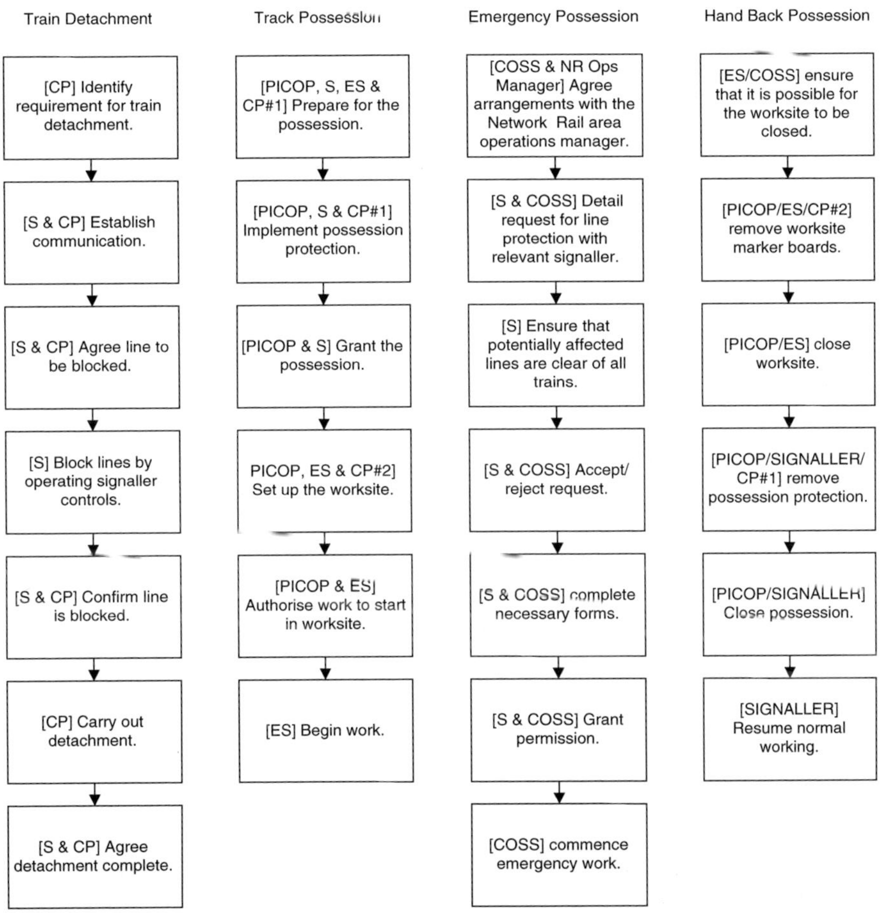


**Figure 1. Task networks for railway maintenance scenarios.**

## Results and discussion

### *HTA (Task networks)*

The task networks are graphical representations of the "plan 0" contained within the HTA's for each scenario. Due to the proceduralised nature of the scenarios, the task networks are in this case sequential, as illustrated in figure one.

### *CDA*

Between 60 to 75% of the tasks within the scenarios involve coordination with other agents. The nature of that coordination is summarised in the Total Coordination score. This score expresses the mean ratings ascribed to the following task elements; communication, situation awareness, decision making, mission analysis, leadership, adaptability and assertiveness. In table one, Total Coordination appears to be highest within the emergency scenario, and lowest within the hand-back of track possession scenario. Closer inspection of the data reveals that the Total Coordination score is comprised of tasks receiving high ratings for communication and situation awareness. The CDA detects that the development and maintenance of situation awareness characterizes much of the collaborative activity undertaken.

### *SNA and CUD*

The results of the SNA and CUD are presented in figure two. The analysis explores the degree of connectivity between agents with two SNA metrics; socio-metric status and centrality. These metrics suggest that the Person In Charge Of Possession (PICOP), Controller Of Site Safety (COSS) and the Signaller are key agents in the scenarios. The prominence of these *three* agents points to a relatively distributed network, expressed graphically in figure two. The implication of varied early research into social networks suggests that there is unlikely to be a single best structure for group performance, rather

**Table 1. Results of CDA analysis, mean total coordination scores.**

| Scenario | Detachment | Planned maintenance | Emergency maintenance | Hand-back possession |
|---|---|---|---|---|
| Mean TC | 1.58 | 1.59 | 1.65 | 1.48 |

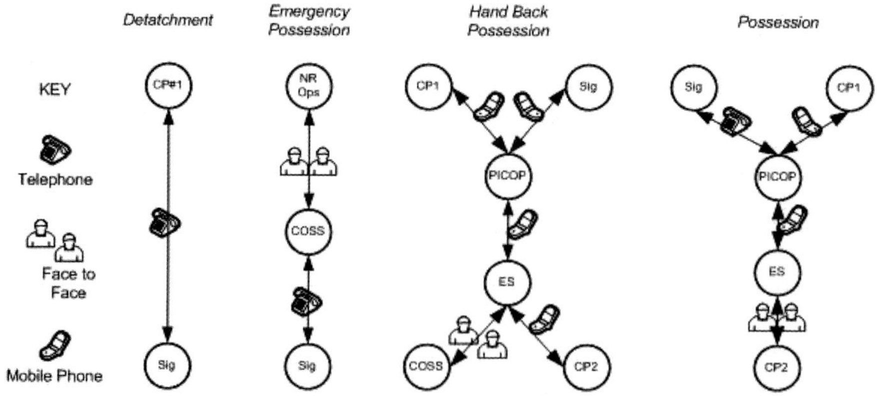


**Figure 2. Social networks annotated with COMMS Media Defined by CUD.**

**Table 2. Task loading table.**

| Agent | PICOP | Signaller | COSS | ES | CP | NRMan |
|---|---|---|---|---|---|---|
| % Total tasks | 32.1 | 26.5 | 7.7 | 13.9 | 19.5 | 0.2 |

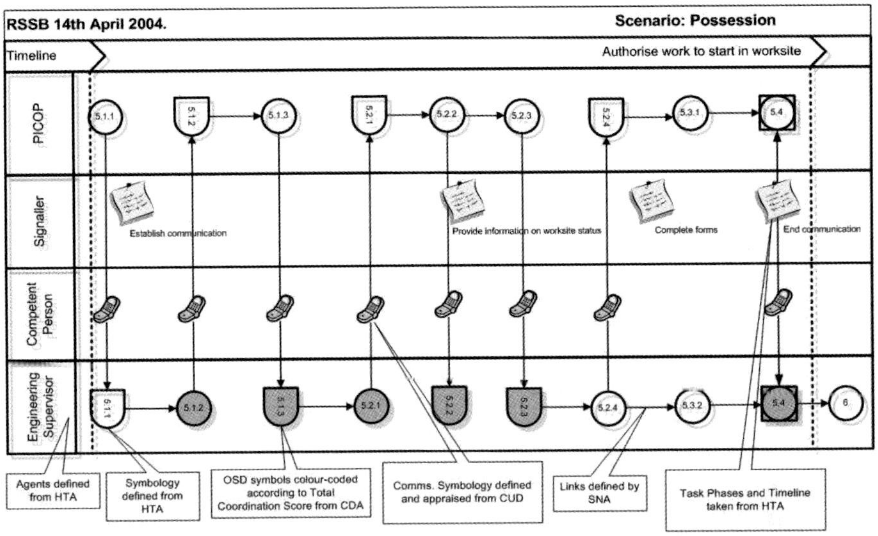


**Figure 3. Enhanced OSD representation.**

the structure of the network interacts with the task loading on members of the group, communications, and various other factors.

Task loading is derived from the OSD (and is presented in table two above) whereas the communications technology that facilitates the links between nodes/agents is annotated on figure two. The facilitation is entirely verbal, although the CUD identifies detailed opportunities for alternatives, including shared situation displays, and text based interfaces combined with greater automation.

*OSD*

Figure three presents a sample of an enhanced OSD and highlights how the preceding methods are integrated with it. The operations loading is presented in table two, showing the Person In Charge Of Possession (PICOP), Signaller and Competent Person (CP) as the most heavily loaded individuals in the network in terms of information flow. This loading is reflected to some extent in the social networks above (figure three) where it can be noted that the PICOP has links to three other agents in the scenario. In addition, it can be further noted that the signaller is also engaged in more decision-making operations, again indicative of their information load.

## Conclusions

The application of the EAST methodology to live railway data briefly illustrates the descriptive capability of the method. The method summarises the task and network structure, and operational context, into a form that enables ready comparisons on key metrics to be drawn between scenarios. The method is also scaleable, and its application by the DTC HFI to a host of military and civil domains is providing input into generic modelling and theories of C4i. The purpose of this wider exercise is to empirically define novel C4i interfaces from Ergonomics first principles.

## Acknowledgement

This work from the Human Factors Integration Defence Technology Centre (HFI DTC) was part-funded by the Human Sciences Domain of the UK Ministry of Defence Scientific Research Programme. The assistance of the Rail Safety and Standards Board is also gratefully acknowledged.

## References

Annett, J. (2003) Hierarchical task analysis. In: D. Diaper & N. A. Stanton. (eds) *Handbook of Task Analysis in Human-Computer Interaction.* Mahwah, NJ: Lawrence Erlbaum Associates.

Burke, C. S. (in press) Team task analysis. In: N. A. Stanton; A. Hedge; K. Brookhuis; E. Salas & H. Hendrick (eds) *The Handbook of human factors and Ergonomics Methods*. Boca Raton, Fl: CRC Press.

Driskell, J. E., & Mullen, B. (in press) Social network analysis. In: N. A. Stanton; A. Hedge, K. Brookhuis; E. Salas & H. Hendrick (eds) *The Handbook of human factors and Ergonomics Methods*. Boca Raton, Fl: CRC Press.

Gibson, H., Walker, G. H., Stanton, N. A., Baber, C., & Salmon, P. (2004). The Event Analysis of Systemic Teamwork (EAST) Methodology and Analyses of Railway Maintenance Tasks. *This Issue.*

Harris, C. J., & White, I. (1987) Advances in command, control & communication systems. (IEE Computing Series 11). London: Peregrinus.

Kirwan, B., & Ainsworth, L. K. (1992) A guide to Task Analysis, Taylor and Francis, London, UK.

Klein, G. A., Calderwood, R., & MacGregor, D. (1989) Critical Decision Method for Eliciting Knowledge. *IEEE Transactions on Systems*, Man and Cybernetics, 19(3), 462–472.

Watts, L.A., & Monk, A.F. (2000) Reasoning about tasks, activities and technology to support collaboration, In: J. Annett & N.A. Stanton (eds) *Task Analysis*, London: Taylor and Francis, 55–78.

# ADOLESCENT ERGONOMICS

**Steve Milanese**

*Centre for Allied Health Evidence, University of South Australia, North Terrace, Adelaide, South Australia 5001*

Adolescence is characterised as a period of rapid development in the areas of physical, cognitive and social development, as the adolescent passes between the relatively stable states of childhood and adulthood. The incidence of adolescent musculoskeletal symptoms ranges from 7% to 40%, yet the determinants of adolescent pain remains unclear. There is a strong suggestion in the literature of a relationship between adolescent and adult spinal pain. When considering the area of Adolescent Ergonomics we need to consider the issues that make this population group so unique.

The World Health Organization (WHO) defines adolescence as the period between 10 and 19 years of age, whilst the Center for Disease Control extends the definition to include ages 10 to 24. The United Nations Population Fund further divides the WHO definition into early adolescence (10–14 years of age) and late adolescence (15–19 years of age).

The variations in definition of adolescence reflect the cultural, racial and regional differences that separate the aging stages. When a person develops from childhood to adolescence and then through to adulthood will vary significantly between different cultural groups, and in different regions of the world. Whilst the use of definitions as a chronological benchmark of population subgroupings facilitates cross professional discussions, it fails to take into account the cultural, racial and regional lines that may differentiate the stages of aging.

In the cultural context, the age set by a given society for the rites of transition to adulthood could be perceived as marking the end of adolescence. Sociologically, adolescence is the transition period from dependent childhood to self-sufficient adulthood. Whilst in a psychological context, it is when an individual has made adjustments that allow progression from child behaviour to adult behaviour in a given society (Muuss, 1962).

A biological definition of adolescence is the span of a person's life between the onset of puberty and the completion of bone growth. The terms adolescence and puberty are often used synonymously, however whilst puberty refers to the physical and physiological changes involved in the sexual maturation of a child, adolescence refers to the stage from the onset of puberty to adulthood, and includes the psychological and cognitive development that accompanies this stage.

At present, using the WHO definition of adolescence, almost one-fifth of the world's population fall within the adolescent grouping, making up more than 1 billion people. In the developing world, more than one-third of the population is under the age of 15.

Adolescence is characterised by a high levels of reported health problems with the incidence of depression, anxiety, drug addiction, and eating disorders significantly increased during this period. It has also been identified that the risk of injury or death is much higher in this period than in any other stage of life (Kelly et al, 2004). Whilst these significant health problems have received justifiable attention, through scientific literature, Government policy and the popular press, the incidence of less obvious health problems receive far less popular attention.

The incidence of musculoskeletal symptoms in the adolescent population has been well documented. Studies have reported that amongst the adolescent population the incidence of musculoskeletal pain is between 7% and 40% (Nissenen et al, 1994). In a recent study of 1466 adolescent students we found reports of symptoms in up to 32% of adolescent students (between 12 years and 17 years of age) in secondary schooling in South Australia. Feldmann et al (2002) found a cumulative annual incidence of neck and upper limb pain of 28.4% in an adolescent population. Of concern are the suggestions that the incidence of low back pain amongst adolescents is approaching that of the adult population (Balague et al, 1988), and that the incidence of spinal pain is occurring at an earlier age than in previous generations (Waddell, 1996). Whether the increasing incidence of symptoms in this population group is reflective of an actual increase in incidence or better reporting mechanisms remains unclear.

It is plausible that the incidence of adult spinal pain is strongly related to adolescent spinal pain. Harreby et al (1995) identified a positive correlation between Low Back Pain in adolescence and continuing pain in the adult population, a situation previously reported with neck and shoulder pain (Hertzberg, 1983). It appears therefore that some of the determinants of musculoskeletal health in the adult population may have developed during adolescence.

The Human-Machine-Environment system (HMES) forms the cornerstone of an Ergonomics approach. Breakdowns in the HMES can be reflected in negative outcomes, such as higher error rates, reduced work efficiency, low morale, and reports of musculoskeletal symptoms, to name just a few. Breakdowns, or mis-matches occur when the demands of the machine, work task or physical environment exceed the capacity of the person (Wilson, 1995).

Before we can utilise the HMES as the model for understanding Adolescent Ergonomics, we need to ensure we are cognisant with all aspects of the system. Traditionally Ergonomics has focussed on the adult as the human in the HMES. This is completely understandable given the historical development of ergonomics, and its application in military, occupational and the political area. However, Adolescents are not "little adults" and the application of adult ergonomic design parameters to the adolescent fails to take into account the specific needs of this population group.

In the USA it has been reported that up to 75% of all students over the age of 15 years of age are in some form of paid employment. Weller et al (2003) and Feldman et al (2002) found significant correlation between the hours worked by adolescents and the reporting of musculoskeletal symptoms. Weller et al, 2003 reported that working more than 20 hours a week increased the likelihood of injury among middle school students. They found that the highest incidence was amongst agriculture workers, followed by restaurant work, construction, and then yard work, whilst Feldman et al (2002) found a higher incidence amongst white collar workers compared to blue collar workers or those adolescents working in childcare. This difference is reflective of the population subgroups selected with Weller et al's study concentrating on lower-income Hispanic South Texas middle school students.

## Adolescent development

Adolescence is characterised as a period of rapid changes in the individual, as they pass between the relatively stable states of childhood and adulthood. Significant advances occur in the areas of physical, cognitive and social development. Apart from the early years of infancy there is no other stage of an individuals life as unsettling as the period of adolescence. Rarely, though do two individual adolescents develop at the same rate across all parameters, with wide variation in an individual's stage and rate of development in each aspect.

Puberty, defined as "the biological changes of adolescence" (Steinberg, 1996), marks the commencement of the physiological changes that the adolescent undergoes, and commences when the hypothalamus is signalled and the pituitary gland activated (Cole & Cole, 1993). It is characterized by the gradual onset of reproductive hormonal activity, which change boys and girls from physical immaturity to biological maturity (Cole & Cole, 1993).

### *Physical changes*

For as many as 3 years during adolescence, the body experiences rapid growth that averages approximately 9 cm a year in females and 10.3 cm a year in males (Muss, 1962). This growth spurt usually occurs in girls about two years before boys. The growth spurt triggers the legs and arms to grow initially, followed by the body's trunk, and ending in the shoulders and chest (Cole & Cole, 1993).

This growth spurt changes the relationship between bone length and musculotendinous units, leading to skeletal vulnerability, decreased flexibility, and increased susceptibility to overuse syndromes. This period may adversely affect the individual adolescents capacity to withstand physical stress during this period.

Body maturation may allow development of a sense of independence, but it may also lead to increased psychosocial stress, as the adolescent grapples with a lack of coordination and a poor understanding of the newfound capabilities.

The effect of the musculoskeletal changes is to alter the capacity of the adolescent to deal with the physical demands placed on them, depending on what stage of development they find themselves in. A standard Ergonomic principle is to design for the 5th%ile to the 95th%ile of the population, based on anthropometric parameters such as strength, height, circumference etc. As individual adolescents are developing at different and differing rates the 5th%ile and 95th%ile members of an adolescent population group will change, at a faster rate than in any other population.

### *Cognitive development*

The development of the thinking and the organizing systems of the brain is defined as cognitive development. The main domains of cognitive development include language, problem solving, memory development, and reasoning. These cognitive changes appear to be a result of learning and brain development. Adolescents and adults have different ways of thinking about different subjects with adults thinking and responding with a more complex attitude due to greater experience and learning.

Steinberg (2004) investigated the influences of age difference on cognitive processes relevant to risk taking and decision making, i.e. risk perception and risk appraisal. The concept of Risk Taking is of importance in ergonomics when considering how an individual interacts with his/her environment. Steinberg (2004) suggested that the increased risk taking amongst adolescents is not due to age differences in risk perception or appraisal, but to differences in intrinsic factors that influence self-regulation. These increased risk taking behaviours may have evolutionary benefits through the development of independence and survival without parental protection, but may also increase the risk of harm (Kelly et al, 2004). Age related development leads to a breakdown between the need for novelty and sensation seeking, which increase dramatically at puberty, and the development of self-regulatory competence, which doesn't develop until early adulthood (Steinberg, 2004). Kelly et al (2004) suggested that neuronal pathways that influence emotional regulation and cognitive function mature at different rates during this transition period, suggesting a psychophysical reason for their risk taking behaviour.

Ergonomic interventions aimed at the adolescent must consider the potential for this inherent increased risk taking potential in this population to occur. The literature suggests that this behaviour is secondary to increased sensation seeking, with lagging development of an adolescent's ability to understand the consequences of their actions. Failure to understand and address these unique cognitive issues in Adolescent Ergonomic programmes potentially dooms any programme to failure.

Adolescents are an important and unique part of our community. Ergonomic interventions aimed at this population group need to understand the uniqueness of this group, and the potential influence of these characteristics on their interaction with their environment. An adolescent's environment includes the playground, the sporting arena, entertainment areas (cinemas, shopping malls), the school environment and even the traditional work environments.

## References

Balague, F., Dutoit, G. and Waldburger, M. 1988, Low back pain in schoolchildren – An epidemiological study. *Scandinavian Journal of Rehabilitation Medicine*, **20**, 175–179.

Cole, M. and Cole, S.R. 1993, *The Development of Children*, Second Edition Scientific American Books, New York).

Fairbank, J., Pynsent, P., vanPoortvliet, J. and Phillips, H. 1984, Influence of anthropometric factors and joint laxity in the incidence of adolescent back pain. *Spine*, **9**, 461–464.

Feldman, D.E., Shrier, I., Rossignol, M. and Abenhaim, L. 2002, Work is a risk factor for adolescent musculoskeletal pain. *Journal of Occupational & Environmental Medicine*, **44**, 956–961.

Galotti, K.M., Kozberg, S.F. and Farmer, M.C. 1991, Gender and developmental differences in adolescents' conceptions of moral reasoning. *Journal of Youth and Adolescence*, **20**, 1–10.

Harreby, M., Neergaard, K., Hesselsoe, G. and Kjer, J. 1995, Are radiographic changes in the thoracic and lumbar spine of adolescents risk factors for low back pain in adults? *Spine*, **20**, 2298–2302.

Hertzberg, A. 1988, Prediction of cervical and low back pain based on routine school health examination. *Scandinavian Journal of Primary Health Care*, **3**, 247–253.

Kelley, A.E., Schochet, T. and Landry, C.F. 2004, Risk taking and novelty seeking in adolescence: Introduction to Part I. *Annals of the New York Academy of Sciences*, **1021**, 27–32.

Muuss, R.E. 1962, *Theories of Adolescence*, Second Edition (Random House, New York).

Niemie, S., Levoska, S. and Kemila, J. 1996, Neckl and shoulder symptoms and leisure activities in high school students, *Journal of Orthopedic and Sports Physical Therapy*, **24**, 25–29.

Nissinen, M., Heliovaara, M., Seitsamo, J., Alaranta, A. and Poussa, M. 1994, Anthropometric measurements and the incidence of low back pain in a cohort of pubertal children. *Spine*, **19**, 1367–1370.

Steinberg, L. 1996, *Adolescence*. Second Edition (McGraw-Hill, Inc, New York).

Steinberg, L. 2004, Risk taking in adolescence: what changes, and why? *Annals of the New York Academy of Sciences*, **1021**, 51–58.

Waddell, G. 1996, Low back pain: A twentieth century health care enigma. *Spine*, **21**, 2829–2825.

Weller, N.F., Cooper, S.P., Tortolero, S.R., Kelder, S.H. and Hassan, S. 2003, Work-related injury among south Texas middle school students: Prevalence and patterns. *Southern Medical Journal*, **96**, 1213–1220.

Wilson, J.R. 1995, A framework and a context for ergonomics methodology, in Wilson J.R. and Corlett. E.R. (eds): *Evaluation of Human Work*, Second Edition (London: Taylor and Francis) 1–39.

# A TASK ANALYSIS OF CRIME SCENE INVESTIGATION

**Paul Smith, Chris Baber, Sandra Wooley, James Cross & John Hunter**

*Department of Electronic, Electric and Computer Engineering, The University of Birmingham, Edgbaston, Birmingham B15 2TT*

The Home Office Science and Technology Strategy identifies the need for criminal investigators to harness new technologies to enhance efficiency during the course of an investigation. Wearable computer technology has developed to provide technological support to many vocational tasks. The aim of the Methods to Support Shared Analysis for Mobile investigators (MsSAM) project is to develop and demonstrate the application of such technology to crime scene investigation. As an initial step in the design process, a detailed task analysis of crime scene investigation has been undertaken. Communicating with individuals involved in crime scene examination and those providing forensic services to police organisations, a theory of technological support to crime scene investigation will be compiled. Using the acquired information, a task analysis of the crime scene examiners duties will be analysed to aid the design specification and creation of a prototype. From this analysis it is possible to divide the process into discrete stages. Each stage has different information requirements, data handling issues and communication demands. Taking into consideration the variables present in crime scene investigation, we explore the interaction of the forensic practitioner with their environment at both the volume crime scene and the major crime scene and the interaction with other forensic disciplines, such as forensic archaeology and fire investigation.

## Introduction

Science and technological developments need to be exploited to equip the police service with the necessary tools to deliver effective policing as part of a modern and respectable criminal justice system (Home Office, 2001). There are many organisations meeting the needs of the police service in areas such as identification, intelligence, criminal justice, force support and the enhancement of forensic services (PITO, 2003). The following paper uses objective and subjective methods of data collection to formulate a Hierarchical Task Analysis (HTA) of the complete process of crime scene investigation, incorporating an understanding of the supporting factors affecting Crime Scene Examiners (CSE). The various tasks that the CSE's are required to carry out in order to achieve their goals at the volume crime scene and the major crime scene will be explored. These data can then be used to formulate a theory of technological support to crime scene investigation, producing design concepts to establish prototypes to support the forensic practitioner.

Integrating technological enhancement within the criminal justice framework requires an understanding of the procedures and recommendations provided by regulatory and police governing organisations, in conjunction with the varying internal force protocols. The purpose of examining a crime scene is to formulate a hypothesis based on all the available evidence, therefore understanding the most likely course of events that has resulted in the observed circumstance (Jamieson, 2004). Criminal investigation has progressively evolved, enhanced by the development of technical and expert services (Davies et al., 1998). However, efficient crime scene investigation must be matched with the best investigative model, which must be undertaken within the parameters of the systems governance of crime scene procedures (Jamieson, 2004). The broad confidence in current and developing forensic technologies can be undermined if the proper procedures are not followed (Townley and Ede, 2004). The role of the investigator has been described as being ambiguous (Wright, 2002); often the direction is torn between the practical demands and official procedural requirements (Hobbs, 1988). There are occasions when procedures and investigative parameters stipulated by governing bodies can conflict with the environmental constraints, and operating practicalities of the investigative process. The practice of the forensic practitioner is increasingly challenged by the court, which has prompted a greater level of transparency in the forensic investigators actions at the scene. Integrating technologies within the criminal justice system requires the designer to have an awareness of, not just the physical process of the task, but the impact on the associated groups involved in the investigative process. Many aspects of the investigation suffer from common and repetitive weaknesses, which can be compounded as the evidence trail progresses. Efforts have been introduced to put into effect tighter managerial control over the investigation, and the introduction of more structured and streamlined procedures (Jacobson, 2003). The introduction of "skills for justice" has set out National Occupational Standards for the police service, in particular a standardised protocol for best practice in the examination of the crime scene for CSE's (HMIC, 2004). All the police authorities in England and Wales are required to support continuous improvement in functionality within their force, enhancing the development of efficiency and effectiveness in an economical manner (Local Government Act, 1999).

## Methods for capturing crime scene investigative practice

For human factors, understanding operator behaviour can be achieved through task analysis; through focusing on an operator's goals and sub goals, a task analysis can supplement the redesign of existing practice with novel and applicable design strategies to improve the efficiency and efficacy of task completion (Shepherd, 2001). Understanding the procedural requirements has helped by providing an overarching template for the CSE's operations but does not account for the interaction of the CSE with the variables of their working environment, for this accompanying the CSE's from different forces and under varying operating conditions was required and has been undertaken.

### *Objective observations*

The basic investigative structure for a typical police force in England and Wales is highlighted in Figure 1. The police response to a crime invariably begins with a call either from a member of the public, or from a police officer, this information is relayed to the call management team, and subsequently the relevant area control desk.

A general overview of the task analysis is shown in Figure 2.

### *Subjective observations*

Three Police Forces are involved in the MsSAM project; these forces provide variation in population, rural/urban distribution, and, most importantly, difference in crime scene investigative operations. The first objective was to visit the Scientific Support Managers (SSM) to understand the specific operational requirements for the relevant force. Once an idea of individual force practice was obtained, time was spent in the company of a Crime Scene Examiner, to record their methods of job allocation, information transfer,

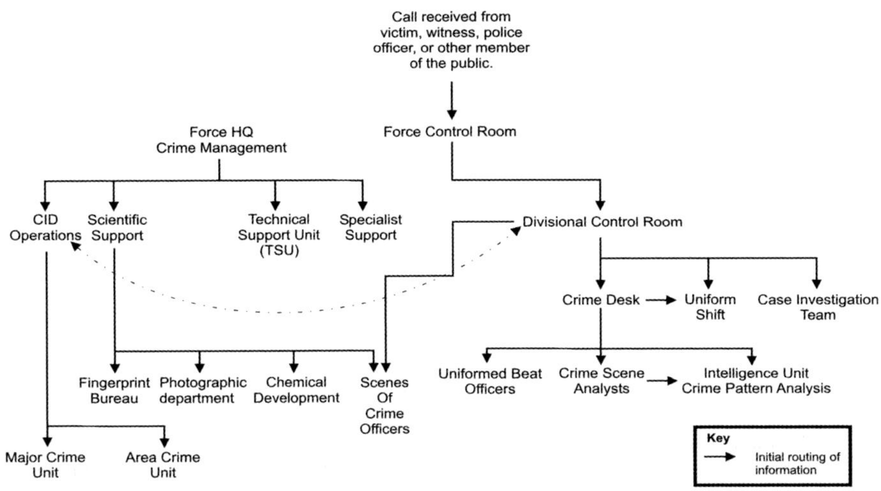


**Figure 1. Basic structure of information flow in a typical police force (Adapted from Townley and Ede, 2004).**

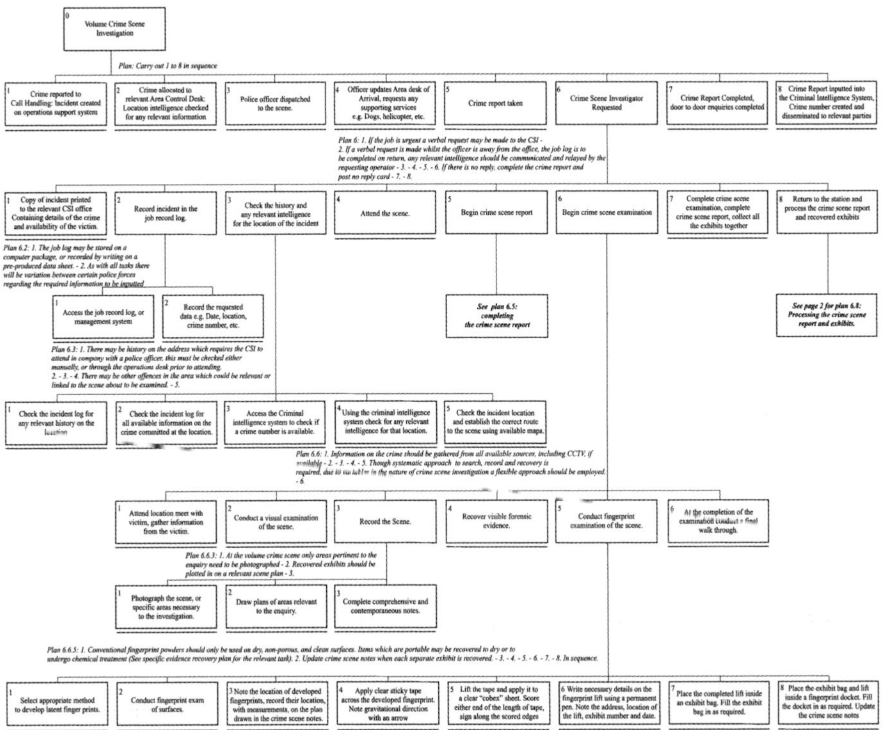


**Figure 2. HTA for crime scene investigation for volume crime.**

**Table 1. Crime scene examiner procedure times.**

| | Force A | Force B | Force C |
|---|---|---|---|
| Record Incident | 3 minutes | 4 minutes | 0 |
| Travelling | 30 minutes | 38 minutes | 0 |
| Undertaking Investigation (average) | 45 minutes | 55 minutes | 0 |
| Completing Paperwork | 20 minutes | 35 minutes | 0 |
| Information Management | 15 minutes | 8 minutes | 0 |

**Table 2. Varying force methods.**

| Force | Incident recording | Investigation | Information management |
|---|---|---|---|
| Force A | Inputted into an evidence management system | Laptop / Hand written report | Evidence Management System |
| Force B | Incident logged on a spread sheet | Hand written report | Criminal Intelligence System |
| Force C | Inputted into an evidence management system. | Laptop/Hand written report | Evidence Management System |

data requirements and information/evidence management. The time spent on each aspect during a typical day was recorded and the information is displayed in Table 1, Force C had no incidents requiring a CSE that day so no timing data was available. The varying methods of each force are highlighted in Table 2.

## Results and conclusion

Established protocols and a training regime operated by Centrex at the National Training Centre for Scientific Support to Crime Scene Investigation have promoted a progression of nationally agreed standards of crime scene investigation. Therefore, a general task analysis for crime scene practice can give a valuable overview of the necessary tasks required to achieve the varying objectives, these overall objectives are predominantly the same for all forces in England and Wales. The apparent differences that exist between police forces have derived from variances in structure and framework of the supporting systems. The main difference may be the hierarchical structure of the scientific support departments. Some forces may employ assistant, or volume crime CSE's in addition to the CSE's that tackle more major crime. Differences may also exist in the dissemination and management of the evidence. Computerised evidence management systems may be used to control and record the continuity of evidence from recovery to examination. As an initial step, monitoring the three police forces has provided valuable data to help evaluate potential integration and compatibility problems in each force. The task analysis not only formulates the overarching framework of task objectives but form an awareness of the subtle differences between police forces, which need to be considered during the design process.

Each stage of the investigation has different information requirements. From the analysis it has become possible to break the process into discrete stages, each demanding differing data handling issues and communication demands. The first stage is invariably the initial request for a scene examination to be carried out, the request coming from the officer in the case relaying the message via the force control room. This message is sent in varying formats to the relevant geographical CSE office, and depending on the type of crime, the CSE records the crime into an incident log, attends the incident, carries out the examination and processes the evidence and data accordingly. However, there are occasions when problems occur restricting the smooth execution of the examination. There can be a number of issues that disrupt the investigation process, including difficulties in communicating with the victim, the incident not being as originally reported, or a history on the address which may compromise the health and safety of attending officers.

In addition to the cognitive elements that affect CSE activities, there are also other elements induced by operating a large organisation, such as staffing levels and sufficient office space for employees, all problems which have been encountered on visiting the various forces. Despite the effects of the above, the CSE's operate to a high standard and achieve numerous identifications leading to an efficient and effective investigative process. Therefore, any technological integration is required to be achieved sustaining the current high levels of practice, whilst improving the operational environment and reducing the probability of error.

## Acknowledgements

The work reported in this paper was supported by EPSRC Grant GR/S85115 Methods to Support Shared Analysis for Mobile Investigators.

## References

Davies, M., Croall, H., and Tyrer, J. (1998) *Criminal Justice: An introduction to the criminal justice system in England and Wales*. Longman, London.

HMIC (2004) *Modernising the police service*. July 2004. Crown Copyright, London.

Hobbs, D. (1988) *Doing the Business*. Oxford University Press, Oxford.

Home Office (2001) *Policing a new century – A blueprint for reform*. Crown Copyright, London.

Jacobson, J., Maitland, L., and Hough, M. (2003) *Reducing burglary initiative: Investigating burglary*. Home Office Research Study 264. Home Office Research, Development and Statistics Directorate. June 2003, Crown Copyright, London.

Jamieson, J. (2004) "A rational approach to the principles and practice of crime scene investigation: 1. Principles." *Science and Justice*. 44 (1) pp 3–7.

Local Government Act (1999) Chapter 27, Section 3 (1). Crown Copyright. London.

PITO (2003) *Forward plan 2003–2008*. Crown Copyright, London.

Townley, L., and Ede, R. (2004) *"Forensic Evidence: its nature, uses, types and Importance"*. In Forensic Practice in Criminal Cases. The Law Society, London.

Wright, A., (2002) Policing: an introduction to concepts and practice. Willan Publishing, Devon.

# DEVELOPMENT OF THE WESTT HUMAN PERFORMANCE ANALYSIS TOOL

**R.J. Houghton[1], M. Cowton[2] & C. Baber[1]**

*[1]HFIDTC, Electronic, Electrical and Computer Engineering*
*The University of Birmingham B15 2TT, UK*
*Email: r.j.houghton@bham.ac.uk; c.baber@bham.ac.uk*
*[2]HFIDTC, Aerosystems International (AeI), Alvington, Yeovil, Somerset BA22 8UZ, UK*

In this paper we describe the development, functionality and applications of the WESTT human performance-modelling tool which represents human and technological activity at multiple levels of analysis; primarily as social networks, propositional networks, and in terms of operational loading. The tool can be used to analyse data from field observations as well as in prototyping new system configurations.

## Introduction

This paper describes the development, functionality and application of the WESTT[1] human performance modelling software. The primary aim of the work was to produce a means of representing, analysing and ultimately modelling team activity, particularly within systems that feature so-called network-enabled capability (NEC). The resulting models allow the exploration of novel system configurations such as those that may be encountered in, for example, combined military or emergency services operations. WESTT represents operational activity at a systems level in which both humans and NEC technologies are represented as agents whose activity can be considered over time within social networks and as propositional networks of knowledge items. WESTT can be used by Human Factors/Ergonomics practitioners in at least two ways; first, to represent observational data from field studies of existing systems to allow their quantitative assessment and qualitative description and second, to provide a reconfigurable representation of systems, i.e., to allow analysts to modify the system description in order to explore the consequences of alternative designs and procedures.

## Overview of WESTT

The organisation of the WESTT software is depicted in Figure 1. The first stage in using the system is to collect data in the field using the CASA application which runs on an iPAQ hand-held device. The analyst observes behaviour within the team and presses marked buttons as different events take place. These data are automatically date and time-stamped by the application creating an overall record that can then be edited in a spreadsheet such as Microsoft Excel (in order to, for example, merge together the observational data from different observers) for eventual import into WESTT itself. Alternatively, an analyst might choose to construct a set of invents directly in Excel on

[1] The acronym WESTT stands for Workload, Error, Situational awareness, Time and Teamwork.

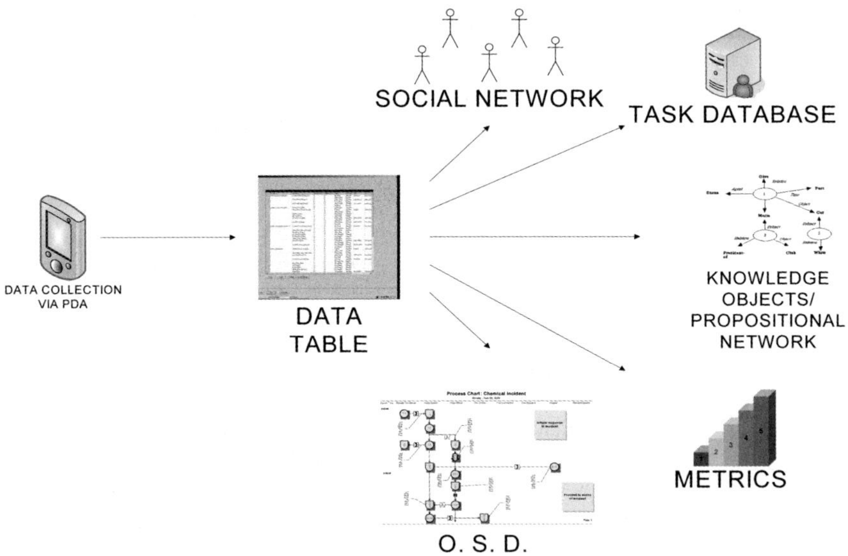


**Figure 1. WESTT architecture.**

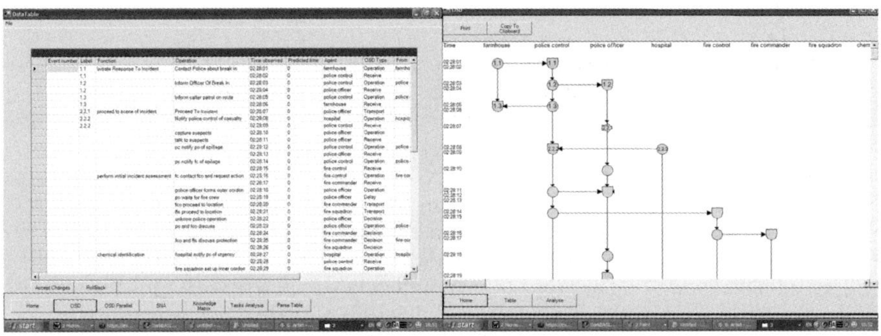

**Figures 2a and 2b. Central data table and operational sequence diagram.**

the basis of interviews, Subject Matter Expert advice, or published doctrine/procedures and so on.

The centre point within WESTT, both in terms of the user interface and in terms of the underlying software itself is the data table (Figure 2a) which holds – in chronological order – the details of the observed activity in terms of what individual actors did and who they communicated with and how. It is from this data that Operational Sequence Diagram (OSD), Social Network Analysis (SNA) and is also used together with the Task Database to calculate performance metrics. The OSD provides the analyst with a graphical flow-chart style representation of operations and agents over time (see Figure 2b). The OSD has long been recognised as a useful aid to interpreting multi-agent activity, but "… the task of drawing a complex OSD can be extremely cumbersome and expensive." (Meister, 1985, p. 68). Various attempts have been made to automate the drawing of OSDs (particularly during the 1980s), although there are few if any commercial products that are currently available to do this. WESTT is able to construct the OSD immediately from the data supplied. Using the observational data, WESTT also tallies up the distribution of

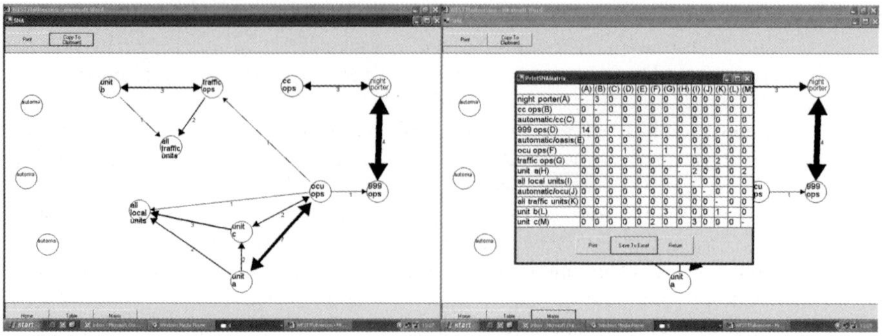


**Figures 3a and 3b. Social network diagram and Social network matrix.**

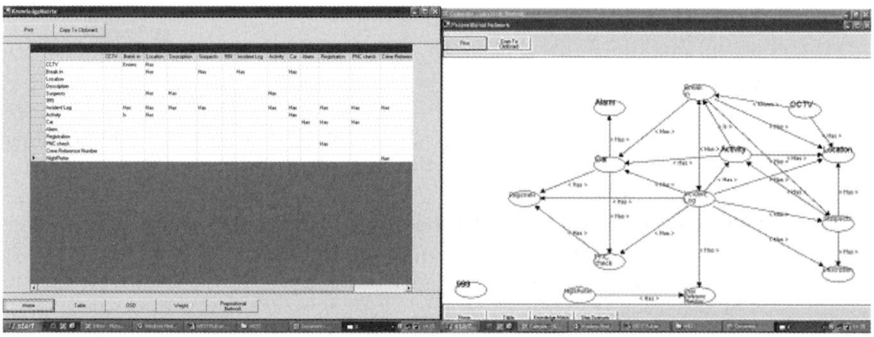

**Figures 4a and 4b. Knowledge objects and the propositonal network.**

different actions and categories of activities by agent, allowing an assessment of team and agent workload. This, integrated with the task database of HEPs developed at Birmingham (Kirwan, Basra & Taylor-Adams, 1997) allows WESTT to generate estimates of task duration and the cumulative chance of error and time-on-task.

The Social Network diagram (Figure 3a) indicates the interconnections between agents within the system. This allows subsequent analysis of the network configuration. As well as giving the analyst a visual representation of the relationships between agents, which can be used to understand the general character of the social network, the data on which it is based can be analysed using statistical techniques. Thus, an analyst can, for example, identify which actors are communications hubs, and whether a given actor is peripheral or central to the social network. Social network theory is a growing area of mathematical interest and WESTT allows the export of the network matrix so that mathematically inclined analysts can deploy techniques as the situation warrants (Figure 3b).

In addition to importing and representing the observation data, WESTT supports the construction of Propositional Networks. The basic approach is for the analyst to collect information from debrief, interviews, procedure manuals and other sources, to define the "knowledge objects" relevant to the mission. A knowledge object is an entity that plays a role in the mission and about which the agents need to acquire and use information. For example, in a road traffic accident, knowledge objects might include {location of accident, vehicles, road conditions, passengers, …, etc.}. In WESTT, the knowledge objects are entered into a matrix which allows the analyst to define the relationship between the objects, e.g., in terms of such terms as "is", "has", "requires", and "knows". (Figure 4a). The terms used to define the relationship between knowledge objects forms the basis for a directed network. This provides a graphical representation of the "space" of knowledge objects that are involved in the mission (Figure 4b). Typically, this network is then presented to Subject Matter Experts (SME) in order to validate the level of detail and the inclusion of specific knowledge objects.

Once the network has reached an acceptable state, it can be subjected to network analysis in a similar way to the Social Networks to identify trends in the relationships between information. It is this network of propositions that embodies the novel conceptualisation of Situation Awareness used by WESTT. Rather than focussing on aspects of "awareness", we are more concerned with defining the "situation". In this work, we view the situation as comprising a collection of objects about which the agents within the system require some knowledge in order to operate effectively. Nodes in a PN can be easily be associated with specific agents (as shown by the colour coding in figure three). This provides an intuitive representation of "who knows what" during the phases of an incident, which can be useful for considering gaps in awareness, requirements for shared knowledge or potential for conflicting interpretations of the same knowledge.

## Discussion

This paper has described the design and use of the WESTT software tool. It is envisaged that WESTT will provide a novel and useful means of representing team activity, and will be particularly beneficial for exploring future configurations in NEC system structure. By supporting analysis at several levels, it is possible to explore the effects of changes to system structure, the introduction or removal of knowledge objects (which might be operating procedures, cultural expectations or tactical information) or the replacement of human agents with technology, on operational effectiveness and system performance. A WESTT analysis can be used to indicate barriers to team activity such as deficits and asymmetries in expertise and situational awareness, pinch-points and sub-optimal social and communication network configurations. Specifically, the analysis is designed to focus on the "system" level, through the use of Social Network Analysis, and on the "agent" level, through consideration of workload, situation awareness, time and human error potential.

## Acknowledgements

This work from the Human Factors Integration Defence Technology Centre was part-funded by the Human Sciences Domain of the UK Ministry of Defence Scientific Research Programme.

## References

Kirwan, B., Basra, G., Taylor-Adams, S.E. (1997). CORE-DATA: A computerised human error database for human reliability support. In D.I. Gertman, D.L. Schurman and H.S. Blackman (eds) *Proceedings of the 1997 IEEE Sixth Conference on Human Factors and Power Plants*, New York: Institute of Electrical and Electronic Engineers

Meister, D. (1985) *Behavioral Analysis and Measurement Methods*, New York: Wiley

# AN ERGONOMICS ASSESSMENT OF POLICE BODY ARMOUR

**Katie Buchanan[1] & Dr Robin Hooper[2]**

*[1]Davis Associates Ltd, Wyllyotts Place, Potters Bar, Hertfordshire EN6 2JD*
*[2]Department of Human Sciences, Loughborough University, Leicestershire LE11 3TU*

Body armour is becoming an essential part of a police officer's personal protective equipment. Officers rely on such garments to protect them against severe injury and at the same time require them to be "wearable" for the duration of their shift. This paper summarises the findings of an ergonomics comparison between the wearability of two types of armour, which was undertaken on behalf of Northamptonshire Police. This investigation initially involved capturing officers' requirements and exploring the wider system issues. User trials were then conducted in both a field and laboratory setting. The trial results showed that the majority of officers preferred the more modern designed armour over their current armour, with a strong preference towards overt wear. However, throughout the trial, officer opinion varied and as such the importance of providing officers with as much individual adjustability was emphasised.

## Introduction

Across the UK, police forces equip their officers with body armour for wear not only in circumstances of heightened risk but for regular patrol. However, the Police Federation (2003) importantly emphasise that "there's no point officers having body armour if they don't wear it". A vitally important concern therefore is striking a balance between providing body armour which gives adequate protection whilst still being comfortable and wearable for the duration of an officer's shift. In 2000 Northamptonshire Police equipped all of its officers with personal issue body armour. However since this introduction the development of body armour fibre technology has been rapid. Today there is the availability of lighter, supposedly more "wearable" armour with only a small compromise in terms of the protection level provided. In 2005 Northamptonshire Police need to decide whether their current armour should be re-certified or if substantial investment in new armour should be made. The purpose of this study was to provide information to help inform this decision through conducting an ergonomics comparison between the current armour and a modern alternative. This study had two interconnected aims:

- To identify the body armour which best suits the needs of Northamptonshire Police's general duty officer.
- To identify officer preference towards style of wear and carriage of police equipment.

## Methodology overview

This investigation adopted a multi-method approach, encompassing qualitative and quantative techniques in field and laboratory settings. This methodological triangulation increased the confidence and ensured the validity of the findings. There were a series of data collection stages:

1. Capturing user requirements and identifying the wider systems issues.
2. Understanding the ergonomics issues surrounding the current armour.
3. Conducting user trials in both a field and laboratory setting.

A sample of eight volunteer Northamptonshire Police officers, four female and four male, were used throughout the investigation. Data collection techniques included semi-structured interviews, observations, activity diaries and a questionnaire.

### *The body armours' characteristics*

The design characteristics of the two armours compared, varied greatly in terms of their weight, flexibility, style of wear and level of protection provided (see Table 1).

One of the most important design principles of the current armour was that it incorporated a carriage system for equipment (see Figure 1). This means that the armour has to be worn overtly but a utility belt is not necessarily required. This armour has a scooped neck shape and is shorter than conventional armour in an attempt to increase comfort whilst seated.

The trialed armour is lighter, more flexible and is primarily designed to be worn covertly; however the armour can also be worn overtly, by placing the armour inserts inside an overt cover. For the purpose of this study the armour was trialed both covertly and overtly in order to investigate officer preference towards style of wear. The trial armour offers a slightly lower protection level than the current armour however, as Hooper (1999) argues, body armour discomfort can lead to officers not wearing their armour, increasing the potential for tragic consequences. Indeed Hooper (1999) states that "smaller, lighter armours may compromise protection, some protection though is better than none".

### *The wider system*

Body armour forms part of a wider system that incorporates the officers, the policing task and the environment in which these tasks are undertaken. The primary role of

**Table 1. Characteristics of the two armours.**

| Design feature | Current armour | Trialed armour |
|---|---|---|
| Style | Overt, 2 piece | Covert or Overt, 2 piece |
| Weight | 7.7 $kg/m^2$ | 6.95 $kg/m^2$ |
| Protection level | KR42 (≈KR2.5) | HG1a / KR1 |

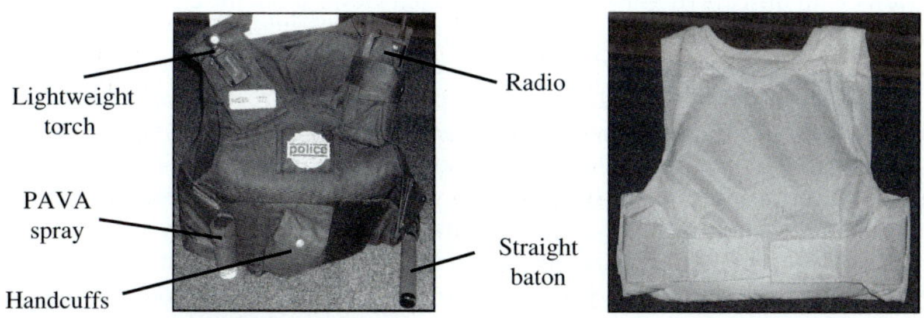


The current armour The trailed armour

**Figure 1. The body armours compared.**

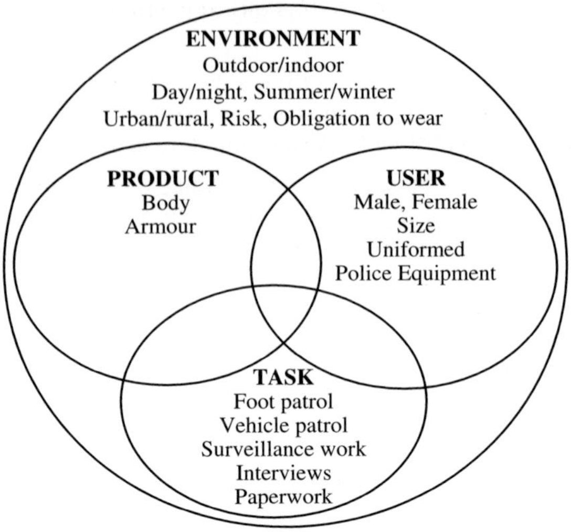


**Figure 2. The wider system.**

body armour is to "save the life of officers wearing it and allow them to use further physical skills to survive the encounter" (Northamptonshire Police, 1996). This role needs to be achieved whilst the officers undertake various activities in an ever changing environment, see Figure 2.

### *Officers' requirements*

In light of officer observations, informal interviews and the knowledge gained of the wider systems issues, the following officers' body armour requirements were identified:

- To provide protection to the vital organs.
- To be wearable for the duration of an officer's 12-hour shift.
- To provide, or be compatible with, an effective carriage of equipment system.
- To be compatible with uniform and other police equipment.
- To not significantly impede everyday activities; walking, running, sitting and standing.
- To be durable and weatherproof.
- To suit a range of body shapes.

### *Issues with the current armour*

Interviews with officers (n = 8) highlighted mixed views about the current armour. Some officers were in the main positive (n = 3) but others were largely negative towards the armour (n = 5). Half of the participants wore the armour all the time regardless of the task being undertaken (n = 4); the remaining participants wore it during the obligation to wear period (n = 2) or alternatively wore it as "little as possible" (n = 2). The officers who wore it all the time highlighted two reasons for this: firstly the fact that equipment is carried on the armour means it is easier to wear all the time and secondly it was considered that "the threat is always there regardless of the time of day". Of the officers who did not wear it all the time, most agreed that they "would wear it all the time but it is so uncomfortable that this is just not possible" (n = 3). Throughout the interviews, officers made numerous comments on the armour and its relationship with other police equipment/uniform, policing tasks and the environment. Table 2 highlights the key issues raised and the relative agreement between officers.

**Table 2. Officer's comments.**

| Comments | Officer Nos. Max (n = 8) |
|---|---|
| Thermally it is a problem in summer months | 8 |
| Incompatible with high visibility jacket | 7 |
| Restricts movement generally | 6 |
| Getting in and out of a car is awkward | 6 |
| The weight of the armour generally causes discomfort | 5 |
| When sitting down the armour cuts into the neck area | 5 |
| Carrying equipment on the vest distributes the equipment weight | 5 |
| Incompatible with utility belt | 4 |
| Association with back problems | 4 |
| Carrying equipment on the vest is far better than on a utility belt | 4 |
| My body is just not long enough to wear armour and belt | 2 |

This table highlights the diversity of issues raised, even though categories overlap. Officers made a number of comments concerned with the inherent weight and rigidity of the garment, which was reported to restrict movements, particularly when running and when entering and exiting a car. Furthermore, there were a number of incompatibility issues with items of police uniform and equipment, notably the high visibility jacket which when fastened up covers the armour and prevents easy access to police equipment.

*User trials with the trialed armour*

Officers wore the trial armour on duty for 3 weeks and undertook a series of exercises in the lab immediately prior to interview. These interviews highlighted different views towards the trialed armour. The majority of officers had a positive reaction to the trialed garment (n = 6) and considered the protection provided by the armour to be appropriate for the policing environment. Officers reported that the armour was considerably more flexible making it much more comfortable when walking (n = 7), sitting down (n = 7), driving (n = 6) and when entering and exiting a police car (n = 7). One officer emphasised "when getting in and out of the car I did not realise I had it on, whereas beforehand it was stiff and would catch you on the chin area". Similarly, one officer stated that the "armour is much easier to run in as it is lighter to wear and makes you more agile".

The laboratory trial corroborates these findings with the trialed armour proving more comfortable whilst undertaking a number of exercises. Interestingly, three of the officers who do not wear the current armour all the time commented that they would wear the trialed armour more often. In contrast, two officers had concerns over the trialed armour as they found it flimsy and were worried over the decrease in blunt trauma (injury incurred when a blunt object hits the body) protection. An assessment of risk and the influence of armour on policing style is necessary to address these issues fully.

Numerous comments were made regarding the comparison between overt and covert wear. Covert wear was considered by some officers to be a "nightmare" (n = 3), with the armour cutting into the neck area when seated (n = 3). This caused two officers so much discomfort that they decided that they could not continue to trial the armour covertly. Officers also commented that this style of wear compounded the heat problem due to the closeness of the armour to the skin (n = 6) and not having the ability to open the armour or take it off (n = 7). From a usability perspective the inability to take the armour off was a major issue (n = 7), with one officer stating that "having to wear it for a twelve hour shift without this benefit is too much". Despite the majority of officers criticising the covert style of wear (n = 7), a number of officers believed that covert style armour may have benefits in terms of image and public perception (n = 3). However, four officers mentioned that image alone should not determine the style of armour selected.

Carriage of equipment was shown to be a complex issue with individual officers carrying equipment differently. This trial showed that the majority of officers in this small sample preferred carrying equipment on the armour, as opposed to on the belt. This might also be a feature of the belt design, not studied here. However, since officer preferences varied, the incorporation of as much individual adjustability as possible in the eventual solution is important.

## Conclusion

On the surface the purchase of body armour may seem relatively simple; however it is in fact a multifaceted issue involving the balance of different priorities and views. This investigation has shown the benefits the advancement in fibre technology has had on armour wearability. The increased flexibility of the trialed armour meant that it was preferred by the majority of officers when worn overtly, which helps to relieve the thermal problems common to all armours. However, two officers were concerned by its flimsy feel and the decrease in blunt trauma protection.

Due to the potentially life saving role of body armour it is important to make recommendations in light of the views of all officers, not only the majority, and based on the risk of injury. With this in mind, it was recommended that the trialed armour should be purchased on the basis that whilst it provides less protection, officers are more likely to wear it, more often. In order to try and accommodate all officers, officers could be allowed to choose which armour they prefer. However, this would have financial, time and effort implications but is the solution adopted by at least one other Police Force. In addition, officers would need to understand risk and their own risk taking behaviour when making their choice.

## References

Hooper, R. 1999, Why Do Police Officers Leave Their Body Armour In the Cupboard? In: *Contemporary Ergonomics*, Taylor and Francis, London

Police Federation, 2003, *Arming the Police*, www.polfed.org/intro/index.html

Northamptonshire Police, 1996. *Protective Vests*

# HOW TO EAT AN ELEPHANT: ACHIEVING HUMAN FACTORS INTEGRATION ACROSS VERY LARGE SYSTEMS

**David Carr**

*Design Services, BAE Systems, 1 South Street, Scotstoun, Glasgow G14 OXN*

The problems facing ergonomics practitioners on large systems projects are very different to those of small-scale equipment design. They typically have to address a wide range of sometimes competing human factors issues across many equipments and sub-systems. Usually systems will comprise off-the-shelf components whose design is outwith the projects scope, and these will be selected by large teams of specialist designers with varying levels of human factors competence. Often ergonomics resources will be relatively scarce and practitioners will be faced with hard choices about where to concentrate their effort.

The best way to eat an elephant is one bite at a time. This paper discusses how techniques such as task analysis can be used to identify critical areas of system activity and hence focus on the parts of the system which require greatest human factors scrutiny. It also discusses approaches, derived from "coal face" experience on how to educate and monitor large design teams towards the achievement of human factors compliance for critical areas. Finally it addresses how ergonomics practitioners can help to ensure that design review and sign-off procedures can be managed to give traceability back to critical human factors requirements.

## Introduction

The Ergonomics Society's website defines ergonomics as "…the application of scientific information concerning humans to the design of objects, systems and environment for human use" (Ergonomics Society). Most typically, the ergonomist is concerned with the design of a single, bounded product – a tool, a system, a workplace – to achieve a

**Figure 1. The best way to eat an elephant is one piece at a time.**

limited number of goals. However, some very large systems comprise many individual equipments, systems and working spaces, which collectively provide the facilities and environment for a large body of personnel carrying out a diverse range of activities. The development of such "systems of systems" provides special challenges to ergonomists working as part of large and diverse design teams.

Examples of such very large systems in whose design ergonomists might be fortunate to become involved include:

- Offshore oil and gas exploration and production facilities which bring together systems concerned with drilling and extraction, maintenance facilities to ensure continued operation, safety-related systems, facilities for embarking and disembarking equipment and personnel and (often) personnel accommodation.
- Transportation facilities (e.g. airports) which include both airside facilities for the handling, preparation and maintenance of aircraft and the groundside facilities for moving and supporting large numbers of passengers and staff.
- Factories and similar production facilities, in which the basic, revenue-earning production tasks are supported by systems for handling and inventory of raw materials and finished goods. In many cases, safety and environmental protection systems also have an essential role.
- Warships, which are highly complex platforms required to transport themselves around the seas while generating power to operate their various detector and weapons systems and maintain themselves at a high state of readiness, even following damage. The flexible nature of naval operations means that large numbers of personnel are required to perform these functions – and these people must be accommodated and sustained over extended deployments.

While these are very different types of installation, they share certain common characteristics that are relevant to the ergonomist. They all are:

- Expensive, thus high performance is needed to deliver a return on investments.
- Operated and maintained by large teams of people.
- Concerned with safety critical areas in their design.
- Comprised by many subsystems to achieve goals, so there are many "stakeholders" to be involved in the design process.

The sheer complexity of such systems – which is magnified by the interactions of their constituent parts – inevitably makes it difficult for the ergonomist to maintain complete control over the whole design. The remainder of this paper discusses this complexity in further depth and suggests some practical techniques for addressing priorities and achieving a high degree of influence.

## The scope of very large systems

Large systems tend to be procured by one customer (e.g., a company; a transportation authority; a government) and "used" by someone else. Here "used" refers to those for whom the system provides a capability (A Company's production division; passengers; military forces), as opposed to the "end user" operators (factory workers; airport staff; sailors). Their design, production, operation and upkeep can typically be contracted out under competitive arrangements. The contractor will, in turn, sub-contract individual systems externally. Increasingly facilities are developed under "*Thin Prime Contract*" arrangements whereby the prime contractor serves solely as an integrator for the various components.

An important issue for ergonomists is that of where the responsibility for an ergonomically sound design lies. Certainly there are areas in which responsibility must be passed

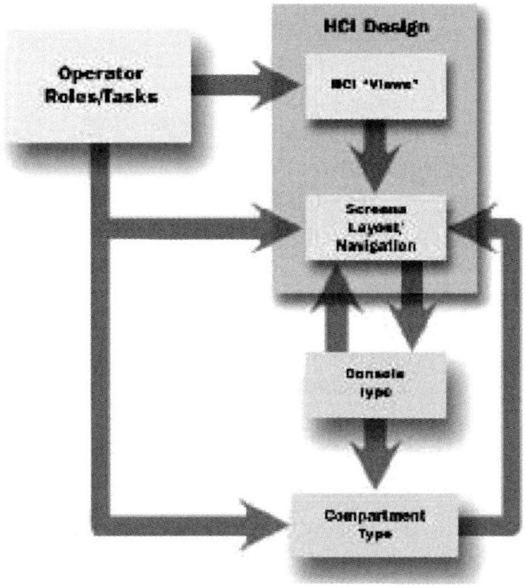


**Figure 2. Some interactions between design elements for a control room and its facilities (*from* David Carr et al., 2003).**

down to the appropriate systems contract – such as the supply of usable man-machine interface in a software-based system. The necessity for arrangements for cascading ergonomics requirements downwards is a strong argument for placing ergonomics expertise at the prime contract level.

In highly complex systems, many ergonomics risks arise from the interaction between systems. To quote from bitter experience: the inappropriate placement of lighting in a control room; the placement of two related controls in separate rooms; the routing of a noisy ventilation trunk through a sleeping cabin and the collocation of a potentially leaky sewage pipe and a food store have all been "problematic". Such risks can only be avoided when attention is given to the totality of the operators' facilities – their workspace, physical equipment and software-based interfaces. A simplified illustration of these interdependencies is shown in Figure 2 – but even this disregards interactions with systems such as cable routings, cooling, lighting, furnishing etc. which can play an important part in delivering an effective working environment. To address these issues, the ergonomist must have ownership of the "human-related system" as a whole, providing further strength to the argument for operating at an integrating level above the various systems.

There is a further systems component which is generally regarded as being outwith the prime contractors scope of supply but which is essential to systems operation and, accounts for a significant proportion of the system's Whole-Life Costs: the end users. The organisation which the system operates is just as much a subject for design as the physical elements. Unless it is attended to, it will be dictated by default. Not only does this risk poor job design, but it also misses opportunities to achieve a cost-effective balance between personnel and hardware costs. The ergonomist's specialist skills in personnel analysis have an important role in integrating organisational considerations into the mainstream of technical design.

## Focusing on priorities

It is axiomatic that pretty much anything that is used by humans can benefit from ergonomics in its design. However, in very large systems, human interfaces are many

| Understands ergonomics ↓ / Cares about ergonomics → | no | yes |
|---|---|---|
| no | *"I'm an engineer. We don't have to worry about that kind of thing."* | *"Yes! The colour of the curtains is very important"* |
| yes | *I'll make sure all the machinery is accessible. But that's just common sense"* | ***"I think you should help me with a Task Analysis before I design the system."*** |

**Figure 3. A "matrix of competence" for system designers.**

and ergonomists few. Difficult decisions need to be made on where to focus attention. An essential tool for this is Task Analysis. Hierarchical Task Analyses provide a convenient means of decomposing high-level system requirements (or "goals") down to the tasks and to the equipment that supports those tasks. Where requirements are traded off at systems level, this can be reflected in the priority and attention given to the various components. Safety-related tasks can be highlighted and the components that influence these can be identified.

## Supporting the design team

Ergonomists must always beware of arrogance. A system will still be built without them (albeit a risky, sub-optimal system), often by a team of engineers who have been building similar systems for years. Assuming, however, there is something to contribute, one way ergonomists can help to ensure other team members address ergonomic issues on our behalf is by proselytisation. Project experience has shown that dissemination of basic information on the value of ergonomics and its role within the project can work wonders in "raising the consciousness" of the development team.

A properly aware team can take a load off the ergonomist's mind. But the emphasis must be on "properly": a little learning is a dangerous thing. An important "sof" aspect to working in teams is assessing where the capabilities lie. Experience suggests that systems designers can vary in both their enthusiasm and understanding of ergonomics (see Figure 3). The ergonomist must be aware of which parts of the team need the closest monitoring. Often a sign of success will be the extent to which one is being sought out to answer ergonomics questions. If the team is uncommunicative, something is going wrong!

## Interpreting requirements

As described above, Ergonomists use task analysis to maintain an overview of who will be doing what with the system and where. During large systems development, much of the design team can be surprisingly unaware of the human needs they are serving. The routing of, say, a hydraulics system may be assigned to draughtspersons whose concern is to connect point A to point B and who may have not been told the space through which they are routing is a main access route for personnel. The ergonomist is in a position to supply this information, which will ensure the requirement "emergency escape shall be permitted" is not compromised.

A successful approach has been to identify the correspondence between the physical spaces of the systems (the rooms and other spaces) which can then be characterised according to their level and frequency of access (e.g. permanently staffed control rooms; spaces visited occasionally for maintenance; unoccupied "void" spaces and the nature and criticality of tasks carried out). Within compartments, working envelopes and access routes can be identified. System designers often think of their jobs as, in part, avoiding "clashes" with other systems. This information allows helps them to avoid clashes with the human component. It also provides clear statements of task requirements which can form the basis for design review and signoff.

## Providing information

It has often been a disappointment that, when confronted with a systems designer eager for a piece of information, one is forced to reach for a book that provides, for example, population dimensions for static, nude people. This is by no means a criticism of excellent publications that supply anthropometric data. Rather it is a reminder that such data need interpretation. An effective approach in the past has been to provide tailored guidance on "frequently asked questions" (how high can a control be; how much passing space is needed around equipment, etc) in an attractive, easy to follow format. Not only does this avoid the workload of answering the same questions repeatedly, but it also – especially if the guidance is updated regularly as new questions arise – serves to maintain a profile for ergonomics.

## Equipment selection

Any large system will inevitably contain a large number of Commercial Off-The-Shelf (COTS) components. The ergonomist's first role in relation to COTS is to assist designers in selecting components that meet basic standards or operability. This includes control/display assessment (Is it legible, comprehensible, consistent with the style guide?) or other basic equipments (Are parts light enough to lift during maintenance? How much strength is needed to operate this lever?). This process yields a potentially huge amount of data, even when the focus is restricted to critical elements. One way of cheating – apart from ignoring the less important components – is to ask the suppliers for assurance they have addressed ergonomics issues. This does not necessarily yield clear answers as few industrial components suppliers will have employed ergonomists. However, careful questioning, (through questionnaires supplied with "requests for quote") can gain insights into the "Human Factors Maturity" of suppliers (Earthy, 1998). The positive answers given may range from "Yes, we've done full task analysis and usability testing" to "We're aware of the Manual Handling Regulations" to "The same equipment is already being used for a similar function in many installations".

## Conclusion: Don't panic!

While very large systems provide particular challenges of complexity, a way forward is provided by familiar, elementary ergonomics techniques and approaches. Task Analysis provides the means of tracing the "context of use" of system elements back to system requirements and highlights design priorities. A focus on the design team members as the "users" of ergonomics information allows the ergonomist to achieve a wide design influence. The rest is simply a matter of a cool head and dogged determination.

## References

Ergonomics Society website http://www.ergonomics.org.uk/ergonomics.htm

David Carr, Iain MacLeod and Lt. Cdr. Mike Cox. (2003). Fighting the Internal Battle: HMI considerations to support teamworking with the Type 45 Destroyer's Platform Management System. *NATO Symposium on Future Developments in Human-Machine Interfaces*. Prague, Czech Republic.

Earthy, J. (1998). Usability Maturity Model: Human Centredness Scale. INUSE Project deliverable D5.1.4(s). Version 1.2. London, Lloyd's Register of Shipping.

# WORKING GROUPS AND USER GROUPS: DIFFERING OBJECTIVES AND FORMS

**Daniel L. Welch**

*Advanced Management Technology, Inc. Arlington, VA 22209, USA*

Human factors practitioners have long known the importance of evoking information from real world users early in the development process. To provide this capability, User Groups function both as sources of important system information and as test beds to evaluate proposed designs.

Most practitioners face additional challenges, however, in coordinating their efforts not only within the system engineering effort, but also among other players in the acquisition bureaucracy and with "external" stakeholders. The Working Group is often employed to achieve this necessary coordination.

This paper discusses the nature of these groups and explores difficulties commonly encountered when the functions of the two groups are combined into a single entity.

## Introduction

The goal of Human Factors in the system development process is to maximize total system performance by optimizing human sub-system performance. This is accomplished by ensuring that the system design supports the performance of those system functions assigned to the human – that the system takes human capabilities and limitations into account to support human task conduct. Studies, analyses, prototyping, mockups, simulations, design guidance and all the other tools and techniques of Human Factors are aimed at supporting the design effort to optimize human performance.

An important source of data for the Human Factors effort is the user. Beyond anthropometric and demographic considerations of the target audience, an essential pool of information which must be used to support the development of a new system includes the user's knowledge of the suprasystem, experience on predecessor or similar systems, understanding of his or her job, cultural and work related stereotypes, and simple user preferences.

A typical way to get at this data has been to form a User Group; a chartered group of users of a similar or predecessor system, or stakeholders in a suprasystem. This group is then involved in a number of data acquisition or knowledge elicitation processes to provide system-related data which is fed-forward into the system design process.

The use of a User Group, however, is typically not sufficient to support all the requirements placed on the Human Factors practitioner. Human Factors is a system engineering sub-discipline; as such it must coordinate its efforts and results with other system engineering disciplines. It must also coordinate with other elements of the acquisition process such as software, hardware, safety, and training, and with "external" stake holders such as requirements generators, regulators, lawmakers, and unions.

In order to accomplish this level of coordination, the establishment of a Human Factors Working Group has often proven useful, and current US Department of Defense and US Federal Aviation Administration acquisition documentation supports the use of a Working Group to respond to the "substantial requirements for frequent formal and informal coordination among organizations, agencies, and individuals with concerns related to" Human Factors (US Department of Defense, 1999).

Thus, the User Group is used as a data source for user-, system- and task-related information and the Working Group is used to coordinate Human Factors activities, concerns and solutions among system engineering, other acquisition entities, and entities beyond the acquisition effort. A difficulty arises, however, when the functions of these two groups are combined. This paper will examine the nature of these groups, their objectives and forms, and the danger in mixing their functions.

## The user group

A User Group is, by definition, a group – i.e., N > 1. A single user is not a User Group. Group size may vary, though generally N < 6 may limit diversity of response (through the impact of a dominant individual) and N > 12 may become unwieldy. Some User Groups may represent a small fraction of the total user population (as for an Automated Teller Machine), while others may encompass all or nearly all potential users (as for a highly specialized system with a small potential user base; e.g., a military weapons system). Regardless of the percentage of the user population represented in the sample, the User Group must represent actual *users*.

Users are people who are currently using an existing system, have been designated or identified as intended users of a new system, or are users of currently existing similar systems. They are not design team members, secretaries to the design team, graduate student interns, or people randomly selected off the street. They are not ex-users and they are not user supervisors. They are not people who "understand the system" but will never use it. They are actual, hands-on, real-life users. A selected group of them is called a User Group. Improperly defining the user can lead to serious and negative impacts on the value of information obtained from the User Group.

The primary functions of a User Group are to serve as a data source and test bed. Data acquisition and knowledge elicitation techniques are employed to probe the users to gain information relevant and important to system design and performance. Developing designs, in whole or part, are user tested on the group members to evaluate design alternatives, establish conformance to performance requirements, determine user preference and acceptance, and identify usability problems. The results of subjective and objective information gathering with the User Group are used to influence system design to optimize human performance.

The User Group can be employed in the full range of Human Factors analysis, design, and test activities, including: function identification, analysis, and allocation; task inventory and analysis; information flow and processing analysis; criticality analysis; critical task analysis; workload analysis; critical design consideration development; alternative design evaluation; usability standards definition; usability metrics development; usability evaluation; field testing; user acceptance testing; and change proposal evaluation.

The early and correct use of a User Group to aid in the Human Factors design support of a developing system can be highly beneficial. Delaying the creation and use of a User Group limits or obviates its potential benefits, and Human Factors improvements identified through its use late in a program are more costly to implement. The improper use of a User Group through the incorrect "case definition" of a user may provide invalid and misleading information on which to make design decisions. Improper use by expecting a User Group to be more than a data source or test bed can lead to catastrophic results and is discussed below.

The management challenges for properly employing a User Group involve avoiding the improper uses listed above. Additional difficulties are encountered when existing or future users are member of an organized labor union. (This is briefly discussed is the next section.)

## The Human Factors Working Group

In order to fulfill its system engineering function, Human Factors must identify Human Factors "issues and concerns" within the developing system and develop solutions which are both technically adequate/elegant and acceptable to the larger system development community. Thus, Human Factors must closely coordinate with other system engineering disciplines (or "ilities"), with other program office disciplines, and with "external" entities which impact the system development effort. The Human Factors Working Group is employed to carry out this function.

The Working Group should be a "chartered" organization; i.e., it should be created by program management through the generation of a formal document of establishment. That document, or charter, should specifically state the mission of the group, define the responsibilities and authorities of the group, establish the initial membership of the group, and broadly outline the procedures to be employed in carrying out its mission. Such "command level" commitment is essential to the success of the Working Group as it both provides legitimacy to the group and demonstrates an understanding of the importance of Human Factors to the eventual success of the developing system.

Since the Working Group is a Human Factors Working Group, it should be led by Human Factors specialists. The Human Factors specialists should set the goals and priorities of the group and be primarily responsible for developing solutions to Human Factors issues and concerns.

The mission of the Working Group obviously should be tailored to the particular system under development, but generally involves assuring that all Human Factors issues and concerns are identified and successfully addressed during development. These issues and concerns are identified from a number of sources, the most obvious being Human Factors studies and analyses, including input from the Users Group. A strength of the Working Group, however, is that the other disciplines and entities represented can also identify Human Factors related issues and concerns raised in their separate activities and bring them to the attention of the Working Group.

The Working Group thus serves as a forum for communications among interested parties and, by extension, as a medium for resolving identified issues and concerns. While a sub-group or executive committee (composed exclusively of Human Factors personnel) might be primarily responsible for both identifying and resolving Human Factors issues and concerns, it is up to the whole Working Group to consider and approve the recommended solutions prior to their implementation. In this manner, the creative process of Human Factors design support is fully integrated into the overall system development process and alternatives obtained to resolve issues, and concerns are fully vetted among interested parties prior to recommendation as a design (or other) solution.

The management challenges for the Working Group are to achieve and maintain proper representation on the group and to facilitate the development of a resolution which answers the issues of all involved parties. (Or at least offends as few as possible.) A management danger is that a single group discipline/entity will come to dominate the resolution process, thus driving solution which are optimized from their perspective but sub-optimized in terms of the total design team and overall system performance.

## The problem with mixed-models

Today, most Human Factors practitioners and organizations involved in system development recognize the value of user input and the power of "concurrent design" achieved through a Working Group. Unfortunately, in an attempt to gain these advantages, the concepts of User Group and Working Group are sometimes combined into a single entity, usually with less than spectacular results. Such a mixed group negatively impacts, or even destroys, the User Group's value as a data source and test bed, and severely impacts the ability of the Working Group to develop appropriate Human Factors solutions, its decision making processes, and its ability to coordinate among disciplines/entities.

A mixed User-Working Group can arise from many sources. Often well-meaning but ill-guided management "involves the user" by simply adding users to the design team and giving them veto power over design decisions. (This can easily occur with unionized users, who often literally do have veto power – "We won't use it.") In other cases, the User Group is simply saddled with the functions and responsibilities of the Working Group, usually with even more disastrous results.

There are many problems with either type of "participatory Human Factors." First, it impacts the Working Groups' function of achieving "good" Human Factors design. The user is generally not trained in the research and experiential data base of Human Factors – you can't ask non-Human Factors experts to make expert decisions on Human Factors issues. What the user wants, while very important, is not always what is best for overall system performance. The user can, and does, make preference-based decisions which sub-optimize for his/her ease of use at the expense of overall system performance. Granting the user veto power over design reduces Human Factors to simply soliciting and defining user preferences and implementing them, thus abdicating and abandoning the value of the entire Human Factors knowledge base.

Additionally, in reality trade-offs sometimes have to be made which negatively impact the usability or user-friendliness of a system. These negative trades may be necessitated by cost, schedule, technical, doctrine, procedural, legal, political or other reasons, but they are simply constraints the Human Factors specialist, as well as the user, has to live with. The Human Factors specialist is more clinically detached than the user and thus better able to "take a hit" for the greater good. Again, the ultimate goal is to optimize overall system performance, not human performance.

The mixed group also impacts the functions of the Users Group. Simply adding one or two users to the Working Group inappropriately restricts the range of the user viewpoint (too small an N isn't a group), thus decreasing its value as a data source. One or two users may sub-optimize not only for the user, but for themselves. This is a problem to the degree the users are not representative of the actual population.

Beyond decreasing the potential for valid user input, limiting user participation to the Working Group also obviates the second function of a User Group and eliminates the ability to employ the User Group as a test bed. Identifying usability problems, evaluating design alternatives, and testing performance requirements require the participation of an objective *group* of *users*. Using the Working Group to user-test a developing system encounters the obvious problem – most of the people on the Working Group aren't users. Data and decisions based on a mixed group violate the basic premise of sampling and are invalid. Relying on the Working Group, even with user representation, to fulfill this function is essentially the same as not having any user representation at all.

These impacts do not imply, however, that users should not be represented on the Working Group: indeed users should be a valuable part of the Working Group. However, they should not be the only users involved in the project; a separate Users Group should still be employed. The Working Group user representatives should be *members* of the Working Group, with voting but not veto rights. They are thus able to represent the views and concerns of the user in proper perspective to the other, often competing, views and needs of the group. Finally, the user representative(s) on the Working Group appropriately may be supervisors or union liaisons, who bring a broader, more detached viewpoint of the job than the individual user.

## Summary

Employing a User Group is today an accepted method of soliciting user input and opinion, and testing developing systems and products. The Working Group is mandated in US military and FAA development efforts as a way of achieving concurrent design. The temptation to realize savings by combining the functions of the two groups into a single entity, however, must be strongly resisted.

Simply adding users to the Human Factors Working Group is not an effective way of achieving user participation and negatively impacts the necessary functions of the two discrete groups, as discussed above. Giving the User Group the role and responsibilities of the Working Group ensures that the functions of the Working Group *will not* be accomplished.

Alternately, employing a separate User Group provides the Working Group, and thus program management, with an invaluable data source and test bed. Adding user representation to the Working Group ensures that the users' viewpoint and input are properly interpreted and disseminated to group members.

It is a consummation devoutly to be wished.

## Reference

US Department of Defense, 1999, *MIL-HDBK-46855A: Human Engineering Program Process and Procedures*, 21

# THE EVENT ANALYSIS OF SYSTEMIC TEAMWORK (EAST): METHODOLOGY AND ANALYSES OF RAILWAY MAINTENANCE TASKS

**Dr Huw Gibson[1], Dr Guy H. Walker[2], Dr Chris Baber[1], Prof. Neville A. Stanton[2] & Paul Salmon[2]**

*Defence Technology Centre for Human Factors Integration (DTC HFI)*
*[1]University of Birmingham, School of Electronic, Electrical and Computing Engineering, Edgbaston, Birmingham B15 2TT, UK*
*[2]Brunel University, School of Engineering and Design, Uxbridge UB8 3PH, UK*

EAST (Event Analysis for Systemic Teamwork) is an analysis methodology which combines seven human factors techniques in order to investigate command, control, communications, computers and intelligence (C4i) (Harris & White, 1987). The methodology has been developed by the DTC HFI. This paper focuses on two methods used in EAST, the critical decision method (CDM) and propositional networks. This paper describes the application of the methods to C4i activities for railway track maintenance.

## Introduction

The C4i activity considered is setting up safety systems required when carrying out railway track maintenance. These systems are required when maintenance personnel work on or near the railway line. They are primarily used to reduce the risks from personnel coming into contact with moving trains and they also help to ensure that trains do not travel over railway infrastructure which is not safe due to ongoing maintenance.

Most of the time, a signaller has the key responsibility of controlling train movements and maintaining safety for an area of railway line. This control occurs remote from the line at a control centre called a signalbox. However, during maintenance, another person takes responsibility for the area of the line in which maintenance is being undertaken. This person is normally on the ground (i.e. on or near the railway line) and they are normally termed the person in charge of possession (PICOP) or controller of site safety (COSS). Communication and coordination is required to transfer responsibility between the signaller and PICOP/COSS. The PICOP/COSS also has to communicate and coordinate with various other personnel, such as those carrying out maintenance within their area of control and personnel assisting in implementing the safety systems.

## Methods

EAST (Event Analysis for Systemic Teamwork) is an analysis methodology which combines seven human factors techniques in order to investigate Command and

Control Activities (C4i). EAST is based upon the collection of live data using a combination of observation and interviews. The methodology has been developed by the DTC HFI. The techniques used in EAST are: Hierarchical Task Analysis of the scenario (Annet, 2003), performing a Coordination Demand Analysis (Burke, in press), applying the Critical Decision Method (CDM), which is used as the basis for a Propositional Network, producing Communications Usage Diagrams and Social Network Analyses (Watts & Monk, 2000). This approach offers a novel form of triangulation, whereby complementary methods offer multiple perspectives on the same data. This paper focuses on outputs related to the CDM and propositional networks. A companion paper (Walker et al.) presents key results for the other techniques and their summarisation in an Operational Sequence Diagram format.

This paper focuses on the application of the CDM and propositional network components of EAST through presentation of example analyses related to rail maintenance. The CDM approach requires assessment of each key decision point in a C4i scenario to be considered from the perspectives of all the actors involved. A slightly modified version of the CDM probes presented in O'Hare et al. (2000) were adopted. The CDM method requires a structured interview to elicit responses against the probe categories: goal specification (what were your specific goals at this decision point?); cue identification (e.g. how did you know that you needed to make the decision?); expectancy (e.g. were you expecting to make this type of decision during the course of the event?); conceptual model (e.g. are there any situations in which your decision would have turned out differently?); uncertainty (e.g. at any stage were you uncertain about either the reliability or the relevance of the information that you had available?); information (e.g. what was the most important piece of information that you used to formulate the decision?); situation awareness (e.g. what information did you have available to you at the time of the decision); situation assessment (e.g. was there any additional information that you might have used to assist in the formulation of the decision?); options (e.g. were there any other alternatives available to you other than the decision that you made?); stress (e.g. was there any stage during the decision-making process in which you found it difficult to process and integrate the information available?). The results are based on information which can be derived from recordings of the communication undertaken during the scenarios and insights gained by the author during observation and discussion with the relevant personnel, but not specifically related to eliciting CDM responses. It should be noted that this approach has been viewed as adequate for the current trial study, but the CDM is fully applied with the relevant system experts outside of this trial application.

While the CDM technique has been applied in a number of areas (Hoffman et al., 1998) a novel extension with the EAST methodology has been translation of the CDM analysis into propositional networks. The propositional network is derived from a content analysis of the CDM transcripts, which takes all the elements defined in a scenario such as people, information and physical objects and links them together. The elements can be viewed as the "knowledge" which makes up the C4i event being investigated. These knowledge "nodes" within the network are then linked through causal paths. The representation can describe both the whole event and also the elements and links associated with a specific decision point in the scenario. The technique provides a description of the required elements of situational awareness at either an event or a decision point level.

The railway maintenance scenarios serve as a source of live data to demonstrate the capability of the method. Data was elicited from recorded transcripts wherein four key maintenance activities were assessed, all involving coordination between distributed teams and individuals.

**Table 1. CDM table excerpt for the request from a COSS for emergency protection to be granted by a signaller.**

| CDM heading | COSS | Signaller |
|---|---|---|
| *Goal specification* (what were your specific goals at this decision point?) | Get signaller to accept request for emergency protection. | Assess whether emergency protection can be granted. |
| *Cue identification* (e.g. How did you know that you needed to make the decision?) | Preceding task step and the emergency event | Call from COSS |

## Results

### *CDM*

CDM analyses were undertaken for all the key decision points in the four railway scenarios investigated. The decision points are defined by the phases in the scenario HTAs (see Walker et al.). Table 1 provides example probe responses to the second key decision point in a process for setting up safety systems to carry out emergency maintenance. A COSS on the ground is requesting that the safety system is set up so that the maintenance can take place and the signaller in the signal box assesses if this is suitable. The key elements of the tasks relevant to the CDM are: the COSS describes the emergency event to the signaller; the signaller has to assess whether the event is suitable for application of the emergency procedures and that the train traffic and railway layout allow the emergency safety system to be implemented.

### *Propositional network*

Figure 1 provides an example of a propositional network. The network as a whole describes all the network elements related to the emergency maintenance. The network elements which are presented in grey are those which are required at the decision point "request from a COSS for emergency protection to be granted by a signaller". The network elements which are contained within an ellipse with a thicker border are those which were activated in the preceding scenario elements. The propositional network therefore provides an overview of all the knowledge elements for a scenario and their relationships. It also allows the knowledge elements related to a specific decision point in the scenario to be described, and the history of previously used knowledge points.

Despite being a relatively simple proceduralised task, the propositional network appears complex. It should be recognised that this complexity does not necessarily reflect complexity in the task itself. For example, some of the knowledge elements are internalised skills (e.g. participant knowledge of the railway rule book or a signaller's knowledge of the current status of the railway) or simple objects (e.g. telephones) which would not provide load or add to the complexity perceived by a skilled participant. It does, however, demonstrate that a large number of related objects are used to accomplish relatively simple tasks. In the context of task redesign, the propositional network provides a prompt which can allow us to question the necessity for knowledge objects and the form those objects can take. In particular, the technique does not differentiate between different types of nodes (e.g. knowledge related to objects, people or decisions) and therefore during re-design, it does not constrain assessments to consideration of existing configurations of people and objects, rather to the required

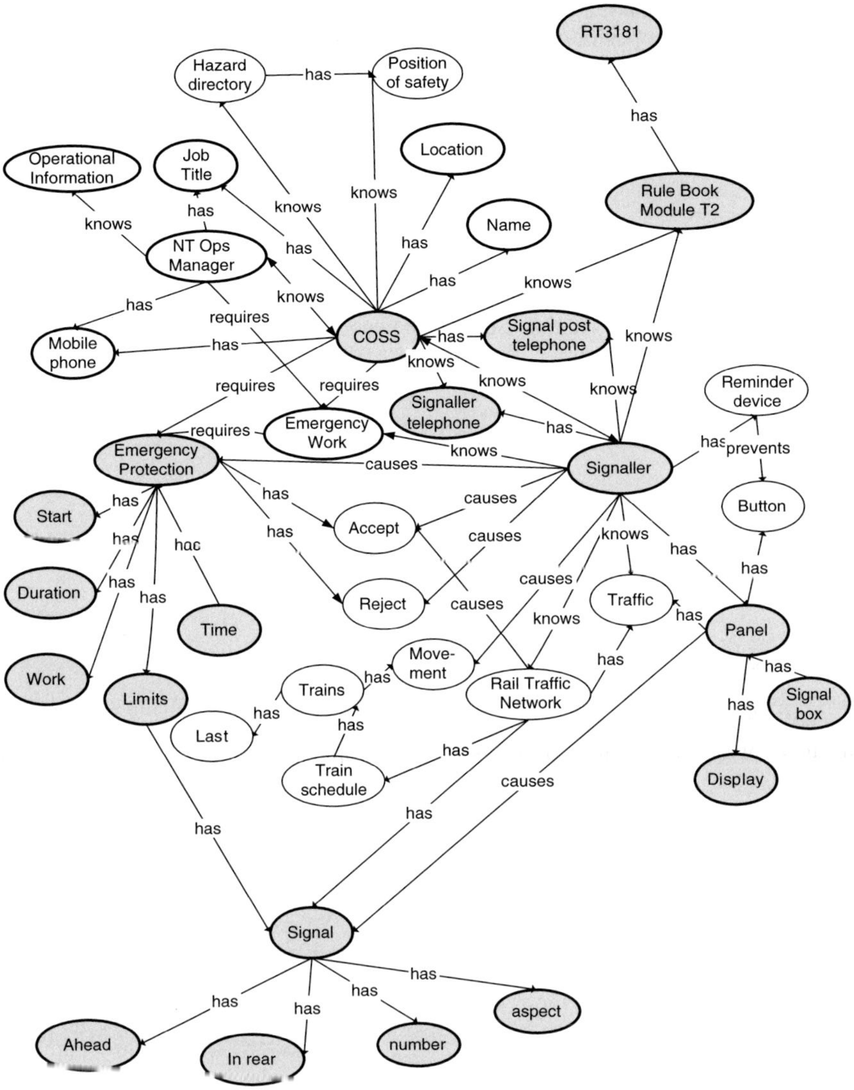


**Figure 1. Propositional network for protection for emergency maintenance.**

knowledge elements associated with a scenario. The network could also be considered as a tool for designing training, which describes the knowledge elements and relationships which someone undertaking the task must have available.

## Conclusions

The EAST methodology integrates a number of methods to support the assessment of C4i. The examples in this paper illustrate two of the methods, and this paper should be considered in the wider EAST methodology context (see Walker et al., this volume).

Based on the CDM data, which allows the key aspects of the decision making process to be defined, the propositional networks provide an opportunity to describe all the key elements related to a scenario or decision point. From these trial applications the methodologies are being applied to military and other civil domains by the DTC HFI. The aspects of the approach related to knowledge objects are also being used by the DTC HFI to develop novel approaches to situational awareness in a systems context.

## Acknowledgement

This work from the Human Factors Integration Defence Technology Centre (HFI DTC) was part-funded by the Human Sciences Domain of the UK Ministry of Defence Scientific Research Programme. The assistance of the Rail Safety and Standards Board is also gratefully acknowledged.

## References

Annett, J. 2003, Hierarchical task analysis. In: D. Diaper and N. A. Stanton. (eds) *Handbook of Task Analysis in Human-Computer Interaction* (Mahwah, NJ: Lawrence Erlbaum Associates).

Burke, C. S. in press, Team task analysis. In: N. A. Stanton, A. Hedge, K. Brookhuis, E. Salas & H. Hendrick (eds) *The Handbook of human factors and Ergonomics Methods* (Boca Raton, Fl: CRC Press).

Harris, C. J., & White, I. 1987, Advances in command, control & communication systems. (IEE Computing Series 11) (London: Peregrinus).

Hoffman, R.R., Crandall, B., & Shadbolt, N. 1998, Use of the Critical Decision Method to Elicit Expert Knowledge: A Case Study in the Methodology of Cognitive Task Analysis. *Ergonomics* 40(2) , 254–276.

O'Hare, D., Wiggins, M., Williams, A., & Wong, W. 2000, Cognitive Task Analyses for Decision Centred Design and Training. In: J. Annett and N.A. Stanton (eds) *Task Analysis* (London: Taylor and Francis) 55–78.

Walker, G.H., Gibson, H., Stanton, N.A., Baber, C., Salmon, P., & Green, D. 2005, The Event Analysis of Systemic Teamwork (EAST); A Novel Integration of Ergonomics Methods to Analyse C4i Activity. *This Issue.*

Watts, L.A., & Monk, A.F. (2000) Reasoning about tasks, activities and technology to support collaboration, In: J. Annett and N.A. Stanton (eds) *Task Analysis* (London: Taylor and Francis), 55–78.

# A GENERAL MODEL OF HUMAN FACTORS STRUCTURED ANALYSIS AND DESIGN METHODS

**J.R. Middlemass & J.B. Long**

*Ergonomics & HCI Unit, University College London,*
*26 Bedford Way, London WC1H 0AP*
*Now the UCL Interaction Centre, Remax House,*
*31–32, Alfred Place, London WC1E 7DP*

A general model of three human factors Structured Analysis and Design Methods is proposed. The model was constructed, following the MUSE method, and represents a high-level process model of activities, occurring during user interface design and evaluation. Further research is suggested.

## Introduction

MUSE, a method for usability engineering (Lim and Long, 1994), is one of a class of methods, supporting a structured approach to design of the user interface (UI). These methods approach design in a "top-down" manner, using information derived from "bottom-up" analyses of existing systems and tasks. In each method, design progresses from the specification of general features of the tasks to be performed (derived from analysis of the user requirements and from existing systems for performing the work), to the specification of the detailed UI. Methods differ in their terminology, level of proceduralisation, inclusion of design knowledge and design products. However, they support similar design transformations, leading to broadly comparable UI specifications. These common design transformations can be represented as a MUSE generalised task model of existing systems.

## Nature of the general model

Lim and Long discuss human factors (HF) contributions to system development in terms of their scope, process and notation. Scope relates to the subject matter, either declarative or procedural; process relates to the timing and provision of HF inputs; and notation relates to the representation of these contributions. Lim and Long further propose that structured HF methods may be likewise considered, both overall and at the level of individual stages. In the context of such methods: scope refers to aspects of the design of the human-computer worksystem, addressed by the product, stage, or method; process refers to the manner in which specifications of the worksystem, required by the scope, are derived (normally, by effecting transformations of information (or previous design products), in line with a set of procedures defined by the method); and notation refers to the representation of these specifications.

The general model, reported here, is a high-level process model. It does not attempt to represent the general properties of the scope or notation of structured HF methods.

### *Derivation of the model*

The general model is derived from three structured HF methods: MUSE (Lim and Long, 1994); STUDIO (Browne, 1994); and GUIDE (Redmonde-Pyle and Moore,1995).

These methods were selected to support research, addressing comparable scope and relationships with software engineering (SE) methods. All three methods result in the production of a graphical user interface (GUI) design, based on a set of user requirements (but not, for example, training or manning requirements). While all define points, at which information is shared with the SE design team, they can be used with a variety of SE methods by virtue of their informal and configurable information cross-checking and exchange procedures.

> *The method used to derive the model was MUSE – Phase 1, the Information Elicitation and Analysis phase. The procedures of each method were used to derive an (idealised) task description for them in structured diagram notation (SDN). SDN is an hierarchical notation, similar to that used in hierarchical task analysis; but employs some additional constructs, for example, to represent selection or iteration. Each diagram was accompanied by a table, providing additional description and commentary on the content of individual nodes. The task descriptions were generified, following MUSE procedures, to construct a general task model (GTM) for each (existing) method, designated GTM(ext). Generification involved: listing the objects and actions in the task descriptions; associating similar terms; assigning a generic descriptor to each group; validating the descriptors by obtaining an independent assessment from someone familiar with the method; and constructing a new (generic) task description, which was similarly validated. Thus, the GTM(ext) products described each method at a level independent of both the particular terminology, employed by the method, and the particular design products, resulting from the method. These GTM(exts) were used to derive a GTM(x), general to the three methods, by further generification, as earlier.*

## Description of methods

An overview of each of the methods, used to derive the general model, follows. Note that the task descriptions, used to model the methods, were at a much lower level of description, and used the lowest level of currently published proceduralisation.

The MUSE method consists of three phases, each divided into stages. The first is the Information Elicitation and Analysis phase and involves the collection and analysis of information to support later design activities. The second phase begins by establishing the HF requirements of the design (that is, performance criteria, likely user problems, and task support). Based on this information and the task models from Phase 1, a conceptual design is produced and agreed with the SE design team. Finally, this conceptual design is split into subtasks to be performed, using the system being designed, from those performed, using other existing devices. Allocation of function is then performed. The third is the Design Specification phase, in which the conceptual design is developed further to a device-specific and implementable specification, including error-recovery procedures. The specification is informed by any UI environment, but also by other earlier products. A number of products detail different aspects of the UI design. The UI is then assessed and refined iteratively. Once the UI is found to be acceptable, the final products are passed to the SE team for implementation.

STUDIO is composed of five stages, each subdivided into steps. Stage 1 – Project Proposal and Planning – is concerned with project justification by cost benefit analysis, and with planning for quality assurance, throughout the design. Stage 2 – User Requirements Analysis – involves the collection of information to support later design and consists of steps, which describe: planning the evidence collection; the collection itself; task analysis; validation; and reporting the findings. Stage 3 – Task Synthesis – uses the information from Stage 2 to produce an initial UI design. Stage 3 consists of five steps: user support (user manual specification); styleguide; task synthesis; design; and formative evaluation. Stage 4 – Usability Engineering – refines the Stage 3 design, by means of iterative prototyping. The first step in Stage 4 is to plan prototyping and evaluation. Following planning, a prototype is constructed and audited by the design team. During prototype construction, evaluation materials are prepared and, when ready, the prototype is evaluated by users. The results undergo impact analysis to prioritise changes, suggested by the results. Once finalised, the design documentation is updated to reflect the Stage 4 modifications. Stage 5 is UI Development. Here, the design is passed to the SE implementers, integration issues are resolved, acceptance testing is performed and termination reporting is conducted.

GUIDE is composed of six processes. Process 1 defines the users and the usability requirements. Descriptions of user classes are constructed, characterising each type, along with usability requirements, containing satisfaction criteria for later testing. Process 2 models and agrees the user tasks and collects scenarios to inform prototyping and later user testing. Process 3 models the "objects in the system" from the user's point of view. User object models are constructed for each class of user, describing: the user's view of the objects; what actions each performs or suffers; the associated object information; and the relationships between the objects. A glossary of user-familiar terms describes the task or system. Process 4 specifies a styleguide, describing the behaviours of each component of any GUI environment used. The styleguide informs later detailed GUI design. Using this information, an initial design is produced in Process 5 – designated Design GUI. A prototype is constructed, evaluated by the design team, together with a user and the scenarios collected earlier, and iteratively refined until found acceptable. Finally, Process 6 evaluates the UI against the usability requirements. Any design revisions are agreed, before full SE implementation commences.

## General model of HF structured methods

The top level of the general model, produced on the basis of analysis of the three methods described earlier, is shown in Figure 1. Additional details of the Prepare Groundwork and Refine Design boxes are shown in Figures 2 and 3 respectively. Details of the Detailed Task Specification, High Level Specification and Implement User Interface boxes are provided next.

The Detailed Task Specification box consists of an iteration over a selection between the following activities: Decompose Tasks; Verify Model with SE team; Allocation of Function; and Communications Level Design. The effect of an iteration over a selection is that each activity becomes optional. If allocation of function was performed during task-level scoping (Figure 2), it would not be repeated here. Similarly, Communications

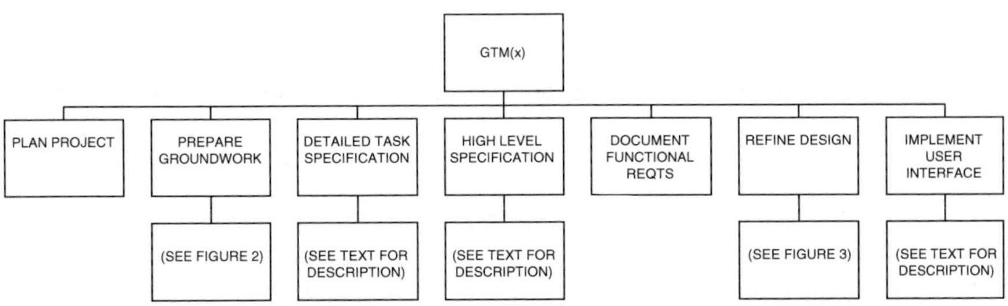


**Figure 1. Top level of the general model of HF structured methods.**

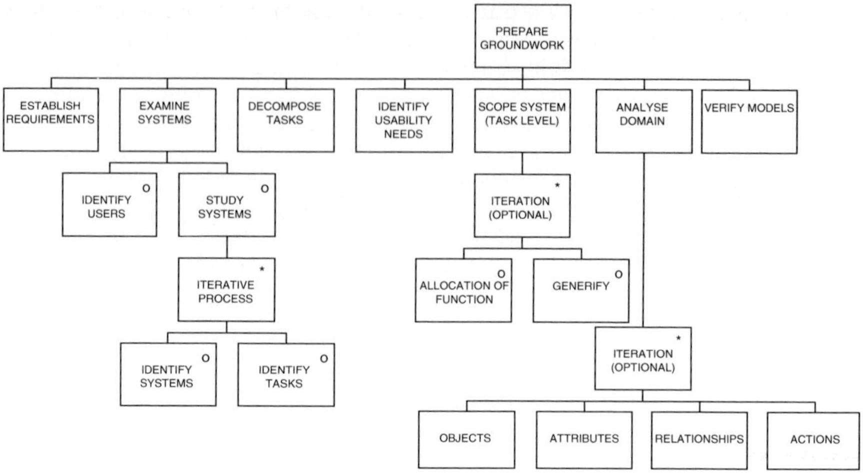


**Figure 2. Detail of the Prepare Groundwork box of Figure 1.**

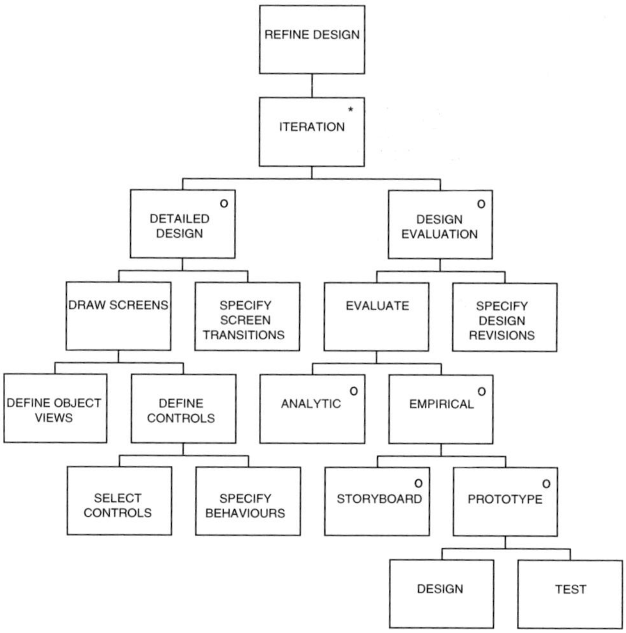


**Figure 3. Detail of the Refine Design box of Figure 1.**

Level Design is optional, because deeper decomposition may have occurred, during task-level scoping or it may be intended to derive the UI design directly from the high-level task and object models.

The High-level Specification box of Figure 1 consists of a sequence of Define Window Hierarchy and Define Styleguide, where the window hierarchy describes the selected styles of windows and their intended rôle in the UI design. The Implement User Interface box is a sequence of: Deliver Specification; Build UI and Test UI; and reflects delivery, implementation and testing of the final design.

## Discussion

The general model, proposed here, is derived from analysis of three HF structured UI design methods and represents a comprehensive, albeit high-level, view of the UI design process. Based as it is on methods claimed to support effective design, a potential use of the model would be for rapid identification of omissions in the more informal design processes, commonly observed in practice. Further research is required to investigate this use. Additional research is also required to incorporate alternative design processes into the model, thus increasing its generality.

## References

K.Y. Lim and J.B. Long, *The MUSE Method for Usability Engineering.* Cambridge University Press. 1994

D.P. Browne, STUDIO: *Structured User-interface Design for Interaction Optimisation.* Prentice Hall International. 1994

D. Redmond-Pyle and A. Moore, *Graphical User Interface Design and Evaluation (GUIDE): A practical process.* Prentice Hall International. 1995

# INTEGRATING HUMAN FACTORS METHODS: THE HF METHODS MATRIX

**Paul Salmon[1], Neville Stanton[1], Guy Walker[1], Damian Green[1] & Chris Baber[2]**

*Human Factors Integration – Defence Technology Centre*
*[1]Brunel University, School of Engineering and Design,*
*BIT-LAB, Uxbridge,*
*Middlesex UB8 3PH.*
*[2]Birmingham University,*
*Department of Electrical and Electronic Engineering,*
*Egbaston, Birmingham B15 2TT, UK*

The application of structured methods has been a feature of the human factors (HF) domain since its origin. Consequently, the HF practitioner now has a plethora of methods at his or her disposal. Despite their great number and increased use, there remains little practical advice available on the effective integration of these methods. As a result of a HF methods review, a methods matrix designed to guide practitioners in the selection, integration and application of HF methods was constructed. The following paper presents an extract of the methods matrix, and also presents the EAST methodology (Baber and Stanton 2004) as an example of an integrated HF approach.

## Introduction

Human factors (HF) methods are used throughout the design lifecycle in order to create, describe and evaluate features associated with man-machine interactions. There are numerous methods available, ranging from task analysis techniques which are used to describe activity, to situation awareness (SA) assessment techniques, which are used to measure operator SA during task performance. Despite the great number of different methods available, there is little evidence of the effective integration of HF methods. Further, there is limited guidance available on how to integrate these methods and also on which methods can be integrated with one another. When analysing activity in complex, socio-technical systems an integrated approach is required in order to ensure comprehensiveness. Kirwan (1998a, 1998b) highlights the need for an integrated approach to human error analysis. In conclusion to a review of thirty-eight existing human reliability analysis (HRA) and human error identification (HEI) techniques, he suggested that as none of the techniques available satisfied all of the fourteen criteria against which they were evaluated, a framework or toolkit approach may be the most suitable approach to error analysis. Kirwan (1998b) suggested that the best approach to predicting human error would be for practitioners to adopt a framework approach to HEI, whereby a mixture of independent HRA/HEI tools are used under one framework. Kirwan (1998b) identified one possible framework comprising the systematic human error reduction and prediction approach (SHERPA) (Embrey 1986), HAZOP, errors of commission analysis (EOCA) (Kirwan 1994), Confusion matrix analyses (Kirwan 1992b), Fault symptom matrix analysis and the skill, rule and knowledge classification system together.

The integration of HF methods in this way is an attractive prospect that offers a number of advantages. Firstly, the analysis becomes much more comprehensive as multiple aspects of the system or task under analysis can be analysed, such as operator activity, workload, situation awareness, error, and performance. Secondly, the analysis becomes more exhaustive as each of the integrated methods act as back up checks for one another. For example, in the case of HEI, any potential errors not identified (or missed) by one method may be picked up by another method within the framework.

## HFI-DTC Human Factors methods review

As part of the design of a novel command, control, communication, computers and intelligence (C4i) system, a HF methods review was conducted in order to identify the most suitable methods for use during the C4i design lifecycle (Salmon et al. 2004a). Initially, a literature review was conducted in order to create a comprehensive database of existing HF methodologies. The literature review was based upon a survey of standard ergonomics textbooks, relevant scientific journals and existing HF method reviews. The result of this initial literature review was a database of over 200 HF methods and techniques, including the following categories of method.

- Data collection techniques
- Task analysis techniques
- Cognitive task analysis techniques
- Charting techniques
- Human error identification (HEI) techniques
- Mental workload assessment techniques
- Situation awareness measurement techniques
- Interface analysis techniques
- Design techniques
- Performance time assessment techniques
- Team analysis techniques

Before the HF techniques were subjected to review, a screening process was employed in order to remove any unsuitable methods. Methods were classed as unsuitable if they were unavailable, software based, inapplicable, or if they were very similar to another method already selected for review. As a result of the method screening procedure, 96 HF design and evaluation methods were selected for the methods review. Each method was then analysed using a set of criteria designed not only to establish which of the techniques were the most suitable for use in the design and evaluation of C4i systems, but also to aid the HF practitioner in the selection and use of the appropriate method(s).

## The Human Factors methods matrix

The methods review highlighted a distinct lack of effective integration of HF methods. Further, in order to ensure comprehensiveness, a need for an integrated approach to be used in the analysis of C4i systems was identified. Consequently, the HF methods matrix was constructed. The methods matrix was constructed in order to allow practitioners to identify possible combinations of methods to suit specific analysis needs. The matrix was also designed to inform the HF practitioner which methods acts as an input into one another. As such, the matrix specifies which of the methods can be used effectively in conjunction with one another (i.e. which methods can be integrated) and

**Table 1. Methods matrix extract.**

| ***Method To → From ↓*** | Obs | HTA | CWA | CDM | OSD | SHERPA | NASA TLX | SAGAT | LA | CDA | CUD | SNA | AF | KLM | PN |
|---|---|---|---|---|---|---|---|---|---|---|---|---|---|---|---|
| Obs | | ▒ ■ | ▒ ■ | ▒ ■ | ▒ ■ | ▒ ■ | ■ | ▒ ■ | ▒ ■ | ▒ ■ | ▒ ■ | ▒ ■ | | ▒ ■ | ▒ ■ |
| HTA | ■ | | ▒ ■ | ■ | ▒ ■ | ▒ ■ | ■ | ▒ ■ | ▒ ■ | ▒ ■ | ▒ ■ | ▒ ■ | ▒ ■ | ▒ ■ | ■ |
| CWA | ■ | ■ | | | | | | | | | | | | | |
| CDM | ■ | ■ | | | ■ | | ■ | ■ | | ■ | ■ | ■ | | | ▒ ■ |
| OSD | ■ | ■ | ■ | ■ | | | ■ | | ▒ | ■ | ■ | ■ | | ■ | ■ |
| SHERPA | ■ | ■ | | | | | | | ■ | | | | ■ | ■ | |
| NASA TLX | ■ | ■ | | ■ | ■ | | | ■ | | | | | | | |
| SAGAT | ■ | ■ | | | | | ■ | | | | | | | | ■ |
| L. An | ■ | ■ | | | | ■ | | | | | | | ■ | ■ | |
| CDA | ■ | ■ | | ■ | ▒ ■ | | | | | | ■ | ■ | | | ■ |
| CUD | ■ | ■ | | ■ | ▒ ■ | | ■ | | | ■ | ■ | ■ | | | ■ |
| SNA | ■ | ■ | | ■ | ■ | | | | ■ | ■ | ■ | ■ | | | ■ |
| AF | | ■ | | | | ■ | | | ■ | | | | | ■ | |
| KLM | ■ | ■ | | | ▒ ■ | ■ | | | ■ | | | | ■ | | |
| PN | ■ | ■ | | ■ | ■ | | | ▒ ■ | | ■ | ■ | ■ | | | ■ |

Key

■ = Can be integrated with

▒ = Acts as an input into

***Glossary***

Obs = Observations

HTA = Hierarchical Task Analysis (Annett 2004)

CWA = Cognitive Work Analysis (Vicente 1999)

CDM = Critical Decision Method (Klien & Armstrong 2004)

OSD = Operation Sequence Diagrams (Baber et al 2004)

SHERPA = Systematic Human Error Reduction and Prediction Approach (Embrey 1986)

NASA TLX = NASA Task load index (Hart and Staveland 1988)

SAGAT = Situation Awareness Global Assessment Technique (Endsley 1995b)

LA = Link Analysis (Stanton and Young 1999)

CDA - Co-ordination Demand Analysis (Burke 2004)

CUD = Comms Usage Diagram (Watts and Monk 2000)

SNA = Social Network Analysis (Driskell and Mullen 2004)

AF = Allocation of functions

KLM = Keystroke Level Model (Stanton and Young 1999)

PN = Propositional Networks (Baber et al 2004)

also which of the methods act as input into other methods e.g. the output of a hierarchical task analysis is used as the input to a SHERPA analysis. Due to its size, the entire matrix cannot be presented within this paper. However, an extract of the matrix is presented in table 1. The extract is based upon a selection of methods from each method category identified during the methods review. Further, an example of a framework or integrated methodology developed from the methods matrix is described below.

## The EAST methodology

The event analysis of systemic teamwork (EAST) methodology (Baber and Stanton 2004) is an example of an integrated HF methods approach extracted from the HF methods matrix. EAST was developed for the analysis of C4i activity. EAST uses a combination of methods to form a framework for analysing C4i activity, including hierarchical task analysis (Annett 2004), co-ordination demands analysis (Burke 2004), comms usage diagram (Watts and Monk 2000), social network analysis (Driskell and Mullen 2004), operation sequence diagrams, critical decision method (Klein and Armstrong 2004) and propositional networks. EAST has been applied in a number of domains, including the fire service (Baber et al. 2004), naval warfare, aviation, energy distribution (Salmon et al. 2004b), air traffic control and rail (Walker et al. 2004) domains in order to analyse C4i activity.

## Acknowledgement

The work reported in this paper was carried out under the UK Defence Technology Centre (DTC) for Human Factors Integration (HFI), funded by the UK Ministry of Defence and the Defence Science and Technology Laboratory (DSTL).

## References

Annett, J. 2004, Hierarchical Task Analysis. In N. A. Stanton, A. Hedge, K. Brookhuis, E. Salas, & H. Hendrick (eds.) *Handbook of Human Factors methods*. Taylor and Francis, London.

Baber, C., & Stanton, N. A. 2004, Methodology for DTC-HFI WP1 field trials. *Defence Technology Centre for Human Factors Integration*, Report 2.1.

Baber, C., Walker, G., Stanton, N. A., & Salmon, P. 2004, Report on Initial Trials of WP1.1 Methodology conducted at fire service training college. WP1.1.1/01, 29th January 2004.

Burke, C. S. 2004, Team Task Analysis. In N. A. Stanton, A. Hedge, K, Brookhuis, E. Salas, & H. Hendrick (eds.) *Handbook of Human Factors methods*. Taylor and Francis, London.

Driskell, J. E., & Mullen, B. 2004, Social Network Analysis. In N. A. Stanton, A. Hedge, K. Brookhuis, E. Salas, & H. Hendrick (eds.) *Handbook of Human Factors methods*. Taylor and Francis, London.

Embrey, D. E. 1986, SHERPA: A systematic human error reduction and prediction approach. *Paper presented at the International Meeting on Advances in Nuclear Power Systems*, Knoxville, Tennessee.

Endsley, M. R. 1995b, Measurement of Situation Awareness in Dynamic Systems, *Human Factors*, Vol. 37, pp 65–84.

Hart, S. G., & Staveland, L. E. 1988, Development of a multi-dimensional workload rating scale: Results of empirical and theoretical research. In P. A. Hancock & N. Meshkati (eds.) *Human Mental Workload*. Amsterdam. The Netherlands. Elsevier.

Kirwan, B. 1992b, Human error identification in human reliability assessment. Part 2: detailed comparison of techniques. *Applied Ergonomics*, **23**, 371–381.
Kirwan, B. 1994, A guide to Practical Human Reliability Assessment, Taylor and Francis, London.
Kirwan, B. 1998a, Human error identification techniques for risk assessment of high-risk systems – Part 1: Review and evaluation of techniques. *Applied Ergonomics*, **29**, 157–177.
Kirwan, B. 1998b, Human error identification techniques for risk assessment of high-risk systems – Part 2: Towards a framework approach. *Applied Ergonomics*, **5**, 299–319.
Klein, G., & Armstrong, A. A. 2004, Critical Decision Method. In N. A. Stanton, A. Hedge, K. Brookhuis, E. Salas, & H. Hendrick (eds.) *Handbook of Human Factors methods*. Taylor and Francis, London.
Salmon, P. M., Stanton, N. A., Walker, G., & Green, D. 2004a, Human Factors Design and Evaluation Methods Review. *Defence Technology Centre for Human Factors Integration*, Report No. HFIDTC/WP1.3.2/1.
Salmon, P. M., Stanton, N. A., & Walker, G. 2004b, National Grid Transco: Switching operations report. *Defence Technology Centre for Human Factors Integration Report.*
Stanton, N. A., & Young, M. S. 1999, A guide to methodology in ergonomics: Designing for human use, London, Taylor and Francis.
Vicente, K.J. 1999, Cognitive work analysis: Towards safe, productive, and healthy computer-based work. Mahwah, NJ: Lawrence Erlbaum Associates, Inc.
Gibson, H., Walker, G., Baber, C., & Stanton, N. 2004, Report on results of WP1.1 EAST methodology for Railway data: Emergency Possession Scenario. *Defence Technology Centre for Human Factors Integration Report.*
Watts, L. A., & Monk, A. F. 2000, Reasoning about tasks, activities and technology to support collaboration. In J. Annett & N. Stanton (eds.) Task Analysis. Taylor and Francis, London.

# OCCUPATIONAL HEALTH AND SAFETY

# MULTI-MEDIA PACKAGING AND ERGONOMIC APPRAISAL

**Chris Fullwood, Orsolina I. Martino & Neil Morris**

*University of Wolverhampton, School of Applied Sciences, Psychology, Millennium Building, Wolverhampton WV1 1SB, UK*

The main aim of a health and safety campaign is to persuade audiences to adopt or avoid certain types of behaviours. Subsequently, it is suggested that the manner in which this information is presented will be one important determining factor in the intervention's success. With the advent of inexpensive multi-media packaging it is possible to import great flexibility in design, combining sounds bites with graphical features. For example, one can use this to present images of celebrities or other icons. This review will consider the ergonomic factors which are important in the design of multi-media packaging.

## Introduction

The need for public safety, for example in the home or in the workplace, requires the dissemination of information on health and safety issues. This information can be distributed to the public using a blend of technical resources so that it is presented in "multiple formats via multiple sensory modalities" (Schnotz and Lowe, 2003, pg 117). This approach is considered useful as it is assumed that it is easier to learn from materials that involve many media, partly because they are more stimulating (Hasebrook, 1995). Although health and safety information can be delivered via the use of more "traditional" media (for example, television), the fact that computers are becoming more widespread (and in turn, so has access to the Internet), there now exists the possibility to make use of a variety of multimedia tools to design and distribute health and safety information easily, effectively and economically. For example, using MS PowerPoint (the most extensively used multimedia tool) one can deliver a variety of media, such as text, pictures, animation and sound bites (Reimann, 2003). Furthermore, this information can be sent to other individuals via email, or uploaded from the Internet.

The success of a health and safety intervention will ultimately be measured by whether or not it can persuade the public to either adopt a set of healthy behaviours or avoid certain risky behaviours. Analysing the impact that health and safety promotions have on the behaviour of the public however is potentially time consuming and difficult to measure in the long term. Therefore, it may be more suitable to consider using comprehension as a measure of effectiveness. In other words, does the audience understand the information that has been transmitted to them? One way to achieve this is by analysing the amount of information successfully remembered (Daneman and Carpenter, 1980). For example, in considering an anti-drugs campaign, this might involve an ability to recognise drugs as being harmful, as recognition of this kind has been correlated significantly with future drug use (e.g. Bachman et al., 1988; Johnston, 1982). Although technological advancements have paved the way for facilitating the promotion of health and safety information, one must consider the design factors, which make multimedia packaging more effective. The following review will reflect on the effects of animation, narrative style, mode of presentation, and the use of celebrities to promote health and safety.

## Using animation

So, how can multimedia materials be designed in such a way as to make them more memorable? According to Downie et al. (1990), information is more likely to be acknowledged and less likely to be forgotten if it is attractive and has instant appeal. Similarly Chebat et al. (1995) noted that advertisements deemed as attractive also tended to be more memorable. One way in which multimedia materials can be made to look more attractive is through the inclusion of animation (Weiss et al., 2002). According to Lowe (2003), in recent years the use of animation (as opposed to static images) for instruction has increased. A number of researchers indicate that animation can have some positive benefits. Animations can be used to serve an attention gaining function. Hong et al. (2004) suggest that "flash" animation used on the Internet helps to attract the attention of users, however this does not help in the retention of the information. Koroghlanian and Klien (2004) also noted that children spent significantly more time using a computer programme about biology when animations were involved as opposed to static images. Large (1996) however suggests that animations generally fail to keep hold of attention once their novelty has worn off. Lowe (2003) suggests a number of reasons why animation may be ineffective: "(a) the imposition on learners of excessive information processing demands (overwhelming) and (b) a reduction in the extent to which learners engage in valuable processing activities (underwhelming)" (Lowe, 2003, pg. 158). Generally it would seem that animations, although impressive on a superficial level, do little to help with learning (Large, 1996).

## Narrative style

The style of the narrative voice is also an important consideration in the design of multimedia materials. Mayer (2003) noted that a conversational style leads to information being processed at a deeper level compared to a formal style. Furthermore, this "personalisation effect" was evident with on-screen text as well as narration. In the context of a health and safety campaign against smoking this might include, for example, referring to "your cigarettes" as opposed to "the cigarettes." Mayer (2003) explains this finding by proposing that when messages are personalised, individuals are primed with a conversation schema, and therefore may be more disposed to believe that they are part of a human-to-human exchange. Furthermore, this disposition would result in the individual trying harder to understand what is being said.

## Mode of presentation

The design of multimedia presentations can be developed further. Nugent (1982) illustrated that in the presentation of hazard information, the mode of presentation is crucial. When text, pictures and narration were combined, information was remembered more easily compared to when either of these individual elements was used to present the information alone. This suggests that presenting information in a number of different formats simultaneously has a positive effect on learning. Mayer and Moreno (2002) however indicate that cognitive load is an important consideration when it comes to designing multimedia presentations, especially if the intention is to remember the information accurately. Cognitive load theory (Chandler and Sweller, 1991; Sweller, 1999) promulgates that the design of an instructional message should reflect the capacity of the learner's cognitive system, and therefore should attempt not to cause overload.

Mayer and Moreno (2002) suggest that information is best presented in words and pictures rather than solely in words (multiple representation principle). When words are presented on their own, an individual is less likely to construct a visual mental representation that connects with their verbal mental representation; however when words and visual material are presented simultaneously, the individual will be more likely to build visual and verbal representations, and make connections between the two (Mayer and Moreno, 2002). Furthermore, the integration of verbal and visual information (i.e., presenting corresponding words and pictures simultaneously rather than separately) results in better retention of the information (Sweller et al., 1998; Mayer and Moreno, 2002). This might include, for example, placing a text box next to a picture (Sweller et al., 1998). The limitations of human working memory are an explanation for this principle of contiguity. When words and pictures are presented at the same time, visual and verbal representations are more likely to be able to be held in working memory (Mayer and Moreno, 2002).

Evidence indicates that words are better presented as auditory narration than on-screen text (modality principle). Essentially, on screen text will be processed through the visual system, along with any animation included in the presentation (Mayer and Moreno, 2002). For this reason, the text and the animation contend for our limited attention, producing a "split attention effect" (Mousavi, Low and Sweller, 1995). However, if narration is used in the presentation, this information can be processed through the verbal channel, meaning that the visual channel is free to process the animation at a deeper level. This modality effect has also been demonstrated by a number of other researchers, and results in lower mental effort (Kalyuga et al., 1999), less time spent solving problems (Jeung et al., 1997), and improved memory for information (Mayer and Moreno, 1998). Findings from these studies may also be explained in terms of a "reduction in visual search" (Tabbers et al., 2004, pg 73).

The amount of information presented to the audience is also an important consideration. Multimedia presentations are better when they include fewer rather than many extraneous words and sounds (coherence principle). An explanation for this is that the irrelevant or extraneous material may fill working memory, and therefore prevent the learner from learning the relevant material (Mayer and Moreno, 2002). Tabbers et al. (2004) advise caution in generalising the findings from such research, especially as they were conducted under experimental conditions in the laboratory, and that virtually all of these studies involved teaching participants about technical information.

## Celebrities

Health and safety promotions often make use of celebrities, for example to promote road safety in Britain (Buckingham et al., 1999), and to speak out against drink driving in the United States (Slater, 1999). There are mixed findings for the effectiveness of using celebrities to promote health and safety behaviours. Borzekowski and Poussaint (2000) noted that celebrity spokespeople for anti-violence messages were less effective than accurate-looking characters in authentic situations. Brown and Basil (1995) on the other hand found that the American basketball player "Magic" Johnson, was an effective spokesperson for HIV/AIDS prevention. The authors concluded that the persuasiveness of the message was caused by the fact that the public felt some form of involvement with the celebrity through media exposure.

For an explanation for this apparent difference, one could look to the advertising literature. Research seems to indicate that attitudes towards products are more favourable when the brand and the celebrity are congruent (Misra and Beatty, 1990; Cronin, 2004). In other words, the celebrity is perceived as more credible if they seem to be an appropriate match with the brand they are attempting to sell. For example, David Beckham would seem a likely candidate to sell football boots, but less congruent with the sale of junk food. The same may also hold true for celebrities promoting health and safety information. Therefore, it would seem that with an already established association between the celebrity and the branding, it is possible build upon an existing schema or link, rather than having to build a new association. This is also explains the findings of Till and Shimp (1998), who noted that attitudes towards fictitious celebrities declined with the introduction of negative information about them, whereas the same did not hold true for actual celebrities.

## Summary

In summary, it would seem that there are a number of design principles that can be employed for effective multimedia health and safety promotions. Information is more memorable if it is attractive, appealing, and includes a conversational style rather than a formal style. Static graphics are preferable to animation. Words and pictures are better than words alone, especially if they are presented simultaneously rather than separately. Auditory information is superior to on-screen text and the use of too many extraneous sounds or images has a negative impact on memory. Furthermore, the use of celebrities to promote health and safety is beneficial but more so if the celebrity and the branding are seen as congruent.

## References

Bachman, J.G., Johnston, L.D., & O'Malley, P.M. (1988). Explaining the recent decline in marijuana use: Differentiating the effects of perceived risks, disapproval, and general lifestyle factors, *Journal of Health, Society and Behaviour*, 29, 92–112.

Borzekowski, D.L.G., & Poussaint, A.F. (2000). Common themes from the extremes: Using two methodologies to examine adolescents' perceptions of anti-violence public service announcements, *Journal of Adolescent Health*, 26, 164–175.

Brown, W.J., & Basil, M.D. (1995). Media celebrities and public health: Responses to "Magic" Johnson's HIV disclosure and its impact on AIDS risk and high-risk behaviours, *Health Communication*, 7(4), 345–370.

Buckingham, D., Davies, D., Jones, H., & Kelley, P. (1999). *Children's television in Britain: History, discourse and policy*, British Film Institute: London

Chandler, P., & Sweller, J. (1991). Cognitive load theory and the format of instruction. *Cognition and Instruction*, 8, 293–332.

Chebat, J.C., Laroche, M., Badura, D., & Filiatrault, P. (1995). Affect and memory in advertising: An empirical study of the compensatory processes, *Journal of Social Psychology*, 135(4), 425–437.

Cronin, J.N. (2004). The effect of celebrity endorsements on attitudes toward advertisements, brands, and purchase intentions, *Dissertations Abstracts International-section A-Humanities and Social Sciences*, 64(7A), 2570

Daneman, M., & Carpenter, P. (1980). Individual differences in working memory and reading, *Journal of Verbal Learning and Verbal Behaviour*, 19, 450–466.

Downie, R.S., Fyfe, C., & Tannahill, A. (1990). *Health promotion: models and values*, Oxford: Oxford University Press.

Hasebrook, J.P. (1995). Learning with multimedia, *German Journal of Educational Psychology*, 9(2), 95–103.

Hong, W., Thong, J.Y.L., & Tam, K.Y. (2004). Does Animation Attract Online Users' Attention? The Effects of Flash on Information Search Performance and Perceptions, *Information Systems Research*, 15(1), 60–86.

Jeung, H., Chandler, P., & Sweller, J. (1997). The role of visual indicators in dual sensory mode instruction, *Educational Psychology*, 17, 329–343.

Johnston, L.D. (1982). A review of the analysis of recent changes in marijuana use by American young people. In *Marijuana: The national impact of education*, New York: American Council on Marijuana, 8–13.

Kalyuga, S., Chandler, P., & Sweller, J. (1999). Managing split-attention and redundancy in multimedia instruction, *Applied Cognitive Psychology*, 13, 351–371.

Koroghlanian, C., & Klien, J.D. (2004). The effect of audio and animation in multimedia instruction, Journal of Educational Multimedia and Hypermedia, 13(1), 23–46.

Large, A. (1996). Computer animation in an instructional environment. *Library and Information Science Research*, 18(1), 3–23.

Lowe, R.K. (2003). Animation and learning: Selective processing of information in dynamic graphics, *Learning and Instruction*, 13, 157–176.

Mayer, R.E. (2003). The promise of multimedia learning: Using the same instructional design methods across different media, *Learning and Instruction*, 13, 125–139.

Mayer, R.E., & Moreno, R. (1998). Split-attention effect in multimedia learning: Evidence for dual processing systems in working memory, *Journal of Educational Psychology*, 90, 312–320.

Mayer, R.E., & Moreno, R. (2002). Aids to computer-based multimedia learning, *Learning and Instruction*, 12, 107–119.

Misra, S., & Beatty, S.E. (1990). Celebrity spokesperson and brand congruence: An assessment of recall and affect, *Journal of Business Research*, 21(2), 159–173.

Mousavi, S.Y., Low, R., & Sweller, J. (1995). Reducing cognitive load by mixing auditory and visual presentation modes, *Journal of Educational Psychology*, 87, 319–334.

Nugent, G. (1982). Pictures, audio and print: Symbolic representation and effect on learning. *Educational Communication and Technology Journal*, 30, 163–174.

Reimann, P. (2003). Multimedia learning: beyond modality, *Learning and Instruction*, 13, 245–252.

Schnotz, W., & Lowe, R.K. (2003). External and internal representations in multimedia learning, *Learning and Instruction*, 13, 117–123.

Slater, M.D. (1999). Drinking and driving PSAs: A content analysis of behavioral influence strategies, *Journal of Alcohol and Drug Education*, 44, 68–81.

Sweller, J. (1999). *Instructional design in technical areas*, Camberwell, Australia, ACER Press.

Sweller, J., Van Merrieboer, J.J.G., & Paas, F. (1998). Cognitive architecture and instructional design, *Educational Psychology Review*, 10, 251–296.

Tabbers, H.K., Martens, R.L., & Merrienboer, J.J.G. (2004). Multimedia instructions and cognitive load theory: Effects of modality and cueing, *British Journal of Educational Psychology*, 74, 71–81.

Till, B.D., & Shimp, T.A. (1998). Endorsers in advertising: The case of negative celebrity information, *Journal of Advertising*, 27(1), 67–82.

Weiss, R.E., Knowlton, D.S., & Morrison, G.R. (2002). Principles for using animation in computer based instruction: Theoretical heuristics for effective design, *Computers in Human Behaviour*, 18, 465–477.

# USING *HOMER SIMPSON* TO PROMOTE SAFETY

**Neil Morris, Chris Fullwood & Orsolina I. Martino**

*University of Wolverhampton, School of Applied Sciences, Psychology, Millennium Building, Wolverhampton WV1 1SB, UK*

With the advent of widely available multi-media presentation packages such as PowerPoint it is possible to develop instructional packages that can be widely, and rapidly, distributed via CD-ROM and the internet. This paper reports the development of a PowerPoint presentation that informs the consumer about chemical hazards in the home and the workplace. Three versions of the presentation were presented to a total of 36 students. One group were presented with screens containing bullet points about the hazards, a second group saw the same material but images of the animated character Homer Simpson appeared with the bullet points, and in a third version Homer Simpson's image and soundbites of his exclamations accompanied the bullet points. A subsequent retention test revealed that the third version, with images and soundbites, resulted in significantly better retention than the other versions.

## Introduction

There is a public health need to provide advice to both adults and children on dangers in the home and the workplace. An obvious example of such a hazard is the range of corrosive, toxic and inflammable chemicals encountered in daily life. The advent of the widespread availability of personal computers with multi-media capabilities has opened up the possibility of providing instructional packages that are stimulating to use and that provide a means of disseminating the necessary knowledge. Software tools such as PowerPoint are particularly attractive because they facilitate easy translation into other languages and can be sent to users via the Internet.

The use of multi-media approaches to disseminate public health information is not a recent innovation. From 1945 to 1970 the US government sponsored a series of short "mental hygiene" films designed to be shown to children in schools and youth clubs. Most of these showed teenagers, or younger children, engaging in activities deemed to be anti-social. An adult voice was dubbed to provide a narrative on the dangers of consuming alcohol or whatever (Smith, 1999). Such an approach might now seem archaic in terms of both content and mode of delivery but there is still a need for public information packages and these might best be delivered from CD-ROM or the Internet.

Multi-media packages designed to have widespread availability of necessity need to be self contained in the sense that the "instructor" is not physically present to elaborate on the information provided. Such packages can only be effective if they can be comprehended without further clarification and a good measure of this is the amount of material that can be recalled after using the software.

Previous research into retention of hazard information has demonstrated that presentation mode is important (Najjar, 2001). In particular, multi-media presentation, (that is, using pictures and narration with text) produced better retention in elementary school children than text, pictures or audio alone (Nugent, 1982). With adults Tindall-Ford et al. (1997) showed better retention when both visual and audio modalities were employed.

Multi-media design can be elaborated further. For example, British television safety promotions have frequently used celebrities that children are familiar with to promote road safety (Buckingham et al., 1999). Advertising research has shown that humorous cartoons can be used to promote certain products. Humour seems to be an appropriate selling device when the implications of purchase are not serious, that is, the purchase is a relatively trivial item in terms of financial investment (Zhang, 1996). Furthermore, Disney characters have successfully been used to promote sales of "fast food" (Schlosser, 2002). Animation was used by Mayer and Anderson (1991) who carried out a study on retention of information about bicycle repair by employing either animated instruction, animated instruction with a narrative soundtrack or just the narration. Subsequent performance was superior following training using the animation with soundtrack.

There are clearly strong indications that an instructional package to educate the user in the hazards associated with chemicals commonly found in the environment should employ both sound and vision and a humorous element might also be efficacious. It was hoped that the package used in the present study would be useful with adults so it seemed inappropriate to employ most cartoon characters as images to hold attention.

However modern animation is not exclusively aimed at children. There are a number of cartoon characters whose actions can be interpreted in a number of different ways. The characters depicted in the animated sitcom "The Simpsons" can be interpreted as rather simple humorous figures and, simultaneously, as providing satirical observations of some sophistication on the nature of American home life (Grimple, 1999). Indeed the pronouncements of the Simpsons have been used to illustrate problems in philosophy (Irwin et al., 2001). Recently Kelley (2002) used magnetic resonance imaging to scan volunteers brains while they watched episodes of The Simpsons. During sequences rated as humorous there was increased processing in the amygdala, an area associated with memory processing. Kelley commented that "you tend to recall the funny bits in sitcoms" (Travis, 2002, p 308).

In The Simpsons the character Homer Simpson is a safety officer in a nuclear power plant and the fact that several story lines have centred on his incompetence makes him an incongruous choice of character to provide safety instruction. However incongruity is an important facet of some types of humour (Lefcourt, 2002) and Homer is an enormously popular cultural icon (Groening and McCann, 2002). Mahon has commented that "We learn from fiction… because it promotes identification. Such identification clearly occurs with characters like Homer and Marge Simpson whose lives bear similarity – albeit caricatured – to those of many viewers" (Mahon, 2001, p 221). Thus the use of Homer Simpson as an icon in a multi-media presentation on chemical hazards may well enhance the impact of the presentation and result in greater retention of the material contained therein. To test this three versions of a PowerPoint were prepared. One version presented the information without Homer's image. The other two versions incorporated Homer but one version had no soundtrack while the other had well known Homer exclamations dubbed on to it. It was hypothesised that retention would be greater when Homer Simpson was present and that performance would be further enhanced in the multi-media version containing Homer's exclamations.

## Method

### *Participants*

Thirty six students from the University of Wolverhampton School of Applied Sciences participated in this study. None of these students were reading chemistry. There were 20 females and 16 males, mean age 23.4 years (sd = 1.2).

### *Materials and procedure*

A list of common hazardous chemicals and their properties were assembled from published sources (Birkin and Price, 1989; Luxon, 1992). Each was allocated to a screen in PowerPoint 2002 so that the name of the substance was presented and six bulletin points providing information on its sensory properties (appearance, smell etc.) and a list of its hazards (for example, flammability, toxicity, corrosiveness) was given. Ten screens were prepared and the information was offset to the right of the screen. A tone indicated the imminent presentation of a screen. A second version was prepared. In this version the blank space on the left was filled by an image of Homer Simpson. Each Homer image reinforced the hazard information. For example, a toxic substance would appear with an image of Homer prostrate and gagging. A substance whose hazards included corrosive by inhalation, for example ammonium hydroxide (household ammonia), would have an image of Homer coughing violently. The images were not animated, they were static images downloaded from large collections of Simpson images on the Internet. This version was then adapted to create a third version in which soundbites of Homer Simpson, again downloaded from internet sites, were dubbed. These soundbites contained no words. They consisted of exclamations (d'oh, urg etc.) which Homer might make when exposed to these chemical hazards. The soundbite occurred at the beginning of the screen presentation and was presented only once. In all conditions each screen was shown for 25 seconds, a pilot study showed that this allowed at least four readings of each screen, and then there was a further five seconds before the next screen was presented.

Twelve students were allocated to each condition. This allocation was determined by seating in a laboratory with individual television monitors and headphones for each participant. All participants wore headphones through which they heard Homer and the pre-trial warning tone if they were in the image plus soundbite condition. In the other two conditions only the warning tone before each screen was heard.

The condition not involving Homer Simpson was run the previous day so that instructions could be adapted to exclude any reference to Homer Simpson. This group were told that they were to be presented with a training presentation for introductory science students. The other two groups were also told this but it was also indicated that Homer Simpson had been added to "lighten" the presentation. The two "Homer" groups were not aware that there were two versions of the Homer presentation.

Participants were given a booklet containing multiple-choice questions. They were not allowed to open this until the presentation was complete. After the presentation participants opened their multiple-choice booklet and answered 10 questions on the presentation. Each question had one correct answer that had to be chosen from a total of 10 options. Participants were requested to guess any answer they did not know. Chance level performance would be approximately one correct answer.

## Results

The multiple-choice test was scored out of ten. Means and standard deviations for each condition are shown in Table 1. The data were subjected to analysis by one-way, independent samples analysis of variance. There was a significant difference between conditions ($F(2,33) = 8.32$, $p < 0.002$) and t-tests showed that the condition employing images of Homer Simpson accompanied by soundbites produced significantly higher scores than the other two conditions ($p < 0.025$, both comparisons). The other two conditions, Homer image present and no image of Homer, did not differ ($p > 0.05$).

**Table 1. Means and standard deviations for multiple-choice test scores (maximum score 10) on a test for knowledge of chemical hazards after a Power Point presentation.**

| Presentation format | Text only | Text plus images of Homer Simpson | Text plus images and exclamations of Homer Simpson |
|---|---|---|---|
| Mean | 1.75 | 2.58 | 4.17 |
| Standard deviation | 1.05 | 1.24 | 1.94 |

## Discussion

These results show significantly better retention when Homer Simpson's image and exclamations accompany a hazard awareness programme. With a single presentation almost 42% of material is retained compared to less than 18% with a list of bullet points. Of particular interest is the facilitatory effect of simply adding linguistically impoverished Homer Simpson exclamations.

The data are consistent with the findings from advertising research. Zhang (1996) argued that when little cognitive investment is involved in the decision to purchase a product then fairly simple humour can be used in advertising to promote the product. This need not mean that humour can only be used to promote the sale of relatively unimportant products. Rather, Zhang argues, it is the extent to which generating an attitude toward a product/service requires intellectual effort, or in Cacioppo and Petty's (1982) terminology, a need for cognition. When little cognitive effort is required to arrive at a decision about whether a product is desirable or unwanted, then humour is a useful device to promote the product. Arriving at the view that it is a good thing to increase one's awareness of dangerous chemicals in the environment is unlikely to require profound thought. If this perspective is valid, then it seems reasonable to employ humour as a component of presentations imparting information almost universally viewed as important for enhancing public safety. We thank Nadia Pervaiz for data collection.

## References

Birkin M and Price B (1989) *C for Chemicals: Chemical Hazards and how to avoid them.* Merlin: London.

Buckingham D et al. (1999) *Children's television in Britain: History, discorse and policy.* British Film Institute: London.

Cacioppo J and Petty R (1982) The need for cognition. *Journal of Personality and Social Psychology,* **42,** 116–131.

Grimple S M (1999) *The Simpsons Forever.* Harper-Collins: New York.

Groening M and McCann J L (2002) *The Simpsons: Beyond Forever.* Harper-Collins: New York.

Irwin W et al. (2001) *The Simpsons and Philosophy.* Open Court: Peru, Illinois.

Kelley W M (2002) The neural funny bone: dissociating cognitive and affective components of humour. *Paper presented at the Society for Neuroscience 32nd Annual meeting,* Orlando, Florida.

Lefcourt H M (2002) Humor. In C Snyder and S Lopez (eds.) *Handbook of Positive Psychology.* Oxford University Press: Oxford. 619–631.

Luxon S G (1992) *Hazards in the chemical laboratory.* Royal Society of Chemistry: London.

Mahon J L (2001) The function of fiction: the heuristic value of Homer. In Irwin W et al. (Eds.) *The Simpsons and Philosophy.* Open Court: Peru, Illinois. 215–232.
Mayer R and Anderson R (1991) Animations need narration: An experimental test of a dual-coding hypothesis. *Journal of Educational Psychology,* **83,** 484–490.
Najjar L J (2001) Principles of educational user interface design. In R Swezey and D Andrews (Ed) *Readings in training and simulation: a thirty year perspective.* Human Factors and Ergonomics Society: Santa Monica. 146–158.
Nugent G (1982) Pictures, audio, and print: Symbolic representation and effect on learning. *Educational Communication and Technology Journal,* **30,** 163–174.
Schlosser, E (2002) *Fast Food Nation.* Penguin: London.
Smith K (1999) *Mental hygiene: Classroom films 1945–1970.* Blast: New York.
Tindall-Ford S et al. (1997) When two sensory modes are better than one. *Journal of Experimental Psychology: Applied,* **3,** 257–287.
Travis J (2002) The brain's funny bone: Seinfeld, The Simpsons spark same nerve circuits. *Science News,* **162,** 308.
Zhang Y (1996) Responses to humorous advertising: the moderating effect of need for cognition. *Journal of Advertising,* **25,** 15–32.

# EMOTIONAL FACTORS IN INITIATIVES TO PROMOTE HEALTH AND SAFETY: EFFECTS OF MOOD STATE ON COMPLIANCE

**Orsolina I. Martino, Chris Fullwood & Neil Morris**

*University of Wolverhampton, School of Applied Sciences, Psychology, Millennium Building, Wolverhampton WV1 1SB, UK*

Ever-increasing awareness of health and safety issues has led to various initiatives to promote healthier and more cautious living. This review considers the most effective approach in encouraging people to not just remember information but also to act upon it, focusing on the influence of mood state on the likelihood of initiating a change and then maintaining that behaviour. Empirical findings in relation to models of positive change are discussed. It is concluded that health and safety information can have an impact on mood regardless of the amount retained; furthermore, enhancing emotional state has important implications for compliance.

## Introduction

Raising awareness of health and safety matters is an increasingly essential requirement of modern living, both at home and at work. "Positive ergonomics" (Martino & Morris, 2004) takes a "prevention rather than cure" approach, highlighting the importance of maintaining well-being and preventing potential hazards from getting to a problematic stage. This is a useful perspective to consider in the promotion of health and safety initiatives, as many of these programmes are designed to inform the public of possible risks in order to reduce the likelihood of accidents or illness. However, the success of such initiatives is not solely determined by the information given. The way in which that information is delivered has significant implications not only for the absorption and retention of the material, but also for the emotional state it evokes. Mood state encompasses a number of contributing factors, with arousal and somatic comfort just as salient as emotion in distinguishing positive moods from negative. Thayer (1996; 2001) proposed a theory of four basic moods, which result from combinations of energy and tension. Whilst calm-energy is almost always a positive mood state and tense-tiredness negative, calm-tiredness and tense-energy can have both positive and negative connotations depending on their context. This paper considers the impact of health and safety information, discussing implications of the resulting mood state in terms of compliance with commencing and maintaining the desired behaviour.

## Stages of change

Models of health behaviour and positive change are used in health psychology to structure effective methods of promoting health-related material. Although the successful application of these models depends to a certain extent on the nature and context of the problem, it is clear to see that key elements can be applied to the issue of health and safety at home and in the workplace. One such model is that of the six "stages of change" – precontemplation, contemplation, planning, action, maintenance and termination (Prochaska & DiClemente, 1992; Prochaska & Norcross, 1994; Prochaska et al., 1994). An influential model in health psychology, the stages of change framework is a useful consideration wherever the adoption of health and safety behaviours is of interest.

The precontemplation stage is typically characterised by denial of the problem in hand; during this stage a person might avoid getting information about this problem and the costs of resolving the issue are overestimated whilst the costs of not addressing the issue are underestimated (see Carr, 2004). Even though the "denial" aspect does not apply as directly to health and safety in the workplace as it does to more personal health matters such as smoking or alcohol consumption, it can nonetheless help determine the mental state needed to progress onto the subsequent stages. Here the presentation of material is important; the attention of the target audience has to be held sufficiently for them to take in and think about the information being given as opposed to dismissing it as unnecessary or irrelevant. In other words, the purpose of the delivery is to facilitate contemplation of the issue and initiate planning to adopt the particular behaviour being encouraged.

## Mood, attention and retention: Contemplating the problem and planning action

Research from health and advertising suggests that a positive mood state tends to correspond to better recall of material and increased compliance. One way in which mood can be enhanced is by using humour, yet findings on the benefits of humour are mixed. Emotional response has been reported to blunt the degree of attention that viewers can pay to advertising (Bushman, 1998; Furnham et al., 1998), with violent and humorous programmes linked with low recall of advertising material. However, Morris et al., 1 (reported in this volume) demonstrated that using images and amusing soundbites of the popular animated character Homer Simpson to accompany bullet points about chemical hazards in the home and workplace resulted in significantly better retention of information than using images alone. Though partly attributable to the familiar figure or "celebrity" influence, one might argue that using humour in the form of Homer's comical exclamations also improved mood.

The discrepancy between this finding and those of Bushman (1998) and Furnham et al. (1998) can be explained in two ways. It is possible that the emotional response generated by the violent and humorous material used in these studies was too extreme, inducing a state of tense-energy rather than calm-energy. This view would be consistent with the major theories of arousal (see Eysenck, 1984), which postulate that although moderate levels of arousal are ideal for optimum cognitive performance, excessive or negative forms of arousal such as anxiety can impair performance. Alternatively, the differences could be a function of the cognitive effort, or "need for cognition" involved in processing the material (see Cacioppo & Petty, 1982). According to Zhang (1996), humour is a useful device to promote a product when little cognitive effort is required to decide whether the product is desirable or unwanted. With certain health behaviours, such as smoking cessation, this may not be appropriate as a great deal of thought is likely to be given to such a decision, but in the context of promoting health and safety this could be a practical strategy to employ. As Morris et al., 1 (this volume) point out, arriving at the view that it is good to increase one's awareness of potential hazards at home and in the workplace is unlikely to require profound thought. Overall these studies suggest that calm-energy is the ideal mood state for attending to and retaining information that has a relatively low need for cognition.

## Mood and compliance: initiating action and maintaining health behaviour

The level of attention paid to health and safety information is not only fundamental to the amount remembered, but also affects the probability of complying with that information. However, the nature of this attention is important. Studies of the relationship between mood and the initiation of health-related changes support the retention findings, indicating that a calm and positive mood state increases compliance and chances of success. Orleans (1985) discussed medical advice given by physicians to aid smoking cessation, asserting that scare tactics were unhelpful – positive approaches, namely outlining the health benefits of cessation, reassuring patients of the temporary nature of unpleasant withdrawal symptoms and enhancing their motives and confidence, were considered to be more beneficial.

An explanation of the cognitive processes underlying these effects can be found in a study by van der Velde & van der Pligt (1991), who examined predictors of sexual health behaviour and behavioural intentions in relation to the protection motivation theory. Protection motivation theory (Rogers, 1985) considers the roles of severity of and susceptibility to a risk, the perceived effectiveness of change, self-efficacy in carrying out that change, and emotional response to education or information about the risk or danger. These components are influenced by the information so that a coping response is generated – this response is either adaptive (e.g. an intention to change behaviour) or maladaptive (e.g. avoidance or denial of the problem). They discovered that although there was a relationship between fear and behavioural intentions, the relationship was less evident when fear levels were high. It was proposed that when experiencing excess fear, attention may be redirected towards reducing anxiety rather than changing behaviour to avoid danger. Again consistent with arousal theory, it seems that bringing about a state of tension rather than calm-energy is more likely to hinder progress to change.

More recently, Morris et al., 2 (reported in this volume) gave participants a short presentation on a fictional disease resembling Ebola. Despite retaining the same amount of information, those who received the symptoms before the prognosis showed a significant positive shift in mood compared with those who received the prognosis first and a neutral control group. It was concluded that health promotions should be structured to maximise mood state rather than evoke fear. Arousing intense emotions can motivate a person into initiating change (Carr, 2004), but it could be argued that the influence of these emotions depends on the situation in which information is given. Emotions such as fear and guilt can sometimes serve as stimuli for maximising health when there is great personal relevance, but for more general health and safety issues one of the difficulties in encouraging compliance may well lie with the fact that self-reference may not be immediately apparent. This fits Zhang's (1996) view that using humour is less suitable for information with a greater need for cognition. Nonetheless, even information with more serious connotations should be treated with caution where inducing fear is concerned, as excess levels are liable to interfere with one's ability to contemplate and tackle the problem.

Now that the issue in question has been acknowledged and acted upon, the final stage is to continue with that behaviour, with "termination" referring to the successful assimilation of the desired routine. Prochaska et al. (1994) examined the pros and cons of change at the different stages of change. In order to progress safely into the "action" stage with a good chance of success, the pros from the precontemplation stage to the action stage must increase by one standard deviation and the cons must decrease by half a standard deviation. As the stages progress further to the maintenance of that behaviour, the pros should continue to increase as the cons decrease. This emphasises the strong subjective component in the effectiveness of any health promotion campaign, and relates back to the role of self-efficacy in protection motivation theory; the individual's belief of their self-efficacy in carrying out the necessary change has considerable implications for the resulting coping response. As Orleans (1985) suggests, people are more likely to respond positively to health advice when their motives and confidence are augmented. Encouraging a calm and positive mood state when disseminating health and safety information is thus expected to be more effective than scare tactics in persuading an audience to adopt and maintain health- and safety-conscious habits.

## Conclusions and recommendations

It is clear that health and safety information can have an impact on mood regardless of how much is actually retained. Nevertheless the resulting emotional state can influence the extent to which information is taken on board, though the degree and nature of this influence depends on the context of the information and the kind of mood it elicits. Whilst emotions such as fear and guilt may sometimes help to motivate a person into initiating a change, extreme emotional responses can blunt attention to information and inducing anxiety may encourage maladaptive coping strategies, hindering progress to change. It is therefore recommended that:

1. Enhancing mood by use of humour may be appropriate when the material being presented does not require complex thought;
2. Using scare tactics to provoke an extreme emotional state should be avoided when disseminating public health and safety information;
3. Health and safety initiatives should be structured in order to maximise mood state, encouraging self-confidence with regards to initiating and maintaining positive change.

## References


Bushman, B.J. (1998). Effects of television violence on memory for commercial messages. *Journal of Experimental Psychology: Applied*, **4**, 291–307.

Cacioppo, J. & Petty, R. (1982). The need for cognition. *Journal of Personality and Social Psychology*, **42**, 116–131.

Carr, A. (2004). *Positive Psychology: The Science of Happiness of Human Strengths.* Hove, East Sussex: Brunner-Routledge.

Eysenck, M.W. (1984). *A Handbook of Cognitive Psychology.* London: Lawrence Erlbaum Associates.

Furnham, A., Gunter, B. & Walsh, D. (1998). Effects of programme content on memory of humorous television. *Applied Cognitive Psychology*, **12**, 555–567.

Martino, O.I. & Morris, N. (2004). Positive Ergonomics: Improving mood before the working day begins. In P.T. McCabe (Ed.) *Contemporary Ergonomics 2004* (pp. 425–429). New York: CRC Press.

Morris, N., Fullwood, C. & Martino, O.I. 1 (this volume). Using *Homer Simpson* to promote safety.

Morris, N., Martino, O.I. & Fullwood, C. 2 (this volume). Manipulating the impact of public health information on mood.

Orleans, C.T. (1985). Understanding and promoting smoking cessation: Overview and guidelines for physician intervention. *Annual Review of Medicine*, **36**, 51–61.

Prochaska, J. & DiClemente, C. (1992). The transtheoretical approach. In J. Norcross and M. Goldfried (Eds.). *Handbook of Psychotherapy Integration* (pp. 300–334). New York: Basic Books.

Prochaska, J. & Norcross, J. (1994). *Systems of Psychotherapy: A Transtheoretical Analysis* (3rd edn.). Pacific Grove, CA: Brooks/Cole.

Prochaska, J., Norcross, J. & DiClemente, C. (1994). *Changing for Good.* New York: Avon.

Rogers, R.W. (1985). Attitude change and information integration in fear appeals. *Psychological Reports*, **56**, 179–182.

Thayer, R.E. (1996). *The Origin of Everyday Moods: Managing Energy, Tension and Stress.* New York: Oxford University Press.

Thayer, R.E. (2001). *Calm Energy.* Oxford: Oxford University Press.

Van der Velde, F.W. & van der Pligt, J. (1991). AIDS related health behaviour: Coping, protection motivation, and previous behavior. *Behavioral Medicine*, **14**, 429–452.

Zhang, Y. (1996). Responses to humorous advertising: The moderating effect of need for cognition. *Journal of Advertising*, **25**, 15–32.

# MANIPULATING THE IMPACT OF PUBLIC HEALTH INFORMATION ON MOOD

**Neil Morris, Orsolina I. Martino & Chris Fullwood**

*University of Wolverhampton, School of Applied Sciences, Psychology, Millennium Building, Wolverhampton WV1 1SB, UK*

Warning the public about the symptoms and prognosis for sufferers of a life threatening disease has psychological consequences. Health promoters need to consider the emotional impact of different modes of presentation. In this study participants were given a short presentation on a fictional disease that closely resembled Ebola. One group was told about the symptoms and then the prognosis, a second group were told about the prognosis and then the symptoms and a third group, the control group, had a presentation on nutrition and diet. Mood was assessed before and after the presentation and a memory test measured retention. There was a significant positive mood shift in the symptoms first group but no change in the prognosis first group or control group. Information retention levels did not differ between groups. The results are discussed in terms of the structuring of health promotions to maximize mood state.

## Introduction

"Health scares" are commonplace (Dalrymple, 1998) and may induce considerable anxiety in individuals who are unlikely to become victims. However, it would be more than remiss to withhold information about potential health threats simply because such warnings are psychologically deleterious. *Potential threats*, as opposed to those that have been realised, for example AIDS, may be averted by responsible public health measures. To this end Morse (1993) examined the dangers of life threatening viral epidemics and included contributions by virologists offering worst case, imaginary scenarios based mainly on extrapolation from smaller outbreaks and likely viral adaptations. Great emphasis was placed on Ebola and Marburg viruses, and in particular, a detailed public health assessment of the impact of a "super-Ebola" virus was offered by Legters et al. (1993). If such a threat appeared to be about to be realised then releasing public health information would be essential. In general, compliance with such advice is greater when the recipient is not unduly frightened by the information (see Martino et al. this volume, for a review).

Filoviruses, for example Ebola and Marburg, can produce symptoms that include massive internal bleeding and major organ failure. It is clear that these symptoms are terrifying. The percent mortality from these infections is very variable. In an outbreak in Zaire mortality was >90% (n = 277) whereas in the USA mortality was 0% (but represented by 4 cases). It is clear that modern medical facilities can result in recovery from Ebola. A public health bulletin on a super-Ebola would need to describe the symptoms but could also reassure that early treatment would massively reduce fatalities. But should we give the good news first or the bad news first? We report a study in which we measured the mood state of the participants before and after they read a leaflet on a fictitious super-Ebola – Hanta Y. For one group we presented the bad news first (symptoms first) and for the other group we presented the prognosis first. A third group had their mood states determined but they read a leaflet about good things to do to avoid obesity. This was to provide a baseline mood change using more neutral materials.

There are a number of reliable instruments for measuring mood changes. The UWIST is particularly useful with British samples and is sensitive to rapid changes in mood. A factor analytic study of the UWIST Mood Adjective Checklist (Matthews, Jones and Chamberlain, 1990), henceforth the UWIST, suggested that three factors dubbed energetic arousal, tense arousal and hedonic tone parsimoniously partitioned the variance and that these are reliable and valid. Two less well defined dimensions, general arousal and anger-frustration, also emerged. An individual scoring high on energetic arousal would report feeling vigorous and alert while high tense arousal scores imply nervousness and perceived stress. Individuals scoring high on hedonic tone report feeling cheerful and happy. "Good" mood states are represented by high scores on energetic arousal and hedonic tone and low scores on tense arousal. General arousal and anger-frustration are not well defined in this instrument and can only be interpreted cautiously, consequently these dimensions will not be used in the present study. However the other dimensions, in addition to having satisfactory psychometric properties, also have good "face validity".

Mood states tend to be rather transient but they may provide the valence of our cognitions (Power and Dalgleish, 1997), that is, they may add to the remembrance of an event an appraisal of whether the event was a positive or negative experience. Thus manipulating the emotional impact of initial briefings on matters of public health, in a way that does not involve misrepresentation, may be an essential first step in promoting a constructive attitude towards potential health hazards. In this study we attempt to influence affective response to a leaflet containing very unpleasant information.

## Method

### *Participants*

78 undergraduates, 56 female and 22 male, from the Psychology division of the University of Wolverhampton School of Applied Sciences participated in this study. The mean age of participants was 22 years (S.D. = 1.3).

### *Materials and procedure*

A brief document, ~500 words, was created that described the symptoms and prognosis after infection with a "super-Ebola" virus as envisaged by Legters et al. (1993). The symptoms were drawn from a range of extant filoviruses (Peters et al., 1993) and the prognosis for suffers treated in a modern hospital and the estimated likelihood of infection from uncontained virus were included. The document was presented as an information leaflet in two forms. One third of participants in this study (26 individuals) received a version presenting the symptoms of the infection followed by the prognosis and likelihood of infection information after this; a further 26 participants encountered the prognosis information before the symptom information. The information was presented on two separate pages, each of which was studied for two minutes before moving on to the next page. A third group, the control group, studied two pages of advice on nutrition and avoidance of obesity.

Participants were told that the virus "Hanta Y" was fictitious (this research was deemed ethical by a research committee committed to discouraging the use of deception) but possessed properties that virologists thought could be found in viruses likely to infect the UK population in the foreseeable future. This "fiction" is based on concerns raised in Morse (1993). The participants were informed that a memory test would be used to examine the educational value of the "pilot form" of the leaflet and that the emotional impact of the release of information about a likely epidemic was to be assessed. They were not told that there was more than one version of the leaflet and the control group were told that the educational value and emotional impact of a brief primer on nutrition and obesity was being assessed. Participants who gave informed consent first filled out the UWIST mood questionnaire and then read through the "leaflet". Following this a second UWIST was completed and then a multiple choice questionnaire on the contents of the leaflet was completed to check that participants had studied the material. The abstract of this paper was placed on an internet page as a debriefing mechanism.

## Results

Scores for each dimension of the UWIST were subjected to two-way, mixed design analysis of variance with group (symptom first, prognosis first and control) as the between subject factor and before and after presentation as a within subject factor. Means and standard deviations for each condition are shown in Table 1. The table legend identifies the acronyms used. The EA analysis revealed no F ratios $>1$ indicating no significant effects. TA showed a main effect for the between subject factor ($F(2, 75) = 6.17$, $p < 0.005$) and *post hoc* analysis indicated that TA scores were significantly higher in the control condition than in the other two conditions ($p < 0.05$). The within subjects factor main effect was not significant ($F(1,75) < 1$) but there was a significant interaction ($F(2,75) = 5.79$, $p < 0.005$) and simple effects analysis showed that this was accounted for by a reduction in TA confined to symptoms first group after the presentation (critical comparison $p < 0.01$). The analysis of HT revealed no significant main effects (both F ratios $<1$) but the interaction was significant ($F(2,75) = 5.41$, $p < 0.01$). Simple effects analysis showed that this was accounted for by a post-presentation elevation in HT confined to the symptoms first group (critical comparison $p < 0.01$). To summarise, there were no changes in EA but TA decreased after the presentation with symptoms first and HT increased after the symptoms first presentation. Prognosis first presentation was statistically indistinguishable from a presentation on nutrition.

The memory test was simply included to ensure that participants attended to the material in the leaflet. For the symptoms first group 65% (S.D. = 4.23) correct recall was found and for the prognosis first group 62% (S.D. = 4.22) of the items were correctly identified. These do not differ statistically ($t(50) = 0.49$, $p > 0.05$). The control

**Table 1. Means and standard deviations for UWIST Mood scale scores before and after a presentation on life threatening diseases or nutrition with either symptoms or prognosis presented first (EA = Energetic Arousal, TA = Tense Arousal and HT = Hedonic Tone).**

| Presentation Format/UWIST scale | | Symptoms first | | Prognosis first | | Control (nutrition) | |
|---|---|---|---|---|---|---|---|
| | | Before | After | Before | After | Before | After |
| EA | Mean | 22.31 | 23.97 | 23.00 | 22.35 | 23.50 | 23.58 |
| | S.D. | 4.02 | 3.86 | 4.08 | 4.12 | 5.38 | 4.28 |
| TA | Mean | 21.92 | 19.62 | 20.73 | 21.96 | 22.42 | 23.00 |
| | S.D. | 2.61 | 2.04 | 2.89 | 2.99 | 3.76 | 2.58 |
| HT | Mean | 22.04 | 24.31 | 23.77 | 22.54 | 23.15 | 23.23 |
| | S.D | 3.34 | 2.11 | 3.34 | 3.40 | 4.00 | 2.45 |

multiple choice test is not strictly comparable and seems to have been a little easier (77%, S.D. = 1.54).

## Discussion

The analysis of mood change is straightforward. There is no change in energetic arousal and in the other two analyses main effects are either absent or modified by interaction. Given that a reduction in tense arousal and an increase in hedonic tone are both "positive" in terms of mood evaluation the most straightforward interpretation is that presenting the symptoms first improves mood on both these dimensions relative to prognosis first and control. These latter two groups do not differ statistically in any analysis. None of the groups differ at first mood test suggesting no sampling problem.

The results are unequivocal. The bad news should be presented first (at least in this scenario). In a short exposure like this it is not the first thing that one is exposed to that determines mood state but the information received just prior to re-testing mood. This suggests that it might be better to rephrase this as mood state is more positive if prognosis is presented last, and what is presented first is probably less crucial. There is a recency effect at work here.

## Recommendations

1. Before creating any bulletin warning of health threats scrutinise the content and identify the most threatening material.
2. Avoid presenting the most threatening material last. If there are effective countermeasures iterate or re-iterate these at the end.
3. Stress the importance of the early symptoms rather than those in the advanced form of the illness, although the latter should not be ignored.
4. Stress that active intervention is possible and high mortality is not inevitable (unless this is untrue) – few people, for example, seem to realise that Ebola is rarely fatal with early treatment.

## References

Dalrymple, T. (1998) *Mass Listeria: The meaning of health scares.* London: Andre Deutsch.

Legters, L.J., Brink, L.H., and Takafuji, E.T. (1993) Are we prepared for a viral epidemic emergency? In S. Morse (Ed.) *Emerging Viruses.* Oxford: Oxford University Press. 269–282.

Matthews, G., Jones, D., and Chamberlain, G. (1990) Refining the measurement of mood: the UWIST Mood-Adjective Checklist. *British Journal of Psychology*, **81**, 17–42.

Morse, S. (1993 - Ed.) *Emerging Viruses.* Oxford: Oxford University Press.

Peters, C.J., Johnson, E.D., Jahrling, P.B., Ksiazek, T.G., Rollin, P.E., White, J., Trotter, R., and Jaxx, N. (1993) Filoviruses. In S. Morse (Ed.) *Emerging Viruses.* Oxford: Oxford University Press. 159–175.

Power, M., and Dalgleish, T. (1997) *Cognition and emotion: From order to disorder.* Hove: Psychology Press.

# TASTE PREFERENCES OF TRANSPORT SAFETY INVESTIGATORS: WHO DOESN'T LIKE SWISS CHEESE?

**Mark S. Young,[1] Steven T. Shorrock[2] & John P.E. Faulkner[2]**

[1]*School of Engineering and Design, Brunel University, Uxbridge, Middx UB8 3PH, UK*
[2]*Department of Aviation, University of New South Wales, Sydney, NSW 2052, Australia*

When it comes to incidents and accidents, the popular template adopted for human factors investigations has been James Reason's "Swiss cheese" model. One of the implications of this has been the tenacious search for latent conditions leading up to the incident. However, it is questionable whether the focus on factors such as management and regulation has gone too far, and whether we should redress some of our efforts back to the human in control and the working environment. This paper uses a recent aviation case study to illustrate the contention that human action at the sharp end can be more significant than has been traditionally assumed. A critique of the Reason model is presented, highlighting a number of key issues with its application. Whilst the organisational model has been an incontrovertible success, we believe it may be time for a rethink.

## "Would you like Swiss cheese with that?"

It is often quoted, particularly in the transport industry, that more than two-thirds of accidents involve "human error" as a major contributory factor (e.g. Boeing, 1996; IATA, 1993). In response, we have been driven to determine why humans are so fallible, and the discipline of human factors has grown from modelling individual cognitive failure to investigating the organisational contribution to accidents (e.g. Perrow, 1999; Reason, 1990; 1997). The popularisation of this way of thinking is largely thanks to the work of James Reason, whose "Swiss cheese" model of accident causation is now adopted as a model (and even the standard) for investigation in many industries.

Reason's distinction between the active, operational errors and the latent, organisational conditions effectively makes human error a contributory factor in 100% of accidents and incidents. Reason asserts that these latent conditions are the true causes of disasters – typically, the operator merely inherits a defective system, and active errors are seen as the consequence, rather than the cause of the accident chain. The term "operator error" became taboo, and it thus became the duty of incident investigators to look at the psychopathology of organisations in the search for clues.

One of the implications of the organisational approach has been the tenacious search for latent conditions leading up to the incident. However, we believe that there can be flaws in such prescriptive implementation. Whilst the importance of analysing human factors throughout the accident sequence is not in question, the dogmatic insistence on identifying the latent conditions could and should be challenged in cases where active errors have played a major part.

Ironically, it is Reason himself who was first to challenge the status quo, warning that "the pendulum may have swung too far in our present attempts to track down possible errors and accident contributions that are widely separated in both time and place from the events themselves" (Reason, 1997). Subsequently, in two separate aviation human factors conferences in late 2003, Reason (2003a, b) reiterated his concerns with the ever-widening search for the upstream or "remote factors" in safety investigation, stating that "maybe we are reaching the point of diminishing returns with regard to prevention". In the present paper, we take this statement as licence to pass a critical eye over the application of Reason's (1990, 1997) organisational model to incident investigations.

We are questioning whether the focus on latent errors has become too strong, and whether we should redress some of our efforts back to the human at the sharp end. It should be made clear at the outset that this is in no way an effort to reapportion blame or change the focus of investigations (i.e. to prevent future accidents). Rather, it is in direct keeping with such philosophy that we are trying to elucidate all of the relevant causes of an accident. We now go on to examine a recent aviation incident where such a front-line action was, in our view, a dominant factor.

## Is too much Swiss cheese a bad thing?

On 23rd September 1999, a Boeing 747 aircraft overran a runway while landing at Bangkok International Airport in Thailand. "The overrun occurred after the aircraft landed long and aquaplaned on a runway which was affected by water following very heavy rain. The aircraft sustained substantial damage during the overrun, but none of the [410 on board] reported any serious injuries. … These events and conditions can be described in many different ways, the most common being the model of organisational accidents as outlined by James Reason and others." (ATSB, 2001; p. v, xii).

Nevertheless, the most critical event in the accident sequence was arguably an active error – the Captain's late and incorrectly handled cancellation of the go-around. Due to an unstable approach, the aircraft was about to land when the Captain instructed the First Officer to go-around. This was a normal decision corresponding with required flight procedures. The next action by the Captain was not normal – some four seconds later, the Captain retarded the thrust levers, "…because he decided to continue the landing rather than go around. The Captain gave no verbal indication of this action or of his intentions and did not take control of the aircraft from the First Officer" (ATSB, 2001; p. 9).

In the sparse discussion of this extraordinary error the report states: "It is very widely accepted that a decision to conduct a go-around should not be reversed … The Captain's rejection of the go-around appeared to be a considered but rapid response to a unique situation". It is not clear why the Captain's actions were "considered", and the situation only became unique when the aircraft ran off the end of the runway. So although the investigation was ably conducted, we would question the extent to which this accident fitted the Swiss cheese model. The fact that this key active error was not sufficiently acknowledged in the report distorted the rest of the findings.

That there were latent factors at work in this accident is not in question. The investigation report (ATSB, 2001) identified deficiencies in company procedures and training for landing on waterlogged runways. However, these latent conditions pale in significance when contrasted with the events at the sharp end. Indeed, every major organisation can be investigated and there will always be room for improvement. The key focus of the inquiry should have been in determining why the Captain acted contrary to his training and experience in cancelling the go-around. This single act precipitated the whole event, but the inquiry sidelined it in favour of the latent conditions.

## The train driver's Swiss cheese sandwich

The Reason model has also been adopted in railway industries around the world as a template for incident investigations. Whilst we maintain that such use of the model could still fall prey to an excessive focus on latent conditions, the causes of many major railway accidents reveal that this industry may actually exemplify the organisational model better than any other. The key systemic deficiencies contributing to railway accidents would appear to lie in design and maintenance. Some of the most high profile fatal accidents in the UK of recent years have been a result of inadequate maintenance (e.g., Clapham Junction, Hatfield, Potters Bar) – classic latent failings.

Not all rail crashes fall into this category either though. The increased public and media concern with signals passed at danger (SPADs) lately has inevitably been the result of major fatal accidents such as that at Ladbroke Grove in 1999. Although there are undoubtedly organisational contributors to such incidents, and it is unjustified to say such drivers were "at fault", SPADs are clearly another category of accidents for which active errors are a necessary and sufficient component in the accident chain.

## Is the Swiss cheese going off?

In light of the above discussion, the remainder of this paper addresses five key issues to ask whether the organisational model is still valid for describing, investigating, and preventing accidents, or if the approach to safety investigation needs to be reworked.

### *Issue 1: Active errors may be the dominant factors*

The Swiss cheese model can lead to the illusion that the roots of all accidents or even errors stem from the organisation's management. This is not the case. Many errors are simply a by-product of normal, adaptive cognitive processes. "Inadequate defences" would make the errors more dangerous, but even then some errors would overcome even well-planned and maintained defences – the Bangkok incident being a prime example.

### *Issue 2: Causal links between latent conditions and accidents are often tenuous*

For many "organisational failings", the establishment of causality is only really evident in hindsight at best, and even then subject to interpretation. Top-down investigations (as advocated by Maurino et al. 1995; and Reason, 1997), working retrospectively from the event outcome, can easily be influenced by knowledge of the consequences. Many "latent conditions" would doubtless seem insignificant to the actors – or even the industry regulators – in the pre-event scenario. It therefore sometimes seems harsh to judge them as organisational failures *post mortem*.

### *Issue 3: Latent conditions can always be identified – with or without an accident*

Reason (1997) stated that distant factors do not discriminate between normal and abnormal states, "… only proximal events – unsafe acts and local triggers – will determine whether or not an accident occurs". Thus it is only the unfortunate occurrence that reveals their pathogenic status (Boston, 2003). Perhaps surprisingly, Reason (1997) further argued that "(t)he extent to which they are revealed will depend not so much upon the "sickness" of the system, but on the resources available to the investigator". It seems that the harder you look, the more latent conditions you'll find.

### *Issue 4: Some latent conditions are difficult to control, or take years to address*

The factors that can be most easily remedied are the local to the task performer – the working environment and supporting processes. Latent or organisational factors are not so amenable to rapid correction. For instance, an organisation's "safety culture" cannot be manipulated easily or rapidly. Again, Reason (1997) declared that our main interest must be in the "changeable and controllable". At the aforementioned 2003 conference, he also argued that remote factors are "out of the control of system managers, and mostly intractable". It is important to consider whether the costs of organisational reform that may be triggered by an investigation are actually justified by the evidence of an accident.

*Issue 5: Misapplication of the model can shift the blame backwards*

Just as the focus of accident investigations has changed over the years, the focus of blame has also changed. It seems that there was swing from a "blame-the-pilot" culture to a "no blame" culture. This over-swing was corrected by the concept of a "just" culture. Somewhere in the midst of this, a "blame-the-management" culture blossomed. Of course, there are many examples of accidents resulting from mismanagement, but there are also accidents that could not have been prevented by better management. Paradoxically, the organisational approach has sometimes tended to focus on a single type of causal factor – "management incompetence" or "poor management decisions".

## Reasons for cutting the Swiss cheese: a healthy balance

It is indisputable that the ultimate and necessary (though not always singly sufficient) cause of all technological disasters relates to human actions – i.e. error. Reason (e.g. Maurino et al. 1995) contends that an error can consist of mostly latent failures, mostly active failures, or a combination of both. As we have argued, though, the accident without a significant contribution from active failures is a relatively rare event. Accidents occur due to varying proportions of predisposing factors and precipitating events, and many require an active "trigger" to keep the window of accident opportunity open.

In summary, the position in this paper is not that Reason's Swiss cheese model should be discarded as a model for accident and incident investigations, despite the seemingly negative tone. On the contrary, since it has clearly proven value in a range of high-risk industries – and perhaps holds most validity in the railways. Our argument is simply that it is sometimes not as applicable as has been thought, and that it can be misapplied in some cases as a prescriptive investigation technique, rather than a theoretical model. The fixation on latent conditions can then result in the sidelining of active errors, which may have had much more direct implications for the outcome. Even in those cases, the search for latent conditions has resulted in recommendations that undoubtedly improve the safety health of the organisations concerned, despite these conditions arguably having only tenuous connections to the actual event.

As we noted earlier, these thoughts have been aired by James Reason himself at two recent conferences (Reason, 2003a; b). Even in his book, Reason (1997) acknowledges the role of active errors, but still argues that "identifying and eliminating latent conditions proactively still offer the best routes to improving system fitness" (p. 237). Again, we cannot argue with this point. Looking only at active errors is a symptomatic approach, and the symptoms of emotional or "irrational" acts are difficult to decipher.

Indeed, the revolution may have already begun. In an analysis of three years' of UK air traffic control (ATC) incidents (Shorrock, 2003), it was found that very few organisational factors were mentioned in the published incident reports. In ATC, it seems, active errors and human performance at the sharp end are currently the major factors contributing to incidents (but see Issue 3 above).

The aim of this paper has not been to criticise James Reason, or to throw his Swiss cheese to the mice. We would just like to see an increased awareness amongst investigators of the spirit of the model, rather than following the letter of Reason's "bibles" so dogmatically. Without wanting to return to the dark ages of "human error" being the company scapegoat for all accidents, there is a balance to be redressed in accounting for the role of active errors. Latent conditions may be significant, but occasionally people really do just slip up.

## References

ATSB – Australian Transport Safety Bureau, 2001, *Investigation report 199904538: Boeing 747–438, VJ–OJH, Bangkok, Thailand, 23 September 1999* (ATSB, Commonwealth Department of Transport and Regional Services)

Boeing Airplane Safety Engineering, 1996, *Statistical Summary of Commercial Jet Aircraft Accidents: Worldwide Commercial Jet Fleet 1959–1995* (Boeing Commercial Airplane Group, Seattle)

Boston, N. 2003, *The Definition and Investigation of Latent Conditions in Aviation Accidents and Incidents*, Unpublished Honours thesis (Department of Aviation, University of New South Wales)

IATA – International Air Transport Association, 1993, *IATA Safety Record 1992* (IATA and Airclaims Ltd., Montreal)

Maurino, D. E., Reason, J., Johnston, N. and Lee, R. B. 1995, *Beyond Aviation Human Factors – Safety in High Technology Systems* (Avebury Aviation, Aldershot)

Perrow, C. 1999, *Normal Accidents: Living with High-Risk Technologies* (Princeton University Press, Princeton, NJ)

Reason, J. 1990, *Human Error* (Cambridge University Press, Cambridge)

Reason, J. 1997, *Managing the Risk of Organizational Accidents* (Ashgate, Aldershot)

Reason, J. 2003a, Keynote Address – Error management: achievements and challenges. Royal Aeronautical Society Human Factors Group Workshop on *Mitigating Human Error*, 15 October 2003, London

Reason, J. 2003b, Keynote Address – Aviation psychology in the twentieth century: Did we really make a difference? *2003 Australian Aviation Psychology Symposium*, 1–5 December 2003, Sydney

Shorrock, S. 2003, *The Retrospective and Predictive Analysis of Human Error in Air Traffic Control*. Doctoral Thesis, The University of Nottingham

# SLIPS, TRIPS AND FALLS – ACCIDENT ANALYSIS

# DIDN'T YOU SEE THE SIGN? RECALL OF WARNING SIGNS FOR SLIP HAZARDS

**Paul Lehane**

*Environmental Health and Trading Standards London Borough of Bromley Civic Centre Stockwell Close, Bromley, BRI 3UH*

Warning signs for slipping hazards are commonplace in shopping malls in the UK providing a visual indication to people that there is a risk of slipping. To be effective they have to have been seen by those at risk, recognised as a slip warning sign and prompt an assessment of the situation followed by appropriate action. One simple way to assess the effectiveness of slip warning signs is to test people's recall of seeing them under field trial conditions. A survey of 110 shoppers at an indoor mall in the South East of London indicates a low level of recall of slip warning signs. Reasons for this are suggested and opportunities for further research outlined.

## Introduction

The prevention of slip injuries is a priority for health and safety professionals in the UK as the number of slip accidents reported to the enforcing authorities under the statutory reporting scheme continues to increase (HSE 2004). In the Borough of Bromley (population 290,000) 47 members of the public slipped in shops in 2003 and were injured seriously enough to be taken directly to hospital.

Most shopping malls in the UK have terrazzo flooring which is slippery when wet and signs warning of slippery floors are widely used during routine cleaning, dealing with spillages or where water might be brought in from the external environment from rain or snow.

The use of warning signs recognises the slipping hazard presented by wet floors and should be applauded as responsible health and safety management by mall managers however in certain situations the effectiveness of slip warning signs may be compromised. In environments where there are high levels of competing visual stimuli such as shopping malls they may not be seen at all. Alternatively they may be seen and recognised but as no important event is associated with them (e.g. no evidence of the hazard or not having slipped) people become habituated to their presence and fail to respond to them.

A typical procedure used in shopping malls for dealing with slip hazards requires the use of warning signs in conjunction with the removal of the contamination of the floor by cleaning or absorption. Warning signs are usually temporary portable free standing three-dimensional objects approximately .65 m high made of brightly coloured material that are placed in the area of the hazard. There are three aspects of a sign that are designed to warn of a hazard – the use of colour, pictorial & written information.

In the UK the Health and Safety Executive have published guidance on reducing the risk of slipping (HSE 1996 a & b and HSE 2003), which refer to the use of barriers and warning signs to prevent pedestrian use until the floor is dry and safe.

To be effective in shopping malls a slip warning sign has to stand out from all the other competing visual stimuli that shoppers are exposed to. To achieve this the sign has to be seen by the shopper and categorised as a warning sign, then the shopper can assess the environment for evidence of the hazard (e.g. a spillage) and respond appropriately.

It is difficult to assess if a person has seen a slip warning sign under field trial conditions unless this results in an observable or measurable response such as the alteration of gait pattern or direction of travel. However the absence of these responses would not necessarily indicate the sign had not been seen. It could indicate that the sign was seen, the situation assessed as safe and no further response was necessary.

In an ongoing study looking at how people use "counterfactual thinking" to undo slip and trip accidents a common response was "If only there had been a warning sign the accident would have been avoided". This implies a belief that the use of slip warning signs are effective in preventing accidents. In the absence of being able to detect or measure other parameters a first stage in assessing the effectiveness of slip warning signs may be undertaken by asking shoppers if they could recall seeing one when visiting a shopping mall.

To be able to recall seeing a warning sign implies that some form of visually based memory exists. Baddely (1997) outlined three levels of visual memory, "iconic memory" lasting milliseconds, "short-term visual memory" lasting several seconds and "long term visual memory" lasting days.

## Experimental procedure

A customer satisfaction survey was carried out by the researcher on behalf of the Manager of the Walnuts Shopping Centre Orpington Kent UK between May and November 2004 during which shoppers were randomly selected and asked a number of questions concerning their shopping experience at the mall. This included asking if they could recall seeing a number of public information signs that were maintained by the mall manager. This included a slip warning sign.

The Walnuts Shopping Centre is a single story covered mall with a simple "L" shape layout. This facilitated the selection of potential respondents walking from one specific direction and the ability to control the number and placement of slip and other information signs that they had passed before reaching the survey point.

Figure 1 shows the mall layout with a coffee shop located in the atrium at the junction of the two malls. The coffee shop has frosted and etched glass partitions to a height of 1.4 m around its perimeter. The mall is over 25 years old and had a minor roof leak near to the coffee shop. After rain a slip warning sign is placed below the leak approximately 4 m from the corner. The survey point was established approximately 4 m further on, so that each respondent had passed a slip warning sign approximately 8 m behind

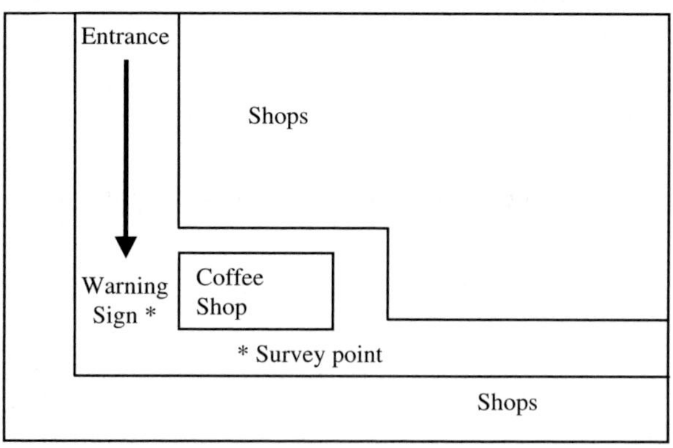


**Figure 1. Mall plan.**

them, but had turned the corner so it was not directly visible. This condition was constant to all respondents, but in wet weather conditions an additional warning sign was placed at the entrance to the mall where water would be walked in.

Having passed at least one wet floor warning sign shoppers were approached and invited to take part in a customer satisfaction survey. One of the questions related to public information signs provided by the mall manager. Respondents were shown five colour photographs of a variety of signs that they had passed between entering the mall and the survey point. These were of two types, general signs such as "opening hours" and safety signs such as "electric shock". Common to both groups was a slip hazard warning sign.

It was hypothesised that people would recall seeing warning signs at a rate greater than would arise from random responses to the question "Can you recall seeing any of these signs during your visits to the mall today", (i.e. greater than 20% (1 in 5) for any one sign). Each respondent was shown one set of signs only. It was also predicted that there would be a greater recall of slip warning signs under conditions where the risk of slipping was increased by the presence of wet underfoot conditions outside and respondents had passed other slip warning signs on entering the mall.

## Results

110 shoppers took part in a face-to-face survey on 5 occasions between May and November 2004. 55 shoppers were shown "safety" signs and 55 were shown "general" signs. 83 respondents were female and 27 were male. 77 responses were obtained in dry external underfoot conditions, 23 damp and 10 wet. The modal age of the respondents was 60 to 70 years (n = 27 representing 24.5% of respondents) who had been shopping at the mall for an average of 16 years.

Tables 1 & 2 present the number of times each sign was recalled and the number predicted by random selection based on a 1 in 5 chance.

**Table 1. Recall of signs from the "safety signs" show card.**

| Sign | Number recalled | Number predicted | Chi square based on 20:80 seen/not seen |
|---|---|---|---|
| 1 Keep Locked | 0 | 11 | Not calculated |
| 2 Keep Clear Automatic door | 8 | 11 | Chi Sq 1.023 1df p = .399 |
| 3 Risk of electric shock | 2 | 11 | Chi Sq 9.205 1df p = .004 |
| 4 No unauthorised person | 1 | 11 | Chi Sq 11.364 1df p = .001 |
| 5 Slip hazard warning | 7 | 11 | Chi Sq 1.818 1df p = .237 |

**Table 2. Recall of signs from the "general signs" show card.**

| Sign | Number recalled | Number predicted | Chi square based on 20:80 seen/not seen |
|---|---|---|---|
| 6 Shop location plan | 10 | 11 | Chi Sq .114 1df p = .867 |
| 7 Opening hours | 4 | 11 | Chi Sq 5.568 1df p = .027 |
| 8 No smoking | 21 | 11 | Chi Sq 11.364 1df p = .001 |
| 9 Keep door shut | 5 | 11 | Chi Sq 4.091 1df p = .061 |
| 10 Slip hazard warning | 11 | 11 | Chi Sq .000 1df p = 1.000 |

The rate at which shoppers recalled seeing slip warning signs was no higher than would have occurred had they been randomly selected from the five signs presented. The recall of three other signs (No.2 keep clear automatic door, No.6 location plan and No.9 keep door shut) was also consistent with random selection. Four signs (No.1 keep locked, No.3 risk of electric shock, No.4 No unauthorised entry and No.7 opening hours) were recalled at rates less than those predicted by random selection. Only one sign (No.8 No smoking) was recalled at a rate exceeding that predicted by random selection.

Slip warning signs were recalled more frequently when presented among more general signs ($n = 11$) than when presented with other safety signs ($n = 7$) however the difference was not significant (Chi Sq 1.063 1df $p = .440$).

No significant differences in the recall of slip warning sign were identified for sex (Chi Sq .897 1df $p = .375$) or for underfoot conditions (chi sq .620 1df $p = .577$). The sample size was too small with regard to males and damp/wet underfoot conditions.

## Discussion

If slip-warning signs were effective in communicating their message the number recalled should have exceed that of a random response. These results indicate that slip-warning signs are recalled less often than expected and their efficiency questioned if people cannot recall seeing them.

If warning signs are not seen, as opposed to not being recalled, it would suggest that a more effective way of preventing slipping accidents would be to use physical barriers to prevent people from gaining access to wet floors until they are cleaned or dried.

There are some possible reasons why shoppers did not recall seeing slip warning signs. Most respondents were frequent visitors to the mall (74% visiting more than once a week) whenever it rained the roof leaked and a slip warning sign was positioned adjacent to the coffee shop. A repeated exposure to a stimulus can lead to a psychological response known as habituation, in which the person stops responding to the stimulus having been repeatedly exposed to it in a situation where nothing important occurs. Shoppers may have become habituated to the signs either specifically at this location or from a more general over exposure in many malls. This effect may be strengthened where shoppers experience slip warning signs in situations where there is no contamination (hazard) but a slip warning sign is still in position.

This research made an assumption that the visual perception of a slip warning sign would bc stored in long term memory for at least several minutes. If however, a person is able to make very quick assessments of situations where there is no risk it maybe possible that such information is only stored in short term visual memory. If this were the case, then the visual trace would decay in the time between perceiving the sign and being asked to recall it. The delay in this experiment was uncontrolled but lasted several minutes while questions were asked about the shopping mall.

Further research in both habituation towards warning signs and the speed at which a person can assess a situation for slip hazards in low and high risk situations would assist in better understanding of the effectiveness of slip warning signs.

The help and assistance of the Manager and Staff at the Walnuts Shopping Centre Orpington Kent is gratefully acknowledged.

## References

Baddely, A. 1997 *Human Memory Theory and Practice*, Revised Edition (Psychology Press, Hove)

Gleitman, H., Fridlund, A.J. and Reisberg, D. 2000 *Basic Psychology*, 5th Edition. John Durbin (ed.) (W.W. Norton & Co, New York)

HSE 1996(a) Slips and Trips – *Guidance for Employers on identifying hazards and controlling risk*, HS(G) 155

HSE 1996(b) *Slips and Trips – Guidance for the Food Processing Industry*, HS(G) 156

HSE 2003 *Preventing Slips and Trips at Work, INDG225 (rev1)*

HSE 2004 *HSC Comprehensive Injury Statistics in support of the Revitalising Programme – Slips and Trips*

# TEXTUAL ANALYSIS OF 459 ACCOUNTS OF SERIOUS FALLS ON THE LEVEL IN OCCUPATIONAL SITUATIONS

**C. Tissot[1] & S. Leclercq[2]**

*Institut National de Recherche et de Sécurité*
*[1]Rue Olivier Noyer – Paris – France*
*[2]Avenue de Bourgogne – Vandoeuvre – France*

459 serious falls on the level, occurring in occupational situations, have been identified in the EPICEA database. The purpose of this work was to detail the circumstances of these accidents and the prevention measures recommended after investigation, using the database text variables constituted by the accident account and the prevention measures. This textual analysis, which follows on from analysis of the same accidents using coded variables, uses effectively the results provided by the previous analysis. It has enabled more accurate identification of these accident cases as falls, completion of fall class descriptions established on the basis of coded variable analysis and summarizing of recommended prevention measures. The advantages and limits of textual analysis emerging from this work are discussed.

## Introduction

Prevention of falls on the level in occupational situations represents a challenge for both individuals and the company. Leclercq and Tissot (2004) described this challenge prior to presenting an analysis of 459 serious falls on the level using coded variables. This first analysis permitted identification of 5 accident classes grouping together cases involving similar circumstances. Textual analysis of accounts of the same accidents and of associated prevention measures allows the present work to complete the fall class descriptions and to summarize the recommended prevention measures.

## Information used

The 459 falls on the level identified in the EPICEA database (Ho et al, 1986) occurred between 1981 and 2000. These "falls on the level" are particularly serious. Indeed, 38% of them were fatal and 37% required hospitalization. For each accident case, the initiated investigation provides systematically certain information. This information appears in the form of modes using coded variables or in the form of numerical variables. It is sufficiently general to describe occupational accidents in their diversity. The general nature of these modes and their limited number results in loss of information. As for the accident account, this is entered as it appears in the investigation report. It contains information that the investigator considered helpful to understanding the accident. Part of the information concerning each accident is therefore included both in the accident account and in the form of coded or numerical variables. On the other hand, prevention measures recommended by the investigator only appear in text form. Analysis of all information unforeseen in coded or numerical variable lists can therefore provide an additional or explanatory perspective to coded or numerical variable analyses. The results of previous coded variable-based analysis of the 459 falls (Leclercq and Tissot, 2004) will be drawn on in order to reveal the complementary nature of results given by analyzing text variables. SPAD© data analysis software used in the previous analysis, has also been used to perform the textual analysis presented in this paper.

## Preparing the set of "*root words*"

Accounts of falls on the level, for all activity sectors combined, describe widely differing accident situations. The vocabulary used by investigators is therefore highly diversified. There are 6,848 words in the 459 analyzed accident accounts. The aim of analyzing accident accounts is to detail their different contexts (Leclercq and Tissot, 2002), so we are interested in sentence content and not sentence structure. This is why "tool" words (of, a, some, etc.) and words such as "casualty" or "accident", which neither allow accidents to be differentiated from other accidents nor fall contexts to be detailed, have been discarded. Words of related meaning have been manually grouped together. For example, the verbs "restrain", "recover", "catch", "grip", "grab" and "catch on" (whether differently conjugated or accorded) have been grouped together under the same root word *"stop_oneself_falling"*. Similarly, the words "pullover", "shirt", "clothing", "oilskin", "suit", "garment", "trousers", "sweater", "cap", "overall", "undergarment", "tee-shirt" and "jacket" have been grouped together under the root word *"clothing"*. When in doubt, the meaning of a word was confirmed by observing its context in the account, this context being given by the software. Finally, root words that appear less than 35 times in all the accounts were discarded, reducing the number of root words to 159. The reason for this was to retain a sufficiently large number of different root words for describing the 459 falls, without weakening the analyses by retaining too many root words. We also memorized 28 segments that were repeated at least 3 times, i.e. word groups that appeared juxtaposed more than 3 times in all the accounts (e.g. *rear running board*).

A second set of 193 root words was similarly built up from the contents of the text variable memorizing recommended prevention measures.

## Analyzing accident account and recommended prevention measure vocabulary

Typology separating the 459 falls on the level according to their circumstances was sought using coded variable analysis (Leclercq and Tissot, 2004). Additional information provided by textual analysis, for 3 out of 5 classes identified is shown in Table 1.

Accounts of falls on the level allocated to the "machinery usage" class contain a wealth of technical descriptions of the equipment and machines concerned. Locations are not described in this class. The accident process is illustrated by notions of incident, malfunction (*failure, problem, anomaly*) associated with the action of wanting to control or intervene by climbing onto the machine, for example. Protection devices are described using terms such as *locking, hood, sensor, screen*, etc. Notions of clothing or part of the body being dragged into a machine entry point or mechanism and violent ejection appear in this class. Notions involving machine operating, shutdown or pressurized state are new, as are actions of climbing onto something and performing maintenance. Words describing machinery reveal frequent involvement of *tanks, liquids* and *belt conveyors*. Recommended prevention measures involve identifying and limiting access to dangerous machine parts. Reference is made to intervention plans and instructions. These measures are mainly aimed at limiting the seriousness of injuries due to contact with the environment.

**Table 1. Complementary information provided by textual analysis for 3 out of 5 classes identified from coded variable analysis.**

| | Information from coded variable analysis reflecting accident circumstances | Information from textual analysis reflecting accident circumstances (in brackets, context-related data leading to better understanding) | |
|---|---|---|---|
| Accident class (size) – name | Variable : class characteristic modes | Some characteristic words* complementing information from coded variable analysis | Some characteristic words* from recommended prevention measures |
| Class 1 (201 accidents – 44%) "machinery usage" | Activity: drive, use, guide, supervise, clean, tidy, adjust, repair, test, measure<br>Activity purpose: installation, machine, vehicle, mobile machine, team, operation<br>Location: workshop<br>Injury: hand, amputation, hospitalization | *Entry_point* (formed by rollers, bobbins, drive belts); *clothing* (pulled into an entry point); *protection_device*; *access*; *operating; shut-down*; *open*; *rotating*; *tank; belt_conveyor*; *liquid*; *pressure; climb; want; want_to_control; maintenance*; *drag; eject_violently; cut* | *Key; device* (protection)*; grill; stop; emergency; access; open; intervene (intervention plan or instruction); operation; movement* |
| Class 3 (34 accidents – 7%) "domestic refuse collection" | Activity: collect, store<br>Activity purpose: refuse, raw material<br>Location: public road, building external traffic route, premises exterior<br>Injury: death | *Rear running board; rear_wheel; reversing; skip; belt_conveyor; lift; reverse; run_over; fall; trunk* | *vehicle; running board; reversing* (reversing signaled); *moving_ forwards; dangerous; rear* (rear-mounted camera, rear of vehicle) |
| Class 5 (70 accidents – 15%) "movement" | Activity: walking<br>Activity purpose: worklocation<br>Location: building internal traffic route<br>Injury: lower member | *Pallet; refuse; ground_ floor_composition; ground/floor; reversing; corridor_passage; room; building; load; see; go_to; go_down* | *Trolley; shoe; passage; ground/floor*; *traffic; pedestrian; pedestrian traffic; instructions; layout; study* |

* A word characterizes a class if its frequency in class vocabulary is higher than its frequency in total vocabulary. Frequency differences are statistically tested.

Accident accounts in the "domestic refuse collection" class include descriptions of *vehicles*, *skips* and also *belt conveyors*. *Rear running boards* and *wheels* are very often quoted. *Reversing* is the most characteristic action in this class. Described locations concern essentially the *road*, *roadway*, *public road*. Accidents in this class are often fatal: an employee who falls from a running board into the roadway or near to vehicle wheels is then run over. Accident context derived from coded variable analysis is enriched by words such as *rear running board, rear wheel* and *reversing*. Notions of running over detail the serious consequences and even the fatal outcome of a fall. Some accidents in this class are characterized by the presence of a belt conveyor or lifting equipment. Recommended measures involve moving skips and vehicles (speed and reversing restrictions, visual or acoustic signaling), improving machine visibility (cameras, rearview mirrors), incorporating and improving running boards, ensuring dustmen wear suitable shoes.

Unlike accident accounts in other classes, those in the "movement" class contain multiple descriptions of various locations and detailing of the ground or floor composition. Verbs concern moving actions (*move towards, walk, arrive*). Characteristic of these accidents, the verb *see* reflects either difficulty in seeing (notion of poor visibility) or the fact of seeing (visual witness, check something). Employee unbalance is caused by colliding with a trolley or part of a machine, by tripping or by slipping. Recommended prevention measures are mainly aimed at trolley and pedestrian movement in passageways. Actions concern ground and floor surfaces (marking, choice of surfacing or covering to ensure slip resistance, cleaning, flatness) and shoes (safety, anti-slip shoes). Training of operators as well as procedures, instructions, traffic layout and station records are all mentioned.

## Discussion – conclusion

As Sorock et al (1997) stress, textual analysis turns out to be time consuming and particularly pertinent after using coded variables associated with the same events. These authors have used occupational accident accounts for prevention purposes in addition to other diagnostic tools. Accident account analysis has, in particular, permitted identification of accident categories or specific dangers that cannot be identified from coded information. McCullough et al (1998) have been able to evaluate accurately the associated costs by identifying a type of accident from textual data. The work presented in this paper, based on accounts of falls on the level occurring in occupational situations, also reveals the complementary nature of information derived from analyzing the text of fall accounts and information derived from analyzing coded variables associated with the same accidents. Two specific characteristics of the textual data provide a partial explanation of this complementarity: the lack of structure in the words drawn from accounts and the preciseness provided by these words.

Words extracted from accident accounts do not, in fact, provide structured information concerning the progression of the accident, unlike coded variables defined within the database. For example, coded verbs in the "activity" variable correspond systematically to the casualty's activity at the time of the accident. On the other hand, verbs of action included in the accident account do not all concern the casualty's activity. It is the textual analysis author who infers a link between a word, or words, and the casualty's activity, in particular with the help of word context. For example, the following notions, which are not revealed from coded variable analysis, were inferred from characteristic words identified by textual analysis: clothing or part of the body being dragged into a machine entry point, and the fact of seeing or not seeing a component of the situation.

Moreover, words contained in the accident account reflect more accurately the accident context than the coded variable lists of predefined modes. The combined modes have to cover the different possibilities during an occupational accident. These modes are therefore general and it is the one that corresponds best that is chosen. On the other hand, textual analysis allows the circumstances of certain falls to be detailed to a degree incompatible with a general accident model. Thus, for certain types of accident, recurring presence of associated rear running board, rear wheel and vehicle reversing or, again, associated vehicle reversing and lack of visibility, was identified. The accuracy achieved by analyzing accident accounts also provides a better picture of the way in which unbalance occurs and thus closer identification of these accidents as falls. Sorock et al (1997), McCullough et al (1998) and Lincoln et al (2004) also observe that accident account analysis leads to greater understanding of accident circumstances. Based on 94 cases of back injury amongst truck drivers, Lincoln et al (2004) have shown that accident accounts provide information on contributing factors, the triggering event and the injury causing event. This information does not appear in coded form but, in the authors' opinion, is essential for classifying accidents and choosing a prevention strategy.

## References

Ho, M.T.; Bastide, J.C. and Francois, C. 1986, Mise au point d'un système destiné à l'exploitation de comptes rendus d'analyse d'accidents du travail, *Le Travail Humain*, **49**, n° 2, 137–146.

Leclercq, S. and Tissot, C. 2004, Serious Falls on the level in occupational situations, In P. T. Mc Cabe (ed.) *Contemporary Ergonomics* (CRC Press, London), 83–87.

Lincoln, A.E.; Sorock, G.S.; Courtney, T.K.; Wellman, H.M.; Smith, G.S. and Amoroso, P.J. 2004, Using narrative text and coded data to develop hazard scenarios for occupational injury interventions, *Injury Prevention*, **10**, 249–254.

Mc Cullough, P.A. and Smith, G.S. 1998, Evaluation of narrative text for case finding: the need for accuracy measurement. *American Journal of Industrial Medicine*, **34**, 133–136.

Sorock, G.S.; Smith, G.S.; Reeve, G.R.; Dement, J. ; Stout, N.; Layne, L. and Pastula, S.T. 1997, Three perspectives on work-related injury surveillance systems. *American Journal of Industrial Medicine*, **32**, 116–128.

# AN INVESTIGATION OF SLIPS, TRIPS AND FALLS IN A UK WATER UTILITY COMPANY

**Anita Scott & Roger Haslam**

*Health & Safety Ergonomics Unit, Loughborough University, Loughborough LE11 3TU, UK*

An investigation of slips, trips and falls (STFs) was carried out in the sewage treatment group of a UK Water Utility company. The main aims of the study were to establish the causal factors for STFs and to provide recommendations to reduce their occurrence. Data was collected from the analysis of STF accident records (n = 226), from site observations (n = 33) and from Operator observations and interviews (n = 13). The main causes of falls were identified as poor grounds maintenance, poor housekeeping and adverse weather. Recommendations were targeted at these areas and included both practical and organisational interventions.

## Introduction

It has been calculated that STFs are responsible for one third of all reported major injuries at work. In addition, the HSE estimate that the cost of slips and trips to employers is £512 million each year (HSE, 2003).

There have been a number of epidemiological studies concerning STFs (See Manning, 1983 and Buck & Coleman, 1985). There are fewer examples however, of studies that address STFs within real organisations and in particular, focus on outdoor workers who are exposed to variable environmental conditions (See Bentley & Haslam, 1998 and Bentley et al., 2003).

For the water utility at the centre of this study, the operations involve a significant amount of outdoor, unsupervised working, this being where most falls occur. In fact, STFs were identified as the main cause of injury for employees in the sewage treatment group. In view of this, the main aims of the project were to:

1. Establish the primary causes of slips, trips and falls within the Sewage Treatment group of the Water Utility.
2. Provide recommendations to reduce their occurrence.

## Method

A multi-faceted approach combining quantitative and qualitative data collection techniques was used to address the project aims.

### *Accident records*

Accident records of STFs (n = 226) from April 2000 to April 2004 were analysed for the Sewage Treatment Group of the Water Utility. The accident records consist of a 2 to 3 sentence description of the event and the prevailing circumstances. The analysis enabled the following information to be derived: temporal trends in accident occurrence, fall initiating event, activity preceding the accident, surface conditions under foot and the underlying causal factors.

*Site observations*

Site observations were conducted by the researcher to determine the STF hazards present on sewage treatment sites, and to also address aspects such as standards of housekeeping and grounds maintenance.

*Operator observation and semi-structured interviews*

The researcher accompanied 13 Sewage Treatment Operators on their daily rounds, enabling observation of their work and semi-structured interviews to be conducted. Interviews addressed issues such as perceived hazards for STFs, potential interventions and also, broader organisational factors associated with accident reporting, training and perceptions towards safety.

## Results and discussion

*Accident records*

Eighty two percent of accidents were non-reportable, the remainder being reportable under the conditions of RIDDOR (injury reporting scheme in the UK for work-related accidents, diseases and dangerous occurrences).

*Time of day*
There were significant differences in accident occurrence over the course of the day ($p < 0.001$). The highest accident rates were between 11:00–12:00 and 14:00–15:00 (Figure 1). These peaks in accident occurrence coincide with the lead up to the lunch period and may be attributed to rushing at this time. Also, peak levels of activity at these times may be influential. The low accident rate between 12:00–13:00 (lunch hour) reflects the lower exposure to STF hazards during the lunch break.

*Time of month*
There was seasonal variation in the occurrence of STFs over the year. The distribution of falls exhibited an approximately U-shaped distribution over the months. The highest fall prevalence occurred in the months December (7.0 accidents per month), January (6.3 accidents per month) and February (6.0 accidents per month) (Figure 2). Given that there are just under 1000 employees in the Sewage Treatment Group, the accident rate over the winter is therefore between 0.006 to 0.007 accidents per person per month. Seasonal variation was especially influential in relation to the occurrence of accidents initiated by slips. The higher rates of accident occurrence during December, January and February may be attributed, in part, to adverse weather conditions with increased prevalence of rain and ice.

*Event initiating accident*
The majority of accidents (60%) were caused by slips whilst 12% were caused by trips. The high proportion of accidents initiated by slips reflects the nature of the employees' work, i.e., working outdoors, under variable weather conditions and walking over surfaces that may be contaminated with sludge, polymer (chemical used to thicken sludge) and general "wetness". The incidents caused by trips were mainly due to obstacles such as uneven slabs, kerb stones, distributor arm wires, abandoned debris and pipework.

*Activity at the time of the accident*
Most accidents (54) occurred when the individual concerned was walking on the level. Other "risky" activities were cleaning (33), ascending or descending stairs (32) and getting into or out of a vehicle (13).

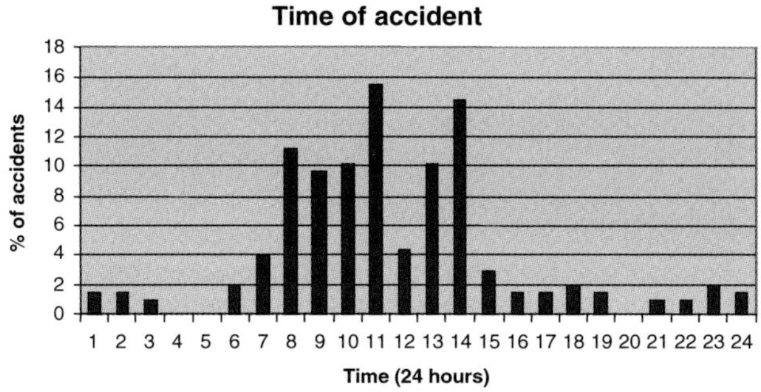


**Figure 1. Time of day when the accidents occurred (n = 206).**

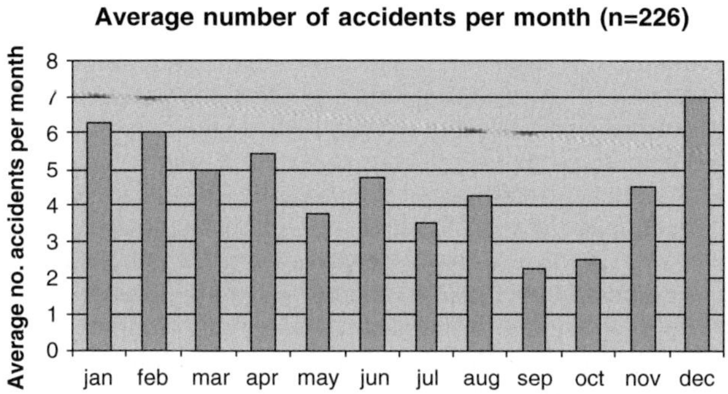


**Figure 2. Average number of accidents per month between April 2000 and April 2004.**

*Underfoot surface conditions*

Common underfoot surface conditions when STFs occurred were: wetness not due to weather (31 accidents), uneven ground (26 accidents), obstacles (permanent) (22 accidents), polymer (21 accidents), ice (20 accidents) and obstacles (temporary) (17 accidents). The accidents initiated by slips commonly involved the contaminants wetness (not due to weather), ice and polymer. The accidents initiated by trips commonly involved both permanent and temporary obstacles and uneven ground.

*Causal factors*

Where possible, the underlying factors thought to be responsible for each accident were identified. The three principle causes of STF accidents were: poor grounds maintenance (44 cases), poor housekeeping (31 cases) and adverse weather (31 cases).

*Site observations*

Thirty three sewage treatment works of varying size, age, type and in three different areas were visited. Site checklists were completed and included details of the types of hazard present and ratings for levels of housekeeping and grounds maintenance.

Common hazards associated with STFs which were observed on many sites were: overgrown grass, uneven ground and pathways, wet floors, sludge spills, polymer residues, abandoned hoses and pipework, raised chamber covers and pathways covered in moss and algae.

The ratings for housekeeping and grounds maintenance varied between sites. These differences can partly be explained by variation in the personal housekeeping standards of each Operator, with some Operators observed to carry out more housekeeping duties than others. There were however, *no* significant differences in housekeeping and grounds maintenance ratings between sites according to regional area, site age or site size.

*Operator observations and semi-structured interviews*

Operator observations (n = 13) revealed that different individuals carry out different amounts of housekeeping duties, such as hosing down and clearing up. Unsafe behaviours were witnessed on occasions, including the taking of short cuts across grassed areas, walking on filter beds whilst the distributor arms are in motion and stepping on to moving equipment.

Interviews with Operators (n = 13) generated data in agreement with that derived from accident records analysis. The hazards perceived to be influential for STFs included polymer, long grass, ice, poor engineering finishings and abandoned debris. Operator suggestions for interventions to decrease STFs included: Increasing hazard visibility, e.g., distributor arm wires and step edges; improve pathway condition; increase the frequency of grass cutting; and introduce designated walkways.

Some organisational issues became apparent during the interviewing process. There was a consensus that accidents were under reported. Reasons for this included reluctance to cause a fuss, embarrassment at the prospect of the accident being read aloud at the team briefing, and potentially being reprimanded if unsafe practice was being undertaken at the time of the incident. When asked about training, Operators explained that the most common approach was to shadow an experienced member of the team. There is therefore the potential for bad habits to be replicated in new team members. Finally, a significant number of Operators believed that the main aim of their work was to maintain water quality standards, a position that reflects the history of the company when bonuses were once given out for maintaining good quality water standards. The company now however state that their main aim is to carry out work to a high standard *and* more importantly in a safe manner.

## Recommendations

A set of recommendations was developed with a view to reduce STFs in the future. Recommendations were mainly targeted to improve levels of grounds maintenance and housekeeping. Some of the recommendations are summarised below:

- Implement a formal system for reporting grounds maintenance work.
- Use concrete as the material for building pathways (slabs are easily damaged).
- Make sure the grass cutting contractors adhere to their agreed schedule.
- Develop a culture where operators take on ownership for the housekeeping and grounds maintenance work of their own sites. Reinforce this with housekeeping audits.
- Provide reels for stowing hosepipes.
- Introduce compounds, e.g., fenced-off areas, for the storage of abandoned equipment away from pedestrian routes.
- Introduce designated walkways and storage areas.
- Monitor weather forecasts and consider gritting prior to icy spells.
- Increase the visibility of distributor arm wires by attaching high visibility material.
- Consider using mentors, trained in best practice to educate new starters.
- Anonymise the individuals involved in accidents.
- Review training for accident investigation.

## References

Bentley, T.A. & Haslam, R.A. (1998) Slip, trip and fall accidents occurring during the delivery of mail. *Ergonomics*. **41**: 1859–1872.

Bentley, T., Moore, D., Tappin. D., Parker, R., Ashby, L., Legg, S., Laird, I. & Hide, S. (2003) Investigating slips, trips and falls in the New Zealand dairy farming sector. COHFE (Centre for Human Factors and Ergonomics), Auckland.

Buck, P.C. & Coleman, V.P. (1985) Slipping, tripping and falling accidents at work: a national picture. *Ergonomics*. **28**: 949–958.

HSE. (2003) Preventing slips and trips at work. INDG225(rev1).

Manning, D.P. (1983) Deaths and injuries caused by slipping, tripping and falling. *Ergonomics*. **26**: 3–9.

# LINKING FALLS AND OTHER LIFE SAFETY CONCERNS IN BUILDINGS: THE WORLD TRADE CENTER AND BEYOND

**Jake Pauls**

*Jake Pauls Consulting Services in Building Use and Safety, 12507 Winexburg Manor Drive, Silver Spring, Maryland 20906, USA*

Insights based on limited research findings from the February 1993 and September 2001 disaster in the World Trade Center in New York, combined with more-general research findings on evacuation and stairway safety, have provided political impetus and technical background for proposed improvements to requirements in codes and standards for design of means of egress/escape in large buildings. Some developments began nationally in the USA. Others, in the City of New York, recommended major regulatory changes for both existing and new high-rise buildings. Some of these improve safety of exit stairways, especially under conditions when lighting fails. Measures needed to reduce the incidence and severity of falls – both in individual and crowd movement conditions – will also have implications for evacuation efficiency. Generally, the design, performance, retrofit and management of evacuation and means of egress/escape have been inadequately examined through systematic research, including disaster investigations. Perhaps one legacy of the World Trade Center disaster will be a major change to this situation. This paper describes some of the developments that could make a difference.

## Introduction

The World Trade Center (WTC) in New York City was the site of two especially notable evacuations (setting aside one due to a bomb scare in 1977). The one following the bombing attack in February 1993 was the larger of the two, involving as many as 20,000 people in each of the two 110-storey towers. The Center's last evacuation on September 11, 2001, involved fewer people – estimated as 7,500 successful evacuees in Tower 1 (the North Tower, the first attacked) and 7,900 successful evacuees in Tower 2 (the South Tower, the first to collapse), but with significant loss of life – 1,434 decedents in Tower 1 and 599 in Tower 2 (NIST 2004).

Although there were small studies of human response in the 1993 incident (Fahy and Proulx, 1995), these did not examine – in detail – evacuation movement in the towers' exit stairways (three per tower). The failure to do such detailed studies of what happened in the stairways has been repeatedly decried by Pauls (1994, 2002, 2003), a long-term student of stairway usability and safety in all conditions including egress.

However, the failure was at least recognized after the final WTC evacuation and, after intensive advocacy work by certain individuals and groups such as the Skyscraper Safety Campaign (one of the "Family Groups" formed by decedents' family members), several million US dollars of US and UK government funding were found and directed to three organizations to examine the WTC evacuation and related human behaviour in detail. The first such study, by the Mailman School of Public Health at Columbia University, started in late 2002 (http://www.mailman.hs.columbia.edu/CPHP/wtc.html). The second, by the US National Institute of Standards and Technology, started in 2003 as part of a larger (US$16 million) comprehensive investigation, under the newly adopted National Construction Safety Team Act, into the towers' collapse and related technical issues including egress (http://wtc.nist.gov/). The third, begun in 2004, titled High-Rise Evacuation Evaluation Database, HEED, is being done by a three-university (Greenwich, Liverpool and Ulster) UK consortium (http://fseg2.gre.ac.uk/HEED/).

All three studies face challenges surveying survivors to obtain detailed, first-person accounts and other information, especially on what happened in the stairwells. Thus the current expectation (at the end of 2004) is that this author will again be disappointed with what little is learned from so many people "committing data" (as first heard from the late John Archea) evacuating down the exit stairways of the WTC.

Despite the paucity, thus far, of comprehensive understanding of what happened in the WTC evacuations, a few efforts are underway to significantly revise requirements for exit stairways in certain high-rise buildings. One spearheaded by the author entails increasing the minimum width of new exit stairways as addressed by the NFPA *Life Safety Code* and the NFPA *Building Construction and Safety Code*. This would amend the traditional 1120 mm (44-inch) minimum nominal width to 1420 mm (56-inch) minimum nominal width for portions serving occupant loads exceeding 2,000 persons. This relatively modest proposal would typically only affect portions of new exit stairways below 14 or more office floors. This is based only slightly on what apparently happened in the WTC stairwells, especially with counter flow of ascending emergency responders and descending evacuees. More-detailed insights largely gained from egress studies directed by Pauls decades ago in Canada, especially on two-abreast movement or overtaking movement, challenged traditional conceptions based on 550 mm (UK) or 560 mm (US) lanes or "units of exit width" for crowd flow (Pauls, 1980, 1984, 1988).

A more drastic effort is occurring in New York City where, as a result of a post-WTC commission study into the City's dated (1968-vintage) building code, Local Law 26 of 2004 is requiring (among many other safety provisions) photoluminescent egress path marking in new and existing high-rise office buildings (75 feet, 23 metres, or more in height) as of July 2006. Of particular relevance to this chapter are the requirements, for photoluminescent marking of stair tread nosings and handrails, that are to be spelled out in technical standards yet to be issued. The author was one of several international experts forming the Egress Marking Task Force, during 2004, advising the New York City Department of Buildings on such standards. While the Task Force dealt with many technical issues, only the tread marking and handrail ones are dealt with in this chapter.

## Risks and countermeasures for stair-related falls during egress

Although the 1993 WTC evacuation was conducted without functioning lighting in each tower's three exit stairwells – with the darkness exacerbated by smoke from the basement-level bomb blast and fire, almost no post-event effort went into documenting how many injuries occurred due to deficiencies, especially of handrails, but possibly also the less-than-desirable step geometry. One small, mostly unpublished study focused on evacuation of people with disabilities (Juillet, 1993) suggested that the installed handrails contributed to occurrence of some falls but this was only anecdotally reported; there was no epidemiological analysis of what injuries resulted from such falls. The stairs' handrails were deficient; the ones on the outside of the dog-leg configuration ended short so that evacuees – who had been told to walk down that side of the stair (to allow ascending emergency responders and descending people requiring assistance to use the shorter inside path) – had an incorrect cue about the bottom of the flights.

Rather than retrofitting the handrails after 1993, to improve their continuity as well as to raise them to a more ergonomically useful, safer height (based, for example, on

Maki et al., 1985, as well as resulting changes in many US building codes), the only stairway retrofit in the WTC involved egress path marking to help future evacuees cope with blackout conditions. However, except for being proposed in New York City regulations, this feature has not been widely employed in other high-rise buildings.

Based on very early epidemiological insights (from Archea et al., 1979), the author estimated that, if the WTC stairways during the 1993 evacuation had a similar risk of injurious falls as did all other stairways in the USA, that risk would be for one injurious stair-related fall (with injury severity warranting treatment in a hospital emergency department) occurring in the evacuation. The 1993 evacuation entailed an estimated 100,000 (or more) person-hours of evacuation behavior, including some five million person-flights of stair use (not counting that by emergency responders (Pauls, 1994). As noted earlier, this amounts to a lot of "data being committed" but not collected!

It would be reasonable to assume that, due to the factors such as the darkness, smoke contamination and complicated usage of the stairways during the 1993 evacuation, the risk of injurious falls would have been higher than expected for all other stairways under all other conditions of use in the USA. But how much higher? A factor of ten? Even more? A confounding effect could be the relatively slow speeds of crowd movement in evacuations; higher speed could be associated with greater risk. On the other hand, the duress of the situation – including poor information on what was happening – plus the unusual, debilitating physical demands of descending as many as about 200 flights of stairs could exact a toll reflected in increased misstep and fall rates.

The bottom line is that exit stairways should generally be designed and maintained to the highest standards reflecting the best available understanding of stairway use and safety ergonomics. This would entail, among other considerations:

- Handrail continuity within and, for at least one handrail, between flights.
- Increased handrail height (to about one metre above step nosings – as recommended by Fenney and Webber, 1994).
- Step geometry to at least the so-called (in the USA) "7-11" standard, with risers no higher than 180 mm and tread depths or goings of at least 280 mm.
- Marking of both the handrails and the critically important nosings of steps (at the leading edges of treads so as to be visible in descent as well as ascent) under all expected lighting conditions including loss of electrical power.

A complete listing of ergonomically justified stairway features for usability and safety, plus a summary checklist, is provided by Pauls (2001).

More generally, it is not sufficient to argue that the risk of falls on exit stairs should be similar to that of falls on stairs generally. That is an unacceptably low standard, especially considering that the cost of injuries on stairways in the USA has been estimated to be of similar magnitude – if not higher than – the cost of stairway construction (Pauls, 2001). This author's suggestion for an acceptable fall risk for exit stairs is on the order of one hospital-treated, fall-related injury for every hundred million flight uses; in other words, a 0.99 999 999 reliability. To put this into a slightly different perspective, by this criterion, the risk of one person being injured in a stair-related injury (and requiring hospital treatment) in an evacuation as large as the with the WTC in 1993 would be about one in 20 and the risk of a particular individual being similarly injured in a WTC-scale evacuation would be on the order of one in a million.

Related reasons for keeping the risk of stair-related falls very low are the potentially devastating effect of a person falling and being seriously injured so that the flow of other evacuees is seriously impacted; moreover, bringing emergency responders to the scene of the fall and moving the injured person would be difficult and very time consuming. The latter might take on the order of hours to accomplish in an emergency situation in a very large building if elevators are not available – as was the case in both 1993 and 2001 situations in the WTC.

## Design and retrofit considerations

As in the current New York City regulatory situation, difficult decisions must be made on the extent and nature of retrofit measures being required. The City's new Local Law 26 of 2004 requires many different measures including improved fire protection and emergency procedures and training.

What is the relative cost effectiveness of various measures? Would provision of more functional handrails be more cost effective than improved egress path marking using photoluminescent materials, aside from the marking specifically on tread nosings and on the upper surface of handrails?

Should a wide band of photoluminescent marking on the wall behind handrails be considered equivalent to, and an option for, a thinner band of photoluminescent marking on the top of the handrails? This was an issue getting a great deal of attention from the task force working on a comprehensive package of recommendations for the City's proposed technical standard. For example, the photoluminescent glow from the band on the wall does not effectively illuminate the handrail and, as a form of glare, the wall band might actually make the handrail even harder to locate in the darkness.

Also, how effective is photoluminescent marking when there is some light coming from other sources, including emergency lights and open exit stair entry doors admitting daylight? Testing to date, such as performed in Canada by Proulx et al. (2000), has not addressed all of such questions. Whether there will be additional testing, especially in the range of stairwells in existing New York high-rise offices buildings that must be retrofitted by 2006, remains to be seen.

Most recent model building codes in the USA do a fairly good job of addressing new stairway design although the NFPA codes and standards tend to be more up to date and responsive to research findings and ergonomic considerations. For example, the NFPA codes and standards are now requiring lighting systems for exit stairs to be capable of providing 100 lux of illuminance at the treads and landings, a light level more appropriate for older workers and residents of buildings. The NFPA documents also do a better job with handrail height, shape, size and clearance to a wall as well as taking very preliminary moves to addressing marking of tread nosings in exit stairwells. NFPA is also currently considering the author's proposals to increase minimum, nominal exit stairway width from 1120 mm to 1420 mm (44 to 56 inches).

What should codes and standards, plus guidance information, be addressing by way of retrofit features? One goal that must be addressed, as a matter of prime importance, is to make sure that existing stairways do not have relatively serious defects such as significant dimensional non-uniformities in step dimensions – either as random, construction-related defects or as systematic, design-caused defects such as combined winding and rectangular treads. If handrails are too low (e.g. less than 860 mm or 34 inches) – as was permitted by older codes, they should be raised. Also traditional, oversized, ungraspable handrails or handrails that do not continue along the entire flight length should be replaced. Damaged tread surfaces, especially those creating trip hazards, should be replaced and be consistent for the stairway system. Generally, assessments of stairway quality and safety should also include estimating the relative effectiveness of the stairways for egress; e.g., what evacuation times do they provide?

Ultimately, improved attention to a building's egress stairways will improve egress efficiency and performance generally while providing reasonable fall protection.

## References

Archea, J., Collins, B. and Stahl, F. 1979, Guidelines for stair safety. NBS-BSS 120, National Bureau of Standards, Gaithersburg, Maryland.

Fahy, R. and Proulx, G. 1995, A Study of Human Behavior During the World Trade Center Evacuation, *NFPA Journal*, **89**, 2, 59–67.

Feeney, R. and Webber, G. 1994, Safety aspects of handrail design: A review. Building Research Establishment Report, BR260, Garston, Watford.

Juillet, E. 1993, Evacuating People with Disabilities, *Fire Engineering*, **146**, 12, 100–103.

Maki, B., Bartlett, S. and Fernie, G. 1984, Influence of stairway handrail height on the ability to generate stabilizing forces and moments, *Human Factors,* **26**, 6, 705–714.

NIST 2004, Progress Report on the Federal Building and Fire Safety Investigation of the World Trade Center Disaster, National Institute of Standards and Technology. Gaithersburg, Maryland, NIST Special Publication 1000–5 (http://wtc.nist.gov/).

Pauls, J. 1980, Building Evacuation: Research Findings and Recommendations, In D. Canter (ed.) *Fires and Human Behaviour,* (John Wiley and Sons), 251–275.

Pauls, J. 1984, The Movement of People in Buildings and Design Solutions for Means of Egress, *Fire Technology*, **20**, 1, 27–47.

Pauls, J. 1988, Movement of People, *SFPE Handbook of Fire Protection Engineering,* National Fire Protection Association, Quincy, MA, 246–268.

Pauls, J. 1994, Vertical Evacuation in Large Buildings: Missed opportunities for research, *Disaster Management*, **6**, 3, 128–132.

Pauls, J. 1999, A Personal Perspective on Research, Consulting and Codes/Standards Development in Fire-related Human Behaviour, 1969–1997 with an Emphasis on Space and Time Factors, *Fire and Materials*, **23**, 265–272.

Pauls, J. 2001, Life Safety Standards and Guidelines Focused on Stairways, In W. Preiser and E. Ostroff (eds.) *Universal Design Handbook,* McGraw-Hill, 23-1-20.

Pauls, J. 2002, Have We Learned the Evacuation Lessons? *Fire Engineering*, **155**, 10, 113–122.

Pauls, J. 2003, Evacuation and Other Movement in Buildings, *Pedestrian and Evacuation Dynamics 2003*, CMS Press, 75–88.

Proulx, G., Kyle, B. and Creak, J. 2000, Effectiveness of a photoluminescent wayguidance system, *Fire Technology*, **36**, 4, 236–248.

# PHOTOGRAPHIC REPRESENTATION OF VISUAL FIELDS JUST BEFORE MISSTEPS AND FALLS INVOLVING STAIRS

**Jake Pauls**

*Jake Pauls Consulting Services in Building Use and Safety, 12507 Winexburg Manor Drive, Silver Spring, Maryland 20906, USA*

Of three key environmental factors in stairway-related falls – visibility, step geometry and handrails – that are often implicated in injurious falls, stair visibility has gotten the least attention in forensic investigation as well as in research. Wide-angle photographs, especially those shot with flash lighting, do not depict adequately what a fall victim saw immediately prior to the critical misstep. A preferred technique is to make a series of many photographs, all taken from a fixed position using fixed exposure settings and a normal-perspective lens. A series of individual photographs is then assembled into a doubly-curved panorama which best represents a complete visual field with preservation of correct perspective for the environment faced by a falling person at the time of the initiating misstep. Presented here is one fall safety expert's experience with the long-term use of such panoramas within a litigation context in North America. The intention is to expand use of such visual documentation techniques in studies of missteps and falls associated with stairs and other pedestrian facilities.

## Introduction

Stairways are a leading consumer product associated with hospital-treated injuries – about one million annually in the USA (recorded by US Consumer Product Safety Commission, CPSC, National Electronic Injury Surveillance System, NEISS) (Pauls, 2000). The problem of stairway safety is not unique to the USA; it is an international concern especially with a growing population of elderly people. However, litigation as a means to examine and document factors contributing to occurrence of missteps and falls (albeit in a relatively small fraction of falls) is most utilized in North America.

Consulting ergonomists are frequently called upon by attorneys, and less frequently by others, to assist in assessing environmental features of stairways relative to the prevention or mitigation of injurious falls or the determination of responsibility for injurious falls once they have occurred. Of three key environmental factors – visibility, geometry and handrails – often implicated in injurious falls, stair visibility has gotten short shrift in forensic investigation as well as in research. However, there have been some notable contributions to the roles of the environment and perception plus cognition by researchers concerned about stairs; they include Archea et al. (1979), Archea (1984), Archea (1985), Cohen and Cohen (1992), Templer (1992) and Clark et al. (1996).

More-general studies of vision, perception and falls include those by Henderson (2003), Lord et al. (2001), Lord and Dayhew (2001) and Lord (2003) who notes: "Recent research has shown that while standard measures of visual functioning such as visual acuity and visual field size are fair predictors of falls, other tests that address more closely the visual functions required for maintaining balance and detecting hazards are superior.... of nine measures of vision, impaired depth perception was the strongest risk factor for multiple falls in community-dwelling older people.... the ability to judge distances accurately and perceive spatial relationships is important for negotiating and avoiding obstacles and hazards in the environment." Thus some of the visual field issues addressed below may not have the same importance as do depth perception and spatial perception generally. Nonetheless, in examining causes of falls, we should pay careful attention generally to accurate documentation of the visual environment encountered people with compromised vision. This includes, for example, the attention devoted to photography without use of supplementary flash lighting that tends to seriously distort important features of the visual environment including contrast issues related to the various walking surfaces encountered, especially where there are steps.

An early attempt to introduce, in an ergonomics conference setting, relatively systematic environmental documentation techniques – based on the three key environmental factors for stairway-related falls, was made by Pauls (1998) and, subsequently (based on one of Pauls' techniques), use of an electronic level to measure stair step geometry in a meaningful fashion has been further explored by ergonomist Daniel Johnson in the USA. A second, somewhat innovative technique, dealing with photographic documentation, is now addressed here in somewhat more detail than in the 1998 presentation, partly in the hope that someone (with better psychology credentials than the author's) explores the technique more intensively, partly with the intention of expanding and improving use of the technique by practitioners who investigate falls.

## Recommended photographic documentation

Although often used to document stairways, perspective-distorting, wide-angle photographs – especially those shot with flash lighting – do not depict adequately what a fall victim saw immediately prior to the critical misstep initiating the fall. (Note that, in accordance with growing convention in ergonomics, the term misstep simply means a departure from normal gait.) A preferred technique is to make a series of many photographs, all taken from a fixed position using fixed exposure settings, with only available light (representative of what was present at the time of the fall) and using a normal-perspective lens such as a 50 mm lens with a conventional 35 mm film camera. The prints are then assembled into a doubly curved, panoramic montage which is viewed at a close distance to replicate the full visual field of the fall victim.

The technique has been employed successfully in the documentation of many stair-fall cases as well as other cases where there are (often-unexpected) elevation changes such as on sidewalks and ramps. This requires use of a tripod to obtain well-focused shots with a large depth of field, e.g., using f22 aperture with relatively long exposures and with the selected exposure setting consistent for all shots and selected as the average for all shots. A series of overlapping photographic frames is shot to capture, for example, 180 degrees of horizontal coverage and 120 degrees of vertical coverage; this might require one or more 24-exposure rolls of film. (Given the relatively good resolution and latitude of film photography as well as extent of manual control and lens quality achievable with conventional SLR cameras, the author has not yet employed digital cameras for this work although the latter are rapidly becoming very acceptable, versatile and affordable for such work.)

To keep the size of the finished panorama manageable, specify 90 mm by 130 mm (3.5 by 5-inch) prints, matt finish and fixed settings for color and density during printing. (Matt finish reduces veiling reflections.) The most labor-intensive activity is assembling the component prints together; the panorama will be a doubly curved montage such as the one depicted in Figure 1.

**Figure 1. Correct viewing of a panorama of a fall site demonstrated by the author.**

As demonstrated in Figure 1 by the author, when this is viewed within several inches of ones eyes (i.e., literally putting ones head into the panorama) one gets a relatively realistic photographic exhibit of what a person would potentially see, for example, within the critical last few steps before encountering a hard-to-perceive, single-riser stair – a relatively hazardous situation. With the reasonably correct preservation of perspective, the small visual size of key features such as a step nosing low in the visual field relative to other features of the environment, directly ahead and to the side, is dramatically clear.

The fall site depicted in the panorama in Figure 1 is a single step at a door between an elevator/lift lobby and a parking garage in a large building used by members of the public meeting with their medical caregivers. In this case the descent direction entails opening and transitioning past a windowless door plus, while visually taking in the environmental features of the occupied parking garage, including a few signs – none of which informed unsuspecting people of the normally hidden step, people were expected to detect the step down and negotiate it without mishap. As with many other single-step situations, the step was not readily detected visually, especially in the descent direction and the elderly lady in this case experienced an air step, landing her leading foot hard on the lower garage floor and falling forward impacting her face on a concrete column shown near the left side of the panorama. Note, for demonstration purposes for the post-fall litigation process, 50 mm (2-inch) wide yellow-black zebra marking has been temporarily applied to the step nosing (the leading edge of the floor surface) for the version of the panorama shown in Figure 1; this marking appears just above the author's thumb. Note also that this particular panorama covers only about one half of the horizontal portion of the normal visual field simply because it spans only what is seen through the doorframe within a step or two of the hidden step.

Despite its size – with a typical panorama occupying a large filing box, it is a very useful, indeed potent, demonstrative exhibit for a deposition and trial. More than anything else this author has seen, such an exhibit helps convey to the trier(s) of fact how a person's perception of an environmental setting could have influenced the inability to detect and assess a danger to avoid a misstep and fall. In the case associated with the situation depicted in Figure 1, the case settled after a deposition given by the author who had been retained by the injured lady's attorney to serve as an expert.

Air steps are one type of misstep (from a set described by Pauls, 2000) in which an unsuspecting person steps forward with a normal gait with the reasonable expectation of the continuation of a level walking surface. When the leading foot drops suddenly through a height less than about 180 mm (7 inches), ones body cannot react quickly enough with any reflexive response or other mechanism to cushion the resulting unexpected, hard impact with the lower walking surface (Jones and Watt, 1971). Such falls can be injurious even if one merely has a hard footfall without falling further; ankles, knees, hips and backs can be injured from the air step alone. Collision injuries to upper limbs as well as knees can occur if the fall involves a complete loss of balance, as was the case in the widely viewed air-step-initiated injurious fall, caught on television, of Cuban President Castro in the autumn of 2004. Falls initiated by air steps are usually good candidates for use of the panorama technique described here.

## Common pitfalls in photographing falls sites

The most common error in photographing falls sites is to use supplementary lighting to improve the exposure of pictures, either film-based or digital (although the latter might be more associated with this error given digital media's need for good light to get good resolutions and given the ubiquitous flash units on many digital cameras, especially compact, "point-and-shoot" models. Fall investigators keen to use flash for photography need to answer one simple question, "How many fall victims carry a light on their heads to supplement the ambient light conditions they encounter on stairways for example?" The photographic documentation should record the environment as it was actually encountered by the fall victim. Time of day, consideration of natural light conditions, even season of the year and sun positions are all important considerations that may affect timing of site inspections. This will generally require that photographs be taken with the camera mounted on a tripod and, to get good focus at all distances, the aperture should be stopped down as much as possible, for example at f22, and exposure time increased significantly. If a step is in a relatively dark area, especially if there are brighter areas close by and in ones visual field, the photographic documentation should record this accurately. This might also require attention to exposure latitude issues, especially in an attempt to record the scene the way the aging eye – with its limited tolerance for differing brightness – sees the scene.

Finally, while there is a concern about documenting a site with the smallest number of individual photographs, the temptation to use wide-angle lenses should be carefully addressed. For incorporation in a report it might be necessary to use some wide-angle photographs to get sufficient coverage of a scene. These should be identified as such so that they are not viewed as representing what a fall victim saw. A normal field of view, encompassing as much as 180 degrees horizontally and 120 degrees vertically cannot be captured realistically by wide-angle lenses although fish-eye lenses capture this with significant geometric distortion. A normal-perspective lens (for a 35 mm film camera) covers a horizontal angle of about 38 degrees; approximately six photos are needed to provide one horizontal strip of the several needed, vertically, for a large full-field panorama. It is important to recognize that using normal-perspective lenses preserves the correct relative size of objects in the environment. A wide-angle, 28 mm lens (for a 35 mm camera) provides about 70 degrees of horizontal coverage. However it distorts relative size of objects and a single photo misses peripheries of a normal field of view on the sides, top and bottom. Space does not permit discussing the serious limitations of video which should be used relatively cautiously in documenting fall sites.

Generally, investigators would be well advised to avoid the pitfalls and to consider careful use of multi-frame, normal-perspective photographic panoramas such as described in this paper. They provide important insights linked to perception and cognition factors that deserve more-careful attention in the missteps and falls field.

## References

Archea, J.C. 1984, Location-specific distraction, *Proceedings of American Society of Safety Engineers Professional Development Conference*, 178–191

Archea, J.C. 1985, Environmental factors associated with falls by the elderly, *Clinics in Geriatric Medicine* (Symposium on Falls in the Elderly: Biologic and Behavioral Aspects), **1**, 3, 555–569.

Archea, J.C., Collins, B.L. and Stahl, F.I. 1979, Guidelines for stair safety. NBS-BSS 120, National Bureau of Standards, Gaithersburg, Maryland.

Clark, M., Jackson, P.L. and Cohen, H.H. 1996, What you don't see can hurt you: Understanding the role of depth perception in slip, trip, and fall incidents, *Ergonomics in Design*, July, 16–21.

Cohen, H.H. and Cohen, D.M. 1992, Walking requires your attention: A human factors analysis of slip and fall accidents, *Forensic Reports*, **5**, 273–282.

Henderson, J.H. 2003, Human gaze control during real-world scene perception, *Trends in Cognitive Sciences*, **7**, 11, 498–504.

Jones, M.G. and Watt, G.D. 1971, Muscular control of landing from unexpected falls in man. *J. Physiology*, **219**, 729–737.

Lord, S.R., Sherrington, C. and Menz, H. 2001, *Falls in Older People: Risk factors and Strategies for Prevention.* (Cambridge University Press, Cambridge).

Lord, S.R. and Dayhew J., 2001, Visual risk factors for falls, *Journal of the American Geriatrics Society,* **49**, 508–515.

Lord, S.R. 2003, Vision, balance and falls in the elderly, *Geriatric Times*, **5**, 6.

Pauls, J. 1998, Techniques for evaluating three key environmental factors in stairway-related falls, *Proceedings of the Human Factors and Ergonomics Society*, 1630.

Pauls, J. 2000, Representation of the Elderly in Premises Liability Cases with a Focus on Falls, *Reference Materials Volume II, Convention of the Association of Trial Lawyers of America*, 2613–2626.

Templer, J.A. 1992, *The staircase: Studies of hazards, falls, and safer design*, (MIT Press, Cambridge, MA).

# SLIPS, TRIPS AND FALLS – BIOMECHANICS

# MINOR VARIATIONS IN GAIT AND THEIR EFFECT ON STAIR SAFETY

**Mike Roys & Michael Wright**

*BRE, Garston, Watford, WD25 9XX*

This paper considers new evidence of a normally distributed variation in the walking gait of stair users between steps (with a standard deviation between 3% and 7.5% of shoe length). Using this information, and an estimate of the average foot position of users on stairs from experimental work, the risk of a large overhang occurring on stairs with different dimensions can be estimated. In addition, the effect of varying the dimensions between two consecutive steps can be determined. An estimate of this risk within an existing stair will enable managers responsible for buildings with stairs, and stair designers, to consider the likelihood of slips occurring on a stair. Further research providing accurate and extensive testing is required to confirm the real values of individual variance and the point where slips on stairs occur under different conditions.

## The risk of slips on stairs

In 2002 in the UK, an estimated 306,000 people had a serious injury in their home involving stairs (DTI, 2003), and 560 people died on stairs in the home (ONS, 2004). In addition, about 125,000 serious injuries occur on stairs when people are involved in leisure activities outside the home (DTI, 2003), and a further 1,000 falls occur on stairs in the workplace. This equates to an incident rate of 1 in 140 people each year, or a serious injury on stairs in the UK every 73 seconds. There may be a further 2 million minor falls on stairs each year where the injuries are less serious and the victim did not seek medical attention (Roys & Hamilton, 2000).

It is known from previous studies that 80% of these injuries occur when users are descending the stairs (Templer, 1995), and often the users report that their injuries occurred because they slipped. Where there is a risk to employees or members of the public, the possibility of slipping on stairs is often considered within a building risk assessment. If the risk is considered to be significant, then action is often taken by adding proprietary nosings to the steps. It is assumed that these proprietary nosings provide a measure of slip resistance to the steps, since the materials used in proprietary nosings often have slip resistant properties when measured under contaminated conditions on the level. If these materials are effective at reducing the potential for slips on the level, then, it is believed, they will also decrease the potential for slips on stairs.

## How is a slip on stairs initiated?

People walk on stairs in a very different way from the way they walk on the level. When walking on the level, the first contact made between the shoe and the flooring material is with the heel. The stride length and the walking speed of the individual determine the angle of contact. Given enough momentum and sufficient lubrication, a squeeze film can be produced at this contact point, generating a slip. Hence rough surfaces, which prevent this squeeze film from being produced, will be effective at reducing slips.

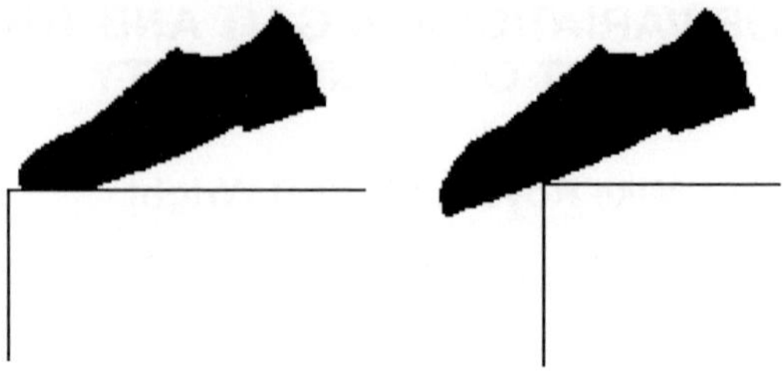


**Figure 1. Position of 300 mm shoe on stairs with 350 mm and 250 mm goings.**

When walking on stairs, both in ascent and descent, the first contact made between the shoe and the flooring material on each step is with the toes or sole of the shoe. The stride length, and hence the potential speed, is controlled by the size of the going and to some extent the size of the rise. However, as the going is decreased it becomes harder to place a substantial part of the foot on the step.

When the going decreases, irrespective of the rise, some users will compensate by rotating their feet, so that a smaller overstep occurs. Other users will maintain a consistent angle between their foot and the walking line regardless of the size of the going. In order to provide positive proprioceptive feedback while descending stairs, it is necessary for part of the metatarsals to make contact with the tread. The metatarsals make up the last 27% of a user's foot, so at a going of 250 mm, the majority of the population, on average, are probably receiving appropriate information, from their muscles, about the stair and where to place their next foot.

Slips on stairs are highly improbable unless the user's foot has a large overhang beyond the edge of the nosing. Consequently, as the going increases the probability of a slip occurring is reduced dramatically, see Figure 1. The effect of going size on slipping is so dramatic that it should be possible to walk with a reduced chance of slipping, even on highly contaminated stairs, if the going is large enough. However, most domestic stairs in the UK have goings of between 220 mm and 240 mm (Roys & Hamilton, 2000). Slips can occur on stairs, but only when a large enough overstep has happened to overcome the frictional characteristics generated at the step/shoe interface. The probability of a slip will therefore depend upon both the variability in the user's walking pattern and the variability in the design of the steps.

## User variability and normal gait

When users first assess a stair before descending they will approach the top landing, placing their leading foot very close to, or just overlapping, the top nosing. Using clues from the visual field, handrails and familiarity with the stair, they start with the first step placing a foot on the first tread. Sensory information from the users' muscles help them to determine the foot position, allowing them to make adjustments to improve it on subsequent steps. Within three steps, most users are able to determine accurately the size of the steps and where to place their feet to descend the stair safely.

Careful observation suggests that the exact position of the foot on the tread varies slightly between steps and between successive descents. It is reasonable to assume that the distribution of foot placements is approximately normal. Using the data reported in Wright & Roys (2005) initial calculations suggest that the standard deviation is in the region of 3% to 7.5% of the shoe length. Using this it is possible to estimate the

**Table 1. Likelihood of large oversteps for different conditions.**

| Going (mm) | (Percentage of foot) SD | Overhang | Overstep | Likelihood Steps | Likelihood on 14 riser stair 5 per day | 2000 per day |
|---|---|---|---|---|---|---|
| 225 | 4.6 | 20.1 | 40.0 | 1:136,873 | 6 years | 5 days |
| 300 | 4.6 | 13.0 | 40.0 | 1:427 million | >1,000 years | 45 years |
| 225 | 6.0 | 20.1 | 40.0 | 1:2,247 | 35 days | 0 days |
| 300 | 6.0 | 13.0 | 40.0 | 1:282,779 | 12 years | 11 days |
| 225 | 6.0 | 20.1 | 50.0 | 1:3,305,090 | 139 years | 127 days |
| 300 | 6.0 | 13.0 | 50.0 | 1:2.7 billion | >100,000 years | 285 years |

*Note*: these figures are based on values calculated for an average male user. It is therefore likely that the risk of slipping due to an overstep is higher than these figures suggest for up to 50% of adult males in the UK.

frequency of foot placements in various positions on treads for any individual. Taking the example of an average British adult male with a shoe size of approximately 300 mm (size 9), we can be 95% confident that the position of the foot in descent is likely to vary by ±17 mm to ±44 mm. We can calculate that when the going is 225 mm (close to an average domestic stair, Roys & Hamilton (2000)) the average size 9 shoe overhangs by 20.1% of its length. Taking a standard deviation of 4.6% of shoe length (the mean from those subjects with size 9 shoes) the foot would overhang by 40%, or more, of its length, once every 136,873 steps, see Table 1. For a 14 riser stair (13 steps) which is descended, on average, 5 times a day, the foot would hang over by 40%, or more, once every 6 years.

From the same data on a 300 mm going, the average size 9 shoe overhangs by just 13.0% of its length. In this case, the foot overhangs by 40%, or more, less than once every thousand years, on a similar 14 riser stair. In a public environment, however, there may be 2,000 individuals using the stair every day, and this relates to a 40% overhang every 45 years, see Table 1.

A slight increase in standard deviation considered in the analysis could dramatically increase the likelihood of an overstep occurring, particularly with smaller goings. For example on a 225 mm going stair, a standard deviation of 6% (the 95 percentile variation for subjects with size 9 shoes) would result in an average foot overhang by 40%, or more, once every 35 days (assuming a 14 riser stair which is descended, on average, 5 times a day). The foot would overhang by 50% once every 139 years, see Table 1. It is unclear from the limited data available what is the correct value for the standard deviation, or what proportion of the foot overhanging is likely to lead to an injury, but clearly the risk is sensitive to small variations in these values. It is essential that accurate measurements are made for foot position in relation to the nosing to determine the real risk associated with various goings.

## Confounding human factors

There are a number of human behaviours that will affect the accuracy with which users can descend stairs. It is likely that, when present, these factors will increase the variation in foot placements and hence increase the probability of a large overstep occurring.

If a user is over-confident, either in his/her own ability or because of familiarity with a particular stair, then they may pay little attention to the stair and take unnecessary risks.

For example, the user may be willing to walk down a familiar stair in low lighting conditions, or while carrying large, heavy or visually obstructive objects. The user might also be distracted while negotiating the stair, either because they are thinking about something else or through external stimuli, like children, pets, variations in lighting or a telephone. Some people will descend a stair while eating, reading, or talking on a mobile phone, all of which could create further distractions. Users may also choose a faster pace than would be suitable for the stair dimensions, to descend the stair. A faster pace will usually result in an increased stride length and hence a higher likelihood of oversteps. It is therefore likely that the standard deviation in the real world will be higher than that estimated under the laboratory conditions in Wright & Roys (2005).

## Sole material and roughness

Slip resistance is dependent upon both the material on the flooring (or in this case the step), and the material on the foot that is making contact with the stair. Often people report slips at home when they are wearing socks, slippers or stockings. It is likely that slips will occur with a smaller overstep if the foot material has a lower slip resistance, e.g. very smooth and hard soles (for example some slippers) or socks. In a similar way, shoe soles with a greater resilience and cleat patterns have been shown, in the laboratory, to allow the user to overstep further without slipping and hence should reduce the likelihood of slips. Grip is also likely to be better with bare feet than most shoes. Consideration should be given to the size of the overstep where slips become likely with different footwear, when calculating the risk.

## Variation in dimensions between steps

It is unusual for the rise and going on any stair to be consistent throughout the whole flight. For some building techniques, variations of 4–6 mm are expected (BSI, 1990), even though consistency is specified in the UK Building Regulations guidance (DETR, 1997). The variation between consecutive steps, in effect, changes the amount of the user's foot that hangs over the nosing in proportion to the change in the rise or going dimension. Differences in going between consecutive steps have be measured as high as 25 mm. Consistency between steps is therefore essential to reduce the potential for slips, since the risk associated with that single step can be as high as a full flight with a much smaller going. The effect of a single step decrease on risk can be seen in Table 2.

**Table 2. Likelihood of large oversteps for non-uniform goings.**

| Going (mm) | (Percentage of foot) SD | Overhang | Overstep | Single step decrease (mm) | Likelihood on 14 riser stair: 5 per day | 2000 per day |
|---|---|---|---|---|---|---|
| 225 | 4.6 | 20.1 | 40.0 | 10 | 2 years | 2 days |
| 300 | 4.6 | 13.0 | 40.0 | 10 | >1,000 years | 8 years |
| 225 | 4.6 | 20.1 | 40.0 | 15 | 299 days | 1 day |
| 300 | 4.6 | 13.0 | 40.0 | 15 | 581 years | 1 year |

## Conclusions

Stairs are dangerous building features; every year in the UK about 290 000 people have a serious injury in their home involving them and more than 500 people die. It is suggested that the small size of goings found in homes leads to infrequent large oversteps, and that this is a major cause of these incidents. A normal distribution in the foot position on a stair can explain why a person suddenly has an overstep incident on a stair they have used safely many times before. The available foot placement data gives a mean overhang of a shoe on the tread of 20% for a 225 mm going, and a standard deviation of around 4.6% of the shoe length, leading to at least a 40% overhang of the shoe once every 6 years for a 14 step stair descended 5 times a day. This estimate does not include any confounding effects such as human behaviour, material of stair covering, footwear, or variation in stair tread dimensions, all of which are likely to increase the probability of an overstep.

The data reported in Wright & Roys (2005) is not sufficiently accurate or extensive to allow testing of the assumed normality. Further research involving accurate measurement of foot placement on a variety of step dimensions is therefore required to determine the potential for slip on various steps.

## References

BSI 1990, *BS 5606:1990 Guide to Accuracy in Building* BSI

DETR 1997, *Building Regulations for England and Wales. Approved Document K: Protection from falling, collision and impact, 1998 Edition* (The Stationary Office)

DTI 2003, *24th (final) Annual report of the Home and Leisure Accident Surveillance System 2000, 2001 and 2002 data*, DTI/Pub 7060/3k/12/03/NP

ONS 2004, *21st Century Mortality- Deaths http://www.statistics.gov.uk/STATBASE/ssdataset.asp?vlnk=6922&More=Y*

Roys, M. and Hamilton, L. 2000, *Stair Questionnaires: A risk assessment*, BRE Client Report 76151

Templer, J. 1995, *The Staircase: Studies of Hazards, Falls and Safer Design*, (MIT press)

Wright, M. and Roys, M. 2005, *Effect of changing stair dimensions on safety*, current proceedings

# EXPERIMENTAL STUDY OF THE BALANCE CONTROL OF STANDING PASSENGERS IN PUBLIC TRANSPORT

**Thomas Robert[1,2]**

[1]*LBMH (Biomechanics and Human Modelling Laboratory)*
*INRETS – Bron (French National Institute Transport and Safety Research)*
[2]*SNCF (French National Railway Company)*
*CIM (Rolling Stock Engineering Centre)*

Experiments were performed on a simulator to observe and measure the balance recovery attempt of standing volunteers after perturbations of their supporting surface. The main particularities of these experiments are the type of perturbation profiles (representative of real deceleration encountered in public transport), and the use of holding devices.

The two main goals are: (1) to understand the volunteers' strategies in restoring their equilibrium with respect to the configuration. (2) to get information about the volunteers' feeling concerning both the level of perturbation and the holding devices.

## Introduction

The falls of people become a more and more important problem in our society because of the general ageing of the population. In the context of public transport, falls are a critical point. Quantitative data are difficult to obtain, because, fortunately, most of the time disequilibrium or falls do not result in serious injuries. Nevertheless, accident studies (Brooks et al., 1978; De Graaf and Van Weperen, 1997; Pereira et al., 2001) are unanimous: standing passenger safety is a major problem for the development of public transport. The lack of knowledge and rules to protect the standing passenger, the development of public transport and the ageing of the population make this point more and more critical and explain the necessity for the study of the standing perturbed equilibrium.

Numerous authors investigated the problem of the standing posture stability, including the question of balance recovery after a perturbation of the supporting surface. However, for some aspects, these studied situations are different from the context of the standing passenger in public transport. The main lacks are the influence of the holding devices, the description of alternatives strategies – only ankle and hips strategies are commonly considered, whereas lots of other strategies are used in unconstraint situation or with higher perturbation level (e.g. Maki and McIlroy, 1997) – and the influence of temporal evolution of the perturbation (De Graaf and Van Weperen, 1997; Bothner and Jensen, 2001).

The intended global aim of this project is to model the balance recovery of standing passengers after a perturbation of their supporting surface. The model will be designed and refined based on experimental observations. This paper presents these experiments and the first obtained observations that can be from them.

## Test set-up

The experimental situation reproduced that of a standing passenger in a hybrid train-tram vehicle submitted to an acceleration pulse resulting from an emergency braking or a light collision. Volunteer subjects took place on a stationary platform connected by a cable and an elastic link to a moving sled launched by the laboratory catapult. The elastic link was designed such that the sled applied a calibrated half sine pulse when the cable came into tension.

A full factorial design of experiment was used. The two independent variables are the initial posture (standing, grasping a vertical bar, the back resting on a seat), and the acceleration profile (low and high). Therefore, six configurations were tested in a random order for each subject.

The two acceleration profiles defined (see Figure 1 below) correspond to an emergency breaking ($\gamma_{max} = 3\,m/s^2$, 300 ms), and a light collision ($\gamma_{max} = 10\,m/s^2$, 200 ms).

Subjects were young male volunteers, close to 50th percentile's morphology (see Table 1 below). They were recruited outside of INRETS (and thus naïve to the results of the experiment) and submitted to a medical examination to ensure that they have never had equilibrium troubles or any medical interventions on the lower limbs. This protocol received the approval from the government ethical committee.

For each test, volunteers were asked to stand comfortably on the moving platform, as if they were on a public transport. They were instructed to use the available devices (vertical bar or bump rest). In order to prevent any fall, the subjects were equipped with a harness which did not interfere with the equilibrium recovery.

In order to avoid as far as possible the anticipation of the perturbation by the volunteers they could not see the start of the moving sled. They also wore a headphone to phonically insulate them from the laboratory and to allow the experimenter to distract their attention by asking them short questions (such as arithmetic calculations or words spelling).

After each test, subjects were asked to value between 0 and 10 the level of the disequilibrium, and the device efficiency and aggressiveness. Due to the difficulties to

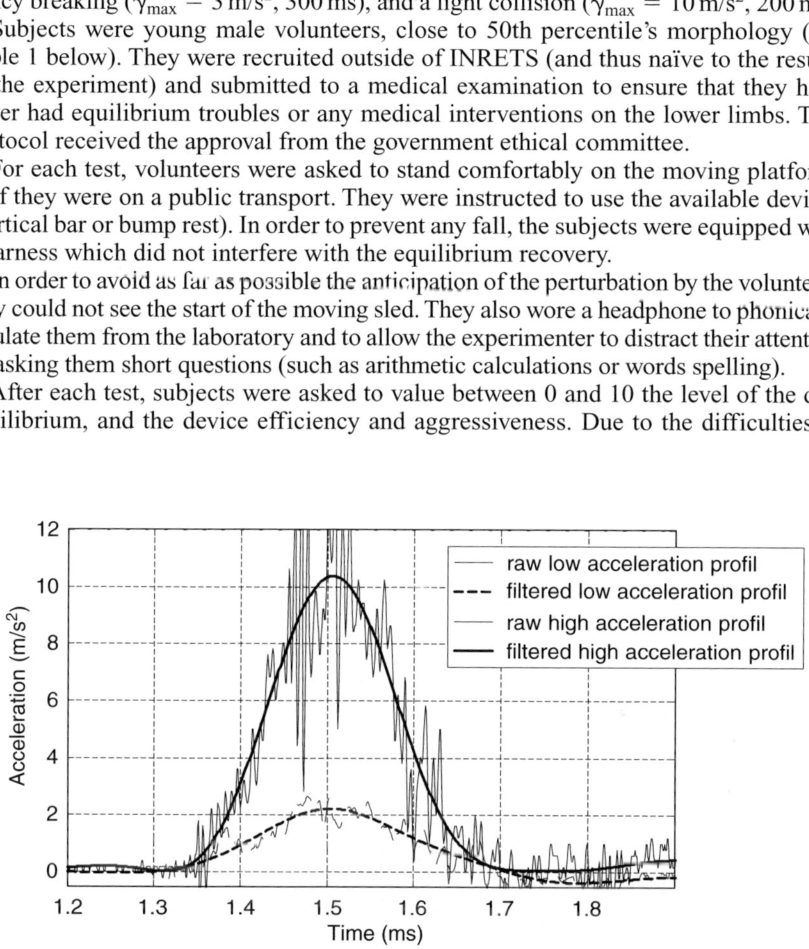


**Figure 1. Measured acceleration profiles.**

**Table 1. Volunteers' main physical characteristics.**

| | Stature (cm) | Weight (kg) | Age |
|---|---|---|---|
| Average | 176.7 | 72.86 | 25.6 |
| S.D. | 2.95 | 5.23 | 2.46 |

evaluate what will be the most favourable (or the worst) configuration, subjects did not use the 0 to 10 scale in the same manner. Therefore a mathematical transformation was applied to the given scores: for each question, the highest score equals 10, the lowest score equals 0 and the middle score (not the average!) equals 5. Subjects' uses of public transports were also investigated.

## Results

### *Balance recovery strategies*

As the subjects were really unconstrained, many different strategies were observed. The predominant strategy was the step forward strategy. Hip strategies (large flexion of the hips) were observed a few times (for two subjects only). It was also observed that arm movements took part in the balance recovery in most of the trials. This depends highly on the subject, and in a smaller part to the level of perturbation and to the previous trials. These arm movements can be divided in two strategies, depending on the objectives: either to help to stabilize the centre of mass over the base of support through inertial effects, or to increase external support through grasping. This is consistent with literature findings (Maki and McIlroy, 1997).

### *Maximum allowable acceleration limit*

Stepping occurs in all high perturbation trials (whatever the initial posture is). For the low level of acceleration configurations, a small recovery step is observed every time in free standing trials, two times with the vertical bar support and none with the back-rest support. This would indicate that this low level of acceleration profile is just above the maximum acceleration allowable limit before stepping. This is consistent with literature data corresponding to our experiments protocol (De Graaf and Van Weperen, 1997): no deceleration phase after the acceleration, no particular instruction given to the volunteer. The differences ($0.2\,m/s^2$ compare to $0.54\,m/s^2$ for De Graaf 1997) can be explained by the higher jerk of our acceleration profile.

### *Configurations' evaluation (see Table 2 below)*

Not surprisingly, the feeling of disequilibrium increases with the perturbation level. For the high level of perturbation, the bump-rest and the vertical bar limit in the same proportion the feeling of disequilibrium. Nevertheless, for the low level of acceleration, the bump-rest appears to reduce the feeling of disequilibrium, whereas it seems that the vertical bar has no particular influence on it. In all cases, the subjects estimate

**Table 2. Configuration evaluation.**

| | | Level of disequilibrium | | Holding device efficiency | | Holding device aggressiveness | |
|---|---|---|---|---|---|---|---|
| Perturbation | Initial posture | Average | S.D. | Average | S.D. | Average | S.D. |
| Low Perturbation | No rest | 2.59 | 1.89 | | | | |
| | Back-rest | 0.00 | 0.00 | 5.45 | 3.76 | 2.17 | 4.16 |
| | Gripping a bar | 2.17 | 1.91 | 8.72 | 2.05 | 2.28 | 3.24 |
| High Perturbation | No rest | 9.56 | 1.41 | | | | |
| | Back-rest | 7.75 | 1.99 | 1.60 | 2.85 | 0.92 | 2.17 |
| | Gripping a bar | 7.77 | 2.12 | 5.68 | 4.48 | 8.33 | 3.60 |

that the vertical bar is more efficient than the bump-rest. For the high level of perturbation, the bump rest seems almost inefficient.

*Influence of the restraint devices*

The feeling of efficiency induced by the restraint devices is not correlated with the feeling of disequilibrium. The impression of efficiency reflects the possibility to stop a fall or to reduce the length needed to restore the balance.

On the other hand, at a low level of perturbation, the feeling of disequilibrium is linked to the initial posture. This is no longer true for a high level of perturbation. A mechanical explanation can be proposed: unlike the standing position (grasping a bar or not), the use of a bump-rest moves the Centre of Mass backward of the Base of Support. The body is then in a stable equilibrium. For the low level of acceleration the perturbation is still not sufficient to disrupt the equilibrium point.

It confirms that these two holding devices act in two different ways:

- The bump-rest tends to modify the initial posture in order to make it steadier, but does not offer the possibility to restore the balance once the falling has started. Therefore, it appears to be highly efficient for the small perturbations, and almost inefficient for the higher one.
- The vertical bar offers the possibility to restore the balance after the start of the disequilibrium, and does not modify significantly the initial posture (the initial steady of the balance). Therefore the vertical bar seems really efficient for the volunteers and reduces the feeling of disequilibrium, even for a high level of perturbation.

This can explain why subjects declare to prefer the use of bump-rests instead of vertical bars: in a classical public transport journey, only small perturbations occur. Using a bump-rest allows passengers to keep a steady balance. Conversely, in grasping a vertical bar, passengers stay in a standing unsteady balance and thus have to continually make efforts to restore their balance.

*Passengers' security*

As mentioned above, stepping forward occurs in most of the configurations tested, even for the low level of acceleration. The stepping distance needed for the subjects to recover their balance could be really big (up to 3 meters). In the context of the public transport's interiors this would lead to secondary collisions between the subject and its environment. Moreover, it has been observed that arm movements were employed in most of the cases by subjects to upgrade their balance recovery.

However, having their arms free to move (as in these experiments) in context of public transport is not the standard: people can read, carry bags, have their hands in their pockets… In these situations the balance recovery would then be less efficient, increasing the risk of injury by falling or secondary collision.

## Conclusion

Experiments have been performed to observe and measure the balance recovery attempt of standing volunteers after perturbations of their supporting surface. This paper described briefly the test set-up and the first observations that could be made from these experiments. The public transport characteristics have been taken into account for the experimental set-up and for the results interpretation.

A global observation has shown the use of different balance recovery strategies by the subjects. Arm movements and change-in-support strategies (e.g. stepping forward) appear to be predominant, due to the high level of perturbation and the absence of constraint.

The analysis of the scores given by the subjects to the different configurations show that the feeling of efficiency of the holding devices appear to be linked to the possibility to restore the balance once the disequilibrium is induced. It also highlights that the feeling of efficiency of a holding device is not correlated to the feeling of disequilibrium. We also showed that the restraint devices act in two different ways: either limiting the risk of loosing balance by increasing the initial posture steadiness, or offering the possibility to restore its balance once the disequilibrium has started.

Further experiments are planned to investigate the influence of the perturbation's temporal characteristics and the importance of the arm movements in balance recovery. The joint kinematics and joint torques are currently being extracted from the 3D measurements, and balance recovery strategies will be characterized. This will allow modelling the subject's reactions to an unexpected movement of their supporting surface.

## Acknowledgment

This work is part of a PhD funded by the SNCF (French National Railroads Company) and is carried out in collaboration with INRETS, INSA Lyon and SNCF. The presented initial results are part of European project SAFETRAM (Passive Safety of Tramways for Europe).

## References

Bothner K.E., Jensen J.L. 2001, How do non muscular-torques contribute to the kinetics of postural recovery following a support surface translation? *Journal of biomechanics*, **34**, 245–250.

Brooks M.B., Edwards H.M., Fraser C.R., Levis J.A. 1978, Passenger problems on moving buses, report no. 3140/32 of the Transport and Road Research Laboratory.

De Graaf B., Van Weperen W. The retention of balance: an exploratory study into the limits of acceleration the human body can withstand without losing equilibrium. *Human Factors* 1997, **30**, 111–118.

Maki B.E., McIlroy W.E. 1997, The role of limb movements in maintaining upright stance: the "change in support" strategy, *Physical Therapy*, **77**, no. 5, 488–507.

Pereira M.S., Hecht M., Segurado V, Sohr S., Uettwiller M. 2001, Passive safety in urban railway system – Light Rail Vehicule (LRV) accident statistics, *Final report*, 94 p.

Runge C.F., Shupert C.L., Horak F.B., Zajac F.E. 1999, Ankle and hip postural strategies defined by joint torques, *Gait and Posture*, **10**, 161–170.

# ANALYSIS OF THE AGE EFFECT IN THE PRINCIPAL POSTURAL-CONTROL PARAMETERS USING THE DYNAMIC POSTUROGRAPHY SYSTEM NED/IBV S.V.E.

**José María Baydal, Ricard Barberà, María Francisca Peydro, Rakel Poveda, José María Gutierrez, Francisco Parra, José Ramiro & Pedro Vera**

*Instituto de Biomecánica de Valencia, Universidad Politécnica de Valencia, edificio 9C, Camí de Vera s/n. E-46022, Valencia, SPAIN*

It is known that falls are a serious problem and the 5th leading cause of death in older adults. Postural control in a standing position is externally reflected by the centre of pressure (COP) displacement. It is necessary to determine COP movement normality patterns in function of the age in order to obtain an accurate balance assessment. A non pathological subject group was studied using NED/IBV s.v.e. measurement equipment. Static trials based on Romberg's test were undertaken. Normality patterns have been obtained for different parameters based on the COP displacement and for different ages. COP movement comparative analysis, related to postural behaviour patterns of non-pathological persons, represents an accurate method to assess postural balance in a standing position. This information enables the design of customised rehabilitation strategies to reduce the risk of falling.

## Introduction

The postural dizziness and in general the disorders of the balance capability are unquestionably the alterations more frequently dealt by the specialists. About 30% of the population will have experimented balance disorders before the 65 years; these patients have to be added to an important number of subjects that experiment falls from this age as, direct or indirect consequence of these alterations and that mean a severe problem, as much economic (24% will need medical assistance, with 5% of fractures, most of them of hip), as social (fear of new falls and loss of autonomy).

In the last years a deep change has taken place in the study of these patients. The technological advances have enabled the development of new techniques of measurement as the Posturography, that have revolutionized and improved the evaluation, the diagnosis and treatment of these patients, allowing knowing the functional condition and detecting the risk of falls. In spite of this, many doctors feel uncomfortable with the arrival of a patient suffering from balance disorders, due to the difficulty to carry out a correct exploration, essential to orientate the treatment of the patient.

The aim of the present work is to present the patterns of the postural control in healthy people, through real measurements of COP displacements and to evaluate the age influence in the balance of healthy people. This information of reference is basic to be able to carry out scales of measurement that help to understand adequately the balance alterations in pathological population and to better orientate the treatment, either medical, surgical or rehabilitator.

## Methodology

### *Subjects*

One hundred and twenty subjects participated in the study, same number of male and female, with no known musculoskeletal injuries or neurological disorders that might have affected their ability to maintain balance. During the period of study none of the subjects took any medication, as sedative ones of the vestibular device, antidepressive, anxiolytic or similar. No subject had carried out the test beforehand. The subjects have been selected in order to obtain a representative sample of all the ages from 20 to 79 years old. The characteristics and age segmentation of the sample of study are reflected in Table 1.

### *Procedures*

In the accomplishment of the measurements it has been followed a methodology similar to the Test of Sensorial Interaction used by Norré (1990), which is a modification of the Test of Sensorial Organization designed by Black et al. (1983), all of them are based on the test of Romberg. In the different conditions of study, the subject had to maintain the balance in a standing position, barefoot, with the heels in contact and with the feet forming an angle of 30°, the arms were extended and close to the body. The time of exploration was 30 seconds (Shumway and Horak, 1986), with an interval of a minute among every test and for every condition of study. The characteristics and the order of the tests were:

1. **Open eyes**: with the stare in a wall situated at 2,5 meters of distance. It represents a usual situation in the daily life in which the subject has his three sensorial systems: vision, somatosensory and vestibular, to maintain the balance.
2. **Closed eyes**: eliminating the vision the subject must keep his balance with the somatosensory and vestibular information.
3. **Open eyes with foam**: with the stare in a wall situated at 2,5 meters of distance, standing on a foam surface with 9 centimeters of thickness and 56.7 $kg/m^3$ of density. The foam surface causes instability due to alteration of the sensorial information of the ankle; therefore the subject must be helped by his vision and vestibular systems.
4. **Closed eyes with foam**: standing on the surface of foam. The vision system is eliminated and the somatosensory information is altered by the foam.

The postural responses were measured with the dynamometric platform NED/IBV s.v.e., developed by the "Instituto de Biomecánica de Valencia". This platform allows to record the position of the centre of pressure (COP).

The COP displacements were made independent from the stature, using a model of inverted pendulum, (Figure 1). This model converts the displacements into a value equivalent to the angular oscillation of the body from the base of support.

**Table 1. Age segmentation of the sample of study.**

| Age (years) | Number | Male | Female |
|---|---|---|---|
| 20–30 | 25 | 12 | 13 |
| 31–40 | 22 | 12 | 10 |
| 41–50 | 27 | 12 | 15 |
| 51–65 | 21 | 10 | 11 |
| 66–79 | 25 | 9 | 16 |
| Total | 120 | 55 | 65 |

The following parameters where calculated.

- Maximum displacements in antero-posterior and midlateral directions.
- Maximum forces antero-posterior and midlateral.
- Sway velocity (Suarez et al., 2001). The average speed was calculated for each sample using the following equation:

$$v = \frac{Frec}{N}\sum_{i=2}^{N}\sqrt{\left(COP_{X_i} - COP_{X_{i-1}}\right)^2 + \left(COP_{y_i} - COP_{y_{i-1}}\right)^2}$$

- Confidence ellipse (95%). The CE is the ellipse which encloses the 95% of the COP samples. The area of the CE is:

$$Area = 2\pi F_{0.05[2,N-2]}\sqrt{\sigma_x^2\sigma_y^2 - \sigma_{xy}^2}$$

Where F0.05[2,N-2] is the Snedecor's F distribution at the 95% confidence level. N is the number of samples and σx2, σy2, σxy2 are the covariance of the coordinates.

## Results

Table 2 shows the results obtained when studying the postural control. In this table is possible to see the angular displacements in midlateral and antero-posterior directions, the

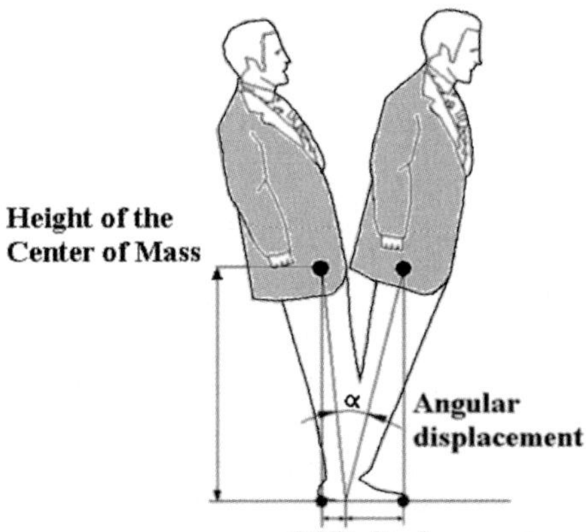


**Figure 1. Model of inverted pendulum.**

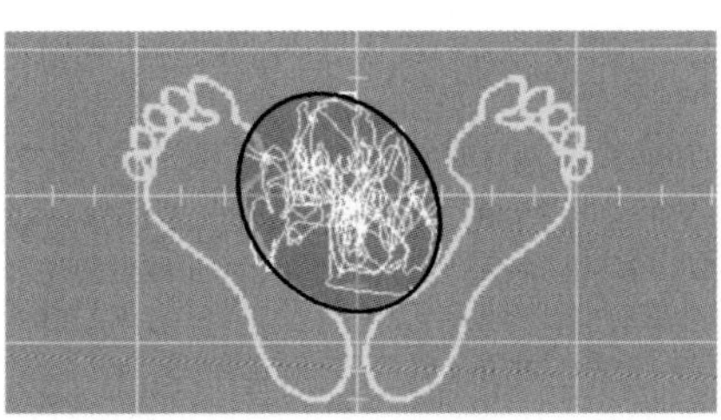

**Figure 2. Area of the Confidence Ellipse (95%).**

CE Area and the mean velocity. All the parameters are bigger as the difficulty of the test increases, being the test of close eyes with foam the most difficult one, since the visual and somatosensorial systems are nullified the postural response loses its efficiency.

It can also be observed that the parameters and consequently the displacements increase with the age of the people, being these differences more pronounced with the tests of greater difficulty. It is very interesting to stand out that the angle displacements in antero-posterior direction are bigger than in midlateral direction, this is due to the fact that the ankle joint is more used to maintain the posture that others joints such as knee or hip.

Table 3 presents a strong correlation of the age effect with the parameters measured in this study with a signify value below 0.01. The parameter with best correlation with the age effect is the Angular displacement in Antero-posterior direction. This indicates that the balance strategy changes from an ankle strategy to a hip strategy while the age is increased. This correlation confirms the fact that the natural posture response is

**Table 2. Postural response obtained in the study.**

| | | Angular displacement ML (°) | | Angular displacement AP (°) | | CE Area ($mm^2$) | | Mean velocity (m/s) | |
|---|---|---|---|---|---|---|---|---|---|
| Test | Age | Mean | Std. | Mean | Std. | Mean | Std. | Mean | Std. |
| Open eyes | 20–30 | 0.87 | 0.31 | 1.08 | 0.26 | 36.63 | 25.17 | 0.024 | 0.003 |
| | 31–40 | 0.92 | 0.26 | 1.09 | 0.28 | 41.20 | 15.59 | 0.026 | 0.003 |
| | 41–50 | 1.00 | 0.32 | 1.24 | 0.37 | 43.83 | 22.43 | 0.027 | 0.006 |
| | 51–65 | 1.12 | 0.27 | 1.25 | 0.40 | 52.63 | 25.32 | 0.028 | 0.008 |
| | 66–79 | 1.14 | 0.32 | 1.44 | 0.46 | 70.13 | 29.39 | 0.029 | 0.009 |
| Close eyes | 20–30 | 1.18 | 0.51 | 1.45 | 0.44 | 66.74 | 35.07 | 0.026 | 0.005 |
| | 31–40 | 1.14 | 0.46 | 1.39 | 0.46 | 67.41 | 38.89 | 0.028 | 0.003 |
| | 41–50 | 1.24 | 0.43 | 1.48 | 0.42 | 72.25 | 44.02 | 0.030 | 0.006 |
| | 51–65 | 1.45 | 0.51 | 1.75 | 0.67 | 83.84 | 57.41 | 0.030 | 0.007 |
| | 66–79 | 1.52 | 0.63 | 1.90 | 0.57 | 125.58 | 73.35 | 0.033 | 0.008 |
| Open eyes with foam | 20–30 | 1.49 | 0.42 | 1.87 | 0.57 | 66.55 | 35.27 | 0.027 | 0.004 |
| | 31–40 | 1.59 | 0.39 | 1.91 | 0.50 | 102.73 | 41.51 | 0.030 | 0.003 |
| | 41–50 | 1.79 | 0.37 | 2.22 | 0.71 | 112.45 | 42.78 | 0.033 | 0.006 |
| | 51–65 | 1.96 | 0.52 | 2.44 | 0.67 | 158.17 | 70.46 | 0.034 | 0.007 |
| | 66–79 | 2.52 | 0.59 | 2.89 | 0.80 | 217.00 | 86.71 | 0.037 | 0.005 |
| Close eyes with foam | 20–30 | 3.19 | 0.82 | 3.56 | 0.80 | 371.96 | 113.27 | 0.051 | 0.011 |
| | 31–40 | 3.40 | 0.98 | 3.58 | 0.78 | 490.29 | 201.19 | 0.054 | 0.011 |
| | 41–50 | 3.74 | 0.96 | 4.20 | 0.92 | 541.91 | 250.08 | 0.055 | 0.011 |
| | 51–65 | 4.32 | 1.09 | 4.45 | 1.05 | 688.93 | 294.66 | 0.060 | 0.014 |
| | 66–79 | 5.47 | 1.68 | 5.90 | 1.22 | 999.61 | 459.04 | 0.072 | 0.015 |

**Table 3. Age correlation with the measured parameters.**

| Parameter | Correlation with the age (p_value) |
|---|---|
| Angular displacement in ML direction (°) | 1.38457E−05 |
| Angular displacement in AP direction (°) | 1.1212E−06 |
| CE Area ($mm^2$) | 7.91591E−05 |
| Mean Velocity (m/s) | 0.005627611 |

becoming deficient as the age increases. The exploration of any subject with balance deficit should have into account the age affect influence and his postural response should be compared with a normal pattern at the same age.

## Conclusions

In all the results we can observe a very significant dependence among the studied parameters and the age, this show in an objective way that the balance system deteriorates with age and as a consequence the risk to suffer falls is also increased.

The quantification of the patterns of postural behaviour in healthy population could have clinical application in the prevention of the risk of falls based on the comparison with these patterns. Using a system of dynamic Posturography, based on the comparison with patterns of normality, the doctor will be able to examine each patient in a simple way and the results will help him to identify the causes of the dizziness or lack of balance and prescribe the most adequate treatment of rehabilitation, allowing an objective valuation. This way the doctor will be able to maintain, to modify or to suppress a treatment of rehabilitation according to its contrasted efficiency.

A system of Posturography based on the comparison with patterns of normality is also useful in the area of the medical-legal expert appraisal, to be able to detect the veracity of an illness of vertiginous origin in an objective way.

## References

Black F. *6. Black FO, Wall C III, Nasher LM*, 1983. Effects of visual and support surface orientation references upon postural control in vestibular deficient patients. *Acta Otolaryngol*, **95**, 199–210.

Norré M. 1990. Posture in otoneurology. *Acta Otorhinolaryngol Belg,* **44**, 55–364.

Peebles L, Norris, B, 1998. *Adultdata. The handbook of anthropometric and strength measurements: data for design safety*. Institute for occupational ergonomics. Department of trade and industry (DTI).

Shumway-Cook, Horak FB, 1986. Assesing the influence of sensory interaction on balance: suggestion from the field. *Phys. Ther.*, **66**, 1548–1550.

Suarez H, Musé P, Suarez A, Arocena M, 2001. Assesment of the risk of fall, related to visual stimulation, in patients with central vestibular disorders. *Acta Otolaryngol,* **121**, 220–224.

# BALANCE ASSESSMENT AND REHABILITATION WITH DYNAMOMETRIC PLATFORMS

**Ricard Barberà, Maria Francisca Peydro, Juan-Manuel Belda, Maria José Vivas, José Montero, Ricardo Dejoz, Carlos Chirivella & Ana Cruz García**

*Instituto de Biomecánica de Valencia., Universidad Politécnica de Valencia, edificio 9c., Camí de Vera s/n. E-46022. Valencia, Spain.*

Rehabilitation has been shown as an effective method to treat balance disorders. A correct evaluation of balance is needed to plan the rehabilitation therapy and to monitor the patient evolution. This paper presents a system that can be used to assess balance disorders and to recommend rehabilitation programs based on a force plate and software. The patient centre of pressure (COP) movement is evaluated and compared to normal patterns. An individual score is provided for each test: visual, somatosensory, vestibular and dynamic systems. After the evaluation a rehabilitation program based on biofeedback is proposed. This system is used satisfactory in hospitals (otorhinolaryngology, neurology, and rehabilitation), insurance companies working on safety and health at work, spas and other clinical resorts.

## Introduction

Balance disorders affects the performance of Activities of Daily Living and functional capacity related to job and entertainment and It supposes negative consequences in the professional and social live of the people suffering from pathologies that affects vestibular or sensory systems. Objective and accurate evaluation of patient functional state is one of the most important problems in the treatment and rehabilitation program of the balance disorders.

The assessment of balance disorders is usually made by means of functional scales, but its limitation is that does not provide objective data in order to perform a clear and reproducible evaluation that can be used in the diagnosis or balance rehabilitation program. The goal of balance rehabilitation is re-establish mechanisms to compensate the alterations between systems. The effectiveness of physical rehabilitation in the treatment of balance disability has been demonstrated. Correct evaluations are necessary in order to program a balance rehabilitation therapy and also to assess the improvement of the patients.

Posturography is currently considered the “Gold Standard” for the postural control study. The American Academy of Otoralyngology-Head and Neck Surgery and the American Academy of Neurology have selected the dynamic posturography as a useful clinical method for the human balance study that isolate and quantify the sensory and motor components that contribute to postural control and allows the assessment of the sensory and motor integration whether in normal subjects or in people suffering from balance disorders. The appropriate tool to study the posturography is a force plate. The IBV (Institute of Biomechanics of Valencia) is working in the use, development and improvement of dynamometric platforms since the beginning of its activity as a RTD centre in 1976.

This work presents a *system to assess balance disorders based on the measurement of the center of mass by means of a force plate and to rehabilitate and improve balance performance based on biofeedback: NedSVE/IBV.*

## Methodology

The system is based on force plates and software both of which have been developed for balance analysis and rehabilitation by IBV.

The goal of patient assessment is to find out the degree of balance disability by means of *testing the different systems for balance maintenance: visual, somatosensory, vestibular, and dynamic*. The visual, somatosensory and vestibular analysis are performed with four different tests (Figure 1):

- ROA: Romberg open eyes. It analyses visual, somatosensory and vestibular systems.
- ROC: Romberg closed eyes. It analyses somatosensory and vestibular systems.
- RGA: Romberg open eyes with rubber foam. It analyses visual and vestibular systems.
- RGC: Romberg closed eyes with rubber foam. It analyses vestibular system.

The system includes gait analysis for dynamical evaluation. It analyses, among other parameters, mediolateral forces in order to detect an abnormal gait (Figure 2).

The *results* of the tests are *compared with the information of a database* that belongs to IBV. The normal punctuation for not disabled people is 100. The numerical and graphical results of the test are reported by means of the NedSVE/IBV software (Figure 3).

The balance evaluation is completed with a test of the *stability limits*. During this test the patient is asked to move its centre of gravity from the centre and maintain it inside a target. The test concludes when the patient complete eight targets around the centre (Figure 2).

The balance rehabilitation by means of NedSVE/IBV has two parts:

- Multisensory rehabilitation program. It takes into account the punctuation of Rombergs in order to propose a program of exercises to be performed by the patient at home. The content of the program depends of the state of each system.

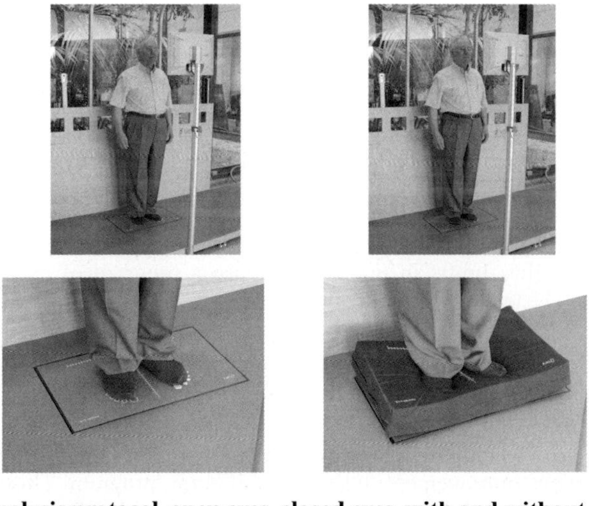

**Figure 1. Analysis protocol, open eyes, closed eyes, with and without rubber foam.**

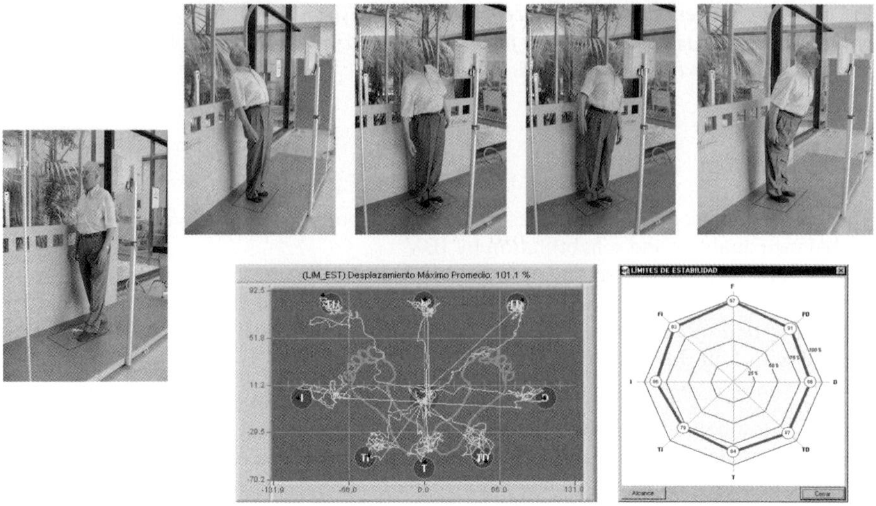

**Figure 2. Dynamic analysis (left) and test of stability limits with graphical results (below).**

**Figure 3. Example of results. Each test and system has and score.**

- Control stability rehabilitation program. It takes into account the punctuation of the test of the stability limits in order to propose a program of exercises based on biofeedback. The tests are usually done moving the centre of gravity to different targets situated on the screen (Figure 4); the patient can see by means of a monitor the representation of his centre of gravity and the target. The degree of difficulty of these exercises will increased by increasing the number of targets, the distance between targets and also decreasing the target area.

Progression can be compared from one session to another. It allows the clinician to evaluate the patient improvement during rehabilitation program. This system can be used with the current configuration or create a new one with different characteristics.

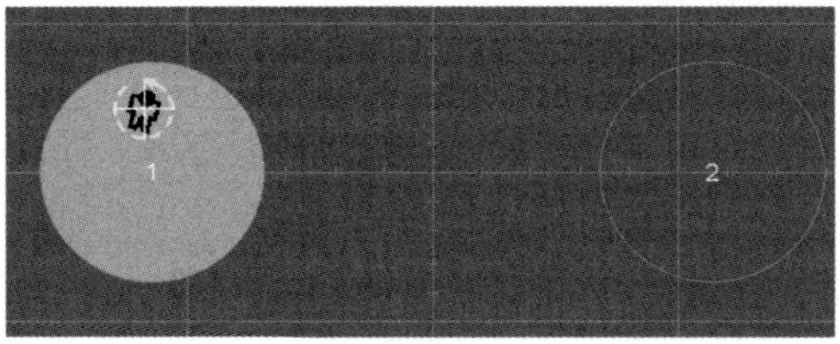


**Figure 4. Example of a low difficulty exercise with two targets. Centre of gravity representation is on target 1.**

## Results

Balance training base on this system will produce an important improvement in altered functions. The main indications focus on:

- Peripheral vestibular lesion: vestibular neuritis, after vestibular surgery, Menière injury, polyneuropathy…
- Central nervous system injury after vascular brain infarction.
- Mixed lesion, central and peripheral, such as cranial traumatism.
- Unbalance of multifactor origin.

## Conclusions

The innovation of the IBV/SVE balance assessment equipment, based on dynamometric platforms and software for balance assessment, lies on the possibility to:

- Objectively evaluate the balance performance in static and dynamic conditions, in comparison to normal patterns. This allows to stablish degrees of alteration of the balance function as well as to detect malingerers.
- Establish adequate training therapies to compensate and reduce balance alteration obtaining better clinical intervention results in balance disability treatments.
- Gather information about the patient evolution and monitor it. This will permit a better treatment selection and an earlier detection of the treatment end.
- Obtain visual biofeedback. The possibility to obtain biofeedback in the training sessions allows the patient to check his progression.
- Customize the exercises depending on the capabilities of the patients and their evolutions by means of the parameters defined in the balance training software.

This system is an important new tool that will provide reliable, objective and quantitative measurements to help specialists working in the field of balance disorders, independently of their backgrounds (geriatrics, labour medicine, sports medicine, neurology, orthopaedics, otorhinolaringology or physical rehabilitation). This type of equipment is **used satisfactory in hospitals** (otorhinolaryngology, neurology, and rehabilitation services), **Insurance Companies** working on safe and health at work, **spas** and other **clinical and health resorts**.

## References

Herdman, S., Cleindaniel, R., Mattox, D., Holidat, M. and Niparko, J. 1995, Vestibular adaptation exercises and recovery. Acute stage after acoustic neuroma resection. *Otolaryngol Head Neck Surg*, **113**, 77–87

Barona, R., Zapater, J., Montalt, M., Armengol, M. and Basterra, J. 1994, Efectos del entrenamiento en plataformas dinamométricas por medio de feedback visual en el mantenimiento del equilibrio en sujetos normales. *Acta Otorrinolaring Esp*, **10**, 161–163

Barona, R. 1994, Programa de rehabilitación vestibular: indicaciones y posibilidades terapéuticas. *Rehabilitación*, **4**, 257–263

Barona, R. (Ed.) 2003, Posturografía: ¿algo se mueve? Rev. Biomecánica, (monographic february) *Biomecanica*, Instituto de Biomecánica de Valencia (IBV)

Baydal, J.M., Barberà, R., Soler-Gracia, C., Peydro, M.F., Prat, J. and Barona, R. 2004, Determinación de los patrones de comportamiento postural en población sana española, *Acta Otorrinolaring Esp*, **55**, 160–169

Prat, J (Ed.) 1993, *Biomecánica de la marcha humana normal y patológica*, Instituto de Biomecánica de Valencia (IBV)

# SLIPS, TRIPS AND FALLS – INJURY PREVENTIONS

# THE USE OF EQUIPMENT BY SHELTERED HOUSING WARDENS TO MANAGE UNINJURED ELDERLY FALLS IN THE COMMUNITY

**Philippa Leggett**

*Tyne Tees Health Consultancy, Boldon Colliery, Tyne and Wear*

Research into elderly falls appears to focus exclusively on prevention rather than management. This study addresses the usability of an emergency lifting cushion (ELK) to manage community falls, in particular those clients who live within the alarm call system. A focused ethnographical approach revealed that not only were the relevant wardens using the equipment, but they considered it indispensable for their manual handling work. The equipment also fitted the usability taxonomy as defined by Stanton. The study also revealed that the wardens were beginning to use the equipment beyond its designated remit, which raises further questions about the overall management of elderly falls in the community.

## Introduction

The United Kingdom currently has an increasing elderly population (National Statistics Office 2001), the majority of whom live in the community. The statistics estimate that 2.5 million uninjured falls may happen annually (Gillespie et al. 2000, Kellogg 1987). Many factors contribute to elderly falls (Connell & Wolf 1997, Lord et al. 2000), but the management of falls appears to have had little or no investigation.

Since 1996 a wardens' manual handling training programme has included the use of an "emergency lifting cushion" (ELK) manufactured by Mangar International Ltd. However, the everyday use of such equipment had never been monitored or assessed, so the aims of the study were:

- To discover whether wardens actually used the equipment in their daily duties.
- To ascertain why they used the ELK rather than other equipment.
- To examine how the ELK is used-with which clients.
- To add to manual handling evidence base.
- To possibly provide recommendations for non-wardens working in similar circumstances.

## Methodology

Ethnography reflects immersion in a subculture in order to understand its structure and mores (Miles & Huberman 1994, Silverman 2000), and in this study the limited study field (manual handling of fallers), restricted time frame, and single researcher created a narrow band of focus. Specific methods used were focus group and participant observation: individuals within focus groups tend to "disclose" (Krueger & Casey 2000) and provide "real life" information about product usability (Page 1998).

Observation within a group's "natural setting" is viewed as fundamental in understanding how that group behaves (Silverman 2000), so the researcher spent approximately forty hours in the field with the wardens. Data gathered through observation offers reliability and validity to what may be poorly replicable data from the focus group (Morse & Richards 2002). Observation is also "the most obvious way of collecting information about a person's interaction with a device" (Stanton & Young 1998).

### *Observational data collection*

Two methods of collecting observational data were used, both constructed from the manufacturer's video:

- A task analysis checklist which would record the level of operator error during use. This was used to define the level of usability of the ELK (Stanton & Baber 1998).
- Recorded timing of the wardens' use of the equipment against that of the video was used to determine their efficiency (Jordan & Servaes 1995).

A basic checklist of seven non-verbal components was also constructed from the video to give some idea whether the wardens were truly comfortable with using the ELK (Morris 1977).

## Results and analysis

Using Strauss' coding paradigm (Strauss 1987), four themes emerged from the transcript analysis which in turn contained sub themes and led to conclusions, including:

- The wardens' remit has changed over the years, with many more people and many more problems to deal with.
- Their protocols involve a discrete distinction between "injured" and "uninjured". "Injured" clients require paramedic attendance.
- Wardens use the ELK because they deem it efficient and effective and it satisfies a specified goal, i.e. getting a fallen person up from the floor with as little risk as possible.
- The wardens view the ELK as indispensable to their work and would not countenance its removal.
- They feel the ELK use confers confidence and competence on them as manual handling problem-solvers.
- The wardens do not consider other equipment (hoists, the "lifting chair") to be as useful as the ELK.
- They feel that other care agencies with which they interact need to update their manual handling knowledge, including ELK use.

### *Observation-results and conclusions*

The wardens used the ELK with ease, confidence and competence. No user errors were committed during the manual handling tasks observed, and at least five out of seven non-verbal components were present at each occasion. Each of the clients was compliant (even the lady who had no previous experience of the equipment) and positive unsolicited comments were made on more than one occasion by the families.

These conclusions appear to support the first seven conclusions drawn from the focus group data.

## Discussion

Internally the wardens' workload now consists of clients with a range of problems that are not only physical but mental and social. This means in turn that many of their clients

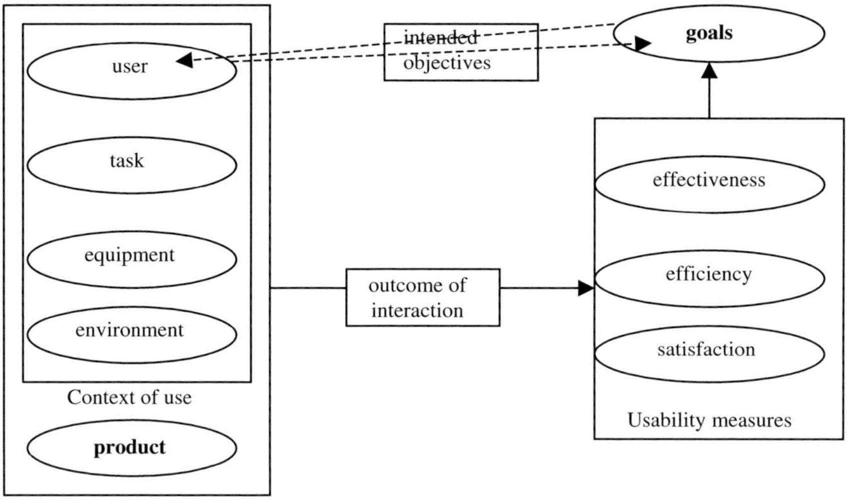


**Figure 1. Usability framework from ISO 9241 (Stanton 1998).**

may be with the system for many years to come and the increase in elderly numbers inevitably means increased workload for them.

External constraints included other carers' attitudes (to manual handling problems), environmental design, client expectations and attitudes, and family expectations and attitudes.

As most of these constraints are unalterable, the usability of the equipment needs to be considered carefully. According to Stanton's usability framework (see Figure 1), the ELK is indispensable to the wardens precisely because it operates within these constraints to achieve the goal they require.

It is noticeable that there is a correlation between the categories in the framework, the constraints identified by the wardens, and the components of a manual handling situation as defined by the MHOR (HSE 1992).

The wardens are so practised at using the ELK effectively that they are beginning to use it outside its designated remit, e.g. with client who have little sitting balance or with clients that are injured. This is indicative of the potential of a device's use (Gaver 1991). They are also using the ELK to perform modified goals, which is "advanced" problem-solving (Bonner 1998).

## Manual handling in the community-the future

There are many more factors regarding the management of elderly fallers in the community than merely the usability of the equipment. The future number of falls will inevitably increase, and the conclusions gathered from this study could possibly be used to inform action that may form some kind of established community manual handling strategy.

Areas that would need further investigation include:

- The structure, frequency, content and impact of manual handling training.
- Comparative usability studies of the new model ELK with the old, and the ELK and the "lifting chair".

- Shift in emphasis from an algorithmic division of "injured"/"uninjured" to "remaining at home"/hospitalised.
- Possibility of "lifting teams" within the community (Hignett et al. 2003).
- The influence of altered criteria for sheltered housing care.
- Empowerment of the wardens to manage change within client's care packages.

## Conclusion

The wardens in this study considered the ELK to be paramount in its usability for the manual handling task of getting fallers from the floor. If a community manual handling strategy were to be considered in the future to deal reactively with the increasing number of elderly fallers, then consideration should be given to factors such as inter-and intra-agency communication, client assessment, environmental design and possible empowerment of a "lifting team" approach rather than merely providing training and equipment. However, more investigation needs to be done into these areas to produce sound reasoning and conclusions before proceeding.

## References

Bonner J.V. (1998) Towards consumer product interface guidelines in Stanton N. (ed.) 1998 *Human Factors in consumer products* chapter 14 Taylor & Francis

Gaver W.W. (1991) in Stanton N. (ed.) (1998) *Human Factors in Consumer Products* Taylor & Francis

Connell B.R., & Wolf S.I. (1997) Environmental and behavioural circumstances associated with falls at home among healthy elderly individuals. Atlanta FICSIT group *Archives Physical & Medical Rehabilitation 78:2 179–86*

Gillespie L.D., Gillespie W.J., Cumming R., Lamb S.E., & Rowe B.H. (2000) Interventions for prevention of falls in the elderly *Cochrane Database Syst. Rev. 2000:2 CD000340*

Health & Safety Executive (1992) *Guidance on Manual Handling Operations Regulations 1992* Her Majesty's Stationery Office

Hignett S., Crumpton E., Ruszala S., Alexander P., Fray M., & Fletcher B. (2003) *Evidence-based patient handling-tasks, equipment and interventions* Routledge

Jordan P.W. & Servaes M. (1995) Pleasure in product use: beyond usability in Robertson S.A. (ed.) *Contemporary Ergonomics 1995* Taylor & Francis

Kellogg International Work Group (1987) The prevention of falls in later life. A report of the Kellogg International Work group on the Prevention of Falls in the Elderly *Danish Me Medical Bulletin 34:supplement*

Krueger R.A., & Casey M.A. (2000) *Focus groups-a practical guide for applied research* 3rd edition Sage publications

Lord, S.R., Sherrington C., & Menz H.B. (2000) *Falls in older people-risk factors and strategies for prevention* Cambridge University Press

Miles M.B. & Huberman A.M. (1994) *Qualitative data analysis-an expanded sourcebook* 2nd edition Sage publications

Morris D. (1977) *Manwatching-a field guide to human behaviour* Abrams New York

Morse J.M., & Richards L. (2002) *Readme first-for a user's guide to qualitative methods* Sage publications

National Statistics Office (2002) Census 2001-England & Wales National Statistics Office(online) http/*www.ageconcern.org.uk/ageconcern/inform_428.htm* (Accessed 05 May 2003)

Page M. (1998) Consumer products-more by accident than design? in Stanton N. (ed.) 1998 *Human Factors in Consumer Products* chapter 7 Taylor & Francis
Silverman D. (2000) *Doing qualitative research-practical handbook.* Sage publications
Stanton N., (ed.) (1998) *Human Factors in Consumer Products* Taylor & Francis
Stanton N., & Baber C. (1998) A systems analysis of consumer products in Stanton N. (ed.) 1998 *Human Factors in Consumer products* chapter 4 Taylor & Francis
Stanton N. & Young M. (1998) Ergonomics methods in consumer product design and evaluation in Stanton N. (ed.) 1998 *Human Factors in Consumer Products* chapter 2 Taylor & Francis
Strauss A.L. (1987) *Qualitative analysis for social scientists* Cambridge University

# FLOOR CLEANING – DROP THAT MOP!

**Robert Shaw, Kevin Hallas, Paul Lemon & Steve Thorpe**

*Health and Safety Laboratory Harpur Hill, Buxton SK17 9JN*

Cleaning regimes are often considered important in the context of hygiene and aesthetics but rarely in terms of slip resistance. Many techniques, though common in the workplace, are not appropriate for the environment in which they are used. The viscosity of the contaminants present and the microroughness of the floor surface are key considerations in determining the relevant cleaning technique for a given area. The Health & Safety Laboratory (HSL) has been studying different approaches to floor cleaning over a wide range of locations in order to ascertain examples of best practice and identify areas in which cleaning regimes could be improved. In many cases the slip resistance of flooring immediately following cleaning could be improved without a great increase in time or money spent, by simple re-evaluation of the situation and application of appropriate techniques.

## Introduction

Slips and trips are the most common cause of major injuries at work and they occur in almost all workplaces; 95% of major slips result in broken bones and they can also be the initial causes for a range of other accident types such as falls from height and workplace transport accidents. In a typical year, slips and trips account for 33% of all reported major injuries, 20% of over-3-day injuries to employees and 50% of all reported accidents to members of the public. It is important to realise that slips and trips are serious and that simple cost effective measures can reduce the numbers of these accidents. Rarely is there a single cause of a slipping accident, and so a holistic approach to reducing the number of slips and trips is required. Obviously the interaction of a pedestrians' shoe with the floor is the major factor in slip accidents, but what properties does the floor surface present? How slip resistant are the shoe soles? And importantly, is there contamination between the two?

Cleaning has been identified as a significant factor in the occurrence of slips for two reasons. Firstly, floor surface contamination is often implicated in slip accidents, and therefore regular cleaning to remove contamination is influential in reducing accidents. Secondly, the process of cleaning can present slip and trip hazards to those working in the area being cleaned. For example, smooth floors left damp by a mop may be extremely slippery and trailing wires from a vacuum cleaner or buffing machine can present a trip hazard. The objective of this study was to observe what cleaning practices are currently in use in different work environments and how appropriate they are to those environments in terms of the amount of contamination removed from the surface and the potential slip risk presented by the technique.

## Methods

A wide range of premises were visited to evaluate the current cleaning regimes in terms of their effects on slip resistance. Premises included several food production facilities, a soft drinks production plant, a car plant, a printing works, a shopping centre, several pub and restaurant kitchens, a butchers shop, supermarket in-store bakery and a school. This has allowed floors with a wide range of roughness levels to be studied in terms of slip resistance, and in many cases a variety of types of contamination on a given type of flooring. Where the duty holders' cleaning was felt to be inadequate further cleaning was undertaken by HSL staff using simple techniques.

Measurements of floor surface "Slip Resistance Values" (SRV, closely related to Coefficient of Dynamic Friction) were made using a calibrated Pendulum slipperiness assessment instrument. Data was generated in [i] the as-found, dry condition, [ii] after application of low volumes of potable water to the flooring by hand-spray, and [iii] after cleaning. Two test slider materials were used as appropriate; Four-S rubber (Standard Simulated Shoe Sole, developed by the United Kingdom Slip Resistance Group (UKSRG) to represent a footwear material of moderate performance), and TRRL rubber (which is used as a secondary standard shoe soling material by HSL and the UKSRG). Data was generated in three complementary test directions where appropriate in order to account for the presence of floor surface directionality. Further testing was undertaken using a calibrated surface microroughness transducer, set to the Rz parameter. These results were interpreted using the UKSRG guidelines and HSE Information Sheet FIS22. Due regard was also given to HSE's Slips & Trips 1 Information Sheet.

## Results

The following is a summary of successful cleaning techniques for a range of types of contamination, in relation to the roughness of the flooring on which they were encountered. The entries in table 1 refer to the advice relating to each technique that follows.

**S/D** – Scrubber dryer machines can be an effective way to clean large areas of most kinds of flooring. Different designs of scrubber dryer lend themselves to different situations. Consideration should be given to the following positive and negative aspects when selecting or using a scrubber dryer:

- The squeegee needs to be wide enough to recover all of the water put down by the scrubber dryer. Single scrubber machines tend to throw water out to one side, and may require an asymmetric squeegee to recover this.
- The squeegee needs to be well maintained to ensure there is no leakage, which may, for example, leave a smooth floor dangerously wet.
- On very rough or profiled surfaces the squeegee may not be malleable enough to allow adequate removal of water from the surface.
- On greasy floors a detergent should be used to remove and hold the oil or grease in the water.

**Detergent** – Detergent is essential if there is any greasy or oily contamination on the floor. Water, whether it is cold or warm is not effective in removing this contamination without the aid of a detergent. The detergent should be left on the floor for an adequate time period to allow effective removal of grease prior to subsequent rinsing (c.f. washing-up, heavily soiled pots and pans require soak time in the detergent). Where detergents are used, the floor should be rinsed with clean water to avoid leaving a detergent residue.

**Fat Solve** – In very greasy environments a fat solvent may prove more useful than detergent.

**Mop** – Mopping is usually only effective on smoother floors. Where smooth floors ($Rz < 20\ \mu m$) are mopped, care must be take to ensure the floor is left completely dry

**Table 1. Successful cleaning techniques.**

| | Microroughness Rz (μm) | | | | | |
|---|---|---|---|---|---|---|
| Contamination | 0–10 | 10–20 | 20–30 | 30–50 | 50+ | Profiled |
| Water | S/D<br>Detergent<br>Mop<br>Wet Vac<br>Dried | S/D<br>Detergent<br>Mop<br>Scouring<br>Pad | S/D<br>Detergent<br>Immersion<br>Mop<br>Scouring<br>Pad | | | S/D Dried |
| Oily/Grease | Fat Solve<br>Detergent<br>Mop | S/D<br>Detergent | Detergent<br>S/D | | Detergent<br>and Hose | |
| Mayo | | | | | Detergent<br>and Brush | |
| Doughy | | | Hose | | | |
| Dry | Vac | Sweep | Mop | | Hose | |

after cleaning, and that only when the floor is completely dry are pedestrians allowed to use the area. To restrict pedestrian access to dangerously wet and slippery floors, the use of physical barriers is strongly recommended, as simple signs are frequently ignored. Consideration should be given to the mopping process – how is the dirt removed from the floor and where does it go? Use a separate dirty water bucket for ringing the mop out to increase the efficiency of dirt removal. Greasy floors require contact time with the detergent solution – use an immersion mopping technique, where the detergent is put down in one stage, and mopped up after a soak time in a second stage. Inappropriate use of mopping techniques to remove spills from smooth floors may simply increase the slip risk by spreading a film of water over a wider area.

**Wet Vac** – Effective at cleaning up liquid spills.

**Dried** – The use of a paper towel or rag to remove small areas of liquid contamination from the floor is the cheapest and most effective method of removing water based spills. This avoids spreading the contamination, or level of risk by mopping a large area. Specialist products are available to achieve this kind of cleaning over a wider area in a short timescale.

**Scouring Pad** – Using a scouring pad or brush to agitate the surface, possibly in combination with a detergent, increases the amount of contaminant removed.

**Dry Vac** – A dry vac can be used to remove dusty contaminants, particularly on rougher floors.

**Sweep** – On a smooth floor sweeping may be adequate to remove dry contaminants

**Hose** – A high pressure water hose can be used to remove dusty or doughy contaminants provided the floor has adequate surface roughness to cope with the water left behind.

**Squeegee** – The situations investigated for this report in which the squeegee has been used have shown it to have a negative effect. If a floor has sufficient roughness to be left wet, the volume of water is not important and the squeegee is unnecessary. Where oily or greasy contamination was present, the squeegee had the effect of spreading a thin layer of the contamination over a wider area, or forcing the contamination into the rough surface, the outcome of both was a surface that was more difficult to clean than when the squeegee was not used.

## Discussion

During the course of this study it was observed that cleaning plays a very important role in the slip resistance of flooring. Use of an appropriate cleaning regime can help maintain a high level of slip resistance by adequately removing contamination from the surface. Conversely if an inappropriate regime is used little or no difference is observed in the level of available slip resistance, and the process itself may even increase the slip risk presented by a floor. A layer of water only two micrometers thick is sufficient to form a squeeze film between the heel of a pedestrian and the floor surface and initiate a slip. Floors with a surface roughness less then twenty micrometers Rz may not be rough enough to break through this squeeze film and are therefore potentially dangerous if left even slightly damp (FIS22, 1999). Application of the correct regime does not necessarily involve a greater investment, either in terms of money or the cleaners time. If proper thought is given to the appropriate technique for the floor being cleaned and contamination present, then safe and effective cleaning should be easy to achieve.

Throughout the site visits undertaken during this study, two overwhelming impressions were gleaned from those working in the cleaning sector. At management level, the actual cleaning processes are often not well thought out. More consideration needs to be given to the type of contaminant, how it can be removed, and whether the cleaning system actually removes the contamination effectively. For example, where grease or oil is present, rinsing with water has very little effect, as neither grease nor oil are soluble in water, regardless of the water temperature. To remove this kind of contamination successfully, the cleaning solution must contain a detergent. Assuming that a detergent is used, then thought needs to be given to contact time for the detergent to act. Time is required for the detergent to release contamination from the surface and mobilise it in the water. Finally, the dirty water has to be removed, as any water left behind will evaporate and allow the contaminant back onto the floor. Guidance on types of detergent for a given contaminant is readily available from the chemical suppliers. Those working on the shop floor, carrying out the cleaning, are low paid and require more effective training and supervision to ensure cleaning is effective. Most cleaners involved in the study clean according to their instructions, and rarely challenge the status quo. We have come across cases where the cleaning supervisors are insufficiently well informed about the chemicals in use, for example, cleaning a greasy floor with disinfectant rather than detergent, when both were available.

The results have been presented in this report in an easy reference matrix format, which should be accessible for cleaning managers. It was decided to plot the contamination versus the microroughness of the floor to identify a scenario without needing hundreds of very specific examples. Although this requires the use of instrumentation to evaluate the microroughness of the floor, it was felt that defining types of floor had the potential to be misleading, especially as surfaces change with time, wear and cleaning. It is suggested that this format could be usefully adopted for future guidance on cleaning in relation to slips. It is suggested that further information to support the guidance would need to be produced for the cleaning operatives, in a simple and easy to follow layout. There are several communications issues when looking at the cleaning industry. The input of industry bodies in the production of the information sheets may be helpful in using terminology and a style that cleaners are comfortable with. At one of the premises visited, there were employees with thirty one different native languages. Where English is a second language, simplicity of the language used can have a significant effect on its comprehension. The use of drawings, cartoons and diagrams that are not language dependant would be more inclusive than text based guidance and may be more effective in real workplaces.

## References

HSE, *The assessment of pedestrian slip risk; The HSE approach*, 2004 (http://www.hse.gov.uk/pubns/web/slips01.pdf)

HSE, *Preventing Slips in the Food and Drink Industries – Technical Update on Floor Specification*, HSE Free Information Sheet FIS22, HSE Books, 1999

UKSRG, *The Measurement of Floor Surface Slip Resistance: Guidelines Recommended by the UK Slip Resistance Group*, Issue 2, 2000. ISBN 1-85957-227-8 (The UK Slip Resistance Group, RAPRA, Shropshire)

# TACTILE PAVING – RIGHT OR WRONG?

**Marianne Loo-Morrey & Steve Thorpe**

*Health & Safety Laboratory, Harpur Hill, Buxton SK17 9JN, UK*

The advent of the Disability Discrimination Act and increasing awareness of the needs of disabled people has resulted in an increase in the installation of tactile paving in public areas. Guidance for the correct installation of tactile paving has been available from the Department of Transport for a number of years, however incorrectly installed tactile paving is commonplace.

A survey of three city centres in England has shown approximately half of tactile paving to be incorrectly installed, either lacking in colour contrast or the wrong pattern in the wrong place. 4% of paving installations used the wrong warning surface, 17% of paving was the wrong colour and 56% of paving had insufficient colour contrast with the surrounding paving.

## Introduction

Tactile paving began to be introduced in 1990 to provide tactile signing for visually impaired individuals to aid their independent mobility. In the years since its introduction tactile paving has become a key design feature in improving the accessibility of public spaces and city centres. The advent of the Disability Discrimination Act with its requirements for inclusive design and equality of access to services is likely to increase the rate at which tactile paving is being installed. At the time of writing there is currently no agreed British standard for tactile paving, but guidance on the use of tactile paving is available (Department of Transport). The aims of this current survey are:

- Determine what types of tactile paving are being installed?
- Gain a better understanding of how tactile paving is being installed.
- What percentage of tactile paving is being installed inline with current guidance?

"Tactile paving" is the term used for a range of paving units that bear distinctive, raised surface profiles which are designed to be detectable by both sighted and visually impaired pedestrians, it can be used to convey important information about the environment to visually impaired people. There are six different patterns of tactile paving currently in regular use in the United Kingdom:

- Blister warning surface – used to indicate a pedestrian crossing point
- Corduroy hazard warning surface
- Off-set blister warning surface
- Lozenge warning surface
- Cycle path warning surface
- Guidance path surface

There are approximately 1 million blind and partially sighted adults in the United Kingdom. Approximately 5% of these people are completely blind, having no sight at all. The remainder have varying degrees of residual sight, which may enable them to function visually to differing degrees. Visually impaired people will either move around independently or with the aid of a sighted person to act as a guide. Individuals who move around independently will do so either solely by using their residual vision or by using a mobility aid.

The most common mobility aid used by visually impaired pedestrians is a long white cane, which is used to scan the ground in front of the person. The cane is swept in an arc from one side to the other just beyond the width of the body. This technique will usually locate potential obstructions and distinct changes in level such as a kerb of step. Once any feature has been located and possibly identified, the pedestrian will decide how to proceed. When navigating around the pedestrian environment, visually impaired people actively seek and utilise tactile information underfoot, particularly detectable contrasts in surface texture. The ability to detect contrasts in texture underfoot can vary between individuals. Visibly impaired people who use mobility aids are likely to have received training in mobility skills, including instruction in the interpretation of tactile paving.

## Experimental

There were a number of considerations that were taken into account in choosing the sites to be surveyed in this study:

- The study sites had to be controlled by different local authorises to allow the effect of differences in policy to be investigated.
- The study sites had to have reasonable geographical separation to reasonably reflect England as a whole.
- The sites had to have different "characters" to determine the effect of considerations such as conservation areas.

Three city centres were chosen for the survey, 1. Sheffield, 2. Chester, 3. Birmingham. 48 examples of tactile paving were identified and photographed. Each use of tactile paving was then assessed to see if had been installed in line with the guidance available from the Department of Transport, 1) Was the correct pattern used? 2) Was the colour correct? 3) Was there good colour contrast with the surrounding paving?

## Results and discussion

Overall the tactile paving was found to be in good condition and well maintained, in only two cases was the condition of the paving a cause for concern. Four different tactile paving patterns were found in use at various locations surveyed: 1) Blister surface for pedestrian crossing points, 2) Corduroy hazard warning surface, 3) Platform edge (on-street) warning surface, 4) Guidance path surface. The results of the tactile paving survey are summarised in Table 1.

The blister surface used to indicate pedestrian crossing points with dropped curbs was the most commonly installed tactile surface, and all the pedestrian crossings examined in this survey had tactiles with the correct blister surface installed. The colour of the warning surface plays a significant role at pedestrian crossings for partially sighted people. Red is used to denote a controlled crossing point and buff coloured tactiles are used to denote uncontrolled crossing points.

All of the controlled crossings assessed in the survey had correctly installed red blister warning surface, however there were numerous cases where the warning tactiles had little or no colour contrast with the surround paving. The colour contrast strip that the guidance states should be used in these circumstances was omitted.

**Table 1. Summary of the results of the tactile paving survey.**

| Number of examples | Type of tactile | % correct pattern useage | % correct colour problems | % colour contrast | % other problems |
|---|---|---|---|---|---|
| *Sheffield* | | | | | |
| 6 | Blister tactile controlled crossings | 100 | 100 | 50 | 0 |
| 8 | Blister tactile uncontrolled crossings | 100 | 100 | 63 | 0 |
| 2 | Corduroy hazard warning tactile | 100 | 0 | 100 | 0 |
| 2 | lozenge platform edge tactile | 50 | 0 | 100 | 0 |
| 2 | Guidance path tactile | 100 | 0 | 100 | 0 |
| *Chester* | | | | | |
| 3 | Blister tactile controlled crossings | 100 | 100 | 0 | 0 |
| 13 | Blister tactile uncontrolled crossings | 100 | 100 | 69 | 4 |
| *Birmingham* | | | | | |
| 3 | Blister tactile controlled crossings | 100 | 100 | 33 | 0 |
| 5 | Blister tactile uncontrolled crossings | 100 | 100 | 20 | 0 |
| 2 | Corduroy hazard warning tactile | 100 | 50 | 50 | 0 |
| 1 | Guidance path tactile | 0 | 0 | 100 | 0 |

(a) Sand stone (b) Inset brass studs

**Figure 1. Picture showing examples of natural sand stone and in set brass studs.**

All the uncontrolled crossings assessed in the survey had tactile paving with the correct blister warning pattern installed. Conventional concrete or block paviour blister tactiles were installed at half of the 26 uncontrolled crossings surveyed. However natural sand stone or in set brass studs were used at the remaining crossing points see Figure 1.

Although it could be argued that both the sand stone is the correct buff colour for an uncontrolled pedestrian crossing there is no contrast at all with the surrounding pavement, nor is there a contrast strip installed. This clearly ignores the advice given in the guidance document, which repeatedly emphasises the use of colour contrast to aid partially sighted individuals and even raises the possibility that the tactile paving may pose a trip hazard to unwary members of the general public.

The current guidance on the installation and use of tactile paving recognises that there may be concerns regarding the use of tactile paving in conservation areas or in the vicinity of listed buildings. It recommends that some relaxation of the colour requirements for blister tactiles at uncontrolled crossings may be acceptable and that in "… these limited circumstances only, the tactile surface may be provided in a colour which is in keeping with the surrounding material."

The decision to use the brass studs in Chester appears to have been an aesthetic one. Given the sensitive nature of the conservation area the brass stud tactile warning surface can be considered to be in line with the more relaxed version of the guidance. However brass studs may pose and unacceptable slip risk in wet conditions, and other solutions that would have been equally aesthetically sensitive to the surrounding environment without posing a potential hazard are possible. Stone blister tactile paving units could have been used to match the existing flags with the contrast strip of a different coloured stone being used the give clear visual cues to partially sighted individuals.

Overall insufficient colour contrast between the tactile areas and the surrounding paving was found to be a problem at 58% of the uncontrolled pedestrian crossing surveyed. In three out of four instances where corduroy hazard warning tactile paving was used there was little or no colour contrast with the surrounding pavement. This is a clear contradiction of the advice given in the guidance document which recommends, "the surface should be in a contrasting colour to the surrounding area to assist partially sighted people." Corduroy hazard tactile paving units are available in a wide variety of colours and materials; alternative tactiles, which provided good colour contrast with the surrounding pavement, could easily have been selected.

Sheffield city centre is served by a tram network and as such has a number of on-street tram platforms. The tram platforms assessed as part of the survey had the correct platform edge (on-street) lozenge tactile warning surface installed and guidance path tactiles to lead visually impaired individuals the platform edge. The contrast between the tactile paving units and the surrounding paving is poor; selection of tactile paving units of a different colour would have been better.

The most serious mistake with tactile paving is the incorrect use of a pattern; this can give misleading and potentially dangerous misinformation to blind and partially sighted people. The survey found two examples of improper use of tactile patterns, see Figure 2.

(a) Incorrect use of lozenge tactile at bus stop

(b) Blister tactile used as guidance path

**Figure 2. Incorrect use of tactile paving at a bus stop and as a guidance path.**

## Conclusions

The survey showed that 4% of tactile paving installed involves the incorrect use of a warning pattern. This is most serious problem that can be encountered with tactile paving as the incorrect use of a pattern can give misleading and potentially dangerous information to blind and partially sighted people. 17% of tactile paving installed is the wrong colour. Either the specified colour was not used, or where a colour is not specified the colour chosen had little of no colour contrast with the surrounding paving. Overall up to 50% of the tactile paving assessed had one or more problems with its installation.

The guidance for tactile paving recommends good colour contrast in association with tactile paving, however 50% of the cases examined lack colour contrast. In many cases the lack of colour contrast appears to be a deliberate design choice that significantly reduces the usefulness of tactile paving to the partially sighted. With the growing emphasis on accessibility of our public spaces it seems odd that designers, architects and specifiers appear to be intent on minimising the visual impact of tactile paving when colour contrast is held to be a key principle in inclusive design.

The current tactile paving guidance has been available for several years, however incorrect installation of tactile paving is widespread. Whether these mistakes are a result of the guidance being knowingly disregarded or through ignorance is unclear. The fact remains though; that mistakes are commonplace and where patterns are misused potentially dangerous misinformation is being given to blind and partially sighted individuals. The DDA will result in increasing use of tactile paving, decisions need to be taken now about how to ensure tactile paving is correctly installed, merely giving guidance on good practise appears to be insufficient and is unenforceable. In order to ensure that the different patterns of tactile paving are reliably and consistently installed and that good colour contrast is provided, it may be necessary to regulate the installation of tactile paving, for example through the building regulations.

## Reference

Department for Transport, "*Guidance on the use of tactile paving surfaces*", dft.gov.uk/stellent/groups/... /pdf/dft_mobility_pdf_503283.pdf

# PREVENTION OF FALL-INDUCED HIP FRACTURES: USABILITY EVALUATION OF HIP PROTECTORS

**K.R. Dunn[1], C.L. Brace[1], T. Masud[2], R.A. Haslam[1] & R.O. Morris[2]**

[1]*Health & Safety Ergonomics Unit, Department of Human Sciences, Loughborough University, LEICS, LE11 3TU*
[2]*Clinical Gerontology Research Unit, Nottingham City Hospital, Nottingham NG5 1PB*

Hip fractures are a common consequence of falls among the older population, with the incidence set to rise as our aging society increases. It has been suggested that the use of hip protectors are instrumental in substantially reducing the occurrence of hip fractures and so reducing the associated risks of disability and death. A serious limitation to their efficacy however, is that of non-use, particularly in the community setting. This paper presents a qualitative interpretation of the usability issues arising from the use of hip protectors and reasons for non-compliance. Two types of hip protectors (soft and hard) were examined to deduce if one was superior to the other on these accounts. The main reasons for non-compliance were: poor comfort; poor self-perceived appearance of the user; and high levels of self perceived safety. The soft hip protectors were found to be superior in terms of usability and compliance, and reported to be significantly more comfortable than the hard hip protectors ($p = 0.009$).

## Introduction

It has been well documented over the years that a third of individuals over 65, and nearly half of those over 80, fall each year. Current trends show evidence of an ageing population and an associated rise in the number of falls sustained (Easterbrook, 2001). This is a major health concern, especially when considering fall-induced injuries and deaths. Some of the most devastating of these injuries are hip fractures. Deterioration in quality of life results after a hip fracture; inability to manage activities of daily living independently and the subsequent reliance on others after a hip fracture are likely causes of this decline. Repression in social activities is likely to prevail and thus further reduce quality of life. Hip fractures are thus a severe threat to the health and well being of the older population. A time trade off study dramatically illustrates this point: the study found that 80% of the elderly population would prefer a shorter life span instead of a bad hip fracture and the consequences that it would bring (Salkeld et al., 2000). Additionally, hip fractures cost the UK's National Health Service between £12,000 and £20,000 per incident (Easterbrook, 2001). These personal and economic dimensions are imminent issues to note and give weight to the importance of assessing hip fracture prevention techniques and to providing improvements to their efficacy.

It has been suggested that the use of hip protectors may be instrumental in substantially reducing the occurrence of fractures and so eliminating the associated risks of disability and death. Hip protectors are garments worn by older people who are at risk of falling and are thus at an increased risk of experiencing a hip fracture. The standard design of hip protectors includes two pads, one sewn into each leg of a pair of briefs. The materials used in the undergarment and hip protector pads differ between makes, and the size and method of protection offered by the pads also vary. Hip protectors have been developed through substantial biomechanical testing and have been found to be technically able to prevent hip fractures (Parkkiari et al., 1995) by protecting the bones, notably the greater trochanter and proximal femur, during a fall episode.

However, recent studies have shown contrasting views on hip protector efficacy: the serious limitation to efficacy is user non-compliance. Compliance is critical to the effectiveness of hip protectors. If the hip protectors are not worn at the time of a fall, they cannot prevent a hip fracture. A study by Lauritzen et al. (1993) reported that less than / of people issued with hip protectors wore them regularly. Previous studies, which have generally been conducted with institutionalised older people, have helped to identify causal reasons for non-compliance: hip protectors have been reported as being uncomfortable, of a poor fit and too tight (Villar et al., 1998; Hubacher et al., 2001). This may give explanation for complaints of bowel irritation and swollen legs (Kannus et al., 2000). Another important issue of concern was reported by Hubacher et al. (2001) who found that 38.9% of dropouts versus 12.7% of wearers reported that their appearance was unattractive when wearing the hip protectors, which may be another reason for non-compliance. A further factor acknowledged by a number of studies to limit compliance is the case of skin irritation which can occur (Cameron et al., 2001).

This paper presents a qualitative interpretation of the usability issues arising from the use of hip protectors by community dwelling older people and their reasons for non-compliance. Two types of hip protectors were examined to deduce if one was superior to the other on these accounts, and issues for usability and design improvements have been summarised.

## Method

Nineteen community-dwelling, fall-susceptible, older people (aged between 65–96 years, mean 80), who had not used hip protectors previously, were recruited from the community setting and involved in a cross over study design to test two types of hip protectors. The Safehip (hard) hip protector is anatomically designed to fit over the greater trochanter and proximal femur. It is made from two hard shell-shaped polypropylene plastic shields that are incorporated into briefs. The shields are fixed into the briefs and positioned so that one covers each hip. The briefs are made from a combination of cotton and Lycra®, and hold the shields in place. The hip protectors work by diverting the impact produced from a lateral fall, away from the hip joint and into surrounding tissues. The Hipsaver (soft) hip protector is an all-soft one piece brief. Sewn into the undergarment are two thin viscoelastic foam pads which are situated over the greater trochanter, one on each hip. The pads are encased in a waterproof and airtight pouch, made from nylon and during a fall the impact on the hip is significantly reduced through both energy attenuation and energy absorbing mechanisms.

A multi-methodological approach was taken and included the use of focus groups (n = 2) and interviews (n = 19) after a period of wear (at least 7 days per garment type). Repeated measures questionnaires centred on usability issues and compliance, and were used to verify and quantify the qualitative data gathering. Additional questionnaires were also used to determine whether wearing hip protectors had an influence on quality of life and fear of falling. The SF36 (Short Form 36) (a questionnaire that has been developed to identify the health status of individuals) and FES (Falls Efficacy Scale) (developed to measure the fear of falling within individuals, Tinetti 1988) were used to determine these respectively.

## Discussion of results

The hip protectors generally seemed to be well accepted, although many participants were concerned about their appearance in the garments and what others thought. This seemed to negatively affected compliance, especially with regard to the hard hip protector. The majority felt that the hard hip protector was more uncomfortable compared to the soft, and this seemed to influence both preference and compliance. This view was similar for comfort at night, with the soft type again being preferred. Toileting was made difficult with the use of the garments and in some instances made participants have to rush to the toilet due to the extra time required to remove additional clothing. Furthermore, many individuals regarded the garment as unnecessary, which also had an adverse affect on compliance. The percentage of participants who reported wearing the soft hip protector for "most days" or more during the study period was 67% compared to 53% of participants wearing the hard design. The main reasons for non-compliance were: poor comfort and fit; poor self-perceived appearance of the user; and high levels of self perceived safety. The soft hip protectors were found to be superior in terms of usability and compliance, and reported to be significantly more comfortable than the hard hip protectors ($p = 0.009$).

It emerged that many of the participants who found the hip protector to be uncomfortable went on to report that they failed to continue wearing the garment. The responses support the natural conclusion that if the garment is uncomfortable, then the likelihood of it being worn is reduced. Comfort was especially noted to influence compliance at night, as one of the reasons presented for non-use at night included the concern that the garment would be uncomfortable to wear at this time and that it may disturb their sleep.

It is important to recognise that the older population for whom the protective devices are essentially being developed is likely to present with physical difficulties. Those individuals with such attributes have been found to have difficulties in getting the hip protectors on and off (van Schoor et al., 2003), as was an issue reported by some of the participants in this study. It has been noted that this difficulty may increase the risk of falling and thus place the wearer at a greater risk of a hip fracture (Oster & Specht-Leible, 1999). Additionally, this usability issue was also of concern to some individuals in an aesthetic sense as the clothing fitted tightly over the protectors, making them more conspicuous to others. For other users, this issue was not of any concern as they felt that safety takes priority and they were more concerned with protecting their hips during a fall than worrying about their appearance. Fear of falling is often a concern among the older population and this lack of confidence can cause further falls or cause a self-imposed restriction on activities. It was suggested by a number of participants that they felt safer when wearing the hip protectors and a feeling of safety was associated with their wear, influencing compliance positively. However, for some, the use of the protectors made them feel more at risk because it raised awareness of their fall risk. It was revealed in the study that individuals who felt more confident in their walking ability were less inclined to comply, the reasons given were that they felt no need to wear the device because they had not had a fall, or hadn't had one for a while, or that through other interventions (such as physiotherapy) they felt as if their mobility was improving and were therefore at less risk of falling (and had less need for protection). As a linear relationship exists between the number of fall risk factors and likelihood of falling, it is not surprising that individuals with a greater risk of falling have a higher compliance rate (Hubacher et al., 2001). Additionally, it is important to understand the psychological effects that the use of hip protectors can have on users. For some, the inability to dress or toilet independently with their use, to feel undignified, or to feel at increased risk of falling, may affect individuals' quality of life.

Interestingly, although the protectors resulted in a lot of negative comments from the participants, very few stated that their design could be optimised, although this may have been due to concerns that changes could reduce their efficacy in fracture prevention. From comments made during the study, it appeared that users may benefit from cooler fabric, less bulky (less visible) and softer material in the design of the protectors, and an easier way to don and doss such garments. The design could be changed although it is likely that a number of participants will not be willing to wear them as they feel they are accepting old age:

> *"I just thought – they are the last straw. You think, why should it happen to me... and it has happened to me, and I don't like it, do I, but who am I to be that proud at 85 – I felt very degraded [wearing the hip protectors]. I just can't visualise having to wear those things every day and perhaps at night... I just feel as though that's the end."*

It may be necessary to employ other additional support to increase the acceptance and subsequent compliance of hip protectors. Literature has suggested that a possible intervention to help improve compliance involves educating the nursing staff caring for the patients about the use of hip protectors and the risks associated with hip fractures. Therefore, the type of information that the staff can give to their patients can be more tailored to suit an individual. Meyer et al. (2003) reported successful adherence with this methodology and a 40% reduction in hip fractures being achieved as a result. A study by Parkkari et al. (1998) supports this as an effective method of achieving good compliance and reported an adherence rate of 91%, with a one hour introduction lesson on the incidence, causes, consequences and prevention of hip fractures, given to the nursing staff.

## Conclusions

The aim of the study was to develop a greater understanding of the usability issues surrounding hip protector use among community dwelling older people, and to identify reasons for non-use. The study aimed to compare two different types of hip protectors, a hard hip protector (Safehip) and soft hip protector (Hipsaver), on issues of usability. The study concluded that the soft hip protector is superior in terms of usability and general compliance.

It can be suggested that there is a requirement to involve the user in the design process in order to improve acceptance and compliance. It can further be concluded that the nature of the information given to the user on allocation of hip protectors impacts on compliance and needs to be tailored to an individual's perception of their own limitations.

## References


Cameron, I.D., Venman, J., Kurrle, S.E., Lockwood, K., Birks, C., Cumming, R.G., Quine, S., Bashford, G., (2001). Hip protectors in aged-care facilities: a randomised trial of use by individual higher-risk residents. *Age and Ageing*, **30**, 477–481.

Easterbrook, L., Horton, K., Arber, S., Davidson, K., (2001). *International review of interventions in falls among older people*. Department of Trade and Industry, London, URN 01/1173.

Hubacher, M., Wettstein, A., (2001). Acceptance of hip protectors for hip fracture prevention in nursing homes. *Osteoporosis International*, **12**, 794–799.

Kannus, P., Parkkari, J., Niemi, S., Pasanen, M., Palvcanen, M., Jarvinen, M., Vuori, I., (2000). Prevention of hip fracture in elderly people with use of a hip protector. *The New England Journal of Medicine*, **343**(21), 1506–1513.

Lauritzen, J.B., Petersen, M.M., Lund, B., (1993). Effect of external hip protectors on hip fractures. *Lancet*, **341**, 11–13.

Meyer, G., Warnke, R., Bender, R., Mühlhauser, I., (2003). Effect on hip fractures of increased use of hip protectors in nursing homes: cluster randomised controlled trial. *British Medical Journal*, **326**, 76–78.

Oster, P., Specht-Leible, N., (1999). Hip fracture with correctly positioned external hip protector. *Age and Ageing*, **28**, 497–500.

Parkkari, J., Kannus, P., Poutala, J., Sievanen, H., Vuori, I., (1995). Energy-shunting external hip protector attenuates the peak femoral impact force below the theoretical fracture threshold: an in vitro biomechanical study under falling conditions of the elderly, *Journal of Bone Res* 1995, **10**, 1437–1442.

Parkkari, J., Heikkila, J., Kannus, P., (1998). Acceptability and compliance with wearing energy-shunting hip protectors: a 6-month prospective follow up in a Finnish nursing home. *Age and Ageing*, **27**, 225–229.

Salked, G., Cameron, I., Cumming, R., Easter, S., Kirrle, S., Quine, S., Seymor, J., (2000). Quality of life related to fear of falling and hip fracture in older women: a time trade off study. *British Medical Journal*, **320**, 341–346.

Tinetti, M.E., Speechley, M., Ginter, R.N., (1988). Risk factors for falls among elderly living in the community. *The New England Journal of Medicine*, **319**(26), 1701–1706.

van Schoor, N., Smit, J., Twisk, J., Bouter, L., Lips, P., (2003). Prevention of hip fractures by external hip protectors, A randomised control trial. *JAMA*, **289**(15), 1957–1962.

Villar, M., Hill, P., Inskip, H., Thompson, P., Cooper, C., (1998). Will elderly rest home residents wear hip protectors. *Age and Ageing*, **27**, 195–198.

# EFFECT OF CHANGING STAIR DIMENSIONS ON SAFETY

**Michael Wright & Mike Roys**

*BRE, Garston, Watford WD25 9XX*

In research conducted for the Office of the Deputy Prime Minister, BRE used a variable going and rise stair rig to investigate how adults use different stair geometries. Sixty adult subjects walked up and down these stairs, and assessed them for safety and comfort using 7-point Likert scales. Each subject tried 10 different goings with two rise heights. During each climb and descent the subjects were observed for hesitations, missteps, use of handrails and glances at feet. In addition, the position of the foot relative to the nosing was determined for each step taken.

The results confirm the effect of an increased going on providing a firm footage for users in descent, and show how this changes for different rises. This is shown both in the subjective and objective data. It is recommended that for private homes the minimum going should be 250 mm.

## Introduction

Stairs in the home are involved in over 500 deaths and over 300,000 injuries in England and Wales each year. There may be a further 1.5 million minor falls on stairs in the home that are not serious enough for the victim to attend A&E (Roys and Wright, 2003). This equates to a minor fall every 20 seconds on average, and an injury every 2 minutes. Previous research indicates that step dimensions play a major part in the risk of an overstep (Roys, 2001), which is a major contributor to slips and falls down stairs. This is of particular importance, considering that 80% of these injuries are likely to occur when users are descending stairs (Roys and Hamilton, 2000).

## Experimental methodology

### *Set up*

In research conducted for the Office of the Deputy Prime Minister BRE developed a stair rig that allows a stair with 7 steps to be adjusted, such that each step could have a going between 200 mm and 425 mm, in 25 mm intervals, combined with a rise of between 160 mm and 210 mm, in 10 mm intervals. This gives 60 different stair sizes. The minimum going size of 200 mm compares with a survey carried out in 1998 (Roys and Hamilton, 2000) of over 1,700 domestic stairs which shows that only 6.3% of the sample had a mean going less than this value. Each step had carpeted treads and risers with a traditional rounded nosing, as seen in most domestic stairs. The stair also had two continuous handrails and at least 100 lux of illumination, throughout the whole flight.

### *Procedure*

Sixty adult subjects walked up and down these stairs, and assessed them for safety and comfort using eleven 7-point Likert scales. The multiple choice questions were phrased to allow an answer on the same scale, ranging from "Very strongly disagree" to "Very strongly agree" with the mid-point labelled as "Neither disagree nor agree". The order of the statements on the questionnaire had a mix of ascent/descent and positive/negative statements as people worked through it. There were more questions concerning descent because a greater number of stair injuries occur in descent.

Each subject tried the 10 different goings in an order based on a Latin square design, for two rise heights. Each flight was ascended and descended twice, equating to 40 climbs and descents for each subject. Twenty different subjects were therefore tested, twice, on each rise and going combination. During each climb and descent the subjects were observed for hesitations, missteps, use of handrails and glances at feet. In addition, each subject wore standardised shoes with an integrated pressure sensor in the sole, so that the position of the foot, relative to the nosing, could be determined on each step. The data collected from the sensors were processed to give the number of cases where the foot was entirely on the tread (thus giving no data about the foot position), the distance along the centre of the sole of the shoe from the toe to the nosing, and the foot angle on the tread.

### *Subject population*

The subjects were volunteers who lived close to BRE, 63% were female. The youngest volunteer was 17 and the eldest was 72 years of age. Both male and female groups were significantly older ($p < 0.01$) than the census data for the age range 16–75 years, which were the limits set on the sample population. This is due to an over-representation of people aged over 55 in the sample.

The subjects varied in height between 150 cm and 186 cm. The female group is significantly taller ($p < 0.001$) than the data presented in Adultdata (Peebles and Norris, 1998), but the male group is not significantly taller. The subjects had shoe sizes that varied between 4 and 11, which is equivalent to foot lengths of between 224 mm and 284 mm. This is not significantly different from the population data presented in Peebles and Norris (1998). The weight of the subjects varied between 56 kg and 113 kg. Both groups were significantly heavier than that presented in Peebles and Norris (1998).

Because not all subjects tested all rises, there may be differences in the subjects which could cause them to experience the stairs differently, giving different subjective opinions and objective measures across the rises. However, carrying out ANOVA tests across the rise groups for the age, height, weight and shoe size did not generate any significant differences. Although this cannot rule out differences in behaviour between the subjects, leading to different subjective opinions and objective measures, it does reduce the chance of significant differences between rises being due to different sample populations.

## Results

### *Subjective opinions*

The data were analysed using univariate analysis of variance, which allows for a two way comparison of data, comparing across both rise and going. This parametric analysis assume that the data are approximately normal, which is not always the case. However, to allow a two way analysis to be carried out, this assumption has to be made. Significant differences based on Tukey post hoc analysis with a significance level of 0.05 were used to distinguish between opinions. Data are presented for ten of the eleven statements in Table 1 and Table 2, using the following key:

| | |
|---|---|
| ✖ (white on black) | Worse than the other rises/goings. |
| ✖ | Worse than the ✓ rises/goings only. |
| ✓ | Better than the ✖ rises/goings only. |
| ✓ (white on black) | Better than the other rises/goings. |

**Table 1. Statistical differences in the subjective data across rise.**

| Statement | Rise (mm) | | | | | |
|---|---|---|---|---|---|---|
| | 160 | 170 | 180 | 190 | 200 | 210 |
| 1. I had to pull myself up the stair using the handrail | | | | | ✖ | ✖ |
| 2. I felt it was hard work to walk up the stair | | | | | ✖ | ✖ |
| 3. I felt out of control coming down the stair | | | | | ✖ | ✖ |
| 4. I relied on the handrail to support me as I came down the stair | | | | | ✖ | ✖ |
| 5. I hardly had to look at the steps as I came down the stair | | | ✓ | | | |
| 6. It's easy to mis-judge a step when descending | | | ✓ | | ✖ | ✖ |
| 7. I felt safe walking down the stair | | ✓ | ✓ | | | ✖ |
| 8. There was enough room to put my foot comfortably on each step | ✖ | | ✓ | | | |
| 9. I found it easy to keep a comfortable stride length on the stair | | ✓ | ✓ | ✓ | | ✖ |
| 10. The stair was too steep | ✓ | ✓ | ✓ | ✖ | ✖ | ✖ |

**Table 2. Statistical differences in the subjective data across going.**

| Statement (as in Table 1) | Going (mm) | | | | | | | | | |
|---|---|---|---|---|---|---|---|---|---|---|
| | 200 | 225 | 250 | 275 | 300 | 325 | 350 | 375 | 400 | 425 |
| 1 | ✖ | | | | | | | | | |
| 2 | ✖ | ✖ | | | | | | | | |
| 3 | ✖ | ✖ | ✖ | | | | | | | |
| 4 | ✖ | ✖ | | | | | | | | |
| 5 | ✖ | ✖ | ✖ | | | | | | | |
| 6 | ✖ | ✖ | ✖ | | | | | | | |
| 7 | ✖ | ✖ | ✖ | | | | | | | |
| 8 | ✖ | ✖ | ✖ | ✖ | | | | | | |
| 9 | ✖ | ✖ | ✖ | | | | | | | |
| 10 | ✖ | ✖ | ✖ | | | | | | | |

**Table 3. Statistical differences in the objective data across rise.**

| Data | Rise (mm) | | | | | |
|---|---|---|---|---|---|---|
| | 160 | 170 | 180 | 190 | 200 | 210 |
| Looking at feet continuously in ascent | | | ✓ | | | ✖ |
| Looking at feet continuously in descent | | ✖ | ✓ | | ✖ | ✖ |
| All the foot was placed on the tread | ✓ | | ✓ | ✓ | ✖ | ✖ |
| Mean proportion of shoe off tread | ✓ | | ✓ | ✓ | ✖ | ✖ |

**Table 4. Statistical differences in the objective data across going.**

| Data | Going (mm) | | | | | | | | | |
|---|---|---|---|---|---|---|---|---|---|---|
| | 200 | 225 | 250 | 275 | 300 | 325 | 350 | 375 | 400 | 425 |
| Number of missteps in ascent | ✖ | | | | | | | | | |
| Number of missteps in descent | ✖ | | | | | | | | | |
| Looking at feet continuously in ascent | ✖ | | | | ✓ | ✓ | | | | |
| Looking at feet continuously in descent | ✖ | | ✓ | ✓ | ✓ | ✓ | ✓ | ✓ | | |
| Continuous handrail use in ascent | ✖ | ✖ | ✖ | | | | ✓ | | | |
| Continuous handrail use in descent | ✖ | ✖ | ✖ | | | | | | | |
| All the foot was placed on the tread | ✖ | ✖ | ✖ | ✖ | ✖ | | | | | |
| Mean proportion of shoe off tread | ✖ | ✖ | ✖ | ✖ | ✖ | | | | ✓ | ✓ |

*Objective results*

- There were few missteps in this experiment, with 4% of ascents and 1.5% of descents including an event.
- There were only 14 hesitations during the 2,400 descents in the experiment.
- About 80% of the sample continuously looked at their feet, both in ascent and descent.
- Most of the sample never used a handrail, neither in ascent (85%) nor descent (79%).
- The percentage of times all of a foot was placed on the stair tread in descent varied widely across the experiment, rising from 4% on a 200 mm going, to 92% on a 425 mm going.

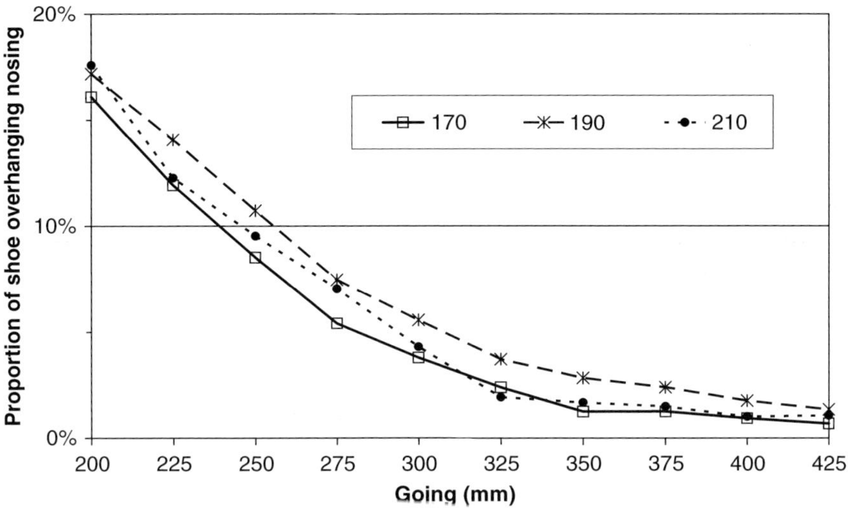


**Figure 1. Variation of proportion of shoe overhanging the front of the stair nosing with going, for three rise heights, 170, 190 and 210 mm.**

- The mean proportion of the shoe overhanging the nosing when descending a stair varied, rising from 1% on a 425 mm going, to 19% on a 200 mm going.

Significant differences based on Tukey post hoc analysis with a significance level of 0.05 were used to distinguish between occurrences of behaviour and foot placement data. Data are presented in Table 3 and Table 4, using the same key as before.

It is interesting to look at the mean proportion of the shoe that was placed on the tread across all goings and some of the rise values. It is likely that the greater the proportion of the shoe that can be placed on the stair tread, the less likely an overstep will occur, which could lead to a slip or other incident. The exact proportion of the shoe that needs to remain on the tread before a slip occurs is unclear, but as explained in the companion paper by Roys and Wright (2005) a smaller mean overhang is likely to reduce the probability of an overstep. The mean values for rises of 170, 190 and 210 mm are presented in Figure 1, the other rise data are similar but not shown for the sake of clarity.

A survey of English domestic stairs had a mean going of 222 mm (Roys and Hamilton, 1998), with a mean rise of 190 mm. However, the distribution is quite narrow, with 88.9% of the sample having a going between 200 and 250 mm. A consequence of this is that only 4.7% of the sample had a mean going of over 250 mm. The current recommendations for private buildings (dwellings) in the Approved Document K (DETR, 1998) are for a minimum going of 220 mm, and a maximum rise of 220 mm.

## Conclusions

Stairs in the home are involved in over 500 deaths in England and Wales each year. It is suggested that a major cause of these incidents is the stair size. This experiment has shown people manage to put a significantly greater part of their foot on a tread for goings 300 mm and above compared to goings of 275 mm or less. The 60 test participants also feel significantly safer for goings 275 mm and above compared to goings of 250 mm or less (see Table 2, statement 7). The subjective and objective data also show that generally rises of 200 mm or larger are less favourable than smaller rises. It is suggested that for private dwellings the minimum recommended going should be increased to 250 mm, and the maximum recommended rise should be decreased to 200 mm.

## References

DETR 1998, *The Building Regulations 1991, Approved Document K. Protection from falling, collision and impact,* (The Stationery Office)

Peebles, L. and Norris, B. 1998, *Adultdata: The Handbook of Adult Anthropometric and Strength Measurements – Data for Design Safety*, (DTI Pub 2917/3k/6/98/NP)

Roys, M.S. 2001, Serious stair injuries can be prevented by improved stair design, *Applied Ergonomics*, **32**, 135–139

Roys, M. and Hamilton, L. 2000, *Stair Questionnaires: A risk assessment*, BRE Client Report 76151

Roys, M. and Wright, M. 2003, *Proprietary nosings for non-domestic stairs*, BRE Information Paper, IP 15/03 ISBN 1 86081 6525

Roys, M. and Wright, M. 2005, *Minor variations in gait and their effect on stair safety*, current proceedings

# REVIEW OF IN-PATIENT FALLS ASSOCIATED WITH HOSPITAL BED RAILS

**Sue Hignett[1], Tash Masud[2] & Sam Jansen[1]**

[1]*Hospital Ergonomics and Patient Safety Unit, Dept. of Human Sciences, Loughborough University, LE11 3TU*
[2]*Clinical Gerontology Research Unit, Nottingham City Hospital NHS Trust, NG5 1PB*

This paper will report on the use of bed rails in a medical directorate of a large acute teaching hospital in the UK. In 1975 a national recommendation was made that bed rails should not be used routinely. This should have resulted in an opt-in culture, whereby the use of rails was subject to the assessment of individual clinical status. A twelve month audit in 12 medical and rehabilitation wards in a UK hospital identified that falls around the bed area comprised the largest category (43%) of falls incidents. A survey was carried out over a 24 hour period to investigate the use of (a) bed rails using observations and (b) the risk assessment procedure using document retrieval. It was found that the average rail use was 29%, with 46% of rails in use between 22.00–01.30 compared with 18% between 10.00–13.00. The reasons for rail use were mostly related to nursing concerns about falls (42%) followed by patient/relative request (22%). Risk assess-ment documentation was available for 71% of the raised rails. Thissuggests that although the rate of rail use is higher than would be expected appropriate assessments had been recorded to support the clinical reasoning.

## Introduction

In 1975 there was a recommendation from the Joint working party of British Geriatrics Society and Royal College of Nursing that bedrails should not be used routinely and that their use should be continually reviewed (Everitt and Bridel-Nixon, 1997). This should have resulted in an opt-in culture with a norm that no bed rails were used unless a full assessment of the individual patient clinical status had been carried out (Frengley, 1999). However approximately 2% of patients fall during their hospital stay, with between 37–90% of falls from the bed occurring with raised bed rails (Mahoney. 1998) and the most common reason for the use of bed rails has been found to be related to a belief in the assistance of rails in the prevention of falls (O'Keeffe et al., 1996).

Donius and Rader (1994) suggested that the use of bed rails can have negative consequences including increasing the distance of falls from the bed, obstructing vision, separating the care receiver from the caregiver, creating noise, causing trauma if the patient's body strikes or becomes entangled in the side rail, dislodging tubes during raising and lowering, and creating a sense of being trapped or jailed. They stressed the need to individualise the use of rails by making repeated assessments for each patient and matching the use of rails to the patient's individual requirements of safety, security and mobility.

There have also been suggestions that bed rails might exacerbate injuries associated with falls from the bed. Van Leeuwen et al. (1994) found, in a retrospective analysis

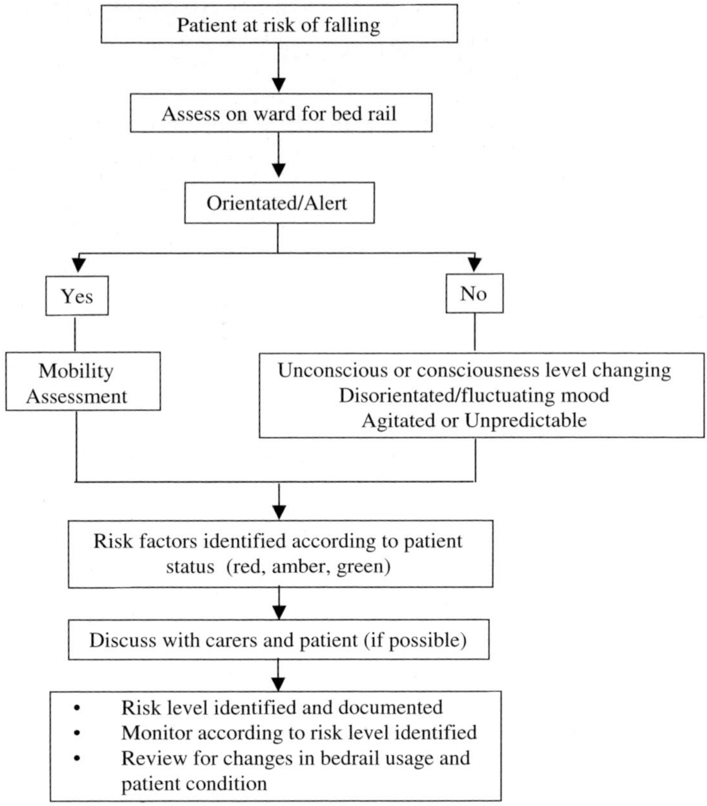


**Figure 1. Risk flowchart for use of bed rails.**

of 419 falls incidents over an 8 year period at one hospital, that the incidence of falls (for all age groups) from beds with raised bed rails was equal to, or higher, than when bed rails were not elevated. Gray and Gaskell (1990) reported that bed rails could raise the height of the bed by 18 inches and described the effect of this "as the conversion of a potential orthopaedic bed (standard fall resulting in a fractured femur) to a neurology bed (fall over cot side resulting in subdural haematoma)". This suggestion is supported by research from Hanger et al. (1999) who evaluated the effect of a change in policy to restrict the use of bed rails in New Zealand. They found that although the intervention was successful in terms of a significant reduction in the use of rails, the falls rate did not change significantly but serious injuries were significantly less common with fewer head injuries.

An audit in 2002/03 at a large acute teaching hospital recorded 955 falls incidents over 12 months, of which 43% were in the bed area (Masud, 2003). Following this audit concerns were raised about falls associated with bedrails which led to the medical directorate producing a risk protocol for (1) patients at risk of falls and (2) the use of bed rails (Figure 1). Risk assessment matrices were produced for both falls and bed rails (Figure 2) and made available on all the wards in the medical directorate.

This study aimed to survey the use of bed rails and associated documentation in the 13 wards of the medical directorate with respect to the risk assessment criteria set out by the hospital.

| Mood/Tolerance (LOW → MEDIUM → HIGH) | IMMOBILE Requires full assistance | LIMITED MOBILITY Requires some assistance i.e. 1-2 nurses | INDEPENDENT Walks frequently |
|---|---|---|---|
| HIGH: Agitated or unpredictable behaviour | BED RAILS MAY BE USED BUT MUST BE REVIEWED EVERY 4 HOURS TO ASSESS REMOVAL OR CONTINUATION | BED RAILS NOT RECOMMENDED | BED RAILS NOT RECOMMENDED |
| MEDIUM: Disorientation fluctuating in mood e.g. anxious or conscious levels | BED RAILS MAY BE USED BUT REVIEW USE EVERY SHIFT i.e. 3 times every 24 hours | BED RAILS MAY BE USED BUT MUST BE REVIEWED EVERY 4 HOURS TO ASSESS REMOVAL OR CONTINUATION | BED RAILS NOT RECOMMENDED |
| LOW: Orientated and alert/or unconscious | BED RAILS RECOMMENDED REVIEW USE EVERY 24 HOURS | BED RAILS RECOMMENDED REVIEW USE EVERY 24 HOURS | BED RAILS MAY BE USED BUT OBSERVE RISK OF UNFAMILIAR EQUIPMENT & ENVIRONMENT FOR PATIENT & REVIEW USE OF BED RAILS EVERY 12 HOURS |

LOW → MEDIUM → HIGH

**Figure 2. Risk assessment scale for use of bed rails.**

## Methodology

The methodology was based on the work of O'Keefe et al. (1996). 12 wards (7 acute and 5 rehabilitation wards, 295 beds) were surveyed over a 24 hour period. A preliminary study was carried out using activity analysis (Jansen, 2002) to identify time trends in bed rail use. It was found that the highest period of use was 22.00–01.00, followed by 16.00–19.00 and the lowest period of use was 10.00–13.00. Ethical constraints limited the collection of data to observable patients. Where curtains were in use or a single room was closed data were not collected. Observations and interviews were carried out on three occasions to collect data on the number of beds occupied, total bed rails in use, subjective reason for use of bed rail from nursing staff and completed risk assessment matrices.

## Results

Bed rails were raised for 29% of the observations (253 from 885 (295 × 3) occasions). The highest proportion of rails were raised between 22.00–01.30 (Figure 3).

The nursing staff were asked why the rails were raised on each occasion. They provided a number of reasons as shown in figure 4. The most likely reason was due to concerns about patient falls (42%) followed by patient/relative request (22%).

The risk assessment documents were checked for the 253 occasions where the rails were raised and found to be available for 180 (71%) of cases.

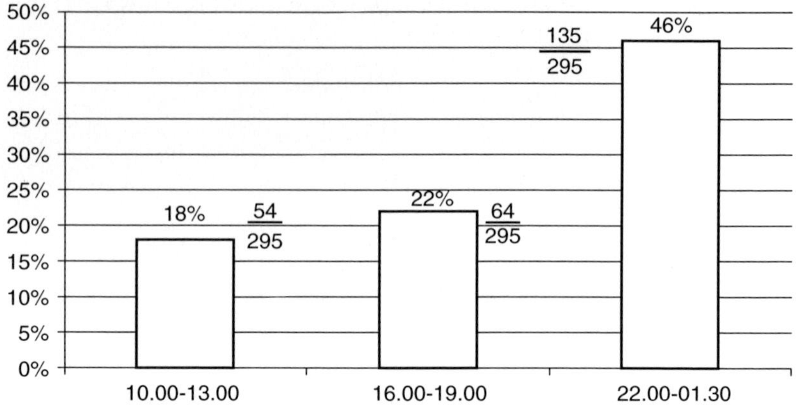


**Figure 3. Frequency of bed rail use by time.**

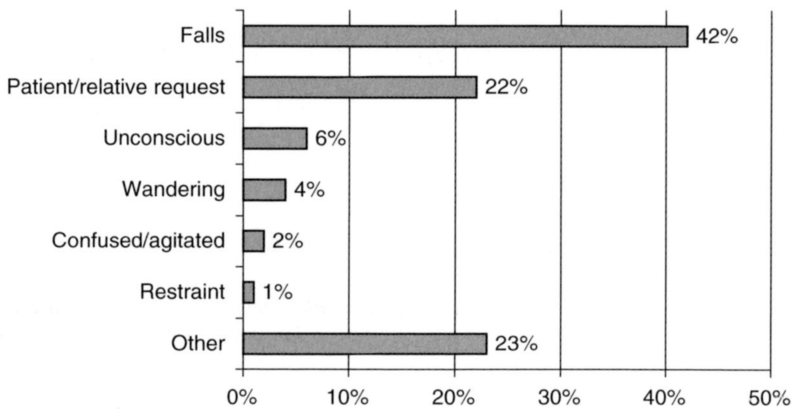


**Figure 4. Subjective rationale for bed rail use.**

## Discussion and conclusion

O'Keeffe et al. (1996) found that 8.4% of rails were raised on the 16 wards in their audit. This was much lower than the rail usage than found in this study (29%). The difference may be due to the first study collecting data on both medical (11%) and surgical (7%) wards whereas this study was only carried out on medical wards. However the O'Keeffe study collected data between midnight and 06.30, the highest usage period for rails whereas this study looked at morning, afternoon and night usage so the difference in rail use is surprising and raises questions about differences in clinical practice and local organisational safety culture. The reasons for rail use given by O'Keeffe et al. are very similar to these findings suggesting that attitudes towards rail use have not changed in the last 10 years.

The risk assessment documentation (either bed rails or falls) was available for 71% of the raised rails. This suggests that although the rate of rail use in this study is higher than would be expected, appropriate patient assessments had been recorded to support the clinical reasoning. In light of recent change to electric beds (Hignett and Lang, 2003) with permanently installed bed rails it would now be interesting to see if there has been a change in the use of bed rails (Hignett and Griffiths, 2005) and whether this has impacted on either the frequency or severity of patient falls.

## References

Donius, M. and Rader, J. 1994, Use of side rails: rethinking a standard of practice. *Journal of Gerontology Nursing*, **20,** 23–27.

Everitt, V. and Bridel-Nixon, J. 1997, The use of bed rails: principles of patient assessment, *Nursing Standard*, **12**, 6, 44–47.

Frengley, J.D. 1999, Bedrails: Do they have a benefit? *Journal of the American Geriatric Society*, **47**, 5, 627–628.

Gray, C.S. and Gaskell, D. 1990, Cot sides: a continuing hazard for the elderly. *Geriatric Medicine*, **20**, 21–22.

Hanger, H.C., Ball, M.C. and Wood, L.A. 1999, An analysis of falls in the hospital: Can we do without bedrails? *Journal of the American Geriatric Society*, **47**, 529–531.

Hignett, S. and Griffiths, P. 2005, Do split-side rails present an increased risk to patient safety? *Quality and Safety in Healthcare* (in press).

Hignett, S. and Lang, R. 2003, Changing to electric hospital beds: A project management case study. *British Journal of Health Care Management*. **9**, 8, 271–276.

Jansen, S. 2003, *Ergonomic investigation into the design, safety and usage of hospital bedrails*. Unpublished M.Sc. Dissertation. (Dept of Human Sciences, Loughborough University).

Mahoney, J.E. 1998, Immobility and Falls. *Clinics in Geriatric Medicine* **14,** 4, 699–726.

Masud, T. 2003, *Audit of Falls in the Medical Directorate*. (Nottingham City Hospital NHS Trust. Internal Report).

O'Keeffe, S., Jack, C.I.A. and Lye, M. 1996, Use of restraints and bed rails in a British hospital, *Journal of the American Geriatric Society*, **44,** 1086–1088.

Van Leeuwen, M., Bennett, L., West. S. et al. 1994, Patient falls from bed and the role of bed rails in the acute care setting. *Australian Journal of Advanced Nursing* **19**, 2, 8–13.

# SETTING UP AN INTEGRATED FALLS AND OSTEOPOROSIS SERVICE IN EAST KENT

**Debbie Janaway[1], Shelagh O'Riordan[1] & Alison Knox[2]**

*[1]Kent and Canterbury Hospital*
*East Kent Hospitals Trust*
*Canterbury, Kent*
*[2]Canterbury and Coastal PCT*
*Chestfield, Whitstable,*
*Kent*

This paper explains how east Kent health and social care providers have addressed standard 6 of the National Service Framework for Older People, to provide an integrated Falls and Osteoporosis Service. The impact of falls and osteoporosis in older people and on the local health economy is discussed. It details the planning, implementation and development processes, illustrating the effect of locally developed screening and assessment tools, multi-disciplinary team working and medically led Falls Clinics. Funding issues and future targets are outlined.

## Introduction

The National Service Framework for Older People (2001), Standard Six: Falls, (NSF) aims "to reduce the number of falls which result in serious injury and ensure effective treatment and rehabilitation for those who have fallen." This standard set out changes needed to reduce the number of falls, prevent fractures through the prevention and treatment of osteoporosis, and improve the diagnosis, care, treatment, rehabilitation and long term support of older people who have fallen.

### *Falls and osteoporosis*

Gibson et al. in 1987 and O'Loughlin, Robitaille, Boivin and Suissa in 1993 define a fall as "an event which results in a person coming to rest inadvertently on the ground or lower level and involving a change of position and lack of intention to do so." 30% of over 65's living in the community and 60% of nursing homes residents fall each year and 40% of admissions to care homes are precipitated by a fall (Cryer and Patel 2001).

Osteoporosis is a progressive, systemic, skeletal disease characterised by low bone mass and microarchitectural deterioration of bone tissue, with a consequent increase in bone fragility and susceptibility to fracture (Royal College of Physicians and Bone and Tooth Society of Great Britain [RCP/BTSGB], 2000). Primary osteoporosis is an age related disease with one in ten men, one in three women over fifty, and half of all women over eighty years affected (NICE, 2004). Each year there are an estimated 200,000 new osteoporotic fractures at a cost of £940 million (RCP/BTSGB, 2000). The most common fracture sites are the wrist, vertebrae and hip. Hip fractures are associated with a mortality rate of 35% over the first year and many sufferers experience long term pain, disability and loss of function and independence (Center et al, 1999).

## Background to the service

In 2001 health and social service leads in East Kent supported the secondment of a nurse post to plan, implement and develop the Falls Service. The NSF describes how agencies and disciplines within primary, secondary and social care, must work together to provide specialist assessment of fall's risk. To reflect the NSF, an inter-agency multi-disciplinary group was formed, representing all four primary care trusts, the acute hospitals trust, social services, the ambulance trust, the mental health trust, local authority and the voluntary and independent sectors. In 2002, a Consultant Geriatrician with a special interest in falls and osteoporosis was also appointed to act as a medical lead.

### *Service objectives*

Research shows that falls in older people may be reduced by 15–60% (Cryer and Patel, 2001). Therefore the objective of the East Kent Falls and Osteoporosis Service is to reduce falls and fractures, particularly hip fractures, by a minimum of 15%. Preventing 15% of the hip fractures that occur locally would save the health economy 1.4 million pounds per annum.

### *Service design*

East Kent has developed four potential levels of screening and assessment, which may be performed on people over 65, who are in receipt of a health or social service.

**Screening Tool 1** – Falls Rapid Assessment Tool (FRAT) is an easily administered, 5 question screening tool (Feder, 2002). A positive response of three or more risk factors indicates a high specificity for a fall in the following six months, and those older people meeting this criteria undergo Screening Tool 2 and the East Kent Osteoporosis Assessment.

**Screening Tool 2** includes questions on the twelve most common risk factors for falling. It addresses risks due to medication, fear of falling, postural hypotension, visual and hearing problems, walking, gait and balance, footwear and foot care and environmental hazards. It states advice that can be given and recommends particular referral options to modify risk factors. People with multiple or unresolved risk factors may be referred for Assessment 3.

**Assessment 3** represents a comprehensive multi-disciplinary assessment incorporating physiotherapy, occupational therapy and nursing components. It is carried out within a Day Hospital setting or by a Community Assessment and Rehabilitation Team within the home environment. A full balance and gait assessment determines risk and may lead to referral to a falls prevention programme. Modification of risk due to home hazards is addressed and referral to a medically led Falls Clinic or Falls Prevention Group occurs when required.

### *Falls clinics*

A study by Close et al. in 1999 entitled the Prevention of Falls in the Elderly Trial (PROFET) showed a 61% reduction in falls, in high risk patients who received a medical assessment and an occupational therapy home visit (OR 0.39 CI 0.23–0.66). In east Kent three geriatricians run Falls Clinics in four hospitals, using Assessment Tool 4. Patients either have multiple unexplained falls and complex problems or are experiencing falls as a result of blackouts (syncope) or dizzy spells. Those with multiple or unexplained causes undergo Assessment Tool 3 prior to attending the clinic.

**Assessment 4** includes a physical examination, falls history, osteoporosis assessment and mobility assessment. Medication is reviewed, concentrating on those drugs known to cause falls, such as, anti-hypertensives, cardiac drugs, anti-depressants, anti-psychotics, sleep inducing drugs and opiates. Causes of the patient's falls are determined, and appropriate referrals, for example, to an opthalmologist for surgery to remove cataracts, are made. Patients may also be referred to a Falls Prevention Group within East Kent Day Hospitals, where they receive a six week course of exercises to improve proximal strength and balance, and education about falls prevention strategies.

It is estimated that 10% of falls in older people are caused by blackouts (Rubenstein et al., 1996). These patients undergo a lying and standing blood pressure check and physical examination, concentrating on the cardiac system. If significant postural hypotension is found medication is reviewed. If no cause is found the patient is referred for carotid sinus massage and head up tilt table testing to assess for carotid sinus hypersensitivity or vasovagal syncope. Positive results are treated by modification of drugs or a pacemaker respectively. Some falls are caused by cardiac arrhythmias and require a 24-hour electrocardiogram or cardio-memo to facilitate diagnosis.

### *Osteoporosis assessment*

A standardised algorithm for investigation and treatment of osteoporosis has been agreed by fund holders, clinicians and pathology and radiology service providers. All patients who progress to Screening Tool 2 undergo an osteoporosis assessment as illustrated below.

### *Service development*

Screening Tool 1 and where necessary further screening and assessment, is used for all people aged 65 and over who are being visited by a community nurse, or who are in patients in an acute hospital, community hospital, or mental health in patient unit. Care homes (nursing) are also included. A pilot with Social Services care managers in one PCT is underway. Screening has been implemented in the local Emergency Care

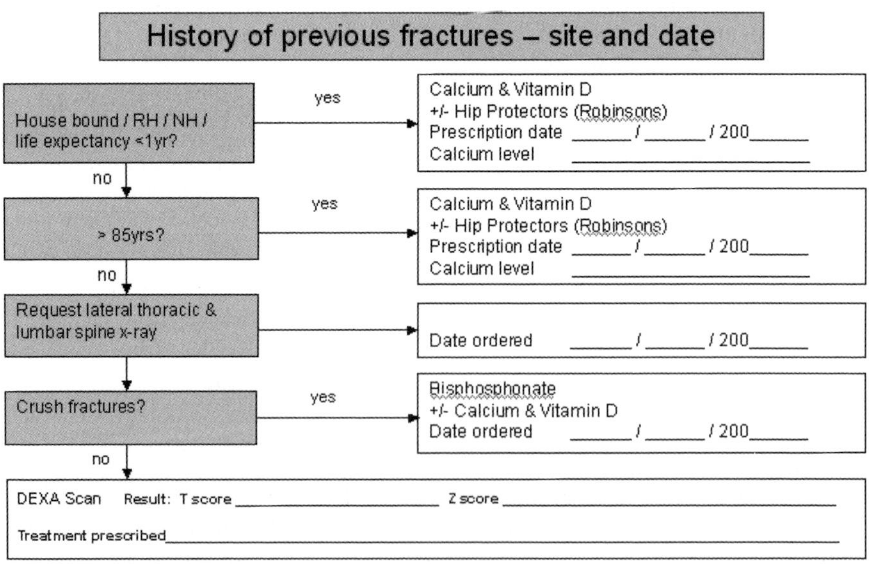

Centres with limited success and attempts are therefore being made to introduce targeted screening for those most at risk. A nurse-led service has facilitated the screening and assessment of appropriate patients attending Fracture Clinics with a fragility fracture. This service includes access to DEXA scanning with recommendation for drug treatment being communicated directly back to the patient's General Practitioner.

### *Hip protector service*

Hip protectors are specially designed garments worn external to underpants. They are designed to divert or absorb the force of an impact on the hip, therefore preventing hip fractures (Cryer and Patel, 2001). The East Kent Falls and Osteoporosis Service are currently developing a hip protector service. The service will offer older people education and demonstrations regarding hip protectors and a non-profitable ordering service. Open access will provide ongoing support for wearers. Providing such a supportive service where usage is actively promoted has been proven to achieve high compliance (Knox, 2002).

### *Developments with the general medical services (GMS) contract*

Service leads have, following negotiation with local GP's, developed an enhanced service for falls and osteoporosis. This includes the compilation of Falls Registers in practices, to highlight older people who have fallen or sustained a fragility fracture and facilitate the use of the available screening tools for targeted groups. An additional benefit of this service is in the maintenance of an electronic record, accessible to other health care workers.

### *Clinical governance*

A clinical governance framework is firmly embedded in all aspects of its service delivery. Each development is evidence based and local research into fragility fractures and patient and public involvement is underway. Education and training includes the development of a local falls prevention module at diploma, degree or masters level and forms part of a unit for the National Vocational Qualification in Care. Information leaflets about falls and osteoporosis and hip protectors are being developed to enhance the provision of information for the local population. A clinical audit programme, assessing the use of the screening tools across the acute trust is in progress and a community audit will follow in 2005.

### *Future implementation within health and social services*

The use of screening tools will be implemented in the future within residential and nursing homes, community specialist nursing teams, older people with mental health needs attending Day Hospitals, the Ambulance Trust, Local Authority and the Voluntary Sector.

### *Future funding*

The service successfully bid for funding for three nurses, working across the acute trust, in 2004. The current application is for the provision of exercise physiologists, to facilitate widespread dissemination of muscle strength and balance training programmes, which is recommended by NICE (2004). This funding bid has involved representatives from the East Kent Falls and Osteoporosis Service, Health Promotion and Public Health.

## Conclusion

Progress to date is significant and meets the milestones laid down in the falls standard of the National Service Framework for Older People, although it is anticipated that full implementation will go beyond April, 2005. The clear message in East Kent for all health and social care workers caring for people over 65, is that careful consideration of the risk of falls and osteoporosis must be made to prevent falls and resultant fractures in older people.

## References

Carter, Y. Close, J. Crown, J. Cryer, C. Donovan, S. Eldridge, S. Feder, G. (Principle Investigator). Nandy, S. Parsons, S. Rashbrook, E. Taylor, S. Underwood, M. Spencer, M. 2002, *Report: G*, Implementation of multi-disciplinary guidelines for the prevention of falls in older people: a pilot study

Close, J. Ellis, M. Hooper, R. Glucksman, E. Jackson, S. Swift, C. 2002, Prevention of falls in older people trial (PROFET): a randomised controlled trial, *Lancet*, **353**, 93–97

Cryer, C. and Patel, S. 2002, *Falls, Fragility & Fractures: National Service Framework for Older people: The case for and strategies to implement a joint Health Improvement and Modernisation Plan for falls and osteoporosis*, (Proctor and Gamble, London)

Department of Health. 2001, *National Service Framework for Older People*

Gibson, M.J. Andres, R.O. Isaacs, B. Radebaugh, T. Worm-Peterson, J. 1987, The prevention of falls in later life. *Danish Medical Bulletin*, **34**, 1–24

Knox, A. 2002, Introducing hip protectors to older people in the residential home setting within care homes. *AGILITY*, Autumn/Winter, 6–9

Leipzig, R. Cumming, R. Tinnetti, M. 1999, Drugs and Falls in older people: A systematic review and meta-analysis: 1 psychotropic drugs. *Journal of the American Geriatric Society*, **47**, 30–39

National Institiute for Clinical Excellence. 2004, *Appraisal consultation document: Secondary prevention of osteoporotic fragility fracture*

National Institute for Clinical Excellence. 2004, *Clinical practice guideline for the assessment and prevention of falls in older people*

O'Loughlin, J.L. Robitaille, Y. Boivin, J-F. Suissa, S. 1993, Incidence of and risk factors for falls and injurious falls among the community dwelling elderly. *American Journal of Epidemiology,* **137**, 342–354

Royal College of Physicians/ Bone and Tooth Society of Great Britain. 2000, *Osteoporosis: Clinical guidelines for the prevention and treatment*

Rubenstein, L.Z. Josephson, K.R. Kenny, R.A. 1996, *Falls in syncope in the older patient.* (Chapman and Hall, London)

Tinnetti, M. Baker, D. McAvay, G.A. 1994, Multifactorial intervention to reduce the risk of falling among elderly people living in the community. New England Journal of Medicine, **331**, 821–827

# INFLUENCE OF PAVEMENT DESIGN PARAMETERS IN SAFETY PERCEPTION IN THE ELDERLY

**Enrique Alcántara, Tomás Zamora, Rakel Poveda, Ricard Barberà, Francisco Matey, Carme Gimeno, Javier Sánchez & Jaime Prat**

*Instituto de Biomecánica de Valencia, Universidad Politécnica de Valencia, edificio 9C, Camí de Vera s/n. E-46022, Valencia, SPAIN*

People modify walking strategy whenever they perceive a floor as slippery independently of real friction properties. However, although functional aspects as friction have been widely studied, there are very few studies on emotional aspects. This paper presents a conjoint analysis of the influence of flooring design elements in safety perception in the elderly. Results show that texture was the most influent parameter. The presence of noticeable joints was revealed as an unexpected factor to transmit security also. Adding these results to the functional aspects it is expected to increase accessibility improving the quality of life and reducing the fear of falling in elderly.

## Introduction

Fatal slips and fall are very frequent in the elderly (Agnew and Suruda 1993). In fact, fall injuries are the main cause of death among people elder than 75 years (National Safety Council 1998, Lockhart Th. et al., 2003). With this respect, many research efforts have been devoted to understand and reduce the risk of falling in the elderly.

It is admitted that there exist multitude and complexity of extrinsic, intrinsic and mixed risk factors for slipping and falling (Grönqvist 1999). Precisely, elderly people are likely to present important intrinsic factors such as limitations in motor control, balance, vision, lower extremity force and others. Considering extrinsic factors, flooring has a clear influence and many researchers have investigated the coefficient of friction as an indicator of flooring safety (Chang et al., 2001). As a result, there are many different testing methods to measure this parameter (Durá JV. et al., 2003).

On the other hand, people has the ability to accommodate to friction conditions in a way that people adopt a gait strategy that mixes force and postural changes for example, reducing or augmenting step length, speed or surface contact (Llewellyn and Nevola 1992), to reduce the demanded friction by decreasing vertical acceleration and forward velocity of the body (Llewellyn and Nevola 1992). Elderly have lower adaptation ability and so are less able to respond when friction conditions are different than expected.

This suggests that people's perception of flooring safety is determinant in the walking strategy adopted for a given floor and thus, influences the risk of falling. However, the influence of flooring design in safety perception has not been addressed before. The purpose of this study was to investigate the influence of several flooring design parameters in elderly perception of walking safety or risk of falling.

**Table 1. Floor parameters and their levels considered for the study.**

| Parameters | Texture | Joints presence | Colour tone | Luminosity | Surrounding ambient |
|---|---|---|---|---|---|
| Levels | Shiny | Noticeable joints | Clear | Good | Bath/Kitchen Pool/Commercial centre/Stairs/sidewalk |
| | Not Shiny | Not noticeable joints | Dark | Poor | |
| | Textured | | | | |

## Methodology

### *Subjects*

One hundred and twenty persons older than 55 (balance distribution of gender, age and living environments) took part in the experiments. People from 3 different living environments were considered for the study: Residence, Day Centres and Assisted Homes. Age was divided into two ranges, between 55 and 75 years old and older than 75.

### *Material*

The Conjoint analysis (Louviere J.J., 1988) was used to quantify the influences of several design parameters in safety perception. Design parameters were extracted from previous experts and users discussion panels. They included texture, joints, colour, profile morphology and the type of ambient where the floor sample was integrated. Levels for each parameter are presented in Table 1. Regarding ambient, bath and kitchen were considered for assisted homes, swimming pool and commercial centre for day centre and walkway and stairs for residence group respectively.

An orthogonal design of experiments was done generating 8 samples for each habitual resident group combining the selected parameters and levels. The conjoint modulus included in the SPSS 11.5 for windows was used. Each combination defined the design of real pictures. Actual elderly volunteers were included in the pictures to stimulate subjects' answers. All pavements in the pictures were virtually modified to randomly include the 8 combinations.

Conjoint analysis issued first the importance of each design parameter in safety perception. Second, the weight of every alternative for each design element is obtained.

Conjoint also computes the regression between perceived security and design variables (level value). Thus, adding the contribution of each parameter (level x importance) and a constant value given by the analysis, it is easy to estimate the perception for any of the possible combinations of design elements.

### *Procedure*

Pictures were shown to the subjects using the method conjoint that asks subjects to order necessarily all the pictures from the best to the worst according to the sensation asked. Subjects were not informed about the differences between each picture even when they asked about. In this sense, answers were based on a global view of the parameters without realising that were combined in a controlled way.

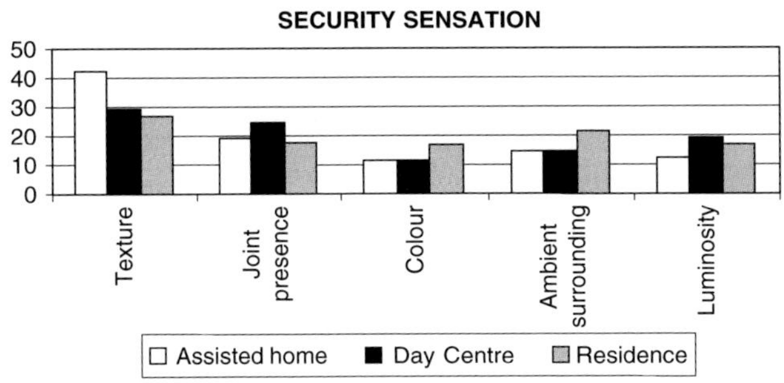


**Figure 1. Importance for each design parameter is presented by distinguishing the 3 selected groups: Assisted home, Residence and Day Centre residents.**

**Table 2. Importance of each design parameter and the influence according to the levels considered. Also it is introduced the constant value necessary to extract the security level perceived for whatever possible combination of parameters considered.**

| | | Assisted homes | | Day Centres | | Residence | |
|---|---|---|---|---|---|---|---|
| | | Import. | Influence | Import. | Influence | Import. | Influence |
| Texture | Not Shiny | 42.28 | −0.184 | 29.31 | 0.159 | 26.57 | −0.0778 |
| | Shiny | | −1.4608 | | −0.9218 | | −0.5861 |
| | Textured | | 1.6448 | | 0.7628 | | 0.6639 |
| Joints | Not Not. Joints | 18.91 | −0.4654 | 24.81 | −0.7538 | 17.52 | −0.575 |
| | Noticeable Joints | | 0.4654 | | 0.7538 | | 0.575 |
| Colour | Clear | 11.65 | 0.0182 | 11.65 | −0.0269 | 17.2 | 0.6667 |
| | Dark | | −0.0182 | | 0.0269 | | −0.6667 |
| Luminosity | Poor | 12.18 | −0.2843 | 19.6 | −0.6808 | 17.16 | −0.5875 |
| | Good | | 0.2843 | | 0.6808 | | 0.5875 |
| Surrounding Ambient | Bath | 14.97 | 0.3528 | – | – | – | – |
| | Kitchen | | −0.3528 | | – | | – |
| | Pool | – | | 14.63 | 0.1923 | – | – |
| | Commercial Centre | | – | | −0.1923 | | – |
| | Stair | – | – | – | – | 21.56 | −0.7458 |
| | Sidewalk | | – | | – | | 0.7458 |
| Constant | | 4.4749 | | 4.4603 | | 4.5144 | |

## Results

Figure 1 shows the level of importance for each design parameter in the three groups. Texture was considered the most important for all the groups whereas colour was the less influent in day centres and assisted homes and luminosity for the residence.

On the other hand, the level of importance differs for the selected groups as can be observed for the texture importance, where assisted home residents gave much more importance to this design parameter when compared to the other groups.

Table 2 includes the value of the influence obtained by each level of the different design parameters. As it can be observed, a level for a given parameter, as for example texture, can have opposite influence for the studied groups. In this case, a not shiny floor had a negative influence for assisted homes and residences whereas it was positive for day centres.

The design parameter called Surrounding ambient reveals interesting results. The importance given to this parameter is higher for the residence residents than for the other groups. Observing the results, stairs are the most negatively perceived ambient followed by the kitchen and the commercial centre. Bath perception registered an unexpected positive influence probably due to the fact that they had been adapted for the elderly. The same happens with sidewalk perception. However, this was contradictory as it is in these two ambient where occur most of the falls reported by the elderly. That could result in an excess of confidence as an important risk factor for suffering a fall.

Observing the table 2, the most secure perceived for assisted homes residents for their baths pavement would be:

| Texture Impor. | Textured level | Joint Impor. | Noticeable joint level | Colour Impor. | Clear Level | Luminosity Impor. | Good level |
|---|---|---|---|---|---|---|---|
| 42.28× | 1.6448+ | 18.91× | 0.4654+ | 11.65× | 0.0182+ | 14.97× | 0.2843+ |

| Const | Result |
|---|---|
| 4.4749= | 87.21 |

While the less secure perceived floor would be:

| Texture Impor. | Shiny level | Joint impor. | NOT Noticeable joint level | Colour Impor. | Dark Level | Luminosity Impor. | Poor level |
|---|---|---|---|---|---|---|---|
| 42.28× | −1.4608+ | 18.91× | −0.4654+ | 11.65× | −0.0182+ | 14.97× | −0.2843+ |

| Const | Result |
|---|---|
| 4.4749= | −57.87 |

## Discussion and conclusions

The results of this work show that design of floorings has an influence in safety perception of elderly in a way that they can induce or not adaptation. In this sense, the floor has not only to be perceived as safe but also be it. In case a slippery floor perceived as safe could be catastrophic as people would not adapt their walking pattern to avoid falling.

From the design elements considered in this study, as it could be expected, texture was the most influent in a way that a shiny floor is perceived always as dangerous whilst an even smoothly profiled texture transmits security. Other parameters did not influence in the same way on each of the selected groups, as for the colour which dark level was considered positive for the day centre residents in opposition to the assisted homes and residence ones revealing that it is important to consider the user and their common ambient. Surrounding ambient also registered important considerations that could help to explain why baths and sidewalks register an increased elderly fallings number.

Finally conjoint methodology is capable to estimate in a quantified manner the safety perception of each possible combination of floor parameters. In this case, (by their importance) a textured surface with noticeable joints, clear tone colours and good luminance levels has been estimated as the most secure perceived. In opposition a shiny finish, a dark colour without noticeable joints and a poor luminance level was estimated as the most dangerous perceived.

## References

Agnew, J. and Suruda, A.J., 1993, Age and fatal work-related falls, *Human factors* 35, 731–736

Chang, W-R., Grönqvist, R., Leclercq, S., Myung, R., Makkonnen, L., Strandberg, L., Brungraber, R., Mattke, U. and Thope, S., 2001, The role of friction in the measurement of slipperiness, Part 1: Friction mechanisms and definition of test conditions. *Ergonomics* 44, **13**, 1217–1232

Cohen, H. and Cohen, D., 1994, Perception of Walking Surface Slipperiness Under Realistic Conditions, Utilizing a Slipperiness Rating Scale, *Journal of Safety Research* 25 **(1)**, 27–31

Durá, JV., Zamora, T., Rosa, D, and Balaguer, E., 2003, The development of a new portable friction testing device, communication for the *Proceedings of IEA 2003*

Grönqvist, R., 1999, Slips and Falls in S Kumar (ed.) *Biomechanics in ergonomics* (London: Taylor & Francis), 351–375

Lockhart, T.E., Woldstad, J.C. and Smith, J.L., 2003, Effects of age-related gait changes on the biomechanics of slips and falls. *Ergonomics*, 12 (**46**): 1136–1160

Lotens, W.A. and Havenith, G. (eds), Proceedings of the Fifth International Conference on Environmental Ergonomics , Maastricht, The Netherlands, 156–157

Louviere, J.J., 1988, Analysing decision making: Metric conjoint analysis, *Sage University Paper* series on Quantitative Applications in the Social Sciences 67, Beverly Hills, California : Sage

National Safety Council, 1998 (NSC), Accident Facts (Itacsa, IL: NCS)

# SLIPS, STRIPS AND FALLS IN THE MARITIME SECTOR

**Brian Sherwood Jones[1] & William Moore[2]**

[1]*Lloyd's Register of Shipping, 71 Fenchurch Street, London EC3Y 4BS, England*
[2]*American Steamship Owners Mutual Protection and Indemnity Association, Inc., 60 Broad Street, 37th Floor, New York, NY 10004, USA*

Slips, trips and falls represent a major risk to seafarers and other ship workers. The poster will set out the current understanding of the hazards in the various operational contexts, and identify risk reduction activities. The objectives are to engage specialists in the slips, trips and falls community with the maritime sector to promote a fuller understanding of both hazards and control measures, and to better inform initiatives within the sector.

## Summary

Roberts (2002) found that trawler fishing and merchant seafaring are the two most dangerous occupations in Britain. Falls contributed significantly to deaths and injuries. Similar patterns are found internationally. Stevedoring is the ninth most hazardous occupation in Britain. Falls contribute significantly here also. Container operations are a particular concern. Slips, trips and manual handling dominate shipbuilding and repair accidents. Guidance exists on aspects of ship design (container operations, stairs, ladders and materials handling), dockside operations, and ship operations such as mooring. Training materials are also available.

## Short bibliography

American Bureau of Shipping 2003, *Guidance Notes for the Application of Ergonomics to Marine Systems* http://www.eagle.org/default.html

American Club 2004, "*Currents*" Newsletter Issue 18. http://www.american-club.com

Australian Maritime Safety Authority 2004, *OH&S (MI) Act Incident Reports for Period 1998/99–2002/03*

Pacific Maritime Association 1998, 2001, Safety Bulletins 98-21, 01-03 *Lasher Safety* http://www.pmanet.org/index.cfm

Roberts S.E 2002, *Hazardous occupations in Great Britain*, The Lancet, Volume 360, Number 9332, pages 543–44

Safer Ports Initiative 2004, *Guide to Preventing Slips Trips and Falls* and *Vessel Design Considerations for Stevedore Safety* Ports Skills and Safety Limited http://www.safeports.org.uk

West Country Club 1994, 1996, *Gangways and Accommodation Ladders* and *Accidents during Mooring Operations* at www.westpandi.com

# SLIPS, TRIPS AND FALLS – TRIBOLOGY

# THE TRL PENDULUM SLIP RESISTANCE TESTER THE REASONING FOR ITS ACCEPTANCE IN THE UK

**Malcolm Bailey**

*Radlett Consultants, 3 Homefield Road, Radlett, Hertfordshire WD7 8PX*

This paper gives the background to the development of the Pendulum and provides the theoretical basis for the correlation which investigators in the UK have found between Pendulum readings and the known accident history of floors. It also considers a number of criticisms of the machine.

## Introduction

The UK has chosen to use the Pendulum Slip Resistance test machine as its main instrument for testing floors to determine their slip resistance characteristics. The UK is not alone in this. Australia and New Zealand have similarly adopted the Pendulum, and a number of CEN standards have already been written around the machine. In this paper some of the reasoning behind their decision is set out.

## History

There are many hundreds of test machines which have been invented to measure friction. In general they give reasonably credible results in the clean dry condition, but as soon as a contaminant such as water is introduced, the machines give a wide range of results for the same floor. More important however is the fact that the majority fail to respond to the presence of the lubricant and give wet results similar to or sometimes even greater than the result in the dry. In any event their wet results generally fail to correlate with the known accident history of the floor.

The Pendulum stands out amongst the majority of the machines in that it clearly does respond to the presence of a lubricant. It was invented by Percy Sigler in the USA (Sigler, 1943, 1948, Sigler et al. 1948) for measuring the slip resistance of floors and footways. It was redesigned by Barbara Sabey (Sabey and Cardew, 1963, Giles et al. 1964) of the UK Transport and Road Research Laboratories in the 1950s and used to measure the Sideways Force Coefficient (SFC) for road surfaces. In the late 1950s/early 1960s the UK version was used by the Greater London Council (GLC) to investigate slipping accidents. They found a good correlation between its results and the known slip history of the floor tested, and adopted it as their primary instrument for testing floors in London.

During the 1970s and early 1980s the Pendulum was used by a relatively small number of investigators of slip accidents in the UK. In the 1980s the UK Ceramic Tile industry introduced the TORTUS, which appeared to be a well designed instrument based on sound technical principles. However it soon became clear that, as with other similar instruments, the TORTUS results frequently suggested that a floor was "safe" where experience had shown that it was not.

Procter at the Health and Safety Laboratories published a paper suggesting that lubrication theory could be the key to the problem. His paper (Procter and Coleman, 1988) suggested that based on this theory both the Tortus and the Pendulum failed to reproduce the characteristics of the lubricating film under the heel when a pedestrian slips. The Tortus developed a critical film thickness which was much lower than that developed by a slipping pedestrian, while the Pendulum developed too thick a film. Unfortunately Procter had used incorrect data for the Pendulum, a fact which was noted by this author (Bailey, 1988). When the correct data was used, the Pendulum's critical film thickness was very close to that developed by a slipping pedestrian. It is for this reason that by pure chance the Pendulum does correlate, as found by the GLC and this author, with the known slip history of the floor surface concerned. The GLC based their results on some 3500 tests, whereas the author can only claim just less than 1000 correlation measurements.

There are two fundamental principles which must be understood when attempting to determine the slip resistance characteristics of a floor surface. The first is that whilst it is tempting to use human test subjects the following factors must be recognised.

Firstly, people as a population require in their natural gait a range of values of slip resistance (Building Research Establishment). Whilst the average is 0.19 (coefficient of dynamic friction) it can vary from as low as 0.08 to as high as 0.35 – it depends on the person. Thus, if a test subject slips in a test all that one can say is that those members of the population who require a similar or greater slip resistance than the test subject when they walk will also be at risk of slipping on that surface.

Secondly, individuals can alter their gait and slip resistance requirements, often subconsciously, as a protective mechanism to prevent a slip.

Thirdly, the development of the lubricating film will depend on how the foot is placed on the floor. Hence pedestrians can sometimes take several strides across a floor seemingly safely and then slip on the next stride.

Fourthly, slipping is a very sudden and unpredictable event and in many cases a subject will have no concept as to how close to slipping over he or she has come.

Based on the above it will be apparent that the only likely means of accurately measuring the slip resistance of human subjects is to arrange for them to slip when their foot lands on a sensitive force plate.

The second fundamental principle is that a test machine must either be such that its parameters, e.g. the loads applied, velocity of slider, area of contact, etc., are identical to that of a pedestrian, that is a scale factor of 1.0, or the scale factor for each parameter is carefully chosen in accordance with the engineering principles used in both the aeronautical and civil engineering fields when testing small scale models. In thesc technologies one usually needs to identify a particular factor which must remain constant in both test and real life situations and adjust the various test scale factors to achieve this. It will usually result in the scale factors applied to, say, the load on the slider being completely different from that used for the velocity of sliding, and indeed some of the factors may need to be greater than 1.0 to compensate for those which need to be less than 1.0.

Because slip resistance test machines are generally required to be portable, it is almost inevitable that at least one of the parameters, for example the vertical load applied during the test, will be significantly less than the real life situation. Hence, model test principles must apply. The reason why the majority of test machines fail to give the right result in wet conditions is that the scale factors have been chosen on the basis of expediency in the overall design of the machine rather than on the principles of model testing.

It is the author's opinion that Procter was indeed correct in suggesting that the critical thickness of the hydrodynamic lubricating film was the key. If the test machine develops the same critical film thickness during its operation as a slipping pedestrian, then it will generate the same level of slip resistance as the pedestrian in wet conditions.

If the hypothesis is correct two things follow. Firstly, the results from the machine should correlate with the known accident history of the floors tested; secondly, two machines designed on different operational principles and with different parameters but with both developing the same critical film thickness, should give similar results in wet conditions.

SlipAlert, a recently developed slip resistance testing machine, was designed on these principles. Although based on different operational principles and using different parameters of load, velocity, slider contact area, etc., it was designed to generate the same critical film thickness as both the Pendulum and a slipping pedestrian. The results of tests carried out by this author, the UK Health & Safety Laboratory and Richard Bowman in Australia show that there is indeed a good correlation with the Pendulum.

There are therefore reasonable grounds for considering Procter's hypothesis to be correct.

## Criticisms of the Pendulum

Various criticisms of the Pendulum have been raised in the sixty years since it was invented and these need to be answered.

### *The Pendulum was designed to measure the skid resistance of a car travelling at 30 mph*

The UK Pendulum was designed to measure the Sideways Force Coefficient which is a totally different concept to the Braking Force Coefficient. It happens to do so quite effectively, but as with the way it develops the correct hydrodynamic film thickness for pedestrian slipping applications it got there more by accident than design. This can be seen by reading any of the relevant papers by TRRL and Percy Sigler.

### *The Pendulum arm sweeps across the surface at a speed far faster than the movement of a pedestrian's heel in a slip situation. It thus cannot represent what happens in a slip*

This is correct but as previously stated, in a "model" situation some scale factors may need to be increased to compensate for the decrease in the scale factor of other parameters. In this case the velocity is increased to compensate for the reduction in vertical force.

### *The Pendulum gives incorrect results for Altro safety floor as published by the GLC in their bulletins. We know Altro flooring is safe, thus the Pendulum must be wrong*

The test results to which this refers are for the "as new" material. It is only when the material wears that the embedded carborundum particles become exposed and provide the significantly higher slip resistance. When tested in this latter state, the Pendulum gives much higher and quite realistic values.

### *The Pendulum does not correlate with other machines, in particular the DIN ramp method*

This criticism presupposes that these other machines are giving the "correct" answer. As far as the author is aware, there has been no authoritative study carried out on the scale of either the GLC work or the author's own work on correlation between the history of the floor and the test results for these other machines or methods. The DIN ramp has the disadvantage that it is laboratory based and it is not normally possible to test the floor as laid and as worn. Whilst the ramp, particularly when used under the UK HSL protocols, can give useful results, there is no fundamental reason to consider it to be the ultimate standard.

*The Pendulum does not always discriminate between two different floorings*

This relates to the perception of slipping as seen by the critic. The Pendulum is a blind tester and may well rate two materials equally. The human problem of perception is that much of the information taken in by the senses is processed subconsciously and the results of a test under blindfold conditions can give very different results in respect of how the user rates the floor compared with when he can see it.

*The Pendulum cannot be used on profiled floors*

With care and with an experienced operator, who understands how profiles interact with a pedestrian heel in a slip situation, the Pendulum can give not only credible results but also valuable information on the directional properties of the profiles.

*It sometimes gives unexpected results*

The problem normally lies in the preconception of what the result should be. This author has found that his own knowledge and understanding of slipping has increased significantly with careful investigation of apparent inconsistencies in results.

In the author's experience the criticisms of the Pendulum generally stem from a lack of understanding of the principles behind slip resistance measurement in wet conditions and are frequently voiced by those with no actual experience of using the instrument. In the UK, following publication of both BS 7976 (the Pendulum Standard) and the UK Slip Resistance Group "Guidelines for the operation of the Pendulum", there is a growing number of users of the instrument who have expressed confidence in the instrument. This has now extended to the UK Tile Association who have published their own document "Slip Resistance of Hard Flooring".

## Conclusions

The Pendulum has survived countless criticisms over the past forty years without any major paper being published in its defence. That in itself says much for the instrument.

While the author would not claim that the Pendulum is the ideal instrument for assessing the slip resistance of a floor/walkway surface, it is virtually the only instrument available which replicates the properties of the lubricating film which is responsible for causing pedestrians to slip on many surfaces in wet conditions.

## References

American Standard for Testing and Materials 1983, *Standard Method for Measuring Surface Frictional Properties using the British Pendulum Tester,* (ASTM Designation: E303-83), 595–599

Bailey, M. 1988, The Measurement of Slip Resistance of Floor Surfaces, *Construction and Building Materials,* **2**

Giles, C.G., Sabey, B.E. and Cardew, K.H.F. 1964, *Development and Performance of the Portable Skid Resistance Tester,* HMSO Road Research Technical Paper No 66, (HMSO, London)

GLC 1971, Bulletin 43 (2nd Series) Item 5 Slip Resistance of Floors, Stairs and Pavings, (GLC Department of Architecture and Civil Design)

GLC 1985, Bulletin 145 Item 2 Sheet safety surfaces for floors, methods for evaluation, *Building Technical File,* **9**, 16–22

Harper, F.C., Warlow, W.J. and Clarke, B.L. 1961, *The forces applied to the floor by the foot in walking, 1. Walking on a level surface,* National Building Studies Research Paper, **32** (HMSO, London)

Procter, T.D. and Coleman, V. 1988, Slipping tripping and falling accidents in Great Britain – present and future, *Journal of Occupational Accidents,* **9**, 269–285

Sabey, B.E. and Cardew, K.H.F. 1963, *The Relation between Coefficient of Friction and Skid Resistance as measured with the Portable Tester,* Road Research Laboratory/Laboratory Note LN/331 (Road Research Laboratory, Crowthorne)

Sigler, P.A. 1943, *Relative Slipperiness of Floor and Deck Surfaces*, US Department of Commerce, National Bureau of Standards, Building Materials and Structures Report BMS 100, (US Government Printing Office, Washington)

Sigler, P.A. 1948, Slipperiness of Walkway Surfaces, *Soap and Sanitary Chemicals,* 148–149, 157

Sigler, P.A., Geib, M.N. and Boone, T.H. 1948, Measurement of the Slipperiness of Walkway Surfaces, *Journal of Research of the National Bureau of Standards,* Research Paper RP1879, **40**, (US Government Printing Office, Washington), 339–346

# COMBINED EFFECTS OF FRICTION AND WEAR BEHAVIORS ON FALL SAFETY MEASURES

**In-Ju Kim**

*School of Sport and Health Sciences, University of Exeter, Heavitree Rd, Exeter, EX1 2LU, UK*

Slip resistance properties are significantly influenced by the joint effects of friction and wear behaviors of shoes and floors. Despite the importance of tribo-physical phenomena, this issue has not been explored in-depth. Main objective of this study is, therefore, to (1) develop a tribo-physical model, which is based on common mechanisms in friction and wear behaviors, (2) validation of the theory model and (3) identify their mutual effects on slip resistance properties. To visualize friction and wear evolutions, three-dimensional images of the shoes and floors are obtained by a stereo scanning electron microscope before and after the tests. All the results clearly identify that the combined effects of friction and wear phenomena were so obvious on both the shoes and floors and made significant effects on the slip resistance performances.

## Introduction

Fall safety analysis between the footwear and underfoot surface is an extremely difficult task in terms of principal understanding on the underlying mechanics and mechanisms of tribo-physical phenomena. Because friction and wear behaviors between the footwear and underfoot surfaces involve a range of sub-mechanisms and continuously change, an averaged format of friction coefficient (COF) measures seem to oversimplify the nature of multi-factorial characteristics of tribo-physical phenomena (Grönqvist, 1995; Kim, 2004). That is, any slip resistance measures (i) have peculiar characteristics to a specific combination of shoe-floor-environments and (ii) irregularly change during entire measuring periods (Kim and Smith, 2000). Moreover, their behaviors are significantly related with surface interactions, which cause to significant changes of both the surface structures. This aspect seems to add more difficulty in the interpretation of slip resistance measurement results. Hence, it is quite vital to thoroughly understand tribo-physical characteristics of the shoes and the floors and their mutual effects on slip resistance properties. To achieve this goal, a new insight may require to systematically identify those factors of friction and wear development. Among various possibilities, geometrical arguments such as surface structures and interlocking mechanism are fully analyzed to recognize major tribo-physical features of the shoes and floors during slip resistance measures. Because surface topographies of the shoes and floors are largely modified throughout the course of wear induced sliding friction, this may largely affect tribo-physical properties and accordingly be one of the most important factors to affect slip resistance performances. In order to claim the argument, this study is focused on exploring main tribo-physical characteristics of slip resistance properties in terms of friction and wear characteristics. Based on this paradigm, a new tribo-physical model is proposed to identifying integrated effects of friction and wear events of the shoes and floors. The model is also verified by a dynamic friction test.

## Theory

### *Background*

It has been generally recognized that there are two basic mechanisms in friction and wear behaviors, namely, abrasion and adhesion. Abrasive friction and wear is a result of ploughing of hard body asperities into a soft body surface whereas adhesive friction and wear is a result of sequential creation and rupture of molecular bonds between two mating bodies. During repeated sliding courses, surface structures of both bodies are largely changed as a result of combined effects of friction and wear phenomena. Based upon those concepts, the following main hypotheses are considered for a model development:

(1) Friction and wear event between the shoes and the floors is elasto-plastic, but nearly plastic deformation,
(2) Deformation is concentrated on the surface areas of shoe heels, whose elasto-plastic modules are ten times or more less than them of the floor surfaces, and
(3) Overall tribo-physical behaviors of the shoe-floor pair are largely influenced by surface structures of the floors.

### *A tribo-physical model*

A traction force between the shoes and floors could be regarded as a tangential force ($\boldsymbol{F_{adh}}$) required to overcome the adhesion at regions of intimate contact between them plus the tangential force ($\boldsymbol{F_{def}}$) required to lift the asperities over each other.

$$\text{Traction Force} = F_{adh} + F_{def} \tag{1}$$

Suppose that initial surfaces of both the bodies have Gaussian height distributions. When two such surfaces are brought into contact, the contour density, ***Dc***, is given by

$$Dc \propto \frac{1}{\sigma^2} \exp\left(\frac{-z^2}{2\sigma^2}\right) \tag{2}$$

where $\sigma^2$ is a variance of the combined height distributions of the two surfaces and $\boldsymbol{Z}$ is a separation of the mean planes of the surfaces. For moderate loads (that is, $\boldsymbol{Z}/\sigma > 1$), the above equation (1) closely represents the density of contacts between the two surfaces. Thus, the number $N$ of contact for a nominal area $\boldsymbol{A_n}$ becomes

$$N \propto \frac{A_n}{\sigma^2} \exp\left(\frac{-z^2}{2\sigma^2}\right) \tag{3}$$

If the average area of each individual contact between the shoe heel and floor surface is $\boldsymbol{a}$, then the real contact area, $\boldsymbol{A_r}$, is deemed to be

$$A_r \propto \frac{A_n a}{\delta^2} \exp\left(\frac{-z^2}{2\sigma^2}\right) \tag{4}$$

From the above model, it could be also assumed that $\boldsymbol{S}$ is the average shear strength per unit area of the junction formed by adhesion between the shoe heel and floor surface and is presumed to be constant, $\boldsymbol{W}$ is the normal force and $\boldsymbol{\theta}$ is the average slope of the asperities measured in the direction in which the tangential force is applied. When gross sliding occurs, the heel surface will undergo both normal and horizontal displacement. This process is explained by assuming that when the shoe heel is sledded from rest it must climb up and pass the forward slopes, that is, the total measured traction force ***F*** from Eqn 1 would be given by

$$F = ArS + W \tan \theta$$
$$= \frac{KAnaS}{\sigma^2} \exp\left(\frac{-Z^2}{2\sigma^2}\right) + W \tan \theta \quad (5)$$

where ***K*** is a constant dependent only on the surface topography. From equation (5), the following possibilities could be considered:

(1) If the shoe and floor are given the same surface topography, that is, ***K*** and ***θ*** are constant, then the dimensionless separation will be the same for contact. The average contact size may also be expected to be constant. Hence, the COF (μ) will be

$$\mu = \frac{K_1}{\sigma^2} + K_2 \quad (6)$$

where $K_1$ and $K_2$ are constants. From equation (6), it could be noticed that traction force would be the sum of a constant term and a term varying inversely as the variance of the combined height distributions of both the surfaces.

(2) Although both the surfaces could have initially same surface profiles, there would be large changes even after a single sliding such as plastic deformation, abrasion and fatigue wear and material transfer. Hence, the nominal contact areas between them will be gradually grown out. In this process, however, the COFs between the shoe heel and floor surface will not exactly vary inversely as the variance of combined height distributions of both the surfaces.

## Experiment

A pendulum-type dynamic friction tester was used for the tests. The normal load of 400 N was applied to the shoe heel and its sliding speed was controlled at a relative speed of 25 cm/sec. Two polyurethane shoes (Nos. 1 and 2) with different sole patterns were rubbed against a ceramic tile. To validate theoretical model and identify wear developments, initial and rubbed heel surfaces were examined by a stereo scanning electron microscope (SEM, XL 30, Philips).

## Results

### *Observations of the wear patterns*

During the repeated sliding, abrasive patterns of the heel surfaces were moved in the sliding direction of each mating floor counterface, as shown in Fig. 1. From the displacement results of the average asperity angles of the shoes, the crack propagation angle was estimated at ***θ***. Figure 2 shows the overall relationships between the calculated asperity angles of the heel surfaces and the rubbing time. The estimated propagating angle in Fig. 2 was about 14°. This means that the crack was initially tended to propagate at a low slope. Although the crack propagation distance was estimated

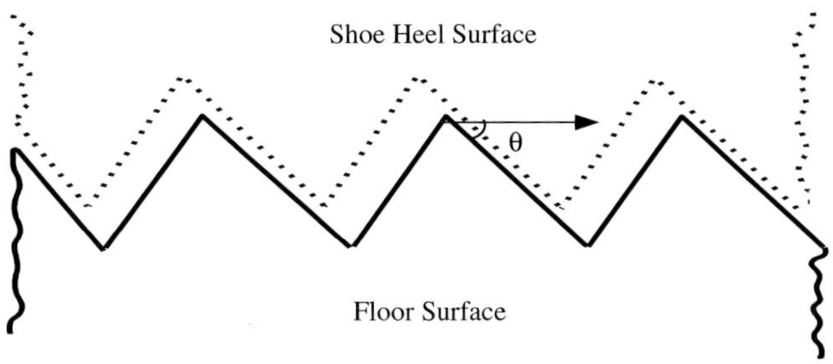


**Figure 1. A schematic illustration of the movement of abrasion patterns of a shoe Heel sliding against a floor surface.**

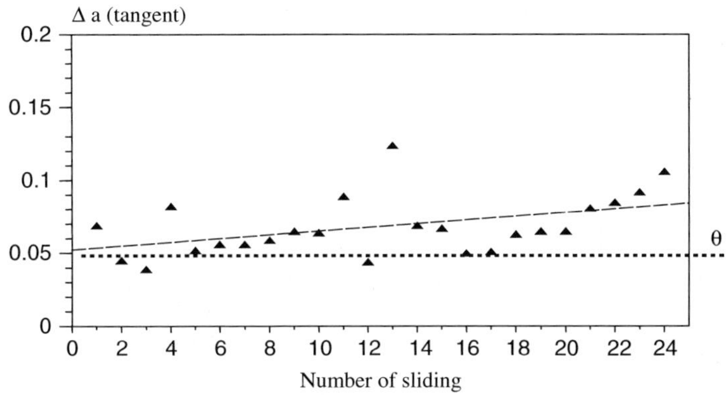


**Figure 2. The relationship between the displacement of the average asperity angles of the shoes and number of rubbings.**

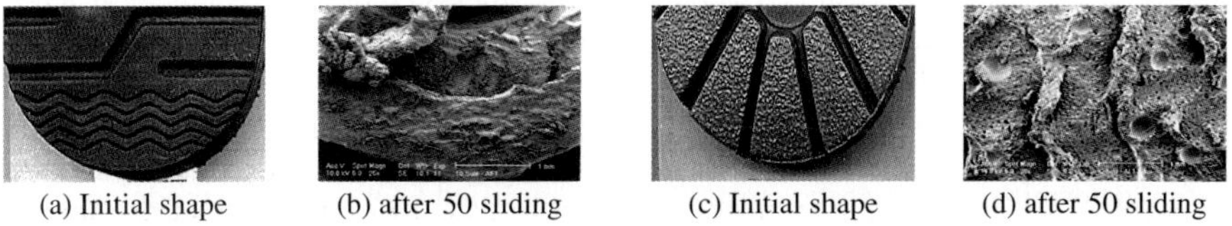


**Figure 3. Micrographs of the wear track for the shoes No. 1 (left) and No. 2 (right).**

numerically, it is probable that cracks of the heel surface do not always propagate with each rubbing.

*Wear observation*

The abrasion patterns of each heel surface were examined by micrographs taken with a scanning electron microscope (SEM). Figure. 3 shows examples of the initial and worn heel surfaces of each shoe, respectively. Initial heel surface of the shoe No. 1 shows clean tread patterns without any protruding shape (Fig. 3(a)) whereas the shoe No. 2 shows shiny and a number of small porous (Fig. 3(c)). After the entire rubbings, however, the heel surfaces were severely damaged so that their distinctive macro- and micro-tread patterns suffered massive changes. The worn heel surfaces of the shoe No. 1 show a number of paralleled tearing traces and micro-layered surface textures (Fig. 3(b)). The width and depth of tearing traces gradually increased with increasing the number of rubbings. In the case of the shoe No. 2, its original micro-porosities were broken up and formed lots of cavities on its surface layer (Fig. 3(d)). This process continuously created new surface structures so that the width and depth of tearing traces gradually increased with increasing the number of rubbing. Here it was also found that overall wear results of the two shoes were severe than expected and occurred from a very early stage of rubbings. The global wear patterns of both the heel surfaces also show severe abrasions, ploughings and deep scratches by fatigue wear.

## Conclusion

An adequate level of traction between the footwear and underfoot surface should be maintained for walking free from falling accidents. The roughened floor surface has a beneficial effect in raising COF values significantly between any shoe heel and floor surface under both dry and wet surface conditions. However, this high-traction surface is problematic from a biomechanical point of view. That is, if the shoe can not be easily rotated on such a high traction floor, then additional torsion stress is placed on the knees and lower back. Whilst such high traction can also result in a major cause of topographic changes of the shoe heel and floor surface from a tribo-physical point of view. This means that surface geometry of both the bodies changes drastically from their initial status and consequently slip resistance performances. This finding highlights the need for developing enhanced methods for reliably characterizing the slip resistance properties. This could be based on in-depth understanding of the nature of frictional and wear phenomena and surface analysis techniques for characterizing the surfaces of shoes and floors and their interactions during slip resistance measures. In the present study, a potential form of tribo-physical model was suggested to identify those mutual effects of friction and wear behaviors on slip resistance properties. From the proposed model, the mutual contributions of friction and wear behaviours were identified by massive structural changes of both the bodies and showed probable explanations on slip resistance results. Even though the suggested model has not yet reached a perfect stage where it would be quantitatively possible to predict the slip resistance property from known tribo-physical characteristics, this paradigm may provide an insight to identifying the complex issues. Therefore, factoring in wear induced material changes with friction measures may provide a way of enhancing the reliability of fall safety determinations over the currently oversimplified COF-index methodology, and make substantial progress towards accounting for one of the mechanics and mechanisms underlying COF variability.

## References

Grönqvist, R., 1995, Mechanisms of friction and assessment of slip resistance of new and used footwear soles on contaminated floors, *Ergonomics*, Vol. 38, No. 2, 224–241.

Kim, I.J. and Smith, R., 2000, Observation of the floor surface topography changes in pedestrian slip resistance measurement, *International Journal of Industrial Ergonomics*, Vol. 26, No. 6, 581–601.

Kim, I.J., 2004, "Development of a new analyzing model for quantifying pedestrian slip resistance characteristics: Part I – Basic concepts and theories", *International Journal of Industrial Ergonomics*, Vol. 33, No. 5, 395–401.

# A NEW UNDERSTANDING ON THE SHOE WEAR MECHANISM AND ITS SIGNIFICANCE ON SLIP RESISTANCE PROPERTY

**In-Ju Kim**

*School of Sport and Health Sciences, University of Exeter, Heavitree Rd, Exeter, EX1 2LU, UK*

One of the most prominent tribological phenomena in slip resistance measures is abrasive wear (AW) of shoe heel surfaces and its continued evolution. This would be a vital factor that affects slip resistance properties and eventually fall safety measures. However, it is very scarce to find any studies on this issue. This paper is, therefore, aimed to investigate principal mechanisms of AW, its propagation and magnitude effects on slip resistance properties. A full cycle of AW is comprehensively assessed by quantitative measures with a number of surface roughness parameters and qualitative analysis using a scanning electron microscope. Both the attempts identify that AW was mainly resulted from crack propagation at the root of a tongue and subsequent cutting off of the tongue when it had reached its largest size. All the results clearly verify a fact that topographic changes of the heel surfaces are a major cause on slip resistance properties.

## Introduction

Slip resistance property between the footwear and underfoot surface is of great importance for evaluating fall safety and has been measured as a form of coefficient of friction (COF). Since friction behaviors are significantly related with the surface structures of shoe heels and their interactions with floor counterfaces (Kim and Smith, 2000), it would be reasonable to expect that slip resistance properties are largely affected by the heel surface geometry and its tribo-physical characteristics (Kim et al., 2001). Although there are evidences on the findings that surface roughness of the shoe soles/heels is one of the major determinants on the slip resistance property, there is still little known about the primary aspects of friction and wear mechanics and mechanisms of the heel surfaces. In particular, it is very scarce to find any measured data on worn areas of the shoe heels and main causes of wear developments over slip resistance measures. This study is, therefore, aimed to investigate tribo-physical attributes of the shoe surfaces during dynamic friction measurements and analyze their effects on slip resistance properties. Surface changes and wear developments are extensively examined by not only for the initial heel surfaces but also for the rubbed ones. To scrutinize the primary aspects of tribo-physical characteristics of the heel surfaces, slip resistance tests are limited to clean and dry conditions in order to eliminate any confounding effects of surface contamination by agents other than shoes themselves. Although it is acknowledged that lubricated environments are potentially more slip hazardous than dry ones, friction and wear behaviors are principally different between two conditions. Because, in fact, we are more exposed to the dry surfaces than the lubricated ones, most shoe heels/soles would be experienced massive surface modifications by repeated friction and wear processes. In this manner, a study on the dry slip resistance property would be a valuable way to understand main tribo-physical characteristics of the shoes.

## Experiment

A pendulum-type dynamic friction tester was used to measure dynamic slip resistance properties between the shoes and floor specimens. It consists of two hydraulic systems, a force transducer, angular displacement transducer, and a PC. This tester simulates movements and loading of the foot during heel-strike and initial-slip, and quantitatively determines slip resistance in terms of a dynamic friction coefficient (DFC). The normal load of 400 N was applied to a shoe heel and its sliding speed was controlled at a relative speed of 25 cm/sec. For test specimens, two polyurethane shoes (Nos. 1 and 2) with different sole patterns were rubbed against ceramic plates. Slip resistance measurements were designed by five sub-groups for the systematic analysis. Each sub-group contained 5, 10, 20, 30 and 50 times of consecutive rubbings, respectively with a new shoe and floor specimen. To obtain full morphological information on the heel surfaces, a range of surface roughness parameters were measured by a laser scanning confocal microscope (LSCM, MRC 600, Bio-Rad). The parameters are centre line average (*Ra*), root mean square (*Rq*), maximum mean peak-to-valley height (*Rtm*), maximum mean height above the mean line (*Rpm*), maximum mean depth below the mean line (*Rvm*), and average asperity slope ($\Delta a$) (ANSI/ASME, 1985). To validate surface roughness data and identify wear developments, initial and rubbed surfaces of the shoes were thoroughly examined by a stereo scanning electron microscope (SEM, XL 30, Philips). The SEM was operated at 10 kV setting to avoid any radiation damages to the naked polymeric surfaces of each shoe heel.

## Results

### *Overall slip resistance results*

Figure 1 shows the slip resistance results between the two polyurethane shoes and ceramic plates. As shown, the DFC values of both the shoes were largely decreased from 1.110 to 0.772 (over 30%) in the case of the shoe No. 1 and 1.079 to 0.646 (over

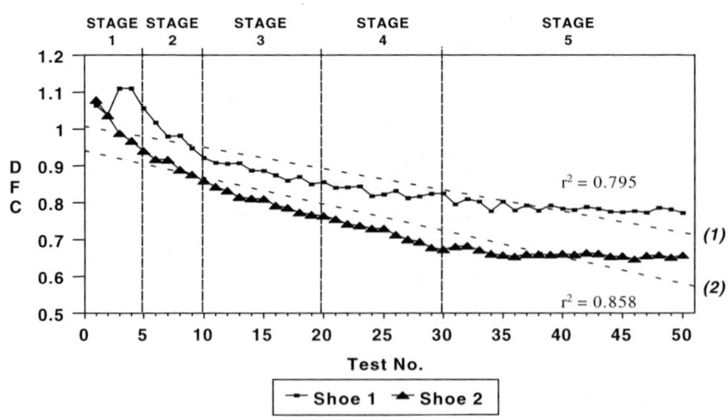


**Figure 1. Slip resistance results between two polyurethane shoes and a ceramic plate.**

40%) in the case of the shoe No. 2, respectively after the entire tests. Although the initial DFC values were very high, overall slip resistance performances of the shoes drastically deteriorated and continuously dropped until the stage 4. Regression results on each shoe also showed significant relationships between the DFC index and number of rubbings ($r^2 \approx 0.80$: the shoe No. 1 and $r^2 \approx 0.86$: the shoe No. 2).

### *Surface analysis*

Figures 2 and 3 summarize the results of surface roughness parameters of each shoe before and after the tests. As shown, all the surface parameters were largely varied during the tests. In particular, they were largely increased after the initial five rubbings. This increment seemed to be caused by abrasive wear that roughened the initial layers of heel surfaces. With further rubbings, however, the roughness parameters were gradually decreased. This was due to removal of the asperity crests in both the shoes. This result implied that new surfaces were evolved in the uppermost areas, whilst at the valley levels original features of the heel surfaces were not severely changed.

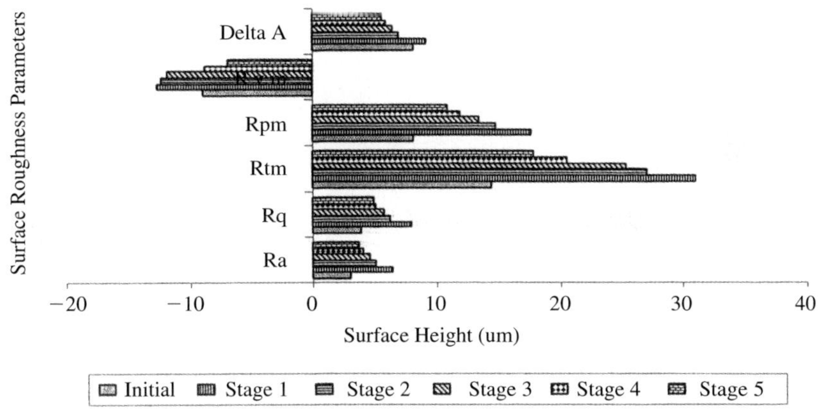


**Figure 2. Summary of surface roughness parameters of the shoe No. 1.**

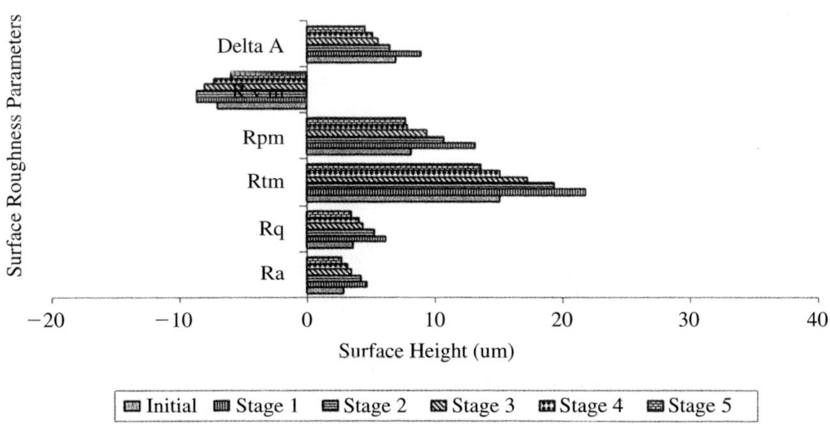


**Figure 3. Summary of surface roughness parameters of the shoe No. 2.**

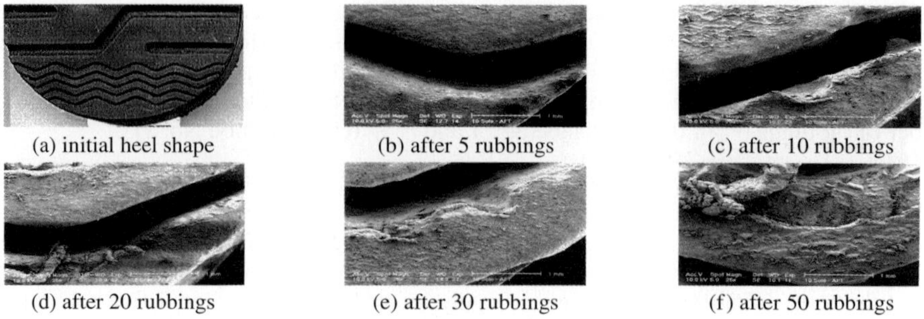

(a) initial heel shape (b) after 5 rubbings (c) after 10 rubbings

(d) after 20 rubbings (e) after 30 rubbings (f) after 50 rubbings

**Figure 4. Micrographs of the wear track for the shoe No. 1.**

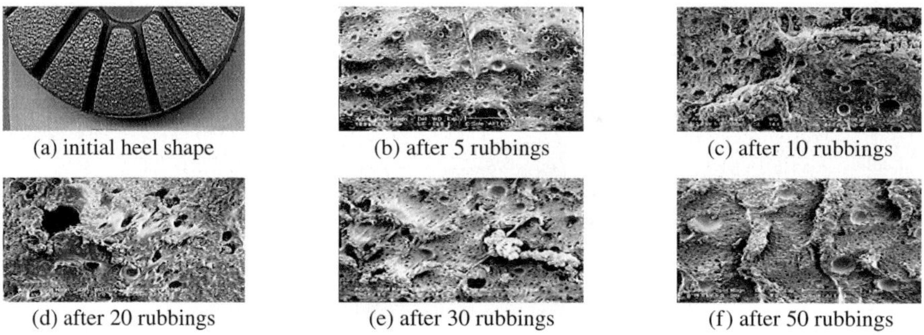

(a) initial heel shape (b) after 5 rubbings (c) after 10 rubbings

(d) after 20 rubbings (e) after 30 rubbings (f) after 50 rubbings

**Figure 5. Micrographs of the wear track for the shoe No. 2.**

*Wear patterns*

Figures 4 and 5 show the micrographs of initial and worn heel surfaces of each shoe. In the case of the shoe No. 1 its surface shows clean tread patterns without any protruding shape whereas the shoe No. 2 shows shiny and a number of small porous. After the entire rubbings, however, the heel surfaces were severely damaged so that their distinctive macro- and micro-tread patterns suffered massive changes. The worn surface of the shoe No. 1 shows a number of paralleled tearing traces and micro-layered geometry. The width and depth of tearing traces gradually increased with increasing the number of rubbings. In the case of the shoe No. 2, its original micro-porosities were broken up and formed lots of cavities on its surface layer. This continuously created new rougher surface so that the width and depth of tearing traces gradually increased with increasing the number of rubbings. Here it is found that overall wear results of the two shoes were severer than expected and occurred from a very early stage of rubbings (stage 1).

*Formation of abrasive wear*

Figure 6 shows an example of abrasive wear formation created on the shoe No. 1 heel surface after 50 times of consecutive rubbings. As shown, the propagated crack marks were clearly observed on the heel surface. On the left-hand side of the figure, a ridge was undercut at the lower front to make a small tongue. It was found that abrasive wear was resulted from crack propagation at the root of a crack tongue and subsequent cutting off of the tongue when it had reached its largest size. Therefore, fine markings of crack propagation should appear on the worn surfaces of each shoe heel.

Figure 7 shows a cross section of a wear scar that was detached from the heel surface. Propagated crack marks are clearly seen in this section. Large tear marks are also seen

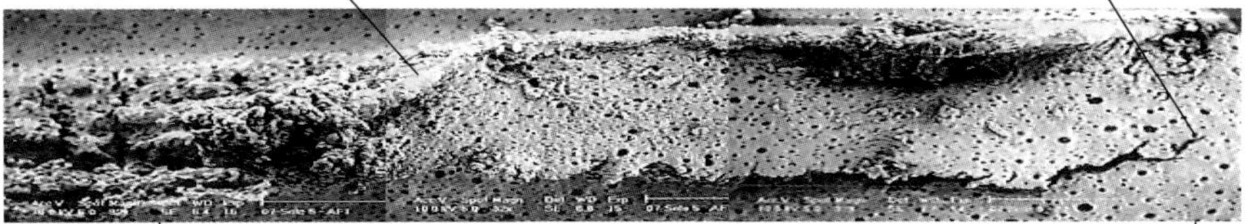


**Figure 6. Formation of abrasive wear at the shoe No. 1 after the tests – x30. It shows a large crack tongue caused by abrasive wear.**

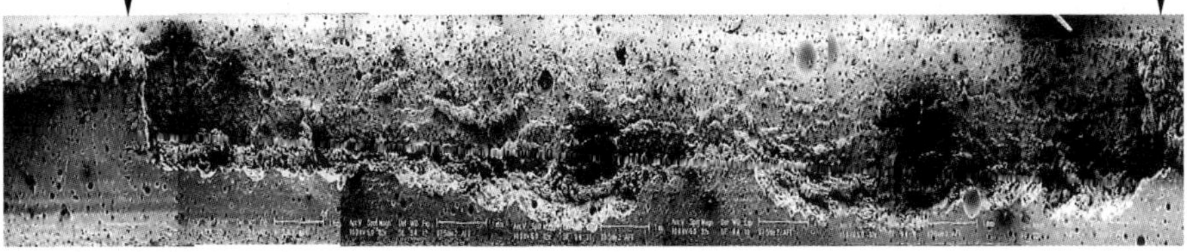


**Figure 7. A section of wear ridges formed at the shoe No. 1 – x30. A large piece of crack tongue was detached from the heel surface.**

amongst the relatively small striations. When the tongue was detached, the crack growth tended to direct itself upwards. Finally, the loose ridge was detached as a loose wear particle. Then the ridge would be undercut once more at the lower part. Thus growth and detachment of the crack tongue occurred repeatedly.

## Conclusion

This study identified fundamental issues on tribo-physical characteristics of the shoe surfaces during slip resistance measures and their effects on slip resistance properties. Repeated sliding friction processes significantly changed the surface structures of the shoe heels and as a result most of the initial surface features were severely damaged by abrasive wear. This aspect was measured by a number of surface roughness parameters showed significant changes on the heel areas of both the shoes. Furthermore, stereo micrographs clearly identified that a series of abrasive patterns appeared on the heel surfaces. The fine marks of propagating cracks were also observed between the abrasive patterns. From the amount of wear and the movement of abrasive patterns, the locus of ridges was obtained and it directed itself downward into the surfaces of shoe heels. Those findings clearly showed that heel surfaces were heavily changed by not only friction and wear evolutions when slip started, but also they significantly affected to the slip resistance properties as slip progressed.

## References

ANSI, 1985, Surface texture: Surface roughness, waviness and lay, *American Standard* ANSI B.46.1.

Kim, I.J. and Smith R., 2003, A critical analysis of the relationship between shoe-floor wear and pedestrian/walkway slip resistance, *Metrology of Pedestrian Locomotion and Slip Resistance*, ASTM STP 1424, ASTM International, 33–48.
Kim, I.J., Smith R. and Nagata, H., 2001, Microscopic observations of the progressive wear on the shoe surfaces that affect the slip resistance characteristics, *International Journal of Industrial Ergonomics*, 28/1, 17–29.

# ASSESSING SLIP RESISTANCE OF WINTRY WALKWAYS WITH PORTABLE SLIP SIMULATOR

**Carita Aschan, Mikko Hirvonen, Erkki Rajamäki & Tarmo Mannelin**

*Finnish Institute of Occupational Health, Department of Physics, Topeliuksenkatu 41 a A, Helsinki FIN-00250, Finland*

In Finland, about 70 000 pedestrian and bicycle falling accidents occur yearly at the street, walkways and courtyards. Around 2/3 of these accidents occur when the walking surface is covered by ice or snow. Slipperiness of different walking surfaces may be assessed by portable devices measuring friction. During winter seasons 2002–2003 and 2003–2004 a study was performed in order to evaluate the usability of the novel portable device, designed and constructed at the Finnish Institute of Occupational Health, for measuring slipperiness of walkways on varying pavement and weather conditions. According to the results, the developed Portable Slip Simulator is applicable for measuring the slipperiness of the wintry walking surfaces *in situ*. Also, the results indicate that the method can be used to evaluate the quality of the winter maintenance as well as anti-slip properties of the footwear in the field conditions.

## Introduction

In Finland, approximately 100 000 people are annually injured in slipping and falling accidents (Penttinen et al., 1998; Grönqvist, 1995), about 70 000 of which take place outdoors at the street, walkways and courtyards (Vuoriainen et al., 2000). Around 2/3 of these accidents occur when the walking surface is covered by ice or snow (Grönqvist, 1995). The total costs of slipping and falling accidents including the indirect costs are estimated to be as high as approximately 420 million euros per year (Vuoriainen et al., 2000). According to the same study, the annual costs due to pedestrian and bicycle falling accidents are estimated to be about 45 million euros in the City of Helsinki. At the same time, winter maintenance causes costs of about 4 million euros per year (Tervonen et al., 2001), i.e. the costs due to slipping and falling accidents are estimated to be tenfold. Amount of slipping accidents as well as the costs due to them could be decreased e.g. by improving the winter maintenance, warning the pedestrians about most slippery conditions and informing the public about the importance of the footwear selection in winter as well as encouraging the use of anti-slip devices, since wintry slipping accidents are mainly due to the inadequate grip between footwear and underfoot surfaces. In general, slip resistance may be defined quantitatively as a coefficient of friction between surfaces. Measured coefficient of friction values are connected to subjective evaluations with the grading system presented in Table 1.

In past years road friction has been assessed in several studies with different types of devices (Wallman and Åström, 2001). Unfortunately, most of the effort has been concentrated on the assessment of slip resistance between vehicle tyres and road. Most devices used in road friction measurements are not suitable for assessing slipperiness experienced by the pedestrian due to the fact that the used parameters differ too much from the human biomechanical parameters (Chang et al., 2001a, b). Also, little attention

**Table 1. The grading system connecting the dynamic coefficient of friction to the subjective evaluations (Grönqvist, 1995).**

| Class | Explanation | Coefficient of kinetic friction |
|---|---|---|
| 1 | Very slip-resistant | ⩾0.30 |
| 2 | Slip-resistant | 0.20–0.29 |
| 3 | Unsure | 0.15–0.19 |
| 4 | Slippery | 0.05–0.14 |
| 5 | Very slippery | <0.05 |

has been paid to the evaluation of slip resistance of footwear in various winter conditions. Therefore, in order to assess pedestrian safety on walkways, a more human-centred approach is needed.

A new portable device measuring friction (Aschan et al., 2003) has been designed and constructed at the Finnish Institute of Occupational Health (FIOH) based on the slip simulator developed for slip resistance measurements at FIOH in the late 1980s (Grönqvist, 1995). The applicable measurement parameters of both devices represent the actual kinematics during critical gait phases. The aim of this study was to evaluate the usability of the novel Portable Slip Simulator for measuring slipperiness of walkways in various pavement and weather conditions in situ. The effects of footwear selection, weather conditions and winter maintenance were also evaluated. In addition, the obtained results are utilized by the Finnish Meteorological Institute in developing a model for predicting pavement conditions in winter in order to give a warning of very slippery days.

## Materials and methods

### *DCOF measurements*

Slip resistance, i.e. dynamic coefficient of friction (DCOF), of various walking surfaces was measured *in situ* (Figure 1) by using the Portable Slip Simulator of FIOH (Aschan et al., 2003). The used measurement parameters of the device were as follows: normal force 170, 250 or 500 N, normal force build-up rate 4.5 kN/s leading to normal force build-up time of approximately 110 ms at $F = 500$ N, horizontal sliding velocity 0.2 m/s and contact angle 5° i.e. heel contact. The result of the measurement was the mean value during the measurement period of 100 ms. Means and standard deviations of at least nine repeated DCOF measurements were calculated.

During the late winter 2003, the measurements were performed with four different footwear: (i) winter shoe found to be slippery in use, (ii) slip-resistant safety footwear (European prestandard ENV 13287:2000), (iii) slippery footwear the outsole of which was made of hard (92 Shore A), unpatterned thermoplastic polyurethane (TPU) which is a commonly used material in women's shoes, and (iv) footwear with soft polyurethane (PU) outsoles the tread pattern of which has been totally removed by using orbital sander. Since the measured DCOF values were similar to each other for footwear (i) and (iii) and for (ii) and (iv), respectively, selection was restricted only to the footwear (ii) and (iii) during the measurements of the second winter period (October, 2003–March, 2004). In addition, measurements were performed with nine pairs of winter footwear for women purchased from the local department store and shoe shops. Footwear samples were selected so that they were estimated in advance to have different slip resistant properties based on their appearance. Slipperiness of the samples were evaluated on trampled snow ($T = -2$°C) and on soft, wet ice ($T = 0$°C).

**Figure 1. The DCOF values of 0,15 for slippery and 0,39 for slip-resistant footwear were obtained when measuring ungravelled trampled snow (T = −5°C).**

*Subjective evaluations*

During the DCOF measurements, the two experienced operators estimated individually the slipperiness of the surface by a subjective rating. The rating was performed by using a gradual scale: (1) very slippery; (2) slippery; (3) neutral/uncertain; (4) slip resistant; and (5) very slip resistant.

*Locations and weather conditions*

The DCOF measurements were started in February 2003. At the time, measurements were performed both at pavements with rather low pedestrian traffic as well as in busy locations, such as the surroundings of the Helsinki Central Railway Station, where many slipping and falling accidents have occurred (Vuoriainen et al., 2000). In addition, the slipperiness of walking surfaces was measured at the locations in which no proper winter maintenance existed. Because the maintenance was found to affect strongly to the measured results, the DCOF measurements of the second winter period were performed at the locations with no regular winter maintenance in order to take better into account the needs of Finnish Meteorological Institute for their weather and pavement condition model development work.

The measurements were performed at different locations in different weather conditions. Air temperature and humidity as well as the surface temperature of a measurement point were recorded. All the locations were photographed, and the description of the surface (ice, snow, slush, etc.) and the used winter maintenance (graveling, ploughing, etc.) were written down.

## Results and discussion

DCOF measurements were performed in different winter conditions, and the walking surfaces were mainly covered by ice, snow or slush. Since the obtained results were similar between the footwear samples i) and iii) as well as between ii) and iv), the results are reported only by using terms "slippery footwear" and "slip-resistant footwear" also indicating the importance of footwear selection in different weather conditions.

Slip-resistant footwear had better grip in all measured pavement conditions being occasionally even more than four times higher than that of the slippery one. On average, the DCOF values measured with slip-resistant footwear were two times higher than those measured with slippery footwear. Equal results with these different types of footwear were obtained when measuring very slip-resistant, well-gravelled surfaces as well as wet, smooth icy surfaces in which the DCOF values, measured even with the slip-resistant footwear, were $<0.14$.

The average value of all measurements performed with the slippery footwear was $0.18 \pm 0.06$ the lowest value being 0.06 (melting ice of 0°C) and the highest 0.42 (trampled newly-fallen wet snow covering trampled snow). Similarly, the average for the slip-resistant footwear was $0.38 \pm 0.12$ the lowest and the highest values being 0.09 (smooth ice covered by water) and 0.63 (trampled newly-fallen wet snow), respectively. The DCOF values measured with the slippery footwear correlated (Pearson's product moment correlation coefficient, r) rather weakly ($r = 0.59$, $p < 0.0001$) with those measured with the slip-resistant footwear.

Statistically highly significant difference ($p < 0.0001$) between the DCOF values of nine pairs of winter footwear for women was revealed by One-Way Analysis of Variance (ANOVA) when measured both on soft, wet ice and on snowy surface. On trampled snow, the measured DCOF values varied from 0.22 to 0.57. When measured on ice, the lower DCOF values (range: 0.11–0.21) were obtained. However, the friction was not so small than normally on wet ice since instead being smooth, the ice was soft and rough. At the same condition, the DCOF value was 0.12 for both the slippery and slip resistant footwear.

Correlation between subjective evaluations and slippery footwear was poor ($r = 0.49$, $p < 0.0001$) whereas a good correlation was found between subjective evaluations and slip resistant footwear ($r = 0.81$, $p < 0.0001$). Better correlation is due to the footwear used in the subjective evaluations: safety footwear with good grip was used by the operators.

## Conclusions

According to the obtained experience, the Portable Slip Simulator is applicable for measuring the slipperiness of the wintry walking surfaces *in situ*. Significant differences between the slip resistance characteristics of different footwear were found indicating the importance of the right footwear selection in prevention of slipping accidents. In addition, the measurement method was found capable for evaluating the quality of winter maintenance in the field conditions.

## Acknowledgements

The authors wish to express their appreciation to Finnish Meteorological Institute and "Jaloin" – project for their good cooperation. The work has been supported by the Ministry of Transport and Communications and the Ministry of Social Affairs and Health who are gratefully acknowledged.

## References

Aschan, C., Hirvonen, M., Mannelin, T. and Rajamäki, E. 2003, Prevention of slipping accidents – development of the portable slip meter II. In *Proceedings of the XVth*

*Triennial Congress of the International Ergonomics Association, August 24–29, 2003, Seoul, Korea*
Chang, W.-R., Grönqvist, R., Leclercq, S., Myung, R., Makkonen, L., Strandberg, L., Brungraber, R.J., Mattke, U. and Thorpe, S.C. 2001a, The role of friction in the measurement of slipperiness, Part 1: Friction mechanisms and definition of test conditions, *Ergonomics*, **44** (13), 1217–1232
Chang, W.-R., Grönqvist, R., Leclercq, S., Brungraber, R.J., Mattke, U., Strandberg, L., Thorpe, S.C., Myung, R., Makkonen, L. and Courtney, T.K. 2001b, The role of friction in the measurement of slipperiness, Part 2: Survey of friction measurement devices, *Ergonomics*, **44** (13), 1233–1261
European prestandard ENV 13287:2000. Safety, protective and occupational footwear for professional use – Test method and specifications for the determination of slip resistance. March 2000, (CEN, Brussels)
Grönqvist, R. 1995, *A dynamic method for assessing pedestrian slip resistance. Ph.D. Thesis. People and work, research reports 2*, (Finnish Institute of Occupational Health, Helsinki)
Penttinen, M., Nygård, M., Harjula, V. and Eskelinen, M. 1998, *Jalankulkijoiden liukastumiset, vaikeimmat kelit ja niiden ennustaminen sekä tiedottamiskokeilu pääkaupunkiseudulla. VTT Research Notes*, (Valtion teknillinen tutkimuskeskus (VTT), Espoo), (In Finnish, English abstract: Pedestrian slipping accidents, detecting and forecasting of hazardous road conditions. Information campaign on the Helsinki region. http://www.vtt.fi/rte/transport/tutkimus/telematiikka/abs34.htm)
Tervonen, J., Hämekoski, K., Myllynen, M., Pihlaja, M. and Penttinen, P. 2001, *Liukkaus, hiekoitus ja katupöly – Yhteiskuntataloudelliset vaikutukset, Helsingin kaupungin Rakennusviraston julkaisut 2001:30*, (Helsingin kaupunki, Rakennusvirasto, Helsinki), (In Finnish)
Vuoriainen, T., Helenius, M., Heikkilä, J. and Olkkonen, S. 2000, *Jalankulkijoiden ja pyöräilijöiden kaatumistapaturmat, Espoo, Helsinki, Jyväskylä ja Oulu, Finnra Reports 48/2000*, (Finnra, road and traffic technology, Helsinki), (In Finnish, English abstract: Incidence of falls by pedestrians and cyclists in Espoo, Helsinki, Jyväskylä and Oulu. http://www.tiehallinto.fi/julkaisut/pdf/3200634.pdf )
Wallman, C.-G. and Åström, H. 2001, *Friction measurement methods and the correlation between road friction and traffic safety. A literature review, VTI meddelande 911A*, (Swedish National Road and Transport Research Institute, Linköping)

# ROLLER COASTER SLIP TESTS PUTTING SLIP TESTING BACK ON THE RAILS!

**Kevin Hallas, Robert Shaw, Paul Lemon & Steve Thorpe**

*Health and Safety Laboratory, Harpur Hill, Buxton SK17 9JN*

There are literally hundreds of slip testing devices available on the market and every year sees new ones launched. The Health & Safety Laboratory has been busy evaluating the latest developments from both academic and commercial organisations. Two Roller-Coaster style slip tests have been studied and compared with the Pendulum on a wide range of installed floor surfaces, in dry, wet and contaminated conditions. The results show some interesting correlations, a few limitations and offer new possibilities for those who would like a quick slipperiness test that actually works.

## Introduction

In a typical year, slips, trips and falls account for 33% of all reported major workplace injuries, 20% of over-3-day workplace injuries and 50% of all reported accidents to members of the public. This means that in the UK there is a slip or trip accident every three minutes. The majority of slip accidents occur in wet contaminated conditions. Amongst other things, the properties of the floor surface can make a significant contribution to the cause or prevention of these accidents, but how is slip resistance measured? There are hundreds of commercially available testing devices for measuring slip resistance of flooring materials, and every year sees new ones launched onto the market. For some time, the Health & Safety Laboratory (HSL) and the Health & Safety Executive (HSE) have condemned sled-type tests for measuring slip resistance values of floor surfaces because they fail to reproduce the correct squeeze film in wet contaminated conditions, leading to inaccurate and indeed misleading slip resistance values for wet contaminated floors. HSL have evaluated two instruments in the past year that were designed with a better understanding of the squeeze film in mind. The first instrument is described by Kirchberg (1997) and was reproduced in the HSL workshop from drawings in the literature. The second instrument, known as SlipAlert, has recently been developed by SLIPALERT LLP.

HSL, HSE and the UK Slip Resistance Group (UKSRG) believe that any valid test for slip resistance in the contaminated condition must establish and manipulate a squeeze film in the same way as the slipping pedestrian. To date, the Pendulum is the only portable test to do this, emerging tests must therefore correlate with the Pendulum if they are to be useful. These "Roller coaster" type tests have the potential to generate a squeeze film on a contaminated floor surface, because the trolley is already in motion as it comes into contact with the floor. The Pendulum data has been supplemented by Rz microroughness measurements, to form a benchmark for comparison of the new instruments.

## Method

The two instruments work in a similar manner, though there are important differences. They are both gravity powered trolleys, rolling down a specially made slope before making contact with the floor surface under examination. The Kirchberg instrument slides on three fixed rubber sliders, the ramp has three different height and therefore speed settings. The distance travelled is determined by tape measure from the point of contact of the rear of the trolley with the floor to the rear of the trolley at rest after testing. The SlipAlert rolls on two front wheels and a single rubber slider towards the rear, mounted at the same angle as the Pendulum slider relative to the ground at impact. It features an inbuilt distance meter and digital display for easy operation.

Preliminary work with the SlipAlert instrument showed the type of rubber and the conditioning before testing to be important in achieving consistent and repeatable results. All of the results detailed in this paper were generated using Standard Simulated Shoe Sole (Four-S) rubber sliders (supplied by RAPRA). The sliders were prepared as follows: for a new slider each trolley was pulled 5 times over P400 grit paper and 9 times over pink lapping film; for a used slider each trolley was pulled 1.5 times over P400 grit paper and 9 times over pink lapping film. These distances are equivalent to the distance travelled by the Pendulum slider over each surface during the relevant conditioning step. As with the Pendulum, eight measurements were taken with each roller coaster test and the final 5 used to determine the coefficient of friction.

The two instruments were evaluated by testing installed flooring materials in dry and water wet contaminated conditions during site visits. Pendulum and microroughness measurements were taken on each floor for comparison. Further limited testing was carried out with other contaminants.

## Results

The surface micro-roughness of each flooring surface was measured, using a Surtronic Duo surface micro-roughness transducer. The mean values of each set of ten measurements are presented in table 1. The Slip Resistance Values (SRV) were generated using the Stanley Pendulum. The slip resistance values of each installed floor are presented in table 1. The CoF results obtained from the last five of eight distance measurements using the Kirchberg and SlipAlert instruments are presented in table 1. The Kirchberg instrument results are presented as low speed (LS), medium speed (MS) and high speed (HS). The Pendulum Slip Resistance Values are approximately 100 × CoF.

## Discussion

The two roller coaster tests were compared with the Pendulum on a variety of installed floors with Rz surface microroughness ranging from 2.1 μm to 23.5 μm. This gives a good basis for comparison of the two tests with the Pendulum in dry and water wet conditions.

### *Kirchberg*

When the results from the Kirchberg roller coaster are ranked in terms of low, moderate and high slip risk according to the UKSRG Guidelines (UKSRG, 2000) the test does not rank the floors in the same order as the Pendulum and does not appear to be able to differentiate clearly between them. It suggests that half of the floors tested would present a moderate slip risk in the as-found dry condition. This result is not supported by the Pendulum, which classifies all the floors as a low or extremely low slip risk in the dry.

In the water wet condition the Kirchberg roller coaster agrees much more closely with the Pendulum results at low values. Although the roller coaster ranks the floors in a similar order to the Pendulum at higher SRV numbers it does not seem able to achieve high CoF values and consistently and dramatically underestimates the level of available

**Table 1. Test results.**

| Flooring | Rz (μm) | Pendulum Dry | Pendulum Wet | SlipAlert Dry | SlipAlert Wet | Kirchberg MS Dry | Kirchberg LS Wet | Kirchberg MS Wet | Kirchberg HS Wet |
|---|---|---|---|---|---|---|---|---|---|
| Terrazzo | 2.1 | 63 | 9 | 0.23 | 0.14 | 0.28 | 0.13 | 0.10 | 0.08 |
| Vinyl Tiles | 2.5 | 58 | 22 | 0.45 | 0.21 | 0.26 | 0.17 | 0.14 | 0.13 |
| Parquet | 3.0 | 56 | 19 | 0.41 | 0.27 | 0.35 | 0.24 | 0.19 | 0.18 |
| Terrazzo | 3.1 | 64 | 16 | 0.55 | 0.20 | 0.39 | 0.16 | 0.13 | 0.12 |
| Linoleum | 4.4 | 64 | 16 | 0.49 | 0.21 | – | – | – | – |
| Linoleum | 6.4 | 49 | 30 | 0.41 | 0.28 | 0.27 | 0.17 | 0.15 | 0.14 |
| Wood | 8.1 | 66 | 20 | 0.61 | 0.34 | – | – | – | – |
| Slate | 8.3 | 56 | 30 | 0.50 | 0.29 | – | – | – | – |
| Parquet | 8.4 | 60 | 20 | 0.62 | 0.32 | – | – | – | – |
| Terrazzo | 10.0 | 56 | 16 | 0.33 | 0.21 | – | – | – | – |
| Ceramic Tile | 11.9 | 59 | 10 | 0.59 | 0.10 | 0.45 | 0.10 | 0.07 | – |
| Quarry Tile | 12.8 | 60 | 20 | 0.70 | 0.18 | – | – | – | – |
| Vinyl | 13.1 | 66 | 15 | 0.53 | 0.15 | 0.44 | 0.18 | 0.12 | 0.10 |
| Safety Vinyl | 13.5 | 64 | 50 | 0.53 | 0.37 | – | 0.28 | 0.28 | 0.22 |
| Quarry Tiles | 14.8 | 64 | 13 | 0.69 | 0.11 | – | – | – | – |
| Mosaic | 15.2 | 57 | 25 | 0.49 | 0.28 | 0.30 | 0.22 | 0.15 | 0.14 |
| Vinyl | 15.6 | 47 | 28 | 0.42 | 0.40 | – | – | – | – |
| Wood | 15.9 | 65 | 19 | 0.63 | 0.28 | – | – | – | – |
| Concrete | 18.2 | 58 | 41 | 0.55 | 0.35 | 0.50 | 0.22 | 0.20 | 0.20 |
| Natural Stone | 19.0 | 60 | 47 | 0.60 | 0.37 | 0.38 | 0.26 | 0.25 | 0.25 |
| Slate | 23.4 | 61 | 33 | 0.63 | 0.41 | – | – | – | – |
| Natural Stone | 23.5 | 55 | 40 | 0.60 | 0.49 | – | – | – | – |

friction. The Kirchberg test is able to distinguish between dry and wet contaminated conditions on each floor tested.

In terms of operation, the Kirchberg instrument proved to be a little difficult at times. The way in which the trolley moves from the ramp to the floor surface requires very careful adjustment of the ramp. If thc ramp is a little high or a little too low the trolley slams down onto the surface in an undesirable manner, and the resulting instability affects the data gathered. When the set up is perfect, the trolley glides across the surface, and shows little tendency to yaw. The ramp can move backwards as the trolley leaves the ramp, so the testing had to be carried out with the ramp braced against an immoveable object (the operators' foot was found to be adequate).

The medium speed (MS) is the most accurate speed to use when determining the coefficient of dynamic friction. At low speed the distance travelled by the trolley is very small and errors in human judgment when measuring the distance can have a larger effect on the final result. At the high speed in contaminated conditions the trolley can travel up to 2 m. This distance can be difficult to measure accurately by tape measure. At the medium speed the trolley travels an acceptable distance to allow ease of measuring and minimisation of potential operator error.

### *SlipAlert*

In the dry SlipAlert is able to achieve much higher CoF values than the Kirchberg roller coaster and mirrors the Pendulum result more closely.

In wet conditions SlipAlert seems more capable of distinguishing between floors and agrees more closely with the Pendulum numbers. It seems incapable of achieving the higher CoF values, limiting its ability to correctly rank floors with a higher coefficient of friction, but it can distinguish between floors which are a low or medium slip risk.

In terms of operation, the SlipAlert is very simple and user friendly. The digital display removes the need to measure the distance travelled, though the number from the display has to be converted to CoF manually using a supplied chart. A certain amount of bouncing has been observed on softer flooring materials, notably the vinyl and safety vinyl. When the contaminant present was changed from water to mayonnaise, the wheels became contaminated and this hindered the subsequent operation of the instrument until a thorough clean was possible. It also allowed the trolley to slide off the ramp sideways, before reaching the end of the ramp. The instrument has a tendency to yaw, especially where there are imperfections in the surface, which can have a dramatic effect on the distance travelled. The surfaces which caused the SlipAlert to yaw also caused the rubber slider to twist relative to the trolley, which was cured by use of double sided tape to supplement the standard screw fixing.

*General comments*

These types of instrument need a large area of flooring to carry out wet and dry evaluations, compared with that necessary for the Pendulum. They lend themselves to large areas of flooring and survey work. For accident investigation several limitations would inhibit their use. It is not possible to measure a small spill, which the Pendulum is ideal for. Floors with variability within a small area are difficult to accurately assess, for example floors consisting of a resin matrix and a fine aggregate that is often distributed by hand, have small areas with very little aggregate, and this is where slips are seen on these floors. The roller coaster tests can only measure the mean slip resistance of a large area. Use of the roller coaster method combined with a roughness meter may allow a more robust study of these inhomogeneous flooring types.

The new test methods lend themselves particularly to training events, where non-experts can easily see the effect on slip resistance of a wet floor, whereas the Pendulum requires a certain degree of understanding of the physics of the test.

## Conclusion

Both roller coaster-type tests have shown the ability to distinguish between wet and dry floors. Although the narrower range of both the Kirchberg and SlipAlert can result in misclassification of dry floors, their main application will be in assessing contaminated surfaces (the conditions in which most slip accidents occur) and therefore the wet results are of more importance.

Of the two tests SlipAlert appears to be able to generate a wider range of results and for this reason produces results that better reflect the Pendulum SRV numbers. Where there is a difference between the CoF values generated and the Pendulum SRV numbers the test generally errs on the side of caution by underestimating the level of available slip resistance. It is therefore unlikely that an unsafe floor would be erroneously classified as safe by SlipAlert.

One factor that must be taken into consideration is the amount of room required to use a roller coaster as opposed to the Pendulum. When the level of available friction is very low the trolley can travel over 2 m. In many areas there will not be sufficient room to correctly use SlipAlert. The trolley is also prone to veer off to one side where there are imperfections in the floor being tested and this can produce misleading results. In order for SlipAlert to be used as a standard test, Four-S sliders would have to be adopted to allow reproducible conditions and results which could be compared with those obtained by the Pendulum. When conditioned and operated in a standardised manner, SlipAlert presents a useful complementary tool to the surface roughness meter, and a possible alternative to the Pendulum for basic slip resistance measurements in contaminated conditions.

## References

Kirchberg, S., Fischer, H., & Reyhl, H. *Untersuchungen zur Optimierung der Prüfparameter für Verhahren zur instationären Messung der Gleitreibung von Fußboden*, Projekt F 1673, 1997, (Bundesanstalt für Arbeitsschutz und Arbeitsmedizin)

HSE, *The assessment of pedestrian slip risk; The HSE approach*, 2004, (http://www.hse.gov.uk/pubns/web/slips01.pdf)

UKSRG, *The Measurement of Floor Slip Resistance – Guidelines Recommended by the United Kingdom Slip Resistance Group*, Issue 2, 2000, ISBN 1-85957-227-8 (The UK Slip Resistance Group, RAPRA, Shropshire)

# A FIELD ASSESSMENT OF POTENTIAL FRICTION VARIATION IN A FAST-FOOD RESTAURANT IN THE USA

**Wen-Ruey Chang, Alfred Filiaggi, Kai Way Li, Yueng-Hsiang Huang & Theodore K. Courtney**

*Liberty Mutual Research Institute for Safety, Hopkinton MA 01748, USA*

Slips and falls are a major problem in occupational injuries in which floor slipperiness is a critical issue. Friction variation is speculated to be a significant contributor to slip and fall incidents, although it has not been quantified and related to a measurement of slipperiness. A friction variation study is more valid when conducted in a field environment. This study investigated friction variation in six major working areas of a fast-food restaurant in the USA. The results indicated that there was slightly more than a 50% chance that employees might experience a friction reduction when stepping from one foot to the other while walking in the major working areas studied. The absolute reduction in friction coefficient had an average and standard deviation of 0.12 and 0.15, respectively. The relative reduction in friction coefficient had an average and standard deviation of 0.19 and 0.22, respectively.

## Introduction

Slip and fall accidents are a serious safety problem in work environments (Leamon and Murphy, 1995). The annual direct cost of occupational injuries due to slips, trips and falls in the USA is estimated to exceed 6 billion US dollars (Courtney et al., 2001). Falls on the same level accounted for approximately 53% of claim costs and 65% of claim cases in total direct workers' compensation for occupational injuries due to slips and falls (Leamon and Murphy, 1995). Roughly 40–50% of same level falls in the USA and European countries are attributable to slips (Courtney et al., 2001). Slips are likely to contribute to many other occupational injuries in addition to falls.

Grease and water contamination on floors is common in restaurant kitchens. Hence, slippery floors, which are a critical factor for falls on the same level, are typical in restaurants (Chang et al., 2003). Leamon and Murphy (1995) reported that slips and falls resulted in the second most frequent workers' compensation injury claims and were the most costly claims within the restaurant industry in the USA. They reported that the incidence rate of falls on the same level over a 2-year period was 4.1 per 100 full-time restaurant employees, resulting in an annual cost of US$116 per employee.

The measurement of friction between the shoe and floor is the most common method used to assess floor slipperiness (Chang et al., 2001b). It is generally assumed that slip and fall accidents are more likely to occur on floors with a low coefficient of friction (COF), and the level of COF is often used to assess the potential risk of these accidents. In addition to the level of COF, friction variation can also play a role in determining slipperiness. People manipulate gait when aware of walking on slippery surfaces, casting doubt on the validity of the level of COF as a sole indicator of slipperiness (Grönqvist et al., 2001). Friction of the floor is highly location dependent (Chang et al., 2003). The potential of slip and fall accidents could be increased due to local friction variation (Strandberg, 1985; Andres et al., 1992; Grönqvist et al., 2001). Discontinuities in friction across floor surfaces may result in a fall due to the unexpected encounter of a low friction area without body posture adjustments.

The Brungraber Mark II, a portable inclinable articulated strut slip meter (PIAST) driven by gravity, is a friction measurement device commonly used in the USA (Chang et al., 2001a). This slipmeter simultaneously applies forces parallel and normal to a floor surface with an impact of a footwear pad on the floor at an inclined angle in order to eliminate the dwell time problem with the static friction measurement. The American Society for Testing and Materials (ASTM) publishes the guidelines for operating this slipmeter (ASTM F1677-96, 2004).

Friction variation was not explicitly addressed in the results published in the literature except that high variability in the friction measured on similar surfaces was mentioned by Leclercq et al. (1997) and significant tile-to-tile variations in friction in the same areas of restaurants were reported by Chang et al. (2003). Friction variation has not been quantified and related to the slipperiness assessments. The main objective of this study was to quantify friction variations over major working areas in a fast-food restaurant in USA as a first step to approach this problem.

## Methods

This study was conducted in a fast-food restaurant in the USA. The floor tiles at the time of the measurements were 9 years old. Six major working areas, including fryer, back vat, grill, sink, front counter and walk through, were identified. The fryer and back vat are the areas for cooking french fries and fried chicken, respectively. The walk through is the area at the entrance of the kitchen where the employees enter and exit the kitchen. The front counter is the area for taking customers' orders and payments, and for delivering food. The sink is used to wash cookware. The grill is used to fry beef patties for hamburgers. Quarry tile with embedded grit was the floor tile in the kitchen and had an approximate dimension of 15 cm by 15 cm (6 in by 6 in).

COF values were measured with the Brungraber Mark II in all six areas immediately after the lunch period on a weekday. Due to a large volume of customers over a short time period and contaminants on the floor surfaces, the floor conditions in this type of restaurant during lunchtime represent one of the worst situations during their daily operation. The measurement protocols recommended by Chang (2002) and Chang et al. (2004) were adopted. A Neolite test liner was used as a footwear pad.

Any loose, gross contaminants on the floor surfaces were removed before taking friction measurements. In the sink area, wet measurements were conducted by applying water to the floor surfaces to simulate floor conditions when washing tasks were being performed. Water was replenished in the footwear pad striking area during repeated strikes under the wet conditions. Besides wet measurements in the sink area and loose debris removal, the surface conditions in other areas were not altered before the measurements.

A line of tiles through each selected area was measured in the walking direction through the area and was at least 30 cm from the wall or edge of the cooking equipment in order to approximate the walk path in the area. This type of selection was adequate for the areas of fryer, back vat, grill and front counter. In the sink area, the actual line of friction measurement was at least 45 cm from the sink since an employee might stand in front of the sink to wash cookware and other employees might not be able to walk within 30 cm from the sink. In the sink, grill, fryer and back vat areas, a tile was measured no more than every 30 cm due to their likelihood of contamination by water and grease. For the front counter, a tile was measured no less than every 60 cm due to a less likelihood of contamination. In the walk through, a tile was measured no more than every 30 cm due to a very limited size of the area. An attempt was made to capture the friction variation when walking from a less contaminated area to a contaminated area such as an employee might experience when approaching each critical area including the fryer, grill, back vat and sink. The area right in front of the cooking equipment was considered the contaminated area. At least two tiles were measured outside of each end of the contaminated area. Dry measurements were conducted on these less contaminated tiles in the sink area. The total number of tiles measured was 53, consisting of 5 in the front counter, 9 in the fryer, 12 in the grill, 9 in the back vat, 14 in the sink and 4 in the walk through areas. On each tile measured, there was one friction measurement for each walking direction before moving on to the next tile. In order to maintain a consistent surface, the Neolite pad was sanded according to the protocol introduced by Chang et al. (2003) before the friction measurements of each tile.

Friction variation was extracted from the measured COF values to simulate the friction changes that an employee might experience while walking through the working areas. Since the distance between two consecutive tiles measured was approximately half of a step length for the areas of sink, grill, fryer, back vat and walk through, the COF values from every other measured tiles in the same direction were used to calculate the friction variation. However, the distance between two consecutive tiles measured was approximately a step length in the front counter area, and the COF values from every two consecutive tiles measured in the same direction were used to calculate friction variation. Andres et al. (1992) speculated that humans could adjust their gait to accommodate different friction conditions in a gait cycle. Therefore, the friction changes in a gait cycle could very likely reflect the potential risk associated with that step. The absolute friction variation of a step was calculated by subtracting the COF value of the preceding tile from that of the subsequent tile. The relative friction variation of a step was obtained by dividing the absolute friction variation by the COF value of the preceding tile.

## Results and discussion

Among the areas measured, the front counter had the highest COF with a mean and standard deviation of 0.82 and 0.063, respectively, while the sink area had the lowest COF with a mean and standard deviation of 0.27 and 0.088, respectively. The mean ($\pm$ standard deviation) COF values for the rest of the areas were $0.74 \pm 0.069$ for the fry vat, $0.75 \pm 0.043$ for the grill, $0.79 \pm 0.063$ for the back vat and $0.76 \pm 0.037$ for the walk through. Despite a lower COF in the sink area, this area had the highest standard deviation. The walk through had the lowest standard deviation despite a high COF.

A paired t test was performed on the measured friction to examine the difference in COF between two measurement directions. The results indicated that there was no statistically significant difference in the COF measured in two different directions. The linear correlation coefficient between two directions from the same tiles across all the tiles measured was 0.95 with $p < 0.0001$. This implies that a friction increase in a pair of tiles in one direction will most likely lead to a friction reduction in the opposite direction.

Friction variation was quantified in the relative and absolute changes. A negative value of friction variation indicates that the subsequent tile in a step has a lower COF value than the preceding tile. All the combinations of friction variation in each area in both directions were calculated and pooled. Among 84 combinations, 44 had a negative value in friction variation and 39 had a positive value with one combination with no change in friction. The probability density function for all the absolute friction variation combinations was constructed and is shown in Figure 1 in the solid line with a Gaussian distribution shown in the dotted line. The mean and standard deviation of the distribution were $-0.002$ and 0.196, respectively. The skewness was 0.090 and the

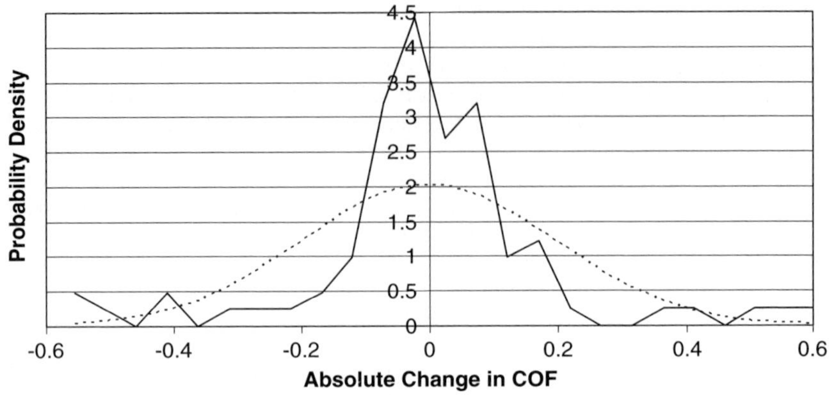


**Figure 1. The distribution of the probability density (shown in the solid line) for the absolute friction variation in all the combinations among the measured tiles and directions in comparison with a Gaussian distribution (shown in the dotted line).**

kurtosis was 3.171, adjusted to reflect the deviation from a Gaussian distribution. As indicated earlier, the paired t test to examine the difference in COF between two different measurement directions indicated no statistically significant difference. The mean and skewness values confirm the results of the paired t test that the distribution is nearly symmetrical with a mean around zero.

A similar distribution was constructed for the relative change in COF. The mean, standard deviation, skewness and kurtosis were 0.121, 0.703, 3.671 and 16.023, respectively. The kurtosis for the relative change in COF reported here was adjusted to reflect the deviation from a Gaussian distribution. Since the relative friction variation of a step was obtained by dividing the absolute friction variation by the COF of the preceding tile, a friction increase usually results in a larger relative change in friction than the absolute change in friction due to a lower friction value on the preceding tile. This is especially more significant in the sink area where the relative change in COF was as high as 4 for changes from wet tiles to dry tiles.

In the change of friction, a friction reduction within a step is of greater interest since a slip and fall incident is more likely to occur. As reported earlier, slightly more than 50% of the friction variation combinations resulted in a friction reduction (44 out of 84). The combinations that resulted in a friction increase were then removed. The absolute reduction in COF had an average and standard deviation of 0.12 and 0.15, respectively. The relative reduction in COF had an average and standard deviation of 0.19 and 0.22, respectively.

## Conclusions

Friction variation is speculated to be a significant contributor to slip and fall incidents and was quantified in this study in six major working areas of a fast-food restaurant in the USA. The results indicated that there was a slightly more than 50% chance that employees might experience a friction reduction in a step in the major working areas studied. The absolute reduction in COF had an average and standard deviation of 0.12 and 0.15, respectively. The relative reduction in COF had an average and standard deviation of 0.19 and 0.22, respectively.

## References

American Society for Testing and Materials F-1677-96, 2004. Standard method of test for using a portable inclinable articulated strut slip tester (PIAST), Annual Book of ASTM Standards, vol. 15.07, West Conshohochen, PA, American Society for Testing and Materials.

Andres, R. O., O' Connor, D., Eng, T., 1992. A practical synthesis of biomechanical results to prevent slips and falls in the workplace, *Advances in Industrial Ergonomics and Safety IV: Proceedings of the Annual International Industrial Ergonomics and Safety Conference*, Edited by S. Kumar, Taylor and Francis, London, 1001–1006.

Chang, W. R., 2002. The effects of slip criteria and time on friction measurements, *Safety Science,* 40, 593–611.

Chang, W. R., Cotnam, J. P., Matz, S., 2003. Field evaluation of two commonly used slipmeters, *Applied Ergonomics,* 34(1), 51–60.

Chang, W. R., Li, K. W., Huang, Y. H., Filiaggi, A., Courtney, T. K., 2004, Assessing floor slipperiness in fast-food restaurants in Taiwan using objective and subjective measures, *Applied Ergonomics*, 35(4), 401–408.

Chang, W. R., Grönqvist, G., Leclercq, S., Brungraber, R. J., Mattke, U., Strandberg, L., Thorpe, S. C., Myung, R., Makkonen, L., Courtney, T. K., 2001a. The role of friction in the measurement of Slipperiness, Part 2: Survey of friction measurement devices, *Ergonomics,* 44(13), 1233–1261.

Chang, W. R., Grönqvist, G., Leclercq, S., Myung, R., Makkonen, L., Strandberg, L., Brungraber, R. B., Mattke, U., Thorpe, S. C., 2001b. The role of friction in the measurement of slipperiness, Part 1: friction mechanisms and definition of test condition, *Ergonomics,* 44(13), 1217–1232.

Courtney, T. K., Sorock, G. S., Manning, D. P., Collins, J. W., Holbein-Jenny, M. A., 2001. Occupational slip, trip, and fall-related injuries – can the contribution of slipperiness be isolated? *Ergonomics,* 44(13), 1118–1137.

Grönqvist, R., Abeysekera, J., Gard, G., Hsiang, S. M., Leamon, T. B., Newman, D. J., Gielo-Perczak, K., Lockhart, T. E., Pai, C. Y. C., 2001. Human-centred approaches in slipperiness measurement, *Ergonomics,* 44(13), 1167–1199.

Leamon, T. B., Murphy, P. L., 1995. Occupational slips and falls: more than a trivial problem, *Ergonomics,* 38(3), 487–498.

Leclercq, S., Tisserand, M., Saulnier, H., 1997. Analysis of measurements of slip resistance of soiled surfaces on site, *Applied Ergonomics*, 28(4), 283–294.

Strandberg, L., 1985. The effect of conditions underfoot on falling and overexertion accidents, *Ergonomics,* 28, 131–147.

# FRICTION MEASUREMENT ON INCLINED SURFACE USING THE BRUNGRABER MARK II SLIPMETER

**Kai Way Li[1,2], Wen-Ruey Chang[1], Chin Jung Chen[2] & John-Chang Wei[3]**

[1] *Liberty Mutual Research Institute for Safety, Hopkinton MA 01748, USA*
[2]*Department of Industrial Management, Chung-Hua University, Hsin-Chu 300, Taiwan*
[3]*Department of Civil Engineering, Chung-Hua University, Hsin-Chu 300, Taiwan*

Slip and fall accidents are a serious safety problem in the workplace. It is a common belief that walking on an inclined surface is more likely to slip as compared to walking on a level surface. While some biomechanical studies concerning slipping on an inclined surface have been reported, tribological investigation on this surface is rare in the literature. Friction measurements are accomplished using slipmeters. One of the slipmeters – the Brungraber Mark II (BMII) is commonly used in the USA. The weight of a cast iron on this slipmeter is used to drive the footwear pad on the floor. On an inclined surface, the measurements of such a slipmeter on the downward direction will be affected by the gravity. Different COF values will be obtained for the same floor with different inclined angles. This paper describes friction measurements conducted on a ramp with different inclined angles using the BMII. The objective of this study is to determine the effects of inclined angle on the measured COF for three footwear materials, three floors, and two contamination conditions using the BMII slipmeter. Friction measurements were conducted on the downward direction of an adjustable ramp. Experimental outcomes and friction measurement problems on inclined surfaces using the BMII are discussed.

## Introduction

Slip and fall accidents are serious safety problems in the workplace (Leamon & Murphy, 1995). Slips may be attributed to inadequate friction between the footwear/floor interfaces. Biomechanical and tribological approaches have been adopted in studying slip and fall phenomena. The former tends to analyze the walking behavior to determine the causes of slip and fall whereas the latter determines the factors affecting the coefficient of friction (COF) between a shoe/floor interface.

Past research, both the biomechanical and tribological ones, has concentrated on the measurements on level surfaces. Investigation and discussion about slips and falls on inclined surfaces were not common. Hanson et al. (1999) conducted a gait experiment on a 1.8 m long ramp covered with either non-waxed vinyl composite tile or a low-loop carpet. The floor conditions included dry, wet, and soapy. Three ramp angles were used: 0°, 10°, and 20°.

The required COF values under their experimental conditions were calculated for downward gait trials. The results indicated that, on either floor, a larger ramp angle resulted in a higher required COF. This implied that slips and falls are more likely to occur on a steeper incline. Similar results were reported by Cham and Redfern (2002).

Determination of COF on an inclined surface has not, however, been reported. Theoretically, the COF of a footwear/floor is determined by the materials, geometrical texture, and contaminants associated with the interface. A floor should possess the same COF whether it is a level one or inclined one. However, this may not be supported by the friction measurement on a ramp using a slipmeter such as the Brungraber Mark II (BMII) because of the gravity effects. This paper describes friction measurements conducted on a ramp with different inclined angles using the BMII. Experimental outcomes and friction measurement problems using the BMII on inclined surfaces are discussed.

## Methods

### *Development and fabrication of an adjustable ramp*

To conduct friction measurements on an inclined surface, a ramp supported by a metal rig was fabricated and used in the study. The ramp encompassed certain features. First, leveling of the ramp was achieved to fulfill the requirement of commonly used slipmeters. Second, the inclined angle may be easily adjusted to set up the ramp angle. Third, the floor on the ramp may be easily changed for measurements of various floor conditions. Finally, a slipmerter is operable on the ramp.

### *Slipmeter*

The friction measurements were conducted using a BMII slipmeter. The operating principle of this tester is to simultaneously apply forces parallel and perpendicular to a floor by impacting a footwear pad on the floor. A weight of 4.54 kg drives an inclined articulated strut to impact the floor surface at an inclined angle to the normal direction of the floor. There is an alumium piece on the lower end of the strut which suspends an alumium shoe holder that accommodates a footwear pad. Both the aluminum piece and shoe holder are connected by a hinge. The aluminum piece is aligned to the strut before the footwear pad strikes on the floor. The hinges allow forward rotation of the piece and the footwear pad upon striking on the floor for a possible slipping movement.

The starting COF for a measurement should be smaller than the COF at which a slip was anticipated and the COF was increased until a slip occurs. A minimum COF of 0 was used as an initial value. The strut was perpendicular to the floor before it struck the floor at this minimum initial value. To operate on an inclined surface in the downward direction, however, the aluminum piece swung forward due to the effects of the gravity and became not aligned with the strut before the footwear pad struck the floor. This created a measurement problem: the striking point of the footwear pad on the floor was very close to the front end of the rig and the slip/non-slip judgment became very difficult. To overcome this problem, two initial COF values were adopted: 0.125 for the 10° and 0.27 for the 20° inclined surfaces. The piece suspending the footwear pad was perpendicular to the floor surface under these two values. The COF was decreased if the footwear pad slipped upon striking on the floor. The COF was increased, on the other hand, if the footwear pad did not slip.

### *Floors, footwear materials, and surface conditions*

Three floors and three footwear materials were tested in the study. The floors included steel, ceramic, and terrazzo with $R_a$ values, also known as the center line average of surface heights (CLA), of 2.81 (±0.38), 6.44 (±0.84), and 1.35 (±0.40) μm, respectively. The footwear materials included Neolite, blown rubber (BR), and ethylene vinyl acetate (EVA) with the densities of 1.28, 0.93, 0.27, and 0.20 g/cm$^3$, respectively. Two contamination conditions were tested: dry and glycerol-contaminated. For the dry conditions, friction measurement was conducted on clean dry floors. For glycerol conditions, two drops (0.06 ml) of glycerol (weight ratio of 98) were applied on the testing area which was identical to the area of the footwear pad. The glycerol was then evenly spread across the measuring area using a flat plastic stick.

Before the friction measurements, all the footwear pads and tiles were wiped with a 50% ethanol solution and were blown dry using a hair drier. For the glycerol-contaminated conditions, glycerol was applied after this cleaning process. In order to maintain a consistent surface, the testing footwear pad was sanded before each measurement using the protocol proposed by Chang (2002).

### *Experimental design and statistical analysis*

The friction measurements were conducted using a four factor (3 footwear materials $\times$ 3 floors $\times$ 2 surface conditions $\times$ 3 inclined angles) randomized experimental design with four replications. A four-way analysis of variance (ANOVA) was performed on the COF values obtained from the friction measurements.

## Results and discussion

The ANOVA results indicated that the main factors, and two-way, three-way and four-way interactions were all statistically significant at $p < 0.0001$. The means and standard deviations of the COF at 0°, 10°, and 20° inclined angles averaged across all other factors were 0.37 (±0.26), 0.30 (±0.23), and 0.13 (±0.27), respectively. On the dry floors, the means and standard deviations of the COF at 0°, 10°, and 20° inclined angles averaged across all the footwear and floor conditions were 0.54 (±0.27), 0.43 (±0.25), and 0.25 (±0.34), respectively. On glycerol-contaminated floors, the means and standard deviations of the COF at 0°, 10°, and 20° inclined angles averaged across all the footwear and floor conditions were 0.19 (±0.11), 0.17 (±0.09), and 0.00 (± 0.00), respectively.

Figures 1 and 2 show the measured COF values under different footwear, floor, and inclined angle conditions on the dry and glycerol contaminated surfaces, respectively. It has been well established that measured COF values are footwear, floor, and contaminant dependent (Grönqvist, R., 1995; Grönqvist, R., 1999; Chang & Matz, 2001; Li, et al., 2004). In addition, the measured COF was dependent on the inclined angle. In general, lower COF values were recorded for floors with a larger inclined-angle.

On dry surfaces (see Figure 1), the highest mean COF on a level floor was recorded in the BR/ceramic combination (0.94). The lowest mean COF was recorded in the Neolite/terrazzo combination (0.27). On the 10° ramp, the BR/ceramic combination also had the highest mean COF (0.89) among all the footwear/floor conditions. The EVA/steel combination, on the other hand, had the lowest COF (0.24). On the 20° ramp, the BR/ceramic combination had again the highest mean COF (0.78). The COF values of five footwear/floor combinations (Neolite/steel, EVA/steel, BR/steel, EVA/terrazzo, and BR/terrazzo) out of nine were zero.

On the glycerol contaminated floors (see Figure 2), the highest mean COF on level floor was recorded in the EVA/ceramic combination (0.34). The lowest mean COF was recorded in the Neolite/terrazzo combination (0.01). On the 10° ramp, the EVA/ceramic combination, again, had the highest mean COF (0.27) among all the footwear/floor conditions. The Neolite on both terrazzo and steel floors had a COF of zero. On the 20° ramp, the COF values of all the footwear/floor combinations were zero.

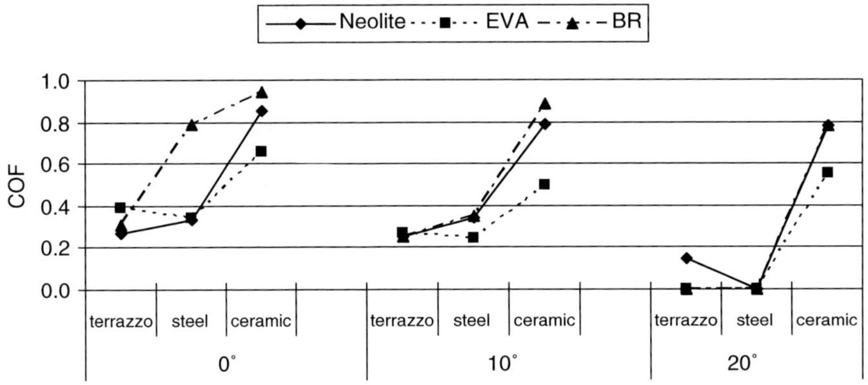


**Figure 1. Measured COF under footwear/floor/inclined angle conditions on dry surface.**

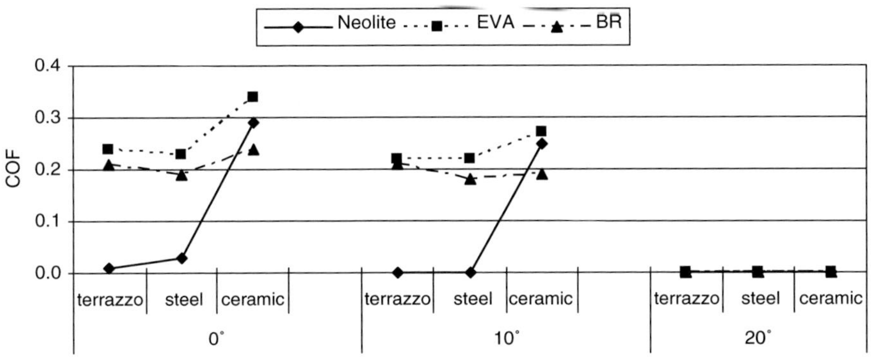


**Figure 2. Measured COF under footwear/floor/inclined angle conditions on glycerol contaminated surface.**

## Conclusion

This research investigated the measured COF on both level and inclined surfaces under 18 footwear material/floor/surface conditions using a BMII slipmeter. While the dependency of the measured COF on the footwear/floor/contaminant conditions has been established in the literature (Grönqvist, 1995; Grönqvist, 1999; Chang, 2003; Li et al., 2004), the present study found that the inclined angle of the floor surface also affects the measured COF. The measured COF using the BMII slipmeter decreased as the inclined angle of the ramp was increased. The results support the commonly held belief that walking downward on a steeper ramp will experience a less slip resistance on the floor.

## References

Cham, R., Redfern, M.S., 2002. Heel contact dynamics during slip events on level and inclined surfaces. *Safety Science* **40**, 559–576.

Chang, W.R., 2002. The effects of slip criteria and time on friction measurements, *Safety Science* **40**, 593–611.
Chang, W.R., Matz, S., 2001. The slip resistance of common footwear materials measured with two slipmeters, *Applied Ergonomics* **32**, 540–558.
Chang, W.R., Cotnam, J.P., Matz, S., 2003. Field evaluation of two commonly used slipmeters, *Applied Ergonomics* **34** (1), 51–60.
Grönqvist, R., 1995. Mechanisms of friction and assessment of slip resistance of new and used footwear soles on contaminated floors. *Ergonomics* **38** (2), 224–241.
Grönqvist, R., 1999. Slips and falls, In: Biomechanics in Ergonomics, Edited by S. Kumar, Chapter 9, 351–375, Taylor & Francis, London.
Hanson, J.P., Redfern, M.S., Mazumdar, M., 1999, Predicting slips and falls considering required and available friction. *Ergonomics* **42** (12), 1619–1633.
Leamon, T.B., Murphy, P.L., 1995. Occupational slips and falls: more than a trivial problem, *Ergonomics* **38** (3), 487–498.
Li, K.W., Chang, W-R., Leamon, T.B., Chen, C.J., 2004. Floor slipperiness measurement: friction coefficient, roughness of floors, and subjective perception under spillage conditions. *Safety Science* **42**, 547–565.

# A SURVEY OF OCCUPATIONAL FALLS AND SLIP RESISTANCE CHARACTERISTICS OF SAFETY FOOTWEAR IN SPAIN

**Juan Carlos González, David Garrido, Amparo Valero, Ricard Barberà, Mario Comín, Mª José Such, Carlos Soler & Pedro Vera**

*Instituto de Biomecánica de Valencia, Universidad Politécnica de Valencia, edificio 9C, Camí de Vera s/n. E-46022, Valencia, SPAIN*

An horizontal study has been performed involving several companies of diverse Spanish sectors where occupational footwear is used. More than 500 questionnaires were completed among different works. Question about number of falls, footwear used, working surfaces and contaminants were included.

Moreover, to complete the study, 19 safety shoes form principal Spanish manufacturers were tested to assess friction properties in order to improve their slip resistance characteristics.

## Introduction

A big effort has been done to establish the role of footwear and floor characteristics and to develop standard friction tests. There is a great variety of methods for measuring slipping behaviour of surfaces (Chang et al., 2001) and diverse standards have been developed to assess friction and safety of floors (DIN 51097, DIN 51130, UNE 127020 to 127024, AS 3661.1) and footwear (EN 13287:2004 and ASTM – F695-96). However, slips, trips and falls are one of the main sources of occupational injuries. Courtney et al. (2001), in a study comparing diverse sources in the USA, the UK and Sweden reported that slip, trip and fall related occupational injuries comprise between 20 to 40% of disabling occupational injuries. The contribution of slips to fall related injuries is between 40 to 50%, percentages that make think about the importance of an improvement in the friction characteristics of safety footwear.

According to the 4th National Survey of Working Conditions of the National Institute of Health and Safety at Work (INSHT) developed from September to December 1999, falls at the same level were the third risk detected by workers, with a frequency of 18.4%. The objectives of this study were, on one hand, the quantification of occupational falls in diverse Spanish industries and, on the other hand, to determinate the influence of slip resistance characteristics of safety footwear in slip related falls. The achievement of these objectives could permit the improvement of safety footwear design and therefore a decrease in the number of falls due to slips and an increase in the safety at work.

## Methodology

Surveys to safety footwear users of different sectors and jobs were developed in order to analyze in depth which are the needs of safety footwear users concerning friction with floor.

To accomplish the surveys three basic premises were taken into account:

- the area of study had to comprise companies of all industry sectors of Spain.
- people interviewed had to be workers of these companies with ages between 18 and 65 years old and current users of safety footwear.
- surveys had to be carried out at the end of the working day.

A total of 16 companies of different sectors, representative of the Spanish industry, were selected. The sectors compressed were: automotion, ceramics, construction, chemical, distribution, food, iron and steel and transport.

The sample size was estimated of at least 457 subjects with an admissible maximum error of 4.5% for a confidence interval to 95%. In order to obtain the as much as possible, 835 surveys were sent to the companies. Finally 501 surveys of men users of safety footwear were received.

It is also necessary to reveal the lacks of safety footwear currently used in order to improve the friction characteristics of this type of footwear. To obtain these lacks objectively, several friction tests were carried out.

So that, as well as the quantification of occupational falls in diverse Spanish industry fields, the aim of this survey was to quantify objectively the friction behaviour of a great sample of occupational shoes. For this reason, 19 safety shoes were selected randomly among the principal Spanish manufacturers in order to measure the Dynamic Coefficient of Friction (DCOF) by means of a slip testing device developed by the Institute of Biomechanics of Valencia. This machine allows to work in different conditions of speed, floor, vertical force and application time simulating, in an accurate way, real conditions of walking. Each sample is placed in an artificial foot and by means of an articulate clamp its contact inclination can be chosen between 0 to 15°.

To perform the tests, a revision of test conditions was made in order to assess which trials have obtained better results in the past. Consequently, according to Wilson (1990), Jung and Schenk (1990) and Gauglhofer (1996) the parameters more commonly used are: displacement speeds of 0.1 and 0.7 m/s, being 0.2 m/s the most extended; vertical force of 500 N for adult man; shoe inclination with floor between 5–10°; floor of stainless steel covered with a thin film of glycerine.

Taking into account this information, the assessed conditions during this survey were: displacement speed 0.25 and 0.5 m/s; Floor steel with glycerine, and dry and wet terrazzo due to their high frequency as industrial floors. Shoe inclination with floor surface was 10° in order to evaluate the most risky condition regarding friction during the first stage of heel strike.

Eventually, the DCOF was obtained in each trial dividing the average value of horizontal and vertical forces in the region of the signal where both remained nearly constant.

## Results

As it is shown in next graph, the main accident related to safety footwear in Spain was falls at the same level (96.8% of workers suffer it at least once a year), followed by bumps on feet, with a frequency of 28.54%.

Studying in depth falls frequency of appearance, 64.7% of workers are falling once per year, whereas 15.49% and 11.60% are falling two or three times per year respectively.

Moreover, through a cross tabulation analysis using ordinal association with Somers' d, there are significant differences in falls due to slips among industrial sectors ($p \leq 0.01$), having food sector a greater incidence of occupational accidents.

The main causes of this falls are the slips and trips, with percentages shared out of 44.14% and 41.44% respectively.

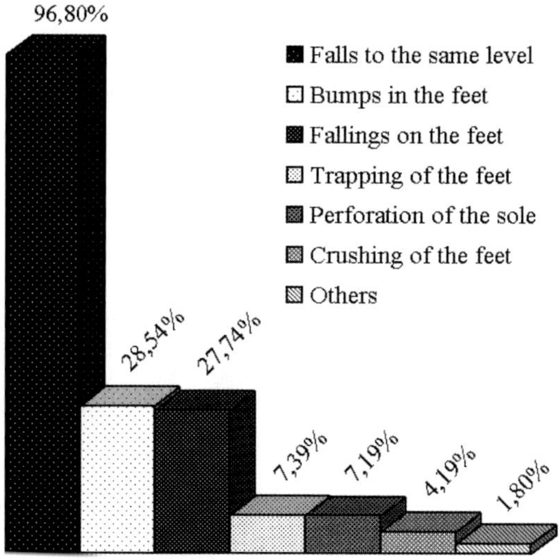


**Figure 1. Accidents.**

**Table 1. Most evaluated aspects.**

| Aspect | Frequency |
|---|---|
| Comfort | 63.5% |
| **Sliperiness over contaminated surfaces** | **20.8%** |
| Flexibility | 10.6% |

Through a cross tabulation analysis using ordinal association with Somers' d it has been observed a significant relationship in frequency and the number of falls due to slips among worker profession ($p \leq 0.01$). In that way, it can be concluded that operators are the group of people with highest percentage of falls.

It was also asked about the most valuated aspects of safety footwear concerning to satisfaction level. In the second place and with a frequency of 20.8% appeared the slipperiness over contaminated surfaces.

Taking into account the subjective perception of workers regarding shoe grip on dry and wet surfaces, it has been found a relationship between the perception and the number of falls due to slips. The analysis was made through an ordinal association with Somers' d and a cross tabulation analysis.

Regarding to the conditions of the working place, 84.03% of workers carried out tasks on surfaces of concrete or cement, although it appeared more types of surfaces like asphalt and terrazzo, with percentages of 27.35% and 21.96% respectively.

The pollutants that appeared with greater frequency were water (49.30%), oil (32.34%) and metallic particles (24.35%), appearing many others in minor measure.

Through a factorial analysis of principal components and a k-Means clustering algorithm, workers were grouped into 8 different combinations according to the type of surface and pollutants commonly used. Significant differences ($p \leq 0.003$) have been found among these groups in the frequency of falls due to slips through a cross tabulation analysis using ordinal association with Somers'd. The group of workers that suffer most frequently falls due to slips are those that walk over concrete/cement in presence of water and oil. Mechanics and operators mainly on food and ceramics sectors, form this group.

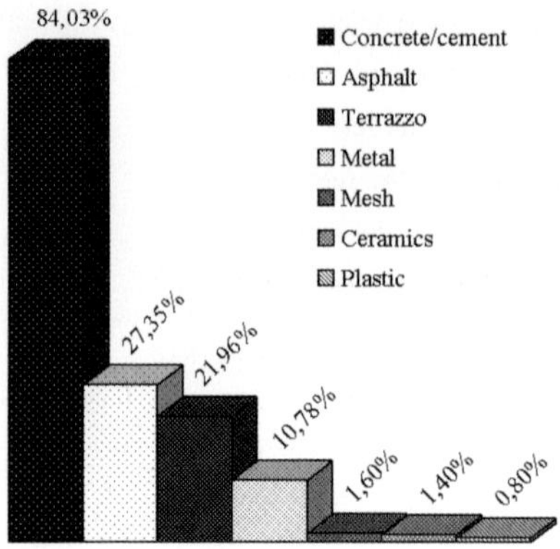


**Figure 2. Type of surface.**

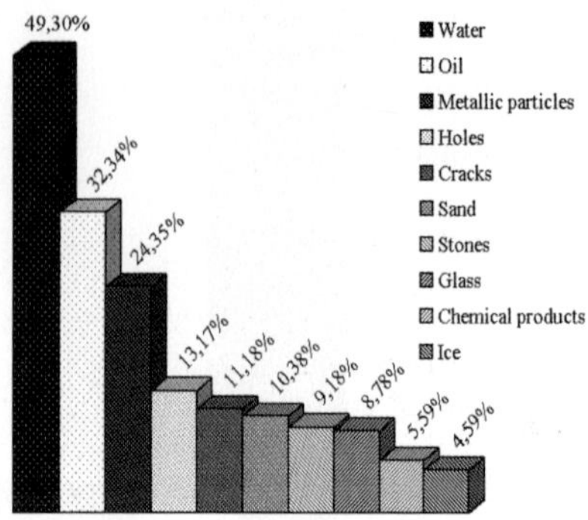


**Figure 3. Type of pollutant.**

**Table 2. DCOF Result.**

| Condition | Displacement speed (m/s) | Mean DCOF | Standard deviation | Max | Min | % of samples under recommended value |
|---|---|---|---|---|---|---|
| Dry Terrazzo | 0.25 | 0.78 | 0.11 | 1.0 | 0.6 | 0 |
| | 0.5 | 0.78 | 0.1 | 1.0 | 0.7 | 0 |
| Wet Terrazzo | 0.25 | 0.43 | 0.08 | 0.6 | 0.3 | 36.8 |
| | 0.5 | 0.42 | 0.09 | 0.6 | 0.3 | 42.1 |
| Steel with | 0.25 | 0.13 | 0.03 | 0.2 | 0.1 | 68.4 |
| glycerine | 0.5 | 0.13 | 0.03 | 0.2 | 0.1 | 73.7 |
| Differences | 0.25 | 0.3 | 0.13 | 0.6 | 0.1 | – |
| dry/wet terrazzo | 0.5 | 0.4 | 0.13 | 0.6 | 0.1 | – |

As a consequence of the results showed above, friction behaviour of the safety shoes has a big responsibility and play an important role in the great number of occupational accidents occurred in Spain.

In order to evaluate how the market answers against that problem, a total of 19 samples of occupational shoes from the principal Spanish manufacturers were evaluated. The nonslip capability of the soles was assessed by means of the DCOF.

In order to quantify the quality of the soles materials and their frictional behaviour the next principles were followed which are based on the European Standard EN 13287:2004 and on our own (IBV) recommendations: the DCOF measured on wet terrazzo should be over 0.4; measured on dry terrazzo over 0.5 and on steel with glycerine higher than 0.15. Next table shows the results of the tested shoes in the surfaces depicted above.

As it is shown in table 2 100% of the samples passed the minimum criterion for dry terrazzo (DCOF $>$ 0.5). According with these results, the frictional behaviour does not depend on the velocity, having all that samples similar characteristics.

On the other hand, for wet terrazzo there are some samples with a DCOF placed bellow the limit. In particular, 36.8% for 0.25 m/s and 42.1% for 0.5 m/s of shoes had a DCOF lesser than 0.4.

As regards a possible change in user walking patter depending on surface conditions (dry/wet), it is observed big differences which could lead to a dangerous situation during walking.

Finally, the values of DCOF on steel with glycerine, highlighting that around 70% do not pass test requirements (DCOF $> 0.15$).

## Conclusions

According to this study, the number of falls due to slips is very high in Spain, in particular operators of food and ceramic sector.

Several factors are involved in the risk of falling due to slips, being the friction characteristics of footwear one of the most important. In that way, this survey has permitted to obtained valuable information regarding the current situation of the Spanish occupational footwear market from a comfort and functionality point of view, which has revealed that most of the safety footwear does not meet the safety standards and this could be one of the reasons for the high incidence of slips.

More effort should be made in the design of footwear in order to improve slip resistance, characteristics of soles and to fit theses characteristics to the floors and contaminants present in the occupational environments.

## References

Courtney, T.K., Sorock, G.S. 2001, Occupational Slip, Trip, and Fall-Related Injuries – Can the Contribution of Slipperiness be Isolated? *Ergonomic,* Vol.44, 13, 1118–1137.

Gray, B., Jablonsky, D. 1990, Biomechanical Measurement: Pedestrian Coefficient of Friction. *American Society for Testing and Materials*, Philadelphia, 124–132.

Manning, D.P., Jones, C. 2001, The Effect of Roughness, Floor Polish, Water, Oil and Ice on Underfoot friction: Current Safety Footwear Solings are Less Slip Resistance than Microcellular Polyurethane. *Applied Ergonomics 32*, 185–196.

National Institute of Health and Safety at Work. 4th National Survey of Working Conditions [on-line] In: http://www.mtas.es/insht/# [10/11/2004].

Gauglhofer, J. (1996), Test method and specification for the determination of the slip resistance of safety, protective and occupational footwear for professional use. Document CENT/TC 161/WG 3, n°22, Rev.1, ECS.

Jung, K., Schenk, H. (1990), An international comparison of test methods for determining the slip resistance o shoes. *Journal of Ocupational Accidents* 13, 271–290.

Wilson, M.P. (1990), Development of SATRA Slip test and tread pattern design guidelines. Slips, stumbles and falls: Pedestrian footwear and surfaces, *ASTM STP* 1103, 113–123.

# TRANSPORT

# MANAGING FATIGUE RISKS

**Claire Dickinson**

*HSE, HMRI, Grove House, Skerton Road,*
*Manchester M16 0RB*

This paper describes the findings from a series of targeted inspections of 8 train/freight-operating companies (TOCs/FOCs) that considered the arrangements and measures for managing fatigue in train drivers. The paper notes the positive points that have emerged, as well as highlighting areas where continuous improvement has been sought. The use of the Fatigue Index has been an asset in carrying out the inspections. It is noted as particularly beneficial in identifying shift turns where there are raised fatigue risks. The presentation will endeavour to share information on what is regarded as good practice in roster design and fatigue management.

## Introduction

In a modern 24-hour society, demanding work schedules are common. However, some work schedules can affect a worker's safety or health. Fatigue can contribute to human error as a key element in many accidents. In round-the clock operations that involve shiftwork or irregular hours fatigue can be common. The symptoms of fatigue include subjective tiredness and a slowing of normal cognitive functions such as recall from memory, decision-making and reaction time. It can lead to reduced vigilance, increased errors, impaired decision making and deterioration in mood and motivation. For safety critical work, the effects of fatigue can give rise to increased risks. HSE has for many years regarded fatigue as a health and safety risk which can and should be managed as any other workplace risk, and produced guidance to assist employers in this duty (HSC, 1996; HSE, 1999, 1999b).

Errors attributed, at least in part, to fatigue are cited as contributory causes in a number of major accidents. Following the Clapham Junction accident Lord Justice Hidden noted there was no effective system for monitoring or limiting excessive working hours and recommended there should be (Dept. of Transport, 1989). Subsequent to the inquiry the rail industry laid down maximum working hour limits for safety critical workers of no more than a 12-hour shift, a rest interval of 12-hours between successive shifts, no more than 72 hours in a week and no more than 13 consecutive days to be worked. These appear in the Railway Group Standard GH/RT 4004, Appendix 1 and are referred to as the "Hidden Limits". The personal safety of workers may also be affected by fatigue. High levels of overtime working were cited as a contributory cause for the death of a lookout working on a busy line at Purley Oaks in 2001 (Croydon Magistrates Court Report, 2003).

HSE's Fatigue Index (FI) is a tool to provide an assessment of short-term, daily and cumulative fatigue risks associated with shift work (HSE, 1999). The FI is based on five main factors known to have an impact on fatigue: shift start time, shift duration, length of time between successive shifts, breaks and the number of consecutive shifts. HSE are particularly concerned about shift turns where the fatigue risk exceeds 30 following an earlier review of fatigue risk in a series of incidents (Lucas, 2003).

## Legal requirements

HMRI is at present part of the Health and Safety Executive, and is responsible for ensuring compliance with health and safety legislation on all UK railways, metros and light transport systems. The general duty to ensure the health and safety of employees is supplemented by a number of more detailed requirements to assess risks, implement control measures and have in place arrangements for ensuring risks continue to be prevented or managed properly. The Railway (Safety Critical Work) Regulations, 1994 regulation 4 requires employers to ensure, so far as reasonably practicable that employees do not carry out work for such hours as would endanger safety. The supplementary Approved Code of Practice (ACOP) on hours of work places a number of further requirements on employers: for carrying out a risk assessment before a material change to existing working time limits/patterns, to set limits for working hours, to provide breaks, monitor working hours, and measures to address problems experienced by individual employees. Railway employers also have legal duties to describe the arrangements for managing working hours to prevent fatigue risks in the safety case they are required to submit in order to be allowed to operate. The Working Time Regulations, 1998 (as amended) (WTR) extend working time measures in full to workers in the railway sector. Whilst the WTR place maximum limits on the amount of time that an employer can ask an employee to work, the provisions of the WTR are not sufficient to prevent work being undertaken by those who may be liable for fatigue.

## The inspections – method

Inspections were carried out by HMRI at 8 train/freight companies (TOCs/FOCs) between November 2002 and November 2004 to establish compliance with the legal requirements. Information was gathered in a standardised way using a question set and interviews were held with train drivers, their supervisors, roster clerks and depot managers. The question set prompted a review of: the arrangements and systems in place for controlling working hours; the steps taken to monitor and implement any changes to working hours; compliance with the Hidden limits; and compliance with good practice in roster design. The rosters were evaluated using the FI. Drivers' work is planned and structured, they follow their assigned roster and the work content and breaks are specified in the diagram provided for that shift.

## Results

### *Arrangements for managing working hours*

The inspections found a high level of consistency between all 8 companies in their arrangements and systems for controlling working hours. This may stem from pre-privatisation days. All companies identified their drivers as safety critical workers and set out arrangements for controlling their working hours. All companies included within their terms and conditions of employment the limits for working hours. In four companies, the drivers recognised improvements in their rosters/working hours (in terms of fatigue) as a result of reductions in their working hours to an average of 35 or 37-hour week. The agreements were restricted in setting out any limits for working patterns to stating a minimum of four days work to be assigned per week. All companies had a very detailed paper-based or computerised system for recording planned and actual hours worked that was linked to the pay system. The Roster Clerks were provided with a base roster twice yearly and used this to prepare a weekly roster and daily sheets that set out the timing of each driver's start and end time, the route they would take and type of train. Planning of working hours is linked to driver's route knowledge and traction knowledge, a lack of either resulting in a change to the weekly roster. In setting out the base roster/weekly roster or daily sheet, the Roster Clerk applies the company's defined limits for working hours. In all 8 companies these corresponded to the Hidden limits. The only deviation that was noted at three companies was the practice of reducing the time of returning to work for any non-safety critical work days e.g. safety briefs, attendance at medicals. The driver's supervisor who booked drivers on/off ensured any late changes to working hours (e.g. in the case when a driver reported in sick and his work was covered by another driver) also ensured compliance with the Hidden limits. Each company defined situations when the limits could be exceeded and where possible made arrangements to relieve the driver in the case that late running was going to result in a shift exceeding 12-hours in length. Best practice was demonstrated by two companies that asked three questions of the driver, in effect a mini risk assessment, to establish if the driver was willing and fit to continue in the event of working for 12-hours.

### *Roster design*

There were elements of good practice demonstrated in the base rosters evaluated for all eight companies. Typically blocks of 4 shifts had a consistent start time, 25% of weekends off, not excessive hours per week. No roster required more than 13-days consecutive days work and included more than 12-hours between successive shift turns. Aspects relating to forward rotation of start time and shift length are areas where improvements were sought. Shifts longer than 10-hours were consistently reported by the drivers as too long and tiring. The number of days rest after a series of night shifts usually was two but in some companies and especially between Saturday and Monday, the length of time provided for rest and recovery was occasionally insufficient. As a result, drivers report early starts on a Monday morning as particularly hard and when they were most tired. Only one company provided a risk assessment to demonstrate they had considered fatigue risks after making a material change to their time-table and driver's working hours. Equally there was less consideration of any factors that influenced the potential for fatigue and control to manage any increases in risk e.g. extensive travelling time, lengthy commuting time and hot working environment.

### *Diagram design*

All companies assigned breaks of sufficient length. In three companies improvements were sought in the timing of breaks. In five companies care had been taken to ensure breaks were assigned 3 hours after booking on and did not occur in the last hour of work. Improvements were sought in 2 company's work patterns that commenced at ~05:00 but provided the first break some 6–7 hours later. The amount of driving demanded within a diagram varied significantly between and within a company's diagrams. Drivers readily highlighted diagrams with very demanding work content. One company had shortened the length of night and early-starting shifts to about 8-hours recognising that drivers can tolerate longer shifts if they are timed during the daytime or evening.

### *Changes to working hours*

All companies demonstrated a controlled, documented system to enable shift swapping. In the event of late changes to the roster e.g. a driver telephoning in sick, a structured system operated to cover the turn. In each case care was taken by the driver's supervisor to ensure consistency of the start times with preceding or subsequent assigned shifts. In all cases compliance with the Hidden limits was adhered to.

Excessive overtime working could be problematic and led to raised fatigue risks as a result of its timing and the cumulative effects of fatigue. In all 8 companies a system of equalisation operated, such that those individuals with the lowest level of additional hours worked over the year were offered overtime in precedence to other people.

*Compliance with Hidden limits*

In all 8 companies the base rosters were compliant with the Hidden limits. Care was taken each day to remain within the Hidden limits, incurring an exceedance only in exceptional circumstances, e.g. diversions due to a tunnel closure. For TOCs, the level of exceedance approximated to 7 per 100 drivers over six months. This figure rose and varied considerably for freight or London-based depots. In two depots, infringements of the Hidden limits were rare, there being no instances in the previous three months. Any infringements for TOCs were invariably due to late-running trains. In 2 companies the reasons for infringements were less than 12-hours between successive shifts i.e. insufficient rest.

*Cultural factors*

Consultation on base rosters with employee representatives occurred at all 8 companies and at all locations was said to have been rushed. In all companies roster changes were made as a result of the employee involvement. Any alterations to diagrams were made less readily. Often problematic diagrams persisted in successive roster changes, especially if they were consistent with the agreements on working hours. One freight company demonstrated best practice by inviting drivers to highlight problem diagrams and assigned some-one to resolve the issues.

*Organisational factors*

All 8 companies had issued a booklet on coping with shiftwork. Drivers from all companies reported it impossible to follow aspects of the guidance ! Companies had issued a briefing to staff on shiftwork and fatigue, thereby demonstrating good practice and consistency with industry guidance (e.g. RSSB, 2003).

*Use of the Fatigue Index*

The FI was found to provide a standardised way of analysing the rosters and identify shifts where fatigue risks were raised. This usually resulted from successive very early, long shifts or a series of long, night shifts. Its use enabled a focused, prioritised way of effecting changes to the roster design. The FI was not relevant to permanent night rosters and was not so readily applied to freight company rosters where drivers work content varied considerably.

*Key deficiencies*

The key points from the inspections of where fatigue risks are raised and errors are more liable are related to roster design, diagram design and organisational/cultural factors.

*Roster design*

- Tiring working patterns/roster design but consistent with Hidden limits
- Rosters including booking on after about 12 hours rest liable to incur more changes to a driver's pattern of working hours
- More than four successive very early starts
- More than four consecutive night turns
- Turning from night to early shifts where insufficient recovery time provided

- Tendency to block 7–9 consecutive days work, not 3–4
- Spare turns – care needed where they are positioned in the roster (i.e. too many spare on late shifts and not enough on early shifts)
- A number of spare turns blocked consecutively are harder because they always end up being longer than assigned
- ±3 hours typical for start time adjustment of spare hours – drivers note this interval as being too much to cope with

*Diagram design*

- Long, early starting shifts (more than 10 hours)
- Long night turns more than 10 hours
- High proportion of driving within a diagram can be very tiring
- Diagrams incorporating a 3–4 hour long break during the night can be difficult to then become alert when resuming driving
- Time given to turn-round times insufficient
- Repetitive routes are problematic e.g. same 9-minute journey driven for 6 hours
- Timing of eating prior to early shifts can be difficult if the assigned break is some 6–7 hours into the shift
- Unfavourable timing of breaks e.g. close to when book on/off so provide little benefit from mitigating fatigue

*Organisational factors and culture*

- Persistent late running trains identified but not addressed in successive rosters
- Need to include training days/safety briefing within the roster
- If permanent nights, scheduling of briefings and training not consistent with working hours
- Position on significant voluntary work/2nd jobs is not always clearly set out in a company policy

## Concluding remarks

The inspections identified good practice for controlling working hours. In TOCs/FOCs control of working hours is primarily by compliance with the Hidden limits. These limits represent what was operational achievable by the rail industry in the late 1980s, rather than being based on technical information on controlling fatigue from working hours. There is an opportunity for the industry to define and agree good practice in roster and diagram design and collectively move away from controlling working hours to a position of managing fatigue risks.

## References

Department of Transport, (1989) *Investigation into the Clapham Junction Railway Accident.* [HMSO:London], ISBN 0 10 108 202 9.

HSC, (2004) *Proposals for new safety regulations for railways and other guided transport systems.* Consultative Document CD199.

HSC, (1996) *Railway safety critical work. Railways (Safety Critical Work) Regulations, 1994.* L50, 2nd edition, ISBN 0 7176 1260 0 [HSE Books : Sudbury].

HSE, (1999a) *Validation and development of a method for assessing the risks arising from mental fatigue.* HSE Contract Research Report CRR 254, 1999.

HSE, (1999b) *Reducing error and influencing behaviour.* HSG48 [HSE Books: Sudbury].
Lucas, D. A. (2003) Developing a fatigue index. Presented at Nodding Off, PACTS Conference, London, 4 February 2003.
RSSB, (1996) *Changes in working hours – safety critical work.* Railway Group Standard GH/RT 4004.
RSSB, (2003) *Feeling tired ? Briefing notes to support the "feeling tired?" powerpoint presentation. "Feeling tired? fatigue and shiftwork"* leaflet, www.rssb.co.uk/factsheets.asp
The Health and Safety Executive v Amec Rail Limited and Primat Recruitment Limited and Amec Services Limited T/A Mapel. Croydon Magistrates Court, 29 October 2003. Reported in the Court of Criminal Appeal Friday 10 March 2000.

The views expressed in this paper are those of the author and may not necessarily reflect HSE policy.

# DRIVING BEHAVIOUR *IN EXTREMIS*

**John R. Bloomfield**

[1]*University of Minnesota, Minnesota, USA*
[2]*University of Derby, Derby, UK*

This paper presents the results of an analysis of fatal and severe injury crashes involving commercial motor vehicles (CMVs). The analysis included 463 crashes that occurred in Minnesota between July 1st 2000 and December 31st 2002. The crashes were assigned to one of eleven categories. Three of these categories – stop sign-controlled intersection crashes; rear-end crashes; and crossing-the-centerline crashes – accounted for 287 (62.0%) of the crashes. Further analysis of these three categories suggests that CMV drivers are far less likely than the drivers of other vehicles to take actions that lead to a fatal or severe injury crash.

## Introduction

Each year there are huge numbers of crashes on highways in the U.S.A. For example, according to the National Highway Traffic Safety Administration (NHTSA), in 2002 there were an estimated 6,316,000 crashes – the crash rate was 223 crashes per 100 million miles traveled (NHTSA, 2004). Further, there were fatalities in 38,309 of these crashes – with a fatality crash rate of 1.35 crashes per 100 million miles traveled (NHTSA, 2004). NHTSA acquires its crash data from two sources. Estimates of the total number of crashes come from the National Automotive Sampling System – General Estimates System (GES), and the data on fatal crashes come from the Fatality Analysis Reporting System (FARS). FARS attempts to obtain data on every fatal crash that occurs in the U.S.A. FARS data are provided by an agency in each state's government. And some states, including Minnesota, collect more information than FARS uses. In an effort to provide a more detailed account of the circumstances in which crashes occur, an analysis of fatal and severe injury crashes involving commercial motor vehicles (CMVs) was conducted. The analysis included 463 crash reports completed by state patrol officers or local police officers. The crashes occurred in Minnesota in the thirty-month period between July 2000 and December 2002. The distribution of fatal and severe injury crashes involving CMVs across months in this time period is shown in Table 1.

The table shows that the more fatal and severe injury crashes occurred in August than any other month. This may be because in August many Minnesotans take driving vacations – and as a result may be traveling on relatively unfamiliar roads.

## Categorization of crash types

Each of the 463 fatal and severe injury crashes involving CMVs that occurred between July 1st 2002 and December 31st 2002 was assigned to one of eleven categories. The distribution of the crashes among these categories is shown in Table 2. Three of these categories accounted for 62% of the crashes – stop-controlled intersections (25.3%); rear-end crashes (17.5%); and crossing-the-centerline crashes (19.2%). Further analysis of these three categories is presented below. [Bloomfield (2003) gives details of the

**Table 1. Number of fatal and severe injury crashes involving CMVs in each month between July 2000 and December 2002.**

| Month | 2000 (July to December) | 2001 | 2002 |
|---|---|---|---|
| January | – | 16 | 8 |
| February | – | 10 | 19 |
| March | – | 15 | 15 |
| April | – | 12 | 12 |
| May | – | 16 | 14 |
| June | – | 17 | 14 |
| July | 24 | 13 | 12 |
| August | 26 | 25 | 21 |
| September | 15 | 14 | 8 |
| October | 9 | 15 | 23 |
| November | 15 | 16 | 10 |
| December | 20 | 12 | 17 |
| Total | 109 | 181 | 173 |

**Table 2. Number of fatal and severe injury crashes involving CMVs between July 2000 and December 2002 in each crash category.**

| Crash category | 2000 (July to December) | 2001 | 2002 | Total |
|---|---|---|---|---|
| Signalized intersections | 7 | 11 | 14 | 32 |
| Stop-controlled intersections | 34 | 38 | 45 | 117 |
| Left turns | 4 | 12 | 6 | 22 |
| Uncontrolled intersections | 0 | 1 | 0 | 1 |
| Exit/entrance ramp | 0 | 1 | 1 | 2 |
| Driveways | 0 | 2 | 4 | 6 |
| Rear-end | 20 | 39 | 22 | 81 |
| Crossing-the-centerline | 19 | 37 | 33 | 89 |
| Crossing-the-median | 0 | 2 | 9 | 11 |
| Run-off-the-road | 10 | 16 | 10 | 36 |
| Other | 15 | 22 | 29 | 66 |
| Total | 109 | 181 | 173 | 463 |

other crash categories – i.e., signalized intersection crashes; left turn crashes; unsignalized intersection crashes; freeway exit/entrance ramp crashes; driveway crashes; crossing-the-median crashes; run-off-the-road crashes; and other crashes.]

## Stop-controlled intersections

The fatal and severe injury crashes involving CMVs that occurred at stop-controlled intersections were categorized further, in terms of the following: (i) whether the vehicle pulling out from the stop sign is the CMV, or the Other vehicle, and (ii) whether the vehicle pulling out from the stop sign strikes the vehicle on the main road, or is struck

**Table 3. Sub-categorization of fatal and severe injury crashes involving CMVs that occurred at stop-controlled intersections between July 2000 and December 2002.**

| | 2000 (July to December) | 2001 | 2002 | Total |
|---|---|---|---|---|
| *CMV thru stop* | 3 | 10 | 9 | 22 |
| CMV hits Other | 2 | 2 | 4 | 8 |
| Other hits CMV | 1 | 8 | 5 | 14 |
| *Other thru stop* | 24 | 25 | 34 | 83 |
| CMV hits Other | 13 | 14 | 21 | 48 |
| Other hits CMV | 10 | 9 | 11 | 30 |
| CMV avoids Other, but overturns | – | 1 | – | 1 |
| Two Others involved before CMV | 1 | 1 | 2 | 4 |
| *Weather-Related* | 3 | 3 | 1 | 7 |
| Weather (CMV thru) | – | 2 | – | 2 |
| Weather (Other thru) | 3 | 1 | 1 | 5 |
| *Incomplete information* | 4 | – | 1 | 5 |
| Total | 34 | 38 | 45 | 117 |

by that vehicle. The reason for the second distinction is as follows. If the vehicle that goes through the stop sign *strikes* the vehicle on the main road when the latter is in the intersection, then it is reasonable to assume that the driver of the latter vehicle would have been unable to take evasive action. On the other hand, if the vehicle that went through the stop sign *was struck* by the vehicle on the main road, then it is possible that, in some cases, the driver of the latter may have been able to take some action to avoid the crash, or that he or she may have been traveling too fast. Table 3 shows the sub-categorization of crashes occurring at stop-controlled intersections.

Excluding the seven weather-related crashes and the five crashes for which there was insufficient information, Table 3 shows that on 22 (18.8%) occasions CMVs went through the stop sign, while Other vehicles went through the stop sign on 83 (70.9%) occasions. The table also shows that 38 times (8 with CMVs, 30 with Other vehicles) the vehicle going through the stop sign hit the vehicle on the main road. And 62 times (14 with the Other vehicles; 48 with the CMVs) the vehicle that went through the stop sign was struck by the vehicle on the main road. It is generally assumed that the driver going through the stop sign is at fault. However, sometimes the driver on the main road may be traveling too fast (either because he or she was speeding, or because the speed limits on the main road are too high). Stop-controlled intersections at which crashes are frequent should be investigated further – it may be that lowering the speed limit on the approaches to some of them may be warranted.

## Rear-end crashes

Table 4 presents additional details of the fatal and severe injury crashes rear-end crashes involving CMVs. Excluding the crashes that involved equipment problems,

**Table 4. Sub-categorization of fatal and severe injury rear-end crashes involving CMVs that occurred between July 2000 and December 2002.**

| | 2000 (July to December) | 2001 | 2002 | Total |
|---|---|---|---|---|
| *CMV driver inattentive* | 9 | 5 | 8 | 22 |
| CMV equipment problem | – | 2 | – | 2 |
| *Other driver inattentive* | 7 | 23 (3 alcohol, 1 medical problem) | 14 (1 alcohol 1 medical problem) | 44 (4 alcohol, 2 medical problems) |
| Weather (CMV hits Other) | – | 3 | – | 3 |
| Weather (Other hits CMV) | 4 | – | – | 4 |
| Misjudgment (CMV hits Other) | – | 2 plus 1 (in which CMV causes crash between three Other vehicles) | – | 3 |
| *Misjudgment (Other hits CMV)* | – | 1 | – | 1 |
| *Stupidity* | – | 2 (1 CMV, 1 Other) | – | 2 |
| Total | 20 | 39 | 22 | 81 |

weather-related crashes, and crashes that involved misjudgments or stupidity, Table 4 shows the CMV drivers were inattentive in 22 (27.2%) of the rear-end crashes, while the drivers of the Other vehicles were inattentive in 44 (54.7%) of them.

## Crossing-the-centerline crashes

Table 5 presents details of fatal and severe injury crossing-the-center crashes involving CMVs. The table shows that 71 (79.8%) of these crashes occurred on straight highway sections, while the remaining 18 (20.2%) occurred on curves. It also shows that, excluding equipment failures, weather related crashes, and one case where a CMV swerved to avoid an animal, crossing-the-centerline crashes occurred 5 times (1 on the curves, plus 4 on straight road sections) when CMVs drivers were inattentive or incapacitated – i.e., on 5.6% occasions. In contrast, crossing-the-centerline crashes occurred 59 times (9 on the curves, plus 50 on straight road sections) when the drivers of the Other vehicles were inattentive or incapacitated – i.e., on 66.3% occasions.

## Comment

Three crash categories accounted for 287 (62.0%) of the 463 fatal and severe injury crashes involving CMVs that occurred in Minnesota between July 1st 200 and

**Table 5. Sub-categorization of fatal and severe injury crossing-the-centerline crashes involving CMVs that occurred between July 2000 and December 2002.**

| | 2000 (July to December) | 2001 | 2002 | Total |
|---|---|---|---|---|
| *Curved Road* | | | | 18 |
| *Inattentive/incapacitated driver (CMV)* | – | 1 | – | 1 |
| *Inattentive/incapacitated driver (Other)* | 2 | 2 | 5 | 9 |
| Excessive speed (CMV) | – | – | – | – |
| Excessive speed (Other) | 1 | – | 1 | 2 |
| Weather (CMV) | – | – | 1 | 1 |
| Weather (Other) | 1 | 3 | 1 | 5 |
| *Straight Road* | | | | 71 |
| *Inattentive/incapacitated driver (CMV)* | – | 3 | 1 | 4 |
| *Inattentive/incapacitated driver (Other)* | 13 | 22 | 15 | 50 |
| Weather (CMV) | – | – | – | – |
| Weather (Other) | 2 | 6 | 6 | 14 |
| Equipment (CMV) | – | – | 1 | 1 |
| Equipment (Other) | – | – | 1 | 1 |
| Animal avoidance (CMV) | – | – | 1 | 1 |
| Total | 19 | 37 | 33 | 89 |

December 31st 2002. In only 49 (22 plus 22 plus 5) of these crashes (17.1%) the drivers of the CMVs went through a stop sign, rear ended the other vehicle, or crossed the centerline; in contrast in 186 (83 plus 44 plus 59) of these crashes (i.e., 64.8%) the drivers of the Other vehicles went through a stop sign, rear ended a CMV, or crossed the centerline. These data suggest that CMV drivers are far less likely than the drivers of Other vehicles to take actions that lead to a fatal or severe injury crash.

## References

Bloomfield, J.R., 2003, Fatal & severe injury crashes involving commercial vehicles (July 2000 to December 2002): Crash types & primary causes, Technical Report submitted to Minnesota State Patrol, Minnesota Commercial Vehicle Section.

NHTSA, 2004, *Traffic Safety Facts 2002: Final Edition*, National Highway Traffic Safety Administration, National Center for Statistics and Analysis, U.S. Department of Transportation Report # DOT HS 809 620; Washington, D.C., January 2004.

# A TEST FLIGHT FOR MALLEABLE ATTENTIONAL RESOURCES THEORY

**Mark S. Young[1] & Greg F. Clynick[2]**

[1]*School of Engineering and Design, Brunel University, Uxbridge, Middx UB8 3PH, UK*
[2]*CAE Australia Ltd., Unit 1 Holker Street, Newington NSW 2127, Australia*

Malleable Attentional Resources Theory (MART) posits that attentional capacity can actually shrink in line with decreases in task demand, thus causing performance decrements in underload scenarios. An experiment was conducted in a medium-fidelity flight simulator to determine whether MART is applicable in the aviation domain, and to plot the decay and recovery curves of attentional capacity under varying mental workload (MWL) conditions. Whilst the MWL manipulation did not produce the expected effects on a secondary task, some intriguing results were still observed. The MART framework is assessed in the light of these findings, and practical implications for pilot MWL and future research are discussed.

## Introduction

Advances in flight deck automation and interface technologies can be paradoxically responsible for causing both overload and underload, which may be equally detrimental to performance (Hancock and Parasuraman, 1992; Leplat, 1978). Attentional resource theories (e.g., Kahneman, 1973; Wickens, 2002a) offer a reasonable explanation of overload, but are not so forthcoming in explaining underload. To fill this gap, Young and Stanton (2002a; 2002b) proposed Malleable Attentional Resources Theory (MART). MART hypothesises that capacity may not actually be fixed, but could shrink with reductions in demand. Underload thus causes performance decrements in situations where capacity has shrunk to such an extent that a sudden increase in demand is no longer within the capacity of the operator. In this way, performance in underload situations is actually resource-limited in the same way as for overload (cf. Norman and Bobrow, 1975).

The present experiment was designed to extend MART into the aviation domain using a flight simulator, while also attempting to chart a "shrinkage curve" of attentional decay and recovery over time, which has to date not been possible. The latter objective is the most significant from both theoretical and practical viewpoints, as it could begin to define the elusive "redline" of MWL for both overload and underload.

## Method

### *Design and procedure*

The experimental task involved a relatively simple flight pattern, involving navigational tracking using the Richmond navigational beacon (NDB), near Sydney.

Eleven student pilots (10 male) faced two conditions in a within-subjects design – one at a reasonable level of task demands (the "normal" condition), and a comparable trial at low demand (the "underload" condition). In the "normal" condition, participants began flying at 3000 ft, with instructions every two minutes to adjust altitude by 1000 ft. At a normal climb/descent rate of 500 ft/min, this meant that participants were in effect continually adjusting altitude for the duration of the trial. In the underload condition, on the other hand, a constant 3000 ft altitude was maintained throughout.

In order to plot both the decay and the recovery of attentional resources as predicted by MART, each trial was divided in two. The first 10 minutes of each trial comprised the experimental manipulation – normal or underload condition. To extend the data collection into attentional recovery, the trial continued for a further 10 minutes at the "normal" level of demand. In total, then, participants faced two trials of 20 minutes each.

Finally, a critical event was introduced to assess the effects of underload on performance. An 80 kt crosswind gradually pushed the aircraft off track – requiring the pilot to make correctional heading changes in order to maintain the nominated track. The crosswind was introduced just prior to the midpoint of the trial for one minute, and again for one minute before the end of the trial to control for effects of fatigue or practice.

Throughout the experiment, participants were required to attend to a visual-spatial secondary task as a measure of spare attentional capacity (and thus indirectly a measure of mental workload). The secondary task was presented on a laptop in the co-pilot's seat, and responses were made via the mousepad buttons.

Dependent variables were therefore reaction time to the crosswind, and the number of correct secondary task responses as well as reaction time to each.

In addition, a low-light video camera was installed in the simulator to record participants' direction of attentional gaze. Time spent on the primary and secondary tasks was recorded throughout the experiment, and this was used to calculate the ratio of time spent on the secondary task against number of correct responses. Young and Stanton (2002b) argued that this ratio reflected the size of the attentional resource pool, and should remain constant under a fixed-capacity model. In an enhancement to previously collected data, though, both attentional gaze and reaction times on the secondary task were recorded continuously throughout the trial (instead of a total at the end), facilitating a plot of attentional capacity for the duration of the trial.

### *The simulator*

The simulator is located at Sydney's metropolitan airport in Bankstown, in the flight training hangar of the Department of Aviation, University of New South Wales. It is a fixed-base, medium fidelity system, with a fully interactive cockpit mock-up and two cathode-ray display screens showing the computer-generated external world and the primary flight instruments. For the purposes of this experiment, the flight model was based on a Cessna 172 single-engine aircraft.

## Results

### *Mental workload*

A visual inspection of the secondary task data revealed a possible reduction in MWL in the normal condition after the critical event. However, this was not borne out statistically in a Friedman analysis of all four conditions (*chi-square*(3) $= 4.86; p = 0.183$).

Nonetheless, further analysis using the Wilcoxon test did produce a significant result for this contrast ($Z = -2.94; p < 0.005$). Although we must treat this result with caution given that the overall Friedman analysis was nonsignificant, the high significance level implies that there may be a genuine difference across these conditions.

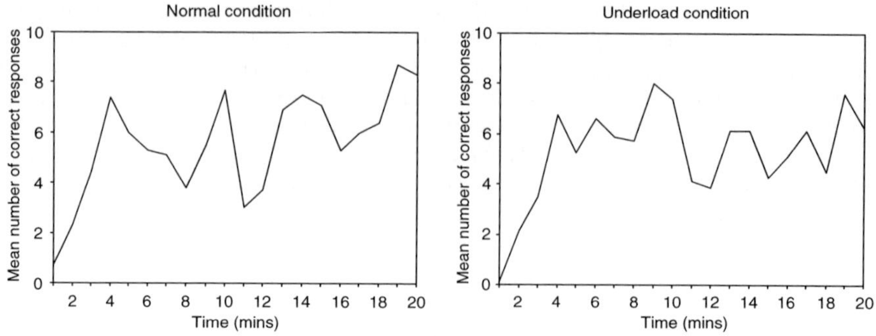


**Figure 1. Timeline plot of mean number of correct responses on the secondary task for each condition.**

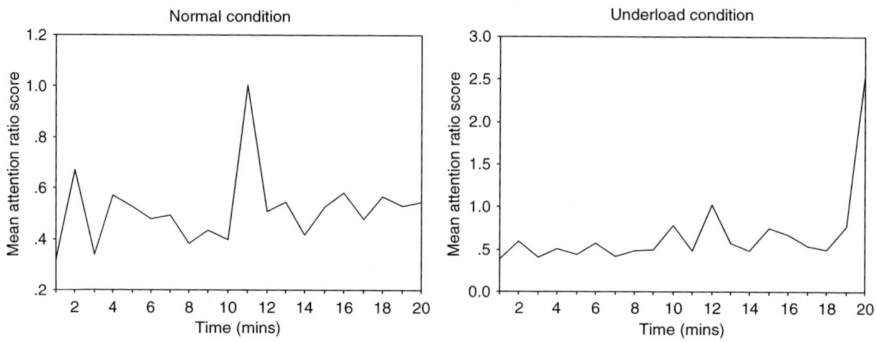


**Figure 2. Timeline plot of attention ratio data for each condition.**

In order to compare MWL data with the plot of attention ratio, the secondary task data were also divided into one-minute segments. Figure 1 shows the pattern of workload variation over each trial.

There was a significant drop in responses (i.e., increase in workload) around the midpoint in each trial – just after the critical event took place between segments 10 and 11 – in the normal condition ($Z = -2.32$; $p < 0.05$), with a marginal result for the underload condition ($Z = -1.69$; $p < 0.1$).

## *Flying performance*

Reaction time to the critical event was the only measure of primary task performance in this study, and Friedman analyses did not reveal any statistical differences across any of the conditions (*chi-square*(3) = 1.35; $p = 0.719$).

## *Attentional capacity*

Although the Friedman analysis across conditions was nonsignificant (*chi-square*(3) = 3.93; $p = 0.269$), post-hoc Wilcoxon contrasts did support the existence of a trend towards a general increase in capacity after the critical event (underload condition: $Z = -2.31, p < 0.05$; normal condition: $Z = -1.96$; $p < 0.1$).

Breaking this analysis down into one-minute segments, a clear difference begins to emerge between the normal and underload conditions (see Figure 2). There is a strong peak in capacity around the midpoint of the trial in the normal condition, whilst in the underload condition a similar peak occurs towards the end of the trial. Compare these graphs with the corresponding plots for secondary task data (Figure 1). Note that both the normal and underload conditions display an increase in workload (i.e., drop in secondary task responses) around the midpoint, yet the attention ratio plot for underload does not increase at this point. Whilst these results are visually compelling, none of the differences for attention ratio reached statistical significance.

## Discussion

Generally speaking, despite great care in the experimental design, the MWL manipulation appeared to be unsuccessful. Thus, conditions that were ostensibly designed as "normal" and "underload" appear to have had little real effect on MWL (although to avoid confusion, the condition labels will be retained for the remainder of this paper).

Indeed, the secondary task indicated a consistent decrease in workload in the "normal" condition following the critical event. This was contrary to expectations, since objective demand was equal in both of these scenarios. Furthermore, pre- and post-event scores in the "underload" condition were not different, again contrary to expectations.

None of these results, though, translated into flying performance effects as measured by reaction time to the crosswind. In the absence of MWL differences, this is actually consistent with MART. The corresponding attention ratio data revealed that attentional capacity was relatively stable (and if anything showed an increase) following the critical event in each condition. Mapping these results onto the MWL data raises problems for MART, as the predictions were unsupported – especially in the normal condition, where a MWL *decrease* was potentially associated with a capacity *increase*. Furthermore, MART would predict performance differences associated with changes in attentional capacity, which clearly did not occur.

The timeline plots shed some further light on this disparity. In the absence of clear MWL differences, the focus instead turned to discrete peaks in the data. There were significant increases in MWL around the midpoint of both conditions, associated with the critical event and the turnaround point. Interestingly, this translated into a similar peak for attentional capacity, but only in the normal condition. These results are more favourable for MART than the overall main effects, since it now appears that the decrease in MWL after the critical event was perhaps due to the trough of secondary task responses at the very start of the trial. Furthermore, the apparent capacity increase in the second half of the trial was probably influenced by the peak at the midpoint.

### *Implications*

From a theoretical perspective, it may be possible that the levels of demand imposed by the two conditions were inappropriately spaced on the traditional inverted-U performance curve. Both tasks could, in actual fact, have presented similar degrees of underload, optimal load, or even overload to the participants. Existing models of overload, which posit that operators adjust their task strategy to manage demands (Farrell, 1999; Wickens, 2002b), could then make it appear that capacity increases with a sudden peak of MWL. However, without primary task data to corroborate these conjectures, it is impossible to be conclusive.

The practical explanations may be more parsimonious, centring around the possibility that MART was not, for some reason, applicable in the present environment. The training regime of pilots encourages a lot of visual scanning of the internal instruments and outside world. This may have negated the MWL manipulation, and certainly could have influenced the attention ratio data. If so, it could have reassuring implications for long-haul commercial pilots, since they might not be so susceptible to underload problems. Alternatively, since this experiment did not use any autopilot systems, MART may have to be revised as an automation-specific theory. Clearly, more research is needed to explore these possibilities.

Notwithstanding the overall drawbacks, there was one key finding in this study that does raise interesting issues for MART and pilot MWL. The lack of a MWL-capacity relation in the underload condition is intriguing, given the associated peaks in the normal condition. Perhaps, as noted above, this provides further insight into the shape of the theoretical curve; alternatively it may represent a performance trade-off with the primary task. Either way, the fact that relation between MWL and capacity appears to depend on task context has both theoretical and practical implications that need to be delineated in future research.

## References

Farrell, P. S. E. 1999, The hysteresis effect, *Human Factors*, **41**(2), 226–240

Hancock, P. A., and Parasuraman, R. 1992, Human factors and safety in the design of Intelligent Vehicle-Highway Systems (IVHS), *Journal of Safety Research*, **23**(4), 181–198

Kahneman, D. 1973, *Attention and effort* (Prentice-Hall, Englewood Cliffs, NJ)

Leplat, J. 1978, Factors determining work-load, *Ergonomics*, **21**, 143–149

Norman, D. A., and Bobrow, D. G. 1975, On data-limited and resources-limited processes, *Cognitive Psychology*, **7**, 44–64

Wickens, C. D. 2002a, Multiple resources and performance prediction, *Theoretical Issues in Ergonomics Science*, **3**(2), 159–177

Wickens, C. D. 2002b, Situation awareness and workload in aviation, *Current Directions in Psychological Science*, **11**(4), 128–133

Young, M. S. and Stanton, N. A. 2002a, Attention and automation: New perspectives on mental underload and performance, *Theoretical Issues in Ergonomics Science*, **3**(2), 178–194

Young, M. S. and Stanton, N. A. 2002b, Malleable Attentional Resources Theory: A new explanation for the effects of mental underload on performance, *Human Factors*, **44**(3), 365–375

# INVESTIGATING TRAIN DRIVERS' PERCEPTIONS OF THEIR VISUAL BEHAVIOUR

**R.K. Smith[1], T. Luke[1], N. Brook-Carter[2], A.M. Parkes[1], A. Mills[2] & E. Grimes[2]**

*[1]TRL, Crowthorne House, Nine Mile Ride, Wokingham, Berkshire RG40 3GA*
*[2]Rail Safety and Standards Board*

This paper presents the findings of a series of verbal reports from a larger study designed to investigate the visual strategies of train drivers. The interviews gave drivers an opportunity to discuss what they believed to be their visual strategies when approaching signals and complex driving scenes and compare this to their measured eye movements. Train drivers' visual strategies are affected by a number of factors including: the signal aspect; complexity of the driving scene; the previous aspect passed; potential hazards; weather and position on route. A number of non-visual strategies were identified, which may have important implications for the future design of train cabs.

## Introduction

Train driving is a highly visual task; requiring the driver to monitor the dynamic scene both inside and outside the cab. Poor performance on this visual task may lead to errors, such as signals passed at danger (SPADs). It is therefore important to understand train driver visual behaviour, particularly in complex visual scenes, or when running on cautionary signals, to prevent future SPADs or rail crashes.

The results reported in this paper are from a larger project investigating train driver visual strategies, funded by the Rail Safety and Standards Board. An eye tracker was used to record the eye movements of drivers as they drove their normal routes for subsequent analysis. In order to form a rich picture of train driver's visual strategies, drivers were shown a video of their own eye tracker footage and asked to report on their perceptions of their visual strategies. Both eye tracker data and verbal report data will be used together in the larger report to better understand train drivers' visual strategies.

Pilot trials were conducted to verify the feasibility of fitting drivers with portable eye monitoring equipment and recording normal behaviour without interference with the normal operating schedule. Initial results showed that drivers' visual strategies are dependant on a number of factors, including position on the route, preceding signal aspects and expectations about hazards.

This paper describes the verbal report method adopted and presents the results in terms of the common visual behaviour and strategies reported by drivers. These include the visual behaviour described when approaching signals and complex scenes; and more general behaviours such as the visual cues drivers' use and differences in visual behaviour in different weather conditions.

## Method

### *Participants*

84 train drivers participated in the main study (77 male, 7 female). Ages ranged from 24 to 63 years, with an average age of 40.

### *Procedure*

Drivers were required to drive one of ten routes wearing an eye tracker to record their visual behaviour. The route was driven in both directions with a short break in between the two journeys. Each of the two journeys lasted for approximately thirty minutes. The drivers were accompanied by an experimenter who recorded the signal aspect and points of interest throughout the route. When the route was completed, drivers were shown the recorded footage of the route that they had just driven. These recordings displayed a crosshair indicating the driver's point of regard on the visual scene, and were useful visual aids assisting the drivers in verbalising their perceived visual behaviour.

### *Equipment*

The eye tracker used during this trial was the ISCAN VisionTrak ETL-500, supplied by Polhemus. The VisionTrak was mounted on a headband that can be adjusted to all head sizes. It used a dichroic mirror to reflect the eye-image onto a highly sensitive head-mounted video camera. The scene camera was mounted below the driver line-of-sight.

The eye tracker consisted of two separate sets of equipment: portable and fixed. The fixed system was set up in the office and possessed the full eye tracking software and functionality. The portable equipment carried onto the train consisted of the eye tracker headband, a Sony Hi8 video recorder, a multiplexing unit and a battery belt.

### *Verbal reports method*

The verbal reports took place after the drivers had completed both of their drives and were back in the office. They were shown a video of a section of the route they had just driven. Sections of the route for review were specifically chosen for having relatively high workloads and contained items such as complex gantry signals and poorly sighted signals. Drivers were shown two clips of their section of route, which lasted for approximately 20 minutes. The eye tracker footage displayed the driving scene from the position of the driver in the train cab. These recordings displayed a crosshair indicating the driver's point of regard to the visual scene.

The drivers were asked to carry out a commentary whilst watching the video of the route they had just driven. The drivers were asked to describe what they were looking at and why they believed they had adopted a particular visual behaviour. A second experimenter was present to write down the drivers' comments throughout the verbal report.

### *Verbal report analysis method*

The verbal reports produced a large amount of raw data. The first stage of processing was to create a spreadsheet containing all comments. Each comment was then printed out on to a separate piece of card.

The method employed to structure the numerous comments was the card sorting technique (Gaffney, 2000). A group of four experimenters sorted the cards in two main stages. The first stage involved sorting the individual cards into the appropriate theme. This was done by grouping together comments relating to a particular theme such as "approaching a station". All comments relating to approaching a station were grouped into one theme. This was repeated with all the cards until all comments were allocated to a theme. During the second stage the experimenters sorted the cards within each of the 15 themes. Within each theme several strategies were identified from the numerous comments.

## Results and main findings

The data collected in this part of the study was purely qualitative and was comprised of comments from the verbal reports. The quality of the reports varied between drivers as some drivers found it easier than others to verbalise their visual behaviour. Due to the nature of the verbal report data collection method and the data itself, no quantitative analysis was carried out on this section of the data. This was used to support the extensive quantitative analysis conducted on the eye tracker data. The main findings from the verbal reports are presented below.

### *General strategies*

Train drivers indicated that they adopted a number of general scanning strategies when driving a train to identify potential hazards on the track. These include looking ahead; looking around; looking for anything out of the ordinary; not focussing ahead all the time; and using peripheral vision to obtain information. General hazards that drivers identified included track workers; people on bridges; looking at other trains for problems; checking the line is clear; and checking that the points were set correctly.

Some general non-visual strategies were also identified. These include "feeling" the location on the route from the vibration from the track; using sound to identify location on the route; and using experience and knowledge to identify location on the route.

### *General observation and checking for signals*

Several of the routes included the approach to a number of signals that were located within a complex environment or were poorly sited. The general visual strategies identified that were related to looking and checking for signals included searching for the next signal; checking the signal; checking and checking again; checking the signal and then looking at other things; looking at signals for other lines; and expecting to see the signal near a particular feature.

Drivers noted anticipating the next aspect based on route knowledge or the previous aspect. Extra vigilance is reported when the aspect is not as would be expected, for example if a particular signal aspect is usually green and on occasion is cautionary, drivers reported being extra vigilant in this case.

### *Identifying the correct signal*

Train drivers reported that it can sometimes be difficult to identify the correct signal until close to it and that they have to be sure it is their signal. Problems can be encountered in complex areas where there are many signals on gantries, or when the signal is located around a bend. Problems may also be encountered when confusing other lights such as street lights or vehicle lights as signals.

The visual strategies used to identify the correct signal included counting across signals; counting across tracks; looking for a specific identifier on the signal; looking at the signal grouping pattern; doing extra checks; looking across the gantry and confirming what line their train is on.

Train drivers reported a number of strategies that they adopt relating to specific types of signals which they reported as being potentially problematic. The visual strategies identified relating to particular signal issues include concentrating on signals in complex areas; concentrating on semaphore signals, looking for obscured signals; looking for poorly sited signals; taking care with bi-directional signals; and taking care with multi-SPAD signals.

*Poor weather strategies*

The two main weather conditions reported by drivers to lead to problems were rain and fog. The non-visual strategies used in poor weather included slowing down, increased vigilance and braking earlier. The visual strategies included searching for specific items in the visual scene such as features close to the track; looking for specific braking points such as bridges; using different braking points; counting signals and searching for sections of the track that were dry, where the brakes could be applied.

*Driver's route knowledge*

As part of drivers training and competency, they are required to have knowledge of the routes that they drive so that they know where they are at any point in the route, and can drive appropriately to ensure safety and maintain train timings. The driver uses features on and around the track as cues to identify where they are. Drivers reported selecting identifiers that were either a fixed feature, or using a feature that was close to the track. These cues included: signals, railway infrastructure, signs, track layout, landscape, stations, level crossings, buildings, bridges, tunnels, train speed, landmarks and sequence of items.

Drivers reported having specific braking points along the route that they used to remind them when to apply the brakes. Braking points used by drivers are similar to those points used as identifiers and included: signals, railway infrastructure, signs, track layout, landscape, stations, level crossings, buildings, bridges, tunnels, train speed, landmarks.

## Discussion

The detail from the verbal reports was extremely useful in understanding drivers' visual and non visual strategies. The verbal reports were not only useful in identifying drivers' perceptions of their visual behaviours but were also important in identifying the reasons why drivers adopt these strategies. This information is not provided with the eye tracker data and therefore was used to validate and better understand the information gathered using the eye tracker.

Eye tracker data only provides data on the point of focus, however, drivers verbally reported taking in information about the driving scene from their peripheral vision. This data was not captured by the eye tracker, and can only be reported from the verbal reports.

Signal identification was an important issue discussed by many drivers. Drivers reported problems in identifying the correct signal in complex areas where there are many signals on gantries. The strategies adopted to identify signals on gantries included counting across all signals and tracks to locate the correct signal on the gantry. Drivers also reported looking for a specific identifier on a complex gantry to locate their signal; this may include a number or a letter. This may have important implications for the current and future labelling of signals on gantries.

The non-visual strategies identified in this study could also have important implications for the rail industry. Drivers reported using the feeling and sound from the track to locate their positioning on the route. It is therefore necessary to ensure that these auditory and kinaesthetic cues are maintained and the driver is not insulated completely from the vibration and sound of the track in future types of train cabs.

Previous research (Groeger, Bradshaw, Everett, Merat and Field, 2001) into train driver's visual strategies identified a number of factors which affect the visual strategies adopted by train drivers when approaching signals, such as the signal aspect, the signal mounting and the aspect of the previous signal passed. Findings from the verbal reports support some of Groeger's findings and suggest that train drivers' visual strategies are affected by a number of factors including: the signal aspect; complexity of the driving scene; the previous aspect passed; potential hazards; weather and position on route.

One limitation identified during the current study was that drivers differed in their ability to verbalise their visual behaviour. In general, more experienced drivers found it harder to verbalise their visual behaviour because the driving task was more automated for them than the less experienced drivers.

Two important results were identified from this study. Signals in complex gantry areas need to have confirmatory information such as numbering to assist drivers in locating their particular signal. Secondly, in bad weather where visual cues normally used may not be available, experienced drivers who are familiar with a particular route can use addition visual cues to identify their location. However, less familiar drivers may not have additional visual cues for bad weather. Therefore, systems should be put into place to support drivers who are less familiar with a route such as in-cab signalling.

With the future plan to introduce in-cab signalling to the UK, many new issues may arise as a result of this. Currently drivers report spending much of their time looking for the next signal, and scanning the track for potential hazards, however, if the signalling system is based in the cab, drivers are likely to spend more time looking in the train cab as opposed to scanning the track for potential hazards ahead. Drivers may have a conflict between looking in the cab for information and scanning the driving scene for potential hazards. Further research into how drivers will distribute their visual attention when using in-cab signalling will be necessary to ensure safety benefits are maximised.

## References

Gaffney, G. (2000). Usability technique series. Information & Design. www.infodesign.com.au

Groeger , J. A., Bradshaw, M. F., Everett, J., Merat, N. and Field, D. (2001). Pilot of Train Drivers Eye Movements, University of Surrey, Guildford, UK.

# PEDESTRIAN FLOW MODELLING – MEETING THE USER'S NEEDS

**Andrew Baker[1], Dr. Ann Mills[2] & Michael Woods[3]**

[1] *Davis Associates Ltd, Wyllyotts Place, Potters Bar, Hertfordshire EN6 2JD*
[2] *Rail Safety and Standards Board, 160 Euston Road, London NW1 2DX*
[3] *Rail Safety and Standards Board, Upper Woburn Place, London WC1H 0HY*

Pedestrian flow modelling is a computer-based technique used to represent and predict pedestrian movement in the built environment. The technique has been used by the rail industry for over 15 years to highlight crowd flow bottlenecks in the assessment of safety at stations, platform train dwell times and the options for greater use of retail space in station concourses.

Whilst offering great potential benefits, in some instances clients have commented that the output did not meet their requirements; the process was not cost effective; or the findings were limited by capabilities of the modelling package used. This paper outlines a suggested process for the use of pedestrian flow modelling. The requirements for each stage in the process are described, sub-divided into: System Goals, Domain Knowledge and Specification Requirements. This process and the associated requirements are intended to support suppliers and clients in the area of station design and operations management. This work is an output from a Rail Safety and Standards Board (RSSB) project, undertaken to research and assess the tools available for crowd management at stations.

## Introduction

The design, operation and popularity of Britain's railway network almost inevitably leads to crowded conditions at stations. Whether caused by service congestion, local events or sheer weight of rush hour traffic, crowds can build quickly. The negative effects of crowds may be more than service delays and unhappy passengers. Crowding can also lead to injuries, whether minor due to slips and trips or more serious incidents at bottlenecks and the platform edge.

Pedestrian flow modelling (PFM) is the computer representation of pedestrian movement within stations and to/from areas such as trains and streets. In a railway context, use of simulation can produce specific measurable findings based on user-definition of the built station environment, passenger population, goals and starting location.

Interviews with rail industry stakeholders highlighted that problems in the application of PFM stemmed not so much from the capabilities of the packages themselves but from poor preparation for their use and difficulties in interpreting the outputs.

Specific concerns included: PFM identifies problems but not solutions; PFM does not consider human interaction, e.g. group behaviour; and the output is only as good as the data used. To address these concerns, a process has been developed for the rail industry for guidance in the use of PFM. The process does not describe how to physically use the packages themselves, but guides the user in defining their requirements, selecting a package, ensuring they get the best value and effective use of the technology.

This work was part of a research study into crowd management at stations on behalf of the Rail Safety and Standards Board (RSSB), which encompassed the following stages:

- Work Package 1: Initial research
  - Literature review and market research
  - Rail stakeholder and domain expert interviews
  - Passenger flow modelling technical review
- Work Package 2: Risk assessment
  - Review data from past incidents
  - Define generic scenarios
  - Document passenger behaviour factors
  - Conduct scenario workshops
- Work Package 3: Pedestrian flow modelling process definition
- Work Package 4: Good practice guide

## Information sources

The PFM requirements and process have been assembled after reviewing information from a number of sources, as illustrated in Figure 1.

## The process framework

Only part of the success of PFM analysis is directly attributable to the software itself. Therefore, we have defined guidelines that encompass both the software and the enabling process that delivers the findings. To do this we have adapted a requirements capture process based on a generic principle of the "World and the Machine" (Jackson,

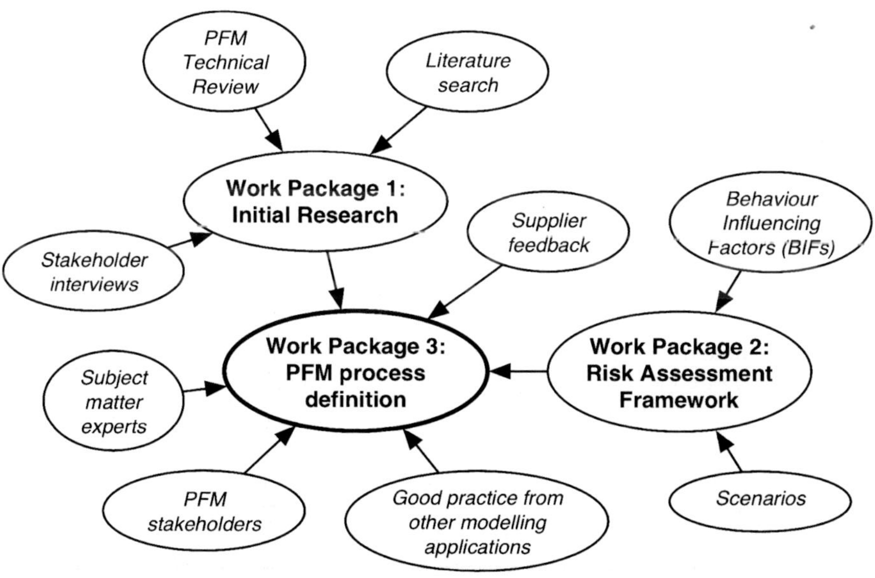


**Figure 1. Inputs to the PFM process definition.**

1995). PFM software (the "Machine") is used to assess and bring about some improvement in the real "World" delivered through a shared interface. From the information sources we have gathered many requirements for successful PFM analysis, subdivided the the following headings:

*System Goal (SG)*

This is something in the real world that we wish PFM to help us achieve. For example, improved safety at a station, service performance or minimised construction costs.

*Domain Knowledge (D)*

This refers to the knowledge of the user outside of the interface that will enable PFM analysis to deliver the System Goals. For example, a clear and objective brief to the pedestrian flow modelling team or a suitable survey of current pedestrian flows.

*Specification Requirement (SR)*

This is something that the PFM software has to have to or be able to do in order to achieve the System Goals. This can only refer to items within the shared interface, i.e. things that the PFM software can actually access and deliver.

It is the combination of the Domain Knowledge outside the interface and Specification Requirements within the interface that together delivers the System Goals. This relationship is illustrated in Figure 2.

It is important to note that this relationship is a "many-to-many" matrix, i.e.:

- Each Specification Requirement and item of Domain Knowledge may contribute to more than one System Goal
- To achieve each System Goal may require more than one Specification Requirement and item of Domain Knowledge

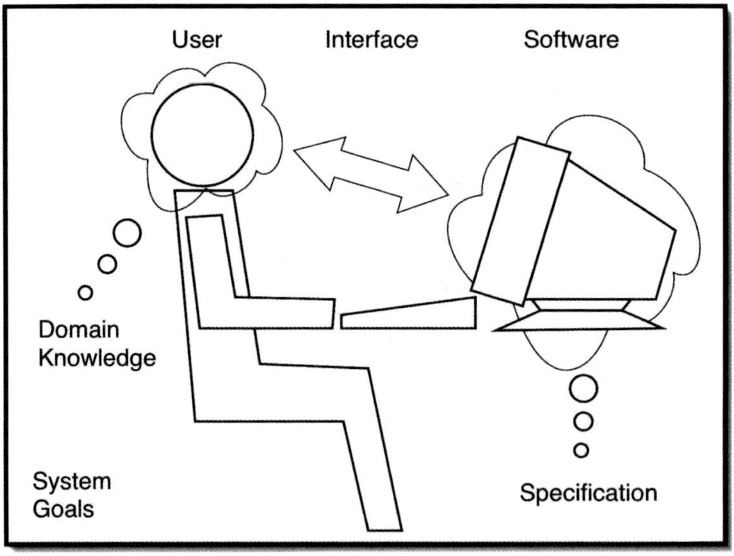


**Figure 2. The World and the Machine.**

## Requirements database

The requirements have been recorded in a database format to enable the efficient search, edit and grouping of items. There are two sub-sections for the database entries:

- System Goals
- Domain Knowledge and Specification Requirements

The database records comprise the following fields:

- Reference (Ref): A record code (System Goal = S, Domain Knowledge = D and Specification Requirement = R) plus a sequential number for that code.
- Creation date: The date the record was first entered into the database.
- Title: The name of the record, expressed as an action to facilitate the targeting of a requirement.
- Stage: The stage of a typical PFM process to which the record applies.
- Type (Domain Knowledge/Specification Requirement records only): The classification of the record according to Domain Knowledge or Specification Requirement.
- Source: The origin of the record, sub-divided into interviews and document references.
- Description: A description of the requirement.
- Rationale: An explanation of why this record has been included and why it is important
- Guidance: How the requirement could be delivered and/or examples of potential responses.
- Enabled by (System Goal records only): Cross-references the System Goal to the Domain Knowledge and Specification Requirements that will enable that Goal to be achieved.

Using the database, a process flow chart was developed to illustrate how the requirements mapped across to project stages.

## Findings and conclusions

The following conclusions are of note arising from this study:

1. This is the first project of its type within the rail industry to study the requirements for the use of pedestrian flow modelling.
2. Through the information review process used, we have captured 50 distinct requirements for the PFM process, encompassing:
   - 16 System Goals
   - 19 requirements for Domain Knowledge
   - 15 PFM software Specification Requirements
3. There were similar responses from various stakeholders, experts and suppliers in certain areas. Particular requirements that received multiple recommendations included:
   - Design for realistic and consistent "worst case" conditions
   - Share the findings with other projects and adopt a proactive approach in assessing opportunities for application of the findings to other projects
   - Define the measurement variables and acceptance criteria at the planning stage
   - Ensure the acceptance criteria suit the application
   - Enlist the services of a subject matter expert before developing the study brief and selecting a PFM supplier
   - Consider the procurement of an in-house package for iterative studies
   - Prepare a set of operational scenarios to focus the study
   - Conduct operational "reality checks" on the findings
   - Demonstrate validation of the data and the process

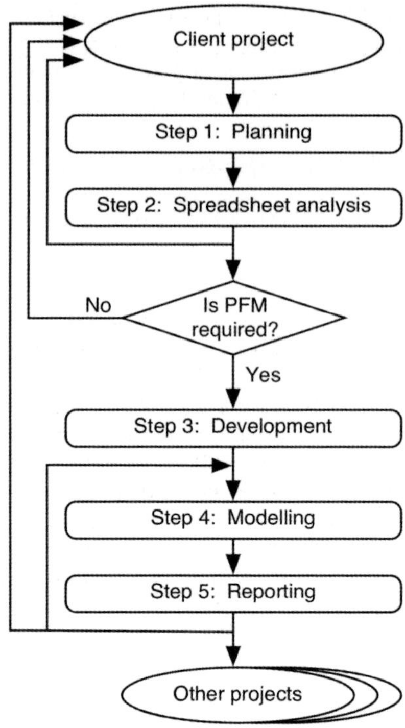


**Figure 3. PFM process flowchart.**

4. A suggested process for PFM work has been developed by this study. This process is sufficiently generic to be applied across a wide range of pedestrian flow studies. The flowchart is shown in Figure 3.

## References

Jackson, M.A. 1995, *The World and the Machine*, Proceedings of the 17th International Conference on Software Engineering, IEEE, 283–292.

Baker, A.J. 2004, *Pedestrian Flow Modelling Process definition*, 30-T161-report-27-May-04, available from: www.rssb.co.uk.

# IT'S NOT ALL PLAIN SAILING. PORT TURN-AROUNDS AND SEAFARERS' FATIGUE: A CASE STUDY

**Paul Allen, Ben Wellens, Rachel McNamara & Andrew Smith**

*Centre for Occupational and Health Psychology, Cardiff University, 63 Park Place, Cardiff CF10 3AS*

A case study is presented from research onboard a mini-bulker illustrating a number of factors which may be critical when trying to account for fatigue on this class of vessel. Whilst the case study highlights that working hours are justifiably considered a key area of concern when addressing fatigue across all ship types, on certain vessels it may be more profitable to consider the challenging issue of crewing in terms of strategically covering for the disparate demands of port and sea work within a competitive market. Evidence from the case study suggests that ship owners may be inclined (and authorised) to favour leaner "open-sailing" crewing arrangements which inevitably struggle to cope when faced with demanding port turn-arounds.

## Introduction

The aim of the seafarers' fatigue project is to investigate the issue of seafarers fatigue across the broadest range of commercial ship types possible. In phase one of the project the offshore oil industry was studied (Smith, Lane and Bloor, 2001), in phase two the focus was upon short sea and coastal trade (Smith, Lane, Bloor, Allen, Burke and Ellis, 2003) and in the current third phase the deep sea and bulk sectors are being investigated. Whilst the seafarers' fatigue project has distinctive onboard and survey elements, the onboard field testing forms the basis of this paper. The onboard field research involves a researcher staying onboard ship for between 10 and 14 days during which time crew members are tested and various environmental recordings are made (see Smith et al., 2001; Smith et al., 2003). In addition to the battery of objective and subjective tests to be completed onboard ship, researcher experience has proved anecdotally valuable in terms of attempting to understand which factors and combinations of factors are critical in terms of accounting for inter-seafarer fatigue differences. This paper will follow the form of a case study in which one onboard visit will be described in detail. This account of a researcher's experience onboard ship is an illustrative means of highlighting the critical role of port turn-arounds when investigating seafarers' fatigue. Furthermore, such an anecdotal account of onboard life serves to illustrate the distinctiveness of the seafarer's working experience which often falls well outside any onshore notions of acceptability.

## The case study

The case study comes from an onboard research visit which took place between 18th November to 2nd December 2003 on a mini-bulker. Bulkers are a versatile class of ship

**Figure 1. View of the vessel from the stern facing forwards.**

**Figure 2. View from the bow looking towards the bridge.**

designed primarily to take bulk cargo such as grain, coal, iron ore and wood pulp with mini-bulkers normally carrying five to seven crew and typically weighing around 3000 deadweight tons (DWT). The vessel involved in the onboard testing carried 6 crew and weighed 3510 DWT. During the two week research trip the vessel visited Holland, Sweden, Germany, Belgium and Portugal and carried cargos of wood pulp and steel coils. The crew consisted of a captain, first officer, chief engineer, deckhand, deckhand/cook and deckhand/motorman who assisted the engineer. Pictures of the vessel can be seen in Figures 1 and 2 above.

The responsibility of navigating the vessel rested solely with the captain and first officer who alternately stood 6 hour watches on the bridge. The 6-on/6-off shift pattern worked by the captain and first officer would be disrupted when coming in and out of port when the first officer would have to oversee cargo loading/discharging operations and the captain would have to be available to deal with officials and requisite paperwork. The rest of the crew worked a more conventional day shift from 8am to 5pm with the chief engineer responsible for getting up in the night to respond to any engine alarms. As with many ships in the mini-bulker market the vessel was working on a "tramp" style charter which meant there was no set schedule with the ship taking cargos from wherever business could be secured on a week by week or even day by day basis. Due to the fact that cargos change regularly on a tramp style charter greater demand is often placed on the crew in terms of preparing the vessel for each new type of cargo that is shipped (e.g. washing coal dust away from the holds ready to take grain).

The crew on the mini-bulker were mostly working 4 months-on/2 months-off however these work/leave periods could be variable with some crew members simply going from one ship to another in search of work. In a 4 month contract the two deck officers (captain and first officer) were unlikely to get any days off unless the ship had a malfunction which required lengthy repair. The standard working arrangement for the deck officers was therefore 12 hours a day, 7 days a week for 4 months without leave. Whilst such a working schedule appears patently excessive by onshore standards, 84 hours a week is actually very much the best case scenario for seafarers working a 6-on/6-off two man watch. Whenever both the captain and first officer were forced to be on duty at the same time it is an inevitable fact that one or both of them was working in excess of their normal 12 hour day. Such "overlap" of watch times consistently occurred when coming in and out of port as the captain and first officer had distinct roles to fulfil simultaneously. The captain in particular would frequently work from the start to the finish of a port visit without sleep, a stretch of as long as 24 hours. With only two officers to cover a 24 hour watch there was no "slack" in the system – if one of the officers was unfit to work the other officer was forced to cover putting an immense burden of duty on both seafarers. The underlying duty-bound principle was therefore "if I rest my colleague suffers". Whilst apart from the deck officers the other crew were also required to work excessive hours in port, the situation was unquestionably worse for the captain and first officer for a number of reasons. Whilst the engineer and deck hands might work a 24 hour day when completing a port turn-around, there was much more opportunity for these crew members to "recover" to some extent with a potential sleep period each night of up to 12 hours compared to 4 or 5 hours for the deck officers. In addition, once the ship was released from the port side the engineer and deck hands were mostly relieved of their duties however the deck officers (and the captain in particular) were responsible for manoeuvring the ship out of port under pilotage which can take a number of hours and requires a high level of alertness and concentration (a situation repeated when coming into port). Finally, the deckhands would occasionally get a day off which was simply an impossibility for the deck officers as previously stated.

Port visits for the ship lasted between 7 and 23 hours depending on the amount and type of cargo to be loaded/discharged. Time in port costs money and therefore the aim was to complete a port turn-around in the shortest amount of time possible. Under such a strict port schedule the crew members were effectively ship-bound with little or no opportunity to take time off to explore any of the places visited. With little or no time to go ashore (beyond the harbour side) the day-to-day living experience of the crew was what most would certainly term "confined" with only one small communal mess room for the crew to relax in apart from their cabin.

## Evidence of fatigue from the case study

From simply observing the working patterns of the crew on the mini-bulker it was apparent that excessive job demands had been assumptively woven into the culture onboard ship. The deck officers on the vessel were expected to work 12 hours a day however this did not account for the fact that paperwork would routinely have to be done in supposedly "off-duty" time for around one or two hours. In addition, the fact that the deck officers in particular would often not sleep during an entire port turn-around makes the concept of a contracted 12 hour day appear naively detached from reality. The level of unreasonable demand can further be judged in relation to the European Union (EU) Working Time Directive which stipulates minimum hours of rest for seafarers of no less than 10 hours in any 24 hour period and 72 hours in any seven-day period (clause 5, 1b). Based on the simple manning requirements of the ship and the frequency with which the vessel was coming into port there is no way the crew could have kept within any such working time regulations. Whilst generalising from one case study is certainly unwise, evidence is provided from the marine accident investigation branch (MAIB) to suggest that many of the problems identified on the research trip are common to smaller vessels in general. The MAIB watchkeeping study looked at accident reports to try and determine which factors are associated with being involved in a marine accident and the two-officer watch system in particular is highlighted as being potentially dangerous, as follows:

"...minimal manning, consisting of a master and a chief officer as the only two watchkeeping officers on vessels...leads to watch keeper fatigue and the inability of the master to fulfil his duties, which, in turn, frequently lead to accidents." (pg. 1)

Whilst preliminary research results from the onboard testing further support the argument that the mini-bulker crew were indeed suffering from fatigue, a basic consideration of operational logistics should be sufficient to conclude that problems are almost inevitable. When an individual regularly works 13–14 hours a day punctuated by periods of 24 hour port work with no recovery time beyond a 4–5 hour sleep period the question of *whether* that individual is fatigued warrants little extended consideration.

## Accounting for fatigue in mini-bulker crews

It could be argued that mini-bulkers are simply a class of ship on which crew members are at a higher risk of suffering from the effects of fatigue. This global conclusion, however, is of limited use when attempting to distil those underlying factors which are critical in terms of causing seafarers' fatigue. Ultimately the class of ship known as a "mini-bulker" represents a constellation of key functional characteristics with these individual characteristics of key interest when examining fatigue across ship types. Using such a deconstructionist approach it is possible to identify a number of factors which come together to make working on a mini-bulker particularly demanding, as listed below:

- **Short port stays.** Small ships carry a small cargo and therefore loading and unloading times are relatively quick. When a port turn-around is completed within 24 hours there is no time for rest or recovery before heading back out to sea.
- **Frequent port visits.** When port turn-arounds are demanding then a high frequency of port turn-arounds only compounds the situation.
- **Changing cargos.** When a vessel changes its type of cargo regularly extra demand is placed on the crew to prepare the ship accordingly.
- **Small crew-2 officers watch.** A small ship can economically only carry a small crew which includes only two officers to cover a 24 hour watch.
- **Longer pilotage.** Small ships can travel further up river and therefore are normally involved in much longer periods of pilotage. Sailing up and down rivers under the guidance of a pilot through locks and narrow waterways is considerably more demanding than sailing in open sea.
- **Unpredictability**. When "tramping" around from port-to-port there is little predictability which can be stressful and makes planning sleep and rest periods difficult.

In terms of causing fatigue the mediating consequence of many of the above listed negative factors is increased working hours. Therefore, for example, short port stays lead to fatigue through the fact that crew members are required to work increased working hours to get the ship out of port as quickly as possible. Similarly, having two officers rather than three requires the officers to work a basic 12 instead of 8 hour day. The most significant point, however, is that if the mini-bulker that was studied had not come into port then few problems would have been evident. Even a 6-on/6-off shift system is arguably acceptable if the pattern is maintained without disruption and without additional hours being added. Returning to the title of this paper, many ships work in two very distinct modes of operation which are largely incompatible when faced with the issue of crewing. If a ship is crewed for work in port then an excessive amount of redundancy will be evident when the ship returns to sea however if a ship is crewed for open sailing then a port visit will stretch the crew beyond their limits. On such a crewing continuum the mini-bulker that was studied was arguably crewed only to cover periods of open sailing however this is perhaps unsurprising considering the economic attractiveness of the smaller crew option. Whilst the obvious answer to the crewing dilemma faced on certain ships is to legally insist on more crew members to cover busier port periods, such a blanket approach takes no account of economic viability. Finding a solution to port under-manning is a relatively simple matter however finding a solution which is not economically crippling to certain classes of ship is considerably more difficult.

## Conclusion

The port/sea crewing dilemma, as with many fatigue-related problems, is only a critical issue on certain types of vessel. Ferries, for example, can not be run purely on the basis of how many crew are needed to sail in open water. In contrast, however, on the mini-bulker in question the issue of coping with two different modes of operation was of critical importance. Whenever a ship which faces the dual-mode crewing dilemma is uncompromisingly crewed in favour of an open sailing arrangement then it is tenable to suggest that fatigue will be a problem unless the demand of a port visit is reduced in some way. In terms of understanding seafarers' fatigue, therefore, where working hours might often appear the obvious culprit a more sophisticated approach is required which sees excessive working hours as potentially symptomatic of a poor solution to the port-versus-sea "crewing conundrum".

## Acknowledgments

This research is supported by the Maritime and Coastguard Agency, the Health and Safety Executive and NUMAST. We would like to acknowledge the contribution made by those who participated in the research. The views expressed here reflect those of the authors and should not be taken to represent the policy of the sponsors.

## References

Marine Accident Investigation Branch (MAIB). 2004, *Bridge watchkeeping safety study.* (Department for transport, London), 1-

Smith, A.P., Lane, A.D. and Bloor, M. 2001. *Fatigue Offshore: A Comparison of Offshore Oil Support Shipping and the Offshore Oil Industry,* (Seafarers International Research Centre, ISBN 1-900174-14-6)

Smith, A.P., Lane, A.D., Bloor, M., Allen, P., Burke, A. and Ellis, E. 2003. *Fatigue Offshore: Phase 2. The short sea and coastal shipping industry,* (Seafarers International Research Centre, ISBN 1-900174-21-9)

Working Time Directive (See: *http://www.dti.gov.uk/er/work_time_regs/exsectors.htm*)

# FATIGUE AT SEA: AMENDMENTS TO WORKING TIME DIRECTIVES AND MANAGEMENT GUIDELINES

**Rachel McNamara, Paul Allen, Ben Wellens & Andrew Smith**

*Centre for Occupational and Health Psychology, Cardiff University, 63 Park Place, Cardiff CF10 3AS*

Seafaring is a physically and mentally demanding occupation, often requiring long periods of sustained attention. The potential for such working practices to cause fatigue and therefore endanger personal, operational and environmental safety is now widely accepted, yet, the most appropriate strategies for combating fatigue are the subject of some debate. The current paper evaluates two distinct approaches: the first relates to the impact of recent changes to working time directives which aim to reduce working hours and therefore minimise fatigue. The second, focuses on the International Maritime Organisation's (IMO) guidelines on management of fatigue at sea in terms of their practical application.

## Introduction

Working at sea has significant potential to cause fatigue for a number of reasons: long tours of duty (up to 6 months in the deep-sea sector) and demanding shift systems (e.g. 6 hours-on, 6-off) reduce leisure time and interrupt normal circadian rhythms (see Smith et al., 2001; Smith et al., 2003). Furthermore, seafaring occupations often require long periods of sustained attention (e.g. those on watch keeping duty) and intense physical labour. Harsh environmental conditions (rough sea, high winds) may also disrupt sleep. The current paper evaluates the impact of two distinct approaches aimed at minimising the fatigue that may result from working within such a unique environment. Firstly, data obtained from a recent survey of seafarers will be used to evaluate the impact of recent changes to working time directives which restrict working hours and prescribe adequate periods of rest. Secondly, the IMO's guidelines on management of fatigue at sea will be discussed with particular reference to feasibility of implementation and their potential to reduce the effects of fatigue.

## The impact of recent changes to working time directives

In the following section, statistics concerning current working practice and the perceived impact of recent working time directives are provided. Data were collected from a cross-sectional survey of three samples: deep-sea (n = 320), offshore (n = 96) and short-sea personnel (n = 132).

### *Current working practice*

Current working time directives (enforced in September 2002) as they apply to seafarers employed by EU countries state that: "…maximum hours of work…shall not exceed either 14 hours in any 24-hour period and 72 in any 7-day period.." (clause 5, 1a) *or* "…minimum hours of rest…shall not be less than 10 hours in any 24-hour period and

77 hours in any 7-day period" (clause 5, 1b). Furthermore, that "Hours of rest may be divided into no more than two periods, one of which shall be at least 6 hours in length, and the interval between consecutive periods of rest shall not exceed 14 hours" (clause 5, 2).

However, it is evident that a minority of seafarers within the current sample report working daily and weekly hours in excess of those laid out in the working time directive (WTD). 2.2% of the total sample work 16 or more hours per day and 2.4% work in excess of 100 hours per week. When asked about rest periods, almost a third of the sample (30.8%) did not regularly have the opportunity to gain 10 hours rest in every 24, and approximately ten percent (11.9%) did not regularly gain at least 6 hours unbroken rest within a 24-hour period. It would therefore seem that nearly a third of all respondents' working hours violate the requirements regarding hours of rest set out in the WTD (clause 5, 1b). It is worth noting that this percentage is much greater than that reporting working hours in excess of maximum levels: it may be the case that respondents felt it was easier to report violations in terms of hours of rest rather than more explicitly in terms of hours worked. Furthermore, 27.6% of the sample reported typically working 15 or more hours continuously, which contravenes the directive laid out in clause 5, 1a. A significant proportion of respondents (21.5%) also reported spending 4 or more hours per day on additional duties.

### *Impact of working hours on personal and operational safety*

The potentially negative impact of working hours on safety is highlighted by the current results: nearly half (46.7%) of respondents felt their working hours presented a potential threat to their personal health and safety, while almost one third (32.5%) felt working hours presented a danger to safe operations onboard their vessel.

### *Impact of recent amendments to working time regulation*

A significant proportion of respondents indicated that working hours had actually increased within the last 5 to 10 years (see Table 1).

Seafarers were also asked more specifically whether recent amendments to working time regulation had altered working practice with regards hours. Results are given in Figure 1.

**Table 1. Changes to working hours in last 5–10 years.**

| | Decreased (%) | No change (%) | Increased (%) |
|---|---|---|---|
| Hours of work changed in last 5–10 years? | 6.8 | 28.4 | 61.5 |

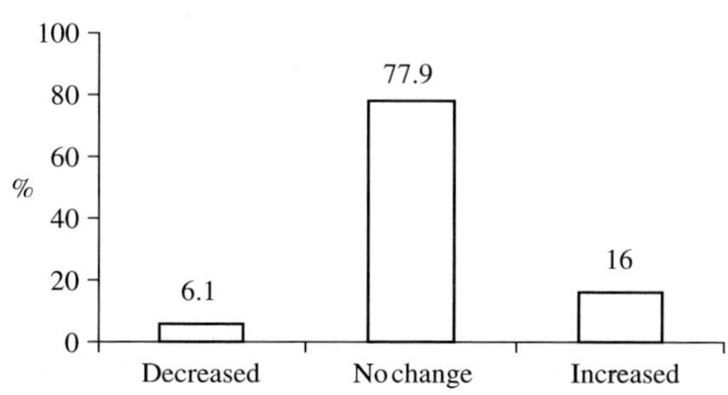


**Figure 1. Have working hours changed as a result of amendments to the WTD?**

*Misrepresentation of working hours*

The WTD also states that records of hours of work and rest must be maintained in order to monitor compliance with the provisions as detailed in clause 5. However, a significant proportion of respondents felt that their actual working hours were at least occasionally under-reported in order to comply with working time regulations: 11.9% reported that their working hours were always or frequently mis-recorded, while a further 28.3% felt this to be the case at least occasionally.

*Knowledge of working time regulation*

The WTD states that regulations should be posted in a highly visible place onboard vessels, yet a significant proportion (15%) of the current sample denied any knowledge of international regulations in place to control their working hours. Furthermore, 7.3% also claimed to have no knowledge of national regulations.

*Sector-specific differences*

Personnel operating in the deep-sea sector seem to be at most risk of working excessively long hours (in violation of working time regulations). The percentage of respondents spending 4 or more hours per day on additional duties was approximately twice that of the offshore and short-sea sectors (28.2% compared with 13.7% and 14.5% respectively). Deep sea respondents were also more likely to report their working hours as a danger to either personal or operational safety. However, few differences were observed across sectors in terms of reported daily and weekly working hours and changes in working practice as a result of amendments to regulations.

*Conclusions*

It is clear from the current data that excessive working hours and inadequate periods of rest are still problematic onboard a range of vessels. Furthermore, hours are likely to be under-recorded, either by management, or by individual seafarers wary of jeopardising their employment by bringing their company under legislative scrutiny. Therefore, auditing of ship records is unlikely to be a sufficient method of ensuring that regulations are adhered to. Better enforcement of existing regulation is needed if excessive working hours and the associated problems of fatigue are to be reduced.

## Appraisal of IMO fatigue guidelines

Given that changes to working time directives appear not to have had the desired effect of reducing working hours (and consequently fatigue), it would seem useful to evaluate current guidelines on fatigue management. The IMO guidelines remain the only comprehensive advice currently available to seafarers. The document provides an informative summary of fatigue, yet has a number of limitations which are subsequently outlined.

*Lack of specific, implementable strategies for reducing fatigue*

The text reads more like a general information document than a set of specific guidelines: for example, working hours and diet are cited as factors influencing susceptibility to fatigue, yet no distinction between the two is made. It is obvious that excessive working hours will have a greater impact on fatigue than diet, although eating well may help to mitigate fatigue effects. Furthermore, general phrases such as "an open communication environment" are used throughout the document: although this is intuitively desirable, there are likely to be many instances where openly communicating that you are too tired to work is not necessarily possible (e.g. within a hierarchical culture and/or one dominated by male bravado).

Suggestions are often made which may be beyond the control of an individual. For example, in a section entitled "Fatigue and the rating" it is suggested that crew members eat regular, well-balanced meals. In practice, "regular" meals are made difficult by anti-social shift systems, and crew on small to medium sized vessels are not likely to have any say in the meals they are provided with. Similarly, it is suggested that seafarers "Make the environment conducive to sleep (a dark, quiet and cool environment and a comfortable bed encourages sleep)" (p.17). Unfortunately, ratings will be assigned a cabin and will have little control over how noisy it is, how comfortable their bed is, or whether light can be adequately shut out.

Management are also advised to consider a number of factors thought to influence fatigue, but no specific information with regards implementation is given. For example, voyage length, time in port, length of service and leave ratios are all cited as important factors to be taken into account when developing fatigue management systems. However, the guidelines do not outline which voyage cycles might be most likely to induce fatigue, how long in port is acceptable for different types of ship, how length of service might impact on fatigue or how long should be spent on leave to achieve optimum recovery. Furthermore, whilst a number of concepts are listed there is little discussion of how the different factors may interact in any cumulative or combined sense.

### *Focus on personal fatigue management strategies*

A distinction can clearly be made between personal and operational/legislative fatigue management approaches. Whilst both forms of approach to fatigue management have obvious strengths and limitations, the IMO guidelines fall indisputably towards the personal side of this continuum. However, many seafarers often find themselves working in situations over which they have little or no control, therefore such an approach is of little value. It would perhaps be more appropriate to concentrate on operational and cultural change if the issue of fatigue is to be tackled effectively.

### *The economic cost of fatigue management*

Advice and best practice cannot compete with economic pressures. There is often little contingency in terms of crew, as many vessels operate at minimum "safe manning" levels and are under pressure to complete port turn-arounds quickly. Under such conditions, it appears unrealistic to suggest fatigue-reducing interventions which do not involve some form of economic trade-off, an issue that is not addressed in the IMO guidelines.

### *Conclusions*

Lengthy, all inclusive guidelines are no substitute for specific and implementable recommendations. Furthermore, the focus of responsibility for fatigue management needs to shift from the personal to the operational. Industry wide, cultural change is needed in order to empower individuals to manage their own fatigue. For example, if provision for extra manning or temporary suspension of operations were allowed for in the design of work schedules, then seafarers would have the option of working shorter hours and gaining more rest when they felt fatigued.

## Discussion

The issue of how best to tackle fatigue at sea is a complex one: the current paper demonstrates that legislation to control working hours and general guidelines on fatigue management are perhaps not sufficient to reduce fatigue. A primary aim of ongoing research on this topic is to identify problem areas within the industry and to make seafarers and their employers aware of how best fatigue can be minimised. Better enforcement of existing regulations and cultural change are needed: the latter may be achieved through implementation and evaluation of novel fatigue awareness and management training schemes, informed by research findings.

## Acknowledgments

This research is supported by the Maritime and Coastguard Agency, the Health and Safety Executive and NUMAST. We would like to acknowledge the contribution made by those who participated in the research. The views expressed here reflect those of the authors and should not be taken to represent the policy of the sponsors.

## References

International Maritime Organization (IMO, MSC/Circ.1014). 2001. *Guidance on Fatigue Mitigation and Management* (IMO, London)

Smith, A.P., Lane, A.D. and Bloor, M. 2001. *Fatigue Offshore: A Comparison of Offshore Oil Support Shipping and the Offshore Oil Industry,* (Seafarers International Research Centre, ISBN 1-900174-14-6)

Smith, A.P., Lane, A.D., Bloor, M., Allen, P., Burke, A. and Ellis, E. 2003. *Fatigue Offshore: Phase 2. The short sea and coastal shipping industry,* (Seafarers International Research Centre, ISBN 1-900174-21-9)

Working Time Directive (See: *http://www.dti.gov.uk/er/work_time_regs/exsectors.htm*)

# COLLISIONS AND COLLISION RISK AWARENESS AT SEA: DATA FROM A VARIETY OF SEAFARERS

**Ben Wellens, Rachel McNamara, Paul Allen & Andrew Smith**

*Centre for Occupational and Health Psychology, Cardiff University, 63 Park Place, CARDIFF CF10 3AS, SmithAP@cardiff.ac.uk*

The human and environmental consequences of collisions at sea are potentially disastrous. For this reason investigating the factors responsible for collisions is important. The MAIB investigates collisions in British waters on a case by case basis but a more general understanding of issues relating to collisions is rarely considered in terms of the average seafarer. Our survey findings demonstrate that a considerable proportion of seafarers have been on vessels involved in collisions. Short sea and coastal vessels and ships working for the offshore oil industry are involved in more collisions than deep sea vessels but it was unclear whether this was due to the increased proximity of potential hazards to these vessels or other differences (e.g. patterns of work and rest). Fatigue was identified as potential contributing factor in collisions worthy of further investigation.

## Introduction

Collisions at sea can have catastrophic consequences; these consequences can be to the health and welfare of a vessel's crew, people onboard the other vessel or installation (if involved) or more generally to the environment. Examples of collisions between vessels at sea causing environmental disasters include the oil tankers *Atlantic Empress* and *Aegean Captain* off Trinidad and Tobago in 1979, the *Sea Star* in the Gulf of Oman in 1972 and *Othello* off Sweden in 1970. Collisions may take place between two vessels or between a vessel and any one of a range of objects including submerged debris or rocks, offshore installations associated with the oil industry, wind farms and even the harbour side. There were 12 collisions under the jurisdiction of the Marine Accident Investigation Branch (MAIB) in the year 2003 involving vessels of 100 gt (gross tonnage) and over; 4 dry cargo vessels were involved, 2 passenger vessels, 2 tanker/combination vessels and 4 categorised as "other commercial vessels" (MAIB, 2004). This was one of the lowest total numbers in the last 10 years and Figure 1 shows a general trend of decreasing collisions (NB: it is worth noting that the MAIB only investigates more serious collisions). The decrease in collisions may be due to technological innovations, improved working practices or other changes in the way that vessels operate (i.e. changes to sailing routes or in the number of vessels at sea); this trend however may be illusory, and an analysis of a greater number of years may show this trend to be a chimera.

It is common sense that collisions can only be investigated after they have occurred; however, it is possible to consider collisions from the point of view of the individual seafarer using retrospective questionnaires rather than investigating each collision separately, for example as in MAIB case studies. Although this approach will not give a necessarily realistic appraisal of the number of collisions that have occurred (NB: some seafarers may have been serving on the same vessels and the collisions may have occurred over many years) it means that issues relating to collisions more generally can be investigated. Of particular interest here is the seafarer's knowledge concerning offshore

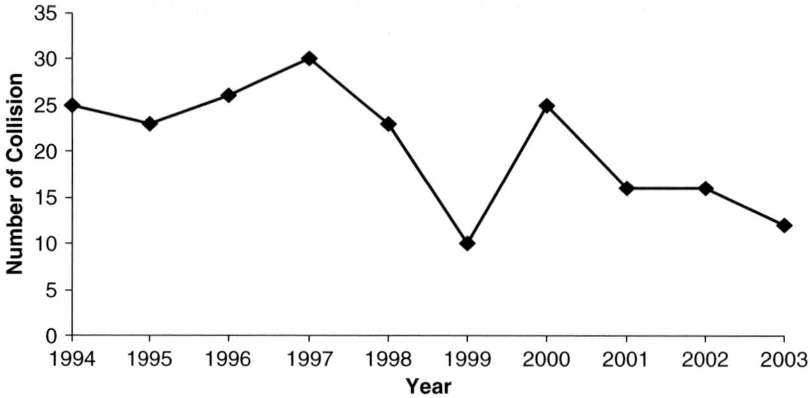


**Figure 1. Collisions at sea investigated by the MAIB 1994–2003.**

installations associated with the oil industry. This is of interest for two main reasons: (1) workers on offshore installations are not seafarers, therefore they may not appreciate the issues for the seafarer; (2) the offshore oil industry is regulated by the Health and Safety Executive (HSE) whereas seafarers are regulated by the Maritime and Coastguard Agency (MCA). It is felt that there is a possibility that the different ways these two industries are regulated may lead to potential collisions between vessels and installations. Offshore installations are aware of this possibility, and engage vessels that come within their exclusion zones; however there is continuing concern in the industry about whether seafarers are aware of the possible consequences of their actions.

## Methodology

This study seeks to gather some data from seafarers on vessels working in the "deep sea" and "short sea and coastal" sectors, as well as data from seafarers on vessels working for the offshore oil installations. These data concern various aspects of collisions and look to see whether there are any systematic differences between the responses from seafarers working in different sectors. This work is best thought of as a pilot study, designed to gather some data about collisions with a view to designing future, bigger studies.

Data were collected from three groups of workers using retrospective questionnaires of self-report measures: deep sea ($n = 302$), short sea ($n = 132$) and offshore ($n = 96$). Initial contact with these sub-samples was through NUMAST, the officers' union. There were some small differences between how the data were collected which influenced the response rates for each sub-sample (i.e. deep sea vs. short sea and coastal vs. offshore); due to these differences the samples were also unlikely to be representative of the wider population of seafarers working in these sectors. Descriptive statistics for the whole sample and comparing the three sub-samples are presented below.

## Some preliminary findings

### *Prevalence of collisions*

Across the sample as a whole, reported collisions with vessels and objects (e.g. fixed installations, harbour side) were relatively high. There were 118 (21.5%) reported

**Table 1. State of vessels at moment of collision.**

| | | Other vessel involved in collision | | |
|---|---|---|---|---|
| | | Sailing (%) | At anchor (%) | Alongside (%) |
| Seafarer's vessel | Sailing | 55.8 | 3.5 | 8.0 |
| | At anchor | 8.9 | 6.2 | – |
| | Alongside | 13.3 | – | 4.4 |

collisions involving another vessel; Table 1 shows that the majority of collisions involved two vessels occurred when both vessels were sailing. Qualitative data indicated that the majority of vessel collisions were attributable to poor weather conditions, in particular strong/gale-force winds and poor visibility; a small number of incidents were directly attributed to fatigue and an even smaller number to equipment and/or mechanical failure. 35.7% of the total sample reported involvement in a collision with another object, the vast majority of which (91.4%) were involved in minor collisions with the harbour side; 4.2% were involved in collisions with fixed oil installations and it is this small number that are of greatest concern. Again, the majority of collisions were attributed to high winds and poor visibility; however, a number of anecdotal reports indicated equipment failure and fatigue and/or human error as elements in a causal sequence leading to the collision.

*Use of offshore installations in navigation*

In terms of the risk of potential collisions, 31.0% of the total sample reported frequently or very frequently sailing close to offshore installations; a substantial proportion of respondents reported using installations as both navigation aids and waypoints (26.5%). 5.2% of respondents often sailing close to installations reported using them as waypoints only. Seafarers were not always aware of the precise location of installations: 28.9% of the total sample of seafarers were at least sometimes surprised by the location of platforms. NAVTEX emerged as the most effective way of alerting seafarers to the position of new installations; 86.9% of seafarers felt this system was adequate whereas 60.9% felt chart corrections were sufficient and 63.4% felt notices to mariners were sufficient.

*Fatigue and collisions*

Nearly half of the total sample (46.9%) felt that being fatigued substantially reduced their awareness of collision risk. 4.0% reported involvement in a collision that they felt was a direct result of personal fatigue. A number of anecdotal reports support the potential involvement of fatigue in collisions though this needs further investigation.

*Sector-specific differences in collision prevalence and related variables*

There were few real differences in terms of the observed collision prevalence; however, unsurprisingly, collisions with installations were more frequent within the offshore sector. Furthermore, collisions directly attributed to fatigue were highest in the offshore sector: 7.7% of the offshore sector reported involvement in an incident they felt occurred as a result of their own fatigue (vs. 2.4% deep sea vs. 4.2% short sea). Differences were also apparent in the extent that routes were formally planned for each voyage: 88.6% of short sea sector seafarers (vs. 99.2% deep sea vs. 98.1% offshore); this may have had a significant impact on collision risk though, as greater proportions of short sea and coastal seafarers reported more collisions with both (a) vessels and (b) objects compared to the deep sea and offshore seafarers.

**Table 2. Sector-specific differences in incidence and perceived risk.**

| | Deep sea (%) | Short sea/coastal (%) | Offshore (%) |
|---|---|---|---|
| Collision with another vessel | 21.7 | 26.2 | 19.8 |
| Collision with object | 34.5 | 46.6 | 43.8 |
| Fatigue decreases awareness of collision risk | 47.6 | 49.3 | 44.5 |
| Involvement in collision where fatigue was an issue | 2.4 | 4.2 | 7.7 |
| Installations are used as: | | | |
| (a) Navigation aids | 68.8 | 50.0 | 62.5 |
| (b) Waypoints | 1.8 | 6.8 | 12.5 |
| (c) Both | 24.1 | 36.4 | 25.0 |
| Surprised by location of installations | 36.2 | 17.7 | 25.5 |
| Adequate to communicate positions of new installations? | | | |
| (a) Notice to mariners | 62.0 | 73.1 | 52.9 |
| (b) Chart corrections | 59.5 | 70.7 | 50.0 |
| (c) NAVTEX | 87.6 | 89.4 | 82.7 |
| Formally planned voyages | 99.2 | 88.6 | 98.1 |
| Planned distance from installations | | | |
| (a) >8 miles | 1.7 | 1.9 | – |
| (b) Between 8–4 miles | 22.7 | 7.5 | 2.0 |
| (c) Between 4–2 miles | 58.0 | 48.1 | 23.5 |
| (d) <2 miles | 17.6 | 42.6 | 74.5 |

## Discussion

Collisions reported in this sample were mostly very minor, causing little damage to the vessel or crew of either vessel; the greater proportion of collisions with objects (vs. other vessels), particularly in the short sea & coastal and offshore seafarers was perhaps due to typically shallower waters and closer proximity to potential dangers compared to the deep sea sector. These data suggest that fatigue be considered in further studies as a potential factor in collisions; the minority of seafarers who reported being involved in a collision where fatigue was an issue was greatest in offshore installation workers and least in the deep sea sector which suggests somewhat intuitively that a closer interface with other objects is a key factor, although seafarers in the deep sea sector also tend to have better work/rest cycles which may be a crucial part of any explanation. The use of installations as navigation aids is encouraged and this is a perfectly safe way to help navigate the oceans; the large proportion of seafarers doing this is promising. There is concern that some seafarers use installations as waypoints, which although not problematic in itself, could possibly be when bad working practices are followed and fatigue is an issue. This is an area that needs further investigation.

It is a major concern that a high proportion of all seafarers were "surprised by the location of installations". That deep sea sector seafarers were sometimes surprised is perhaps understandable as they would tend to cover greater distances; it is peculiar that offshore workers were sometimes surprised as they work within the industry that is responsible for placing the installations where they are. Regarding how to communicate the position of new installations it is clear that NAVTEX is felt to be the most effective; NAVTEX is the automatic broadcast of localised maritime safety information. The "planned distances" from installations that seafarers in the different sectors typically travel demonstrate that the offshore sector work closer to the installations (as their work often involves ferrying people and equipment to and from the installations); deep sea seafarers tend to stay further away, perhaps partially due to these vessels typically being larger with greater manoeuvring problems.

Future research should elaborate on these identified issues. It is important to try and get largely representative samples in future and to consider the plausibility of longitudinal diary studies that will provide a better understanding of causal relationships between key variables. Triangulation of outcome measures (i.e. looking at self-report measures, objective measures of health and fatigue and archived records of working schedules for example) would add further weight to any findings. It also may be worth considering the working practices of other seafarers such as fishermen, who are routinely (somewhat unfairly) regarded as villains concerning collisions at sea, when near the offshore oil installations.

## Acknowledgments

The research detailed in this article is supported by MCA, HSE and NUMAST. The views detailed in this paper represent those of the authors and do not necessarily reflect the policy of the sponsors.

## References

Marine Accident Investigation Branch. 2004, *Annual Report 2003*, MAIB: Southampton.

# THE EFFECT OF SIGNAL INFRASTRUCTURE ON TRAIN DRIVER VISUAL BEHAVIOUR ON SIGNAL APPROACHES

**T. Luke[1], N. Brook-Carter[1], R. Smith[1], A.M. Parkes[1], A. Mills[2] & E. Grimes[2]**

[1]*TRL, Crowthorne House, Nine Mile Ride, Wokingham, Berkshire, RG41 3GU, Email: tluke@trl.co.uk Tel: 01344 770495*
[2]*Rail Safety and Standards Board*

This paper presents the results of a study carried out by TRL on behalf of the Rail Safety and Standards Board. Data for 86 train drivers on ten different routes were collected using a state-of-the-art eye tracking system. Data for 603 individual signal approaches were analysed to examine the effect of different types of signal infrastructure on visual behaviour. Signals mounted on posts, gantries and cantilevers were examined.

Results suggest that gantry signals are looked at more often, for longer and for a greater percentage of the time overall than post signals. This result is explained in terms of the increased complexity of gantry signals. The results have implications in terms of signal identification support in complex areas and the emphasis of route knowledge in driver training.

## Introduction

Following a number of recent incidents on the railways, which have been attributed to errors in human performance, an increased effort has been focussed on conducting research into understanding the behaviour and actions of the train driver. One of these areas of research concerns the visual behaviour of train drivers.

Train driving is primarily a visual task. Train drivers are required to monitor the dynamic scene visually, both outside and inside the train cab. Poor performance on this visual task may lead to errors, such as signals passed at danger (SPADs). For this reason understanding the visual behaviour of train drivers and the strategies they employ when monitoring and searching the visual scene is important.

Currently, only one study has been carried out in the UK investigating the visual behaviour of train drivers. This was a small pilot study involving 10 train drivers (Groeger et al., 2001). The research identified some factors in the visual scene which influenced the visual behaviour of train drivers. They found that signals mounted on gantries were looked at earlier and for a greater duration than signals mounted on simple posts. They also found that cautionary signals were looked at earlier, and for more time overall.

This work built upon that pilot study and extended the methods used. One aim of this work was to examine a larger set of data to determine the effect of signal infrastructure on the visual behaviour of train drivers. Analysis of the data focused on the approach to signals, as this was the primary focus of the research.

## Method

### *Participants*

Eighty six train drivers participated in the main study (79 male, 7 female). Ages ranged from 24 to 63 years, with an average age of 40.

### *Equipment*

The eye tracker used during this trial was the ISCAN VisionTrak ETL-500, supplied by Polhemus. The VisionTrak was mounted on a headband that can be adjusted to all head sizes. It used a dichroic mirror to reflect the eye-image onto a highly sensitive head-mounted video camera. The scene camera was mounted below the driver's line-of-sight.

The eye tracker consisted of two separate sets of equipment: portable and fixed. The fixed system was set up in the office and possessed the full eye tracking software and functionality. The portable equipment carried onto the train consisted of the eye tracker headband, a Sony Hi8 video recorder, a multiplexing unit and a battery belt.

### *Procedure*

Drivers were required to drive one route wearing an eye tracker to record their visual behaviour. The route was driven in both directions with a short break in between the two journeys. Each of the two journeys lasted for approximately thirty minutes. Drivers were accompanied by an experimenter who recorded the signal aspect and points of interest throughout the route. When the route was completed, drivers were shown the recorded footage of the route that they had just driven. These recordings displayed a crosshair indicating the driver's point of regard on the visual scene, and were useful visual aids to assist the drivers in verbalising their perceived visual behaviour.

### *Data coding*

The data recorded in the trials was in the form of video footage with a crosshair overlaid on the scene to depict the driver's point of regard in each frame. 603 fifteen-second signal approach sections of this footage were coded. The coding process identified and recorded the driver's point of regard for each individual frame of video. The data was coded into a number of different glance targets; signal, track, in-cab, ahead, landscape and other. Figure 1 shows a representation of the coding scheme.

Data was classified as a "glance" if five or more subsequent frames (0.17 seconds) were observed with the same point of regard. Only glances were considered in the analysis. Other eye movements which did not dwell on a particular target for five or more frames were excluded. The frame by frame classification was converted into seconds and used to calculate measures of visual behaviour for each target in the scene. The measures used in analysis were: *glance count*, the number of glances during the 15-second section; *glance duration*, the average duration of a glance in seconds; *total time on target*, the

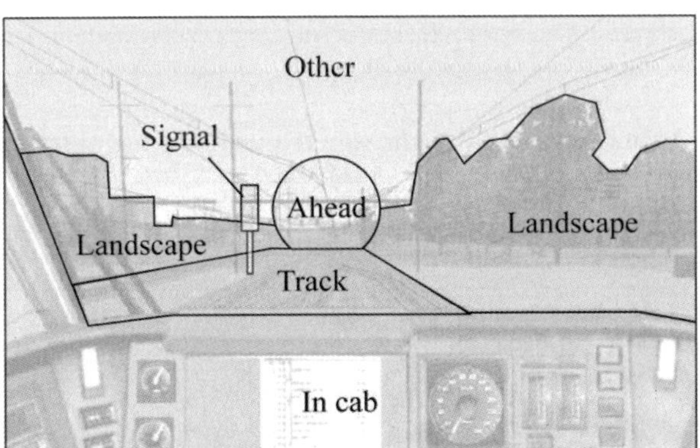


**Figure 1. Representation of the data coding scheme.**

percentage of the 15-second section on the particular target and *time of first glance*, the time between the beginning of the analysis section and the first glance at the signal.

## Results

One-way ANOVA was used to compare the visual behaviour of drivers on approach to different types of signal. Post, gantry and cantilever signals were compared. Post-hoc comparisons were conducted using a Bonferroni correction.

There was a significant effect of signal type on glance count $\{F(2) = 38.179, p < 0.001\}$; glance duration $\{F(2) = 4.280, p = 0.015\}$ and percentage time on target $\{F(2) = 29.613, p < 0.001\}$ for glances to the **signal**. Post-hoc comparisons demonstrated that train drivers looked at gantry signals more often, made longer glances and for a greater time overall than post signals (see Table 1).

There was also a significant effect of signal type on the time of the first glance to the *signal* $\{F(2) = 5.890, p = 0.003\}$. Drivers looked at gantry signals earlier than post signals.

The effect of signal type on visual behaviour to other targets in the driving scene was also examined. Drivers were found to make significantly longer glances *ahead* $\{F(2) = 13.099, p < 0.001\}$ and spend a greater percentage of the time overall looking *ahead* $\{F(2) = 5.746, p = 0.003\}$ when approaching gantry signals compared to post signals. Drivers also made fewer glances at the *landscape* $\{F(2) = 19.779, p < 0.001\}$ and

**Table 1. Glance behaviour to different signal types.**

| Visual Behaviour to the *signal* target | Signal type: Post | | | Gantry | | | Cantilever | | |
|---|---|---|---|---|---|---|---|---|---|
| | Mean | S.D. | N | Mean | S.D. | N | Mean | S.D. | N |
| Total glance count | .96 | 1.45 | 403 | 2.49 | 2.41 | 166 | 1.00 | 1.56 | 19 |
| Mean glance duration (s) | .47 | .33 | 403 | .59 | .42 | 166 | .63 | .62 | 19 |
| Total time on target (%) | 3.35 | 7.28 | 403 | 9.37 | 10.05 | 166 | 3.82 | 6.17 | 19 |
| Time to first glance (s) | 10.43 | 4.34 | 403 | 8.95 | 4.41 | 166 | 11.27 | 2.49 | 19 |

**Table 2. Visual behaviour to the signal according to gantry complexity.**

| | | Number of signals on gantry: 1.00 | 2.00 | 3.00 | 4.00 | 5.00 | 6.00 |
|---|---|---|---|---|---|---|---|
| Total glance count | Mean | 1.00 | 1.98 | 1.57 | 3.00 | 3.25 | 3.80 |
| | S.D. | 1.55 | 1.65 | 2.21 | 3.70 | .96 | 3.07 |
| | N | 6 | 89 | 14 | 7 | 4 | 46 |
| Time to first glance (s) | Mean | 12.80 | 10.25 | 10.16 | 7.88 | 6.13 | 6.59 |
| | S.D. | 1.59 | 4.00 | 4.10 | 5.33 | 5.16 | 4.04 |
| | N | 6 | 89 | 14 | 7 | 4 | 46 |

spent less time overall looking at the *landscape* {$F(2) = 12.370$, $p < 0.001$} when approaching gantry signals.

There were very few differences between post and cantilever signals except in visual behaviour to the *track* target. Drivers were found to look more often {$F(2) = 4.366$, $p = 0.013$} and spend more time overall {$F(2) = 4.502$, $p = 0.012$} looking at the *track* when approaching a cantilever signal compared to when approaching a post signal.

The effect of gantry complexity on visual behaviour was also examined. The number of signals mounted on the gantry architecture was used as a measure of complexity. Complexity was found to have an effect on the number of glances made to the *signal* {$F(5) = 4.539$, $p < 0.001$}. Post-hoc tests revealed that gantries with 6 signals on were looked at more often than gantries with only 2 signals. There was also a significant effect of complexity on the time of first glance at the *signal* {$F(5) = 6.668$, $p < 0.001$}. Gantries with 6 signals were glanced at earlier than gantries with only 2 signals (see Table 2).

## Discussion

The results of this study indicate that train drivers demonstrate different visual behaviour in response to different types of signal infrastructure. Drivers tended to look at gantry signals more often, for longer and for more time overall, than post or cantilever signals. The data also indicated that the first glance at the signal was earlier on approaches to gantry signals than on approaches to post or cantilever signals.

These results are consistent with the findings of the Groeger et al. (2001) study which showed that drivers look at gantry signals earlier and for longer than other types of signal.

However, there are several limitations of the data in this study to consider. The first glance observed in the analysis may not represent the actual first glance made by the train driver. The human eye is capable of resolving objects at a much greater distance than the scene camera used in this study. A related consideration is that there may be a systematic bias towards gantry signals in the data. Gantries are much bigger than posts or cantilevers and are therefore easier to see in the video footage. For this reason more frames may have been classified as "signal" on gantry approaches and it may have been possible to identify gantry signals earlier on the recorded sections.

Despite the limitations, the finding is supported by data from interviews with the drivers who took part in the trials. Drivers described strategies that they use to identify the correct signal on gantries. These strategies included counting across the signals on the gantry to find the correct signal or looking for a specific identifier such as a letter or number attached to a particular signal. These tactics would require the drivers to spend longer looking at gantry signals and possibly to look at them earlier to allow time for identification.

The results for signal complexity are also concurrent with the explanations provided by drivers. There is a general trend which suggests that drivers look at gantries with 4 or more signals more often and earlier than gantries with 3 or less signals.

Overall, the results for signal type seem to suggest that train drivers direct more attention to the signal when approaching a gantry signal. This conclusion is supported by the finding that drivers spent more time looking ahead and less time looking at the landscape when approaching a gantry signal compared to a post signal. This implies that drivers are more focused on searching for the correct signal for their line when approaching a gantry. Drivers also appear to attend more to gantries with an increased number of signals.

This can be interpreted in terms of Wickens (1992) model of visual search speed, which proposed that visual search is mediated by factors in the environment and the higher the number of elements to be searched the more time required. Cluttered environments that contain distractions, particularly those that are similar to the target, require attention to each individual element, therefore resulting in greater visual sampling. Areas that contain complex train signalling are an example of a cluttered environment and multiple signals mounted on a single gantry are similar stimuli from which the driver must select and interpret the appropriate signal.

The findings of this work suggest that gantry signals require more of the driver's attention in order to obtain the necessary information. This has implications for infrastructure design. It may be that providing the driver with identification support, such as labelling signals with large numbers or letters, will reduce the difficulty of identifying the signal, especially where there are four or more signals on a gantry.

The results for the signal type comparisons also suggest that drivers compensate for the difficulty on approach to gantry signals by looking ahead more. It is assumed that this strategy is advantageous in providing early identification of the signal. This is evidence that route knowledge, awareness of current location and important features ahead, is extremely important in train driving. The evidence from the driver interviews suggests that knowledge of the sequence of signals, in terms of the order and position of signals, the type of signal and any unusual or difficult signal features, is particularly important. This has implications for the design of driver training programmes, and suggests that the emphasis on route knowledge, and signalling in particular, should be maintained.

## References

Groeger , J. A., Bradshaw, M. F., Everett, J., Merat, N. and Field, D. 2001, Pilot of Train Drivers Eye Movements, University of Surrey.

Wickens, C. D. 1992, Engineering psychology and human performance. Second Edition, HarperCollins Publishers Inc. New York.

# USING AN ANIMATED SIMULATION TO ASSESS THE VISIBILITY OF AERODROME INFORMATION DISPLAYS

**Kee Yong LIM & Eugene KHOO**

*Center for Human Factors and Ergonomics, School of Mechanical and Production Engineering, Nanyang Technological University, 50 Nanyang Avenue, Singapore, 639798*

This paper reports an analytic evaluation of the effect of 3 design factors on the pilot's field of view (FOV), namely cockpit design, aerodrome design and pilot stature. The case study information used for the analysis is motivated by the SQ006 crash on 31 Oct 2000 at Chiang Kai Shek Airport, Taipei, Taiwan. In particular, the study examines design parameters comprising the cockpit design of a Boeing 747-400, the aerodrome design of the Runway 05R. A 3-dimensional scaled virtual environment is created to simulate a subset of the take-off conditions prevailing on the night of the accident. The model is then used to assess the FOV afforded to the pilot in the cockpit, taking due account of the cockpit structure and nose cone of the Boeing 747-400. An animated simulation of the airplane taxiing towards Runway 05R is also created to support visualization of the dynamically changing FOV experienced along the route. In this way, first pass inferences may be made concerning the visibility afforded to the pilot by the design parameters under assessment. The appropriateness of the specification of these design parameters may be highlighted for more detailed address in aviation safety.

## Introduction

With the increase in international air traffic, New Large Aircrafts (NLAs) would be introduced to the aerodromes to meet this need. Prior to their introduction, the adequacy of aerodromes needs to be verified, for instance runway length and separation and terminal traffic (Barros and Warisinghe, 1997). However, the scope of verification often excludes a review of the visibility of aerodrome displays and markings. These visual displays are vital in helping the pilot's task of taxiing and navigating the aircraft to the designated runway for take off, and for berthing at the terminal. Critical to these tasks is the Field of View (FOV) afforded to the pilot seated in the cockpit.

Taxiing navigation towards the runway would require the pilot and co-pilot to check route features, signs and other visual aids against a map as necessary, to maintain the required situational awareness of the aircraft position and aerodrome conditions. It is to be expected that the maintenance of good situational awareness becomes more difficult under such conditions of low visibility.

Presently, it is postulated that certain design parameters of the aircraft-aerodrome system may also affect the FOV afforded to the pilot; for instance cockpit design and aerodrome design (pilot stature is excluded as a factor if personnel selection and appropriate adjustment to the sitting reference point are assumed). In this respect, International Civil Aviation Organization (ICAO) Annex 14 has been established to standardize the design of aerodromes. However, not all international airports adhere to the requirements stipulated by the Annex. Where discrepancies prevail, the omissions might become contributing factors to an accident, especially when compounded by other factors such as inclement weather conditions. This paper reports an analytic evaluation of the effect of these design factors on the pilot's field of view, when considering the introduction of the NLA.

## Method

Using AutoCAD 2005 and Discreet 3D Studio MAX 6, a scale model of the spatial environment of the aerodrome leading to runways 05L and 05R at Chiang Kai Shek Airport, Taipei, Taiwan, is constructed. The aircraft model assumed is the Boeing 747-400 Domestic (Boeing Commercial Airplanes, 2002). Dimensions corresponding to a no-load condition are assumed to approximate the worst visibility condition; i.e. highest possible plane height leading to maximum obscuring of the runway by the cockpit windscreen and aircraft nose cone. The specified dimensions (Boeing Commercial Airplanes, 2002) assume that the centre of the pilot was 0.53 m from the center of the aeroplane and that the seated height of the pilot is 1.01 m as shown in Figure 1.

The aerodrome layout and specifications used corresponded to the threshold of runways 05L and 05R of Chiang Kai Shek Airport (CKS) (Aviation Safety Council, 2001) as shown in Figure 2. The separations between center lights were set at 30 m for straight routes and at a denser 7.5 m for curved routes, while the edge lights were spaced at 60 m intervals from the threshold of the runway. Both center lights and edge lights used in the simulation were set to a high intensity, omni directional, where in reality it was bi-directional. This was due to the limitation of the software. The omni-directional assumption for the lights would address this in part in respect of the best case for visibility of the lights. Other residual effects associated with the assumption, for example glare and distraction from stray lights, did not constitute main concerns of this project. The radius of curvature of the centerline was calculated to be 49.5 m on the basis of a continuous turn from NP to 05R (a separation distance of 110 m between a 45 m wide non precision runway and a 30 m wide taxi-way). The specifications were extracted from the Taiwanese Air Safety Council's Aircraft Accident Report (Aviation Safety Council, 2001).

To simulate the pilot's visual field from the cockpit, two omni-directional lights were used to project the view from the cockpit window onto the external aerodrome, to represent the pilot and co-pilot's views from their eye level and positions.

The simulation also modeled two boundary conditions when a 90° turn was made, namely Wing Gear Tracking and Nose Gear Tracking (see Figure 3). Nose Gear Tracking

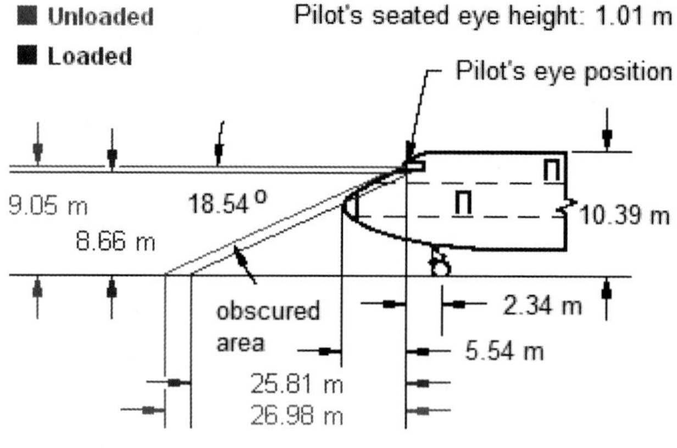


**Figure 1. Eye level reference under loaded and unloaded conditions.**

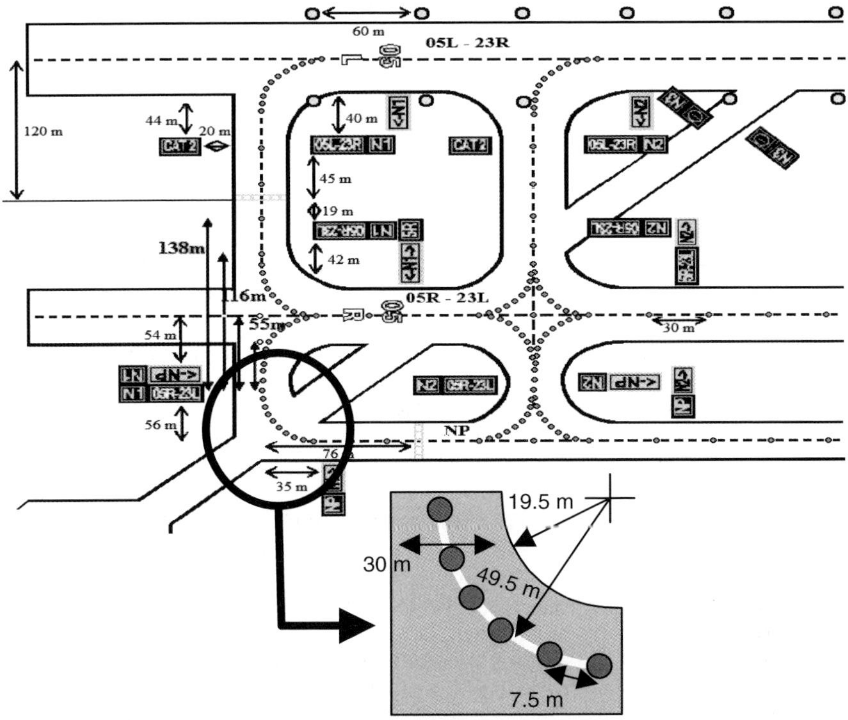


**Figure 2. Aerodrome dimensions assumed for the simulation.**

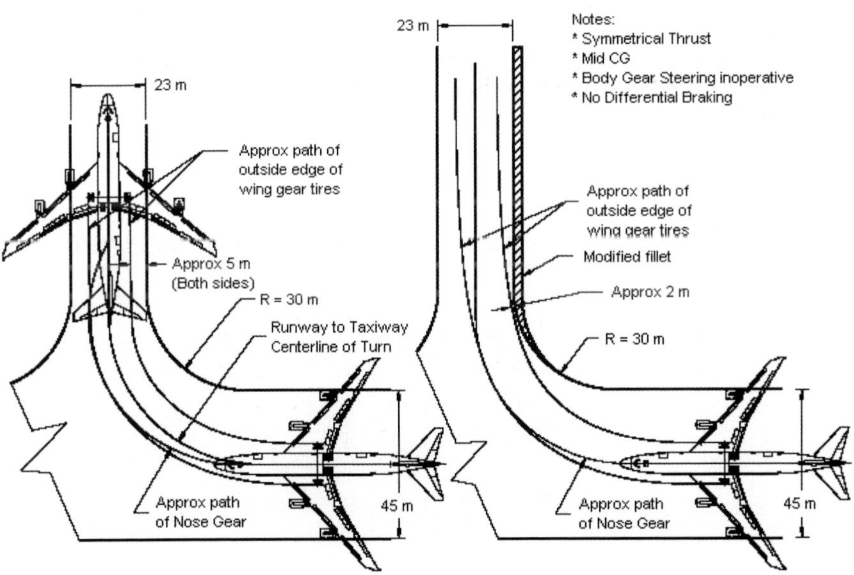


**Figure 3. Wing gear tracking and nose gear tracking.**

implied that the nose gear would be aligned with the centerline during a turn. Wing Gear Tracking implied that the wing gear (or rear wheels) would run parallel to the taxiway centerline irrespective of the trajectory of the nose gear (or front wheel). According to feedback from a pilot, Wing Gear Tracking enabled runway length to be maximized.

## Results and discussion

Figure 4 shows the field of view afforded to a Boeing 747 pilot during taxiing. The different colored areas corresponded respectively to the field of view afforded to the pilot (yellow area) and co-pilot (turquoise area). Areas visible to both pilots are shaded in green. Obscured areas are indicated using shadows cast by the light source. As an A380 has a lower and more forward cockpit and a shorter nose cone, as compared with a Boeing 747. Thus, different NLAs are expected to present the pilot with different FOVs.

Figure 5 shows that upon completing the turn, the 05R runway marking of CKS airport in the SQ 006 case (simulated according to FAA stipulations, 1999) would be

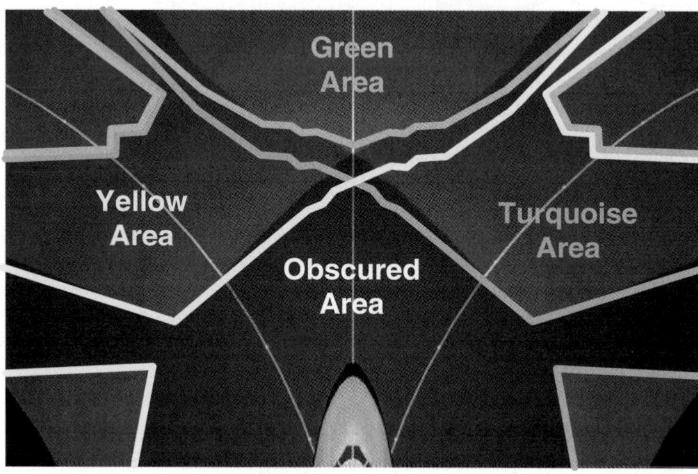


**Figure 4. Field of view from a Boeing 747 cockpit.**

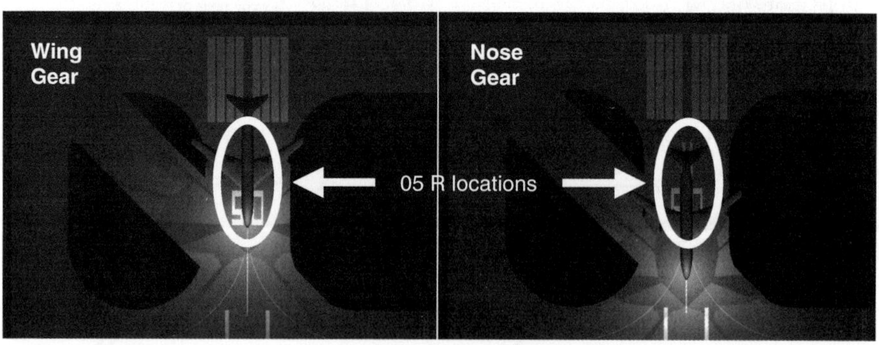


**Figure 5. Aircraft position for wing and nose gear tracking at runway line-up.**

obscured by the Boeing 747 aircraft. While it may be highlighted that the pilots would be able to see the runway designation while executing the turn, this might be inadequate, since the pilot or co-pilot may be preoccupied with other tasks while executing the turn (this situation can happen as evidenced from the SQ006 case). Pilots have also indicated that while turning, they would be looking at the turning reference point rather than at runway designation markings. Consequently, it is presently argued that runway designation markings should also be made visible to the pilots at final line up to serve as an additional safety defense (as opposed to being visible only during turning), as it presents the final opportunity for a visual check before a critical take-off decision is taken. It is possible that the SQ006 crash might have been averted had the runway designation been visible after the aircraft had lined up for take-off, since the pilots would have been prompted at the last minute that they have entered the wrong runway. Presently, simulations are being constructed for forthcoming NLAs to examine how differences in FOV and turning radius might impact the visibility of aerodrome displays.

## Conclusion

In conclusion, the animated simulation was found to be useful in revealing specific issues and concerns associated with aerodrome designs when implementing NLA. In the case of runway designation markings, it was found that compliance with ICAO Annex 14 might not be adequate to comprehensively address the safety critical concerns. Further, it was asserted that existing specifications of runway designations and other visual aids should be reviewed thoroughly, whenever new aircraft models were being introduced to an existing aerodrome. To this end, a similar animated simulation as reported in this paper could be used conveniently to perform preliminary checks on the adequacy of the existing aerodrome design, in respect of the accommodation of the NLA. Follow-up work to extend the simulation to other NLAs will be reported in future papers.

## References

Boeing Commercial Airplanes, 2002, *747-400 Airplane Characteristics for Airport Planning*, Seattle, Washington, USA, 29–34, 111–116.

Aviation Safety Council, 2001, *SQ006 Aircraft Accident Report*, Taipei, Taiwan, Republic of China, 37–71.

International Civil Aviation Organization, Aerodromes Annex 14, 1999, Volume 1 to the Convention on International Civil Aviation, Montreal, Quebec, Canada.

Alim K.M., Fern M.J., Khoo C.S., Chan D, 2000, Non-Conformity to ICAO Standards and Recommended Practices Annex 14: A Case Study of Chiang Kai Shek Airport Taipei, *Internal Report*, Nanyang Technological University, Singapore.

Barros, A.G. & Wirasinghe, S.C., 1997, New aircraft characteristics related to airport planning, *First Air Transport Research Group Conference June 1997*, 1–17.

# MEASURING PILOT WORKLOAD IN SINGLE AND MULTI-CREW MILITARY AIRCRAFT

**Fiona Sturrock & Christian Fairburn**

*BAE SYSTEMS, Warton Aerodrome, W427J, Preston, Lancs, PR4 1AX*

This paper reviews the workload assessments undertaken by Human Factors (HF) Engineers at BAE SYSTEMS on the Nimrod MRA 4 and Typhoon projects. It shows how the traditionally subjective approach, of relying mainly on pilot opinion, has been made more objective by the application of *red-lines*, or tolerance limits, distinguishing between acceptable and unacceptable workload levels. Three complementary workload techniques (BRS, ISA and NASA TLX) have been employed and *red-line* criteria have been developed for each of these techniques. By adopting a more objective workload measurement approach it has been possible to implement design changes that improve the overall HMI of these multi-million pound aircraft, improve human performance and ultimately mission success.

## Introduction

The measurement of human mental workload is notoriously controversial, traditionally relying on subjective ratings. HF Engineers are often then left with the unsavoury task of trying to transform these subjective ratings into "objective" facts that can be used to influence and/or validate the design of a product or system. So, why do we continue to make life so difficult for ourselves?

The fact is that measuring workload within today's complex systems is particularly difficult, as the role that the human takes has evolved to become predominantly mental or cognitive. The humans within these complex systems frequently behave as system managers and tacticians, they "watch over" a system as much as physically interact with it. For military aviation pilots, workload is about performing planned and unplanned tasks in a controlled way that maintains flexibility and respects the temporal demands imposed by the operational environment. From an analytical perspective, the trouble is that as workload shifts away from the physical and toward the cognitive, it becomes both less observable and more difficult to conceptualise. Reduced observability means that subjective data to some extent becomes a necessary "evil". However, data collected in this way must be grounded in a concept of cognitive workload that is shared between HF Engineers and aircrew. Ensuring, and proving, that pilots can cope with the cognitive workload imposed by sophisticated aircraft systems continues to be one of the key challenges facing HF Engineers at BAE SYSTEMS.

Through examples from the Nimrod MRA 4 and Typhoon projects this paper will summarise the ways in which three subjective workload tools can be made more objective by the application of theoretically well-founded, but conceptually simple *red-lines*, distinguishing between *acceptable* versus *unacceptable* workload levels. The paper will then go on to give a flavour of the way in which results have been used to justify design recommendations and then qualify that the workload imposed by the final design is within acceptable limits. By adopting a more objective approach to subjective workload measurement it has been possible to implement traceable design changes that improve the overall HMI of these multi-million pound aircraft, improve human performance and ultimately mission success.

## Development of a subjective workload toolset

Bedford Rating Scale (BRS), the Instantaneous Self Assessment (ISA) technique and NASA Task Load indeX (NASA TLX) were the three workload tools employed during recent workload assessments at BAE SYSTEMS. This multi-tooled approach allowed complementary data sources to be gathered therefore providing a more coherent workload picture.

- BRS (Roscoe, 1987), also known as the Pilot Workload Rating Scale, is a decision tree derived from the Cooper-Harper scale. The user is required to make subjective judgements about workload based on their ability to complete tasks and the amount of spare capacity they have available.
- ISA (Leggatt & Castle, *in print*) is a unidimensional five point rating scale for measuring mental workload. Data are collected electronically via a keypad in the cockpit; ratings are requested at regular intervals indicated by a flashing LED. The tool was originally designed by the National Air Traffic Service but has been further developed by BAE SYSTEMS HF Engineers.
- NASA TLX (Hart & Staveland, 1988) focuses on six dimensions of workload (mental demand; physical demand; temporal demand; performance; effort; frustration level). The user is required to rate each dimension on an incremented bipolar scale. Weightings were not applied to the results.

Together, these measures form a complementary battery of tools, as they all ask the pilot slightly different questions about his/her workload (Figure 1).

A key asset of BRS is the way in which the scale arrives at a quantative rating through a qualitative decision tree. The quantitive rating is useful as a shorthand communication tool between HF engineers and project mangers. Meanwhile, the qualitative decision tree constitutes a usefully "plain English" approach to workload with a built in criterion of acceptability. This criterion relates to the subject's perception of spare capacity. Bearing in mind the goals of military aircraft workload assessment, this conceptualisation of workload is highly appropriate.

The ISA offers the ability to gather real-time data in a minimally intrusive way. It offers the analyst freedom to gather data at specific points in an assessment in a consistent and precise way. In addition, because ISA is administered via an electronic keypad in the flight deck/cockpit it allows workload data to be collected more frequently than would perhaps be possible using the typically paper-based BRS and TLX tools.

NASA TLX is primarily used to compute an overall workload score, but its structure also affords diagnostic use. The six factors within the method can be used to unpack the "spare capacity" conceptualisation of workload and dig down towards the origin of workload peaks in a way that informs design. For example, a lack of spare capacity captured through BRS or ISA could be attributed through NASA TLX to "time pressure" or "mental effort". A "time pressure" diagnosis could direct the designer to look towards

| |
|---|
| BRS - "Was your workload acceptable?" |
| ISA - "What is your workload now?" |
| NASA TLX - "What was your workload?" |

**Figure 1. Three complementary workload tools.**

**Table 1. Red-Lines employed during Nimrod MRA 4 workload assessments.**

| Tool | Development/Risk reduction workload assessments | Qualification workload assessments |
|---|---|---|
| NASA TLX | 0–60: Acceptable<br>61–80: Investigate further<br>81–100: Unacceptable – probable design change | 0–80: Acceptable<br>81–100: Investigate design change |
| BRS | 1,2,3: Acceptable<br>4,5,6,7,8: Investigate further<br>9–10: Unacceptable – probable design change | 1–8: Acceptable<br>9–10: Investigate design change |
| ISA | 1–4: Acceptable<br>5: Investigate further | 1–4: Acceptable<br>5: Investigate further |

the modification of, for example, an interface menu structure or interaction strategy. A "mental effort" diagnosis may prompt the designer to consider offloading tasks to automation, simplifying or modifying the chosen mode of data presentation.

As with the majority of subjective workload tools there are few guidelines to distinguish between *acceptable* and *unacceptable* workload. The next section summarises the way in which HF Engineers at BAE SYSTEMS have developed explicit *red-lines* for each of the three workload tools described.

### *Red-line criteria*

An extensive literature review revealed very few existing workload *red-line* criteria. Consequently, the development of such criteria for the Nimrod MRA 4 project was based on the results of workload studies reported in the literature, Nimrod Mk 2 workload studies and experience of high workload situations in the aviation industry.

Although these *red-lines* were initially developed for the Nimrod MRA 4 project they have been, in part or completely, employed on other projects such as Typhoon. *Red-lines* were identified for each tool depending on the stage in the life cycle that the assessment was being performed. For example, during the early/development phases of the product life cycle a **tri-band** *red-line* process was adopted in order to flush out as many design issues as possible. During the latter stages of qualification however a **dual-band** *red-line* process was adopted. Table 1 details the *red-lines* used for each of the three workload tools.

## Nimrod MRA 4 and Typhoon assessments

Both the Nimrod and the Typhoon workload assessments were carried out on bespoke high fidelity flight deck/cockpit simulators capable of "flying" realistic operational missions.

### *Data collection & analyses*

During the assessments the workload tools were administered at different, pre-defined points during each mission. Links were made between each tool and different mission elements. For example, in the case of Nimrod MRA 4 NASA TLX was used at the highest level to gather data about the different mission phases flown by the pilots. BRS was adopted at the mission segment level and finally ISA was employed to gather task-level data.

The administration points for each workload tool had to be carefully selected to be certain that there was as little interference with the pilots' primary and secondary tasks as possible. Both the Nimrod MRA 4 and the Typhoon presented different challenges in this area. A typical mission for a Nimrod MRA 4 (such as Search and Rescue) will generally be significantly longer in duration and of a different form than a typical Typhoon mission. During the workload assessments the Typhoon missions were flown by single pilots who completed their workload ratings after the end of each mission but used mission elements to refer back and differentiate between different periods of time or aspects of work. During the Nimrod pilots' workload assessments however a more interruptive administration technique was employed, i.e. at predefined points in the mission the aircrew were interrupted in order to gather workload ratings. It should be noted that these missions were not "frozen" during the data gathering in order to preserve situation awareness (as far as possible) as one of the pilots could maintain control of the aircraft whilst the other completed their workload ratings.

Once the workload data had been collected the analyses process was made quick and easy by the application of the *red-line* criteria. In the case of Nimrod, this allowed the data to be colour coded and workload *hot-spots* (areas of unsatisfactory or unacceptable workload) were immediately apparent. These *hot-spots* were then used to drive the crew debriefing sessions during which aircrew, together with HF engineers investigated whether the cause of unsatisfactory/unacceptable workload was directly attributable to HMI design deficiencies.

In the case of Typhoon, the aim of the data analysis was to identify and understand any "consistent workload peaks". The origins of these peaks were studied in order to clarify the interaction between system design and the demands placed on aircrew. Analysing the data in this way allowed for prioritisation and clarification of any HMI issues that emerged.

It is important to note that for both projects the decision about whether there was a design issue was not based on workload data alone, but on the combined results of all the data collected during the assessments (i.e. workload, aircrew performance and situation awareness). This integrated and rigorous process led to a set of recommendations for design changes, substantiated by objective data. The majority of these highly traceable recommendations were accepted by each project's senior management and have been embodied into the baseline designs.

## Discussion

Although the *red-line* criteria developed for the Nimrod and Typhoon projects have not yet been formally validated they have been used on numerous occasions throughout each product's life cycle. The results from these assessments have so far produced meaningful results both in terms of supporting design changes that the pilots considered to be essential but also in helping to reduce, or filter out, the inevitable *requirements-creep* that are a result of individual's preferences and biases. For although the nature of the systems under study and their effect on pilot workload makes the use of subjective data invaluable in our work it is essential that design changes are not implemented based on personal preferences but on "hard" evidence. The integration of several data sources with predefined *red-line* criteria provides clear and traceable evidence for each design change proposed and/or implemented. In some cases the identification of consistent workload peaks also allows the analyst to understand the way in which a single HMI issue may be responsible for more than one instance of unacceptable workload. In this way, HMI issues may be prioritised and the cost-benefit of HMI changes more clearly illustrated.

As evaluators of a complex system design, military aircraft HF Engineers are frequently faced with the task of making an *absolute* judgement with regard to workload. In both case studies reported here, the question being asked related to whether the workload encountered by an operator of the aircraft would be likely to exceed human capacity in a mission situation. The acceptability question would probably have been much easier to answer if the judgement was *relative* (which of these different designs produces the optimal level of workload in a mission situation?). The main problem addressed in this paper relates to having a theoretically valid benchmark of absolute subjective workload acceptability. The *red-line* approach applied during the Nimrod MRA 4 and Typhoon assessments goes some way towards addressing this issue and will continue to be developed in order to support future design requirements.

## References

Hart, S.G. and Staveland, L.E. 1988, Development of NASA TLX (Task Load indeX): Results of empirical and theoretical research. In P.A. Hancock and N. Meshkati (eds.), *Human Mental Workload*, (Elseveir: Amsterdam), 139–183.

Roscoe, A.H. 1987, In-flight assessment of workload using pilot ratings and heart rate. In A.H. Roscoe (ed.), *The practical assessment of pilot workload*, AGARDograph No. 282, AGARD: Neuilly Sur Seine, France, 78–82.

Leggatt, A. and Castle, H. *in print*, Validation of the ISA (Instantaneous Self Assessment) Subjective Workload Tool.

# VIOLENCE IN THE WORKPLACE – DESIGNING OUT THE PROBLEMS

**Suzy Broadbent, Zoë Mack & Lucy Swinson**

*CIRAS, WS Atkins House, Birchwood Boulevard, Birchwood, Warrington WA3 7WA*

The Railway Group Safety Plan 2003/4 states that "88% of staff say they have experienced assault in the past six months" and that many of these incidents go un-reported. A considerable amount of reports received by CIRAS (The Confidential Incident Reporting and Analysis System for the UK Rail Industry) relate to assault or the abuse of staff by members of the public. These events vary in severity and also in the factors underlying them. When these events are analysed for root causes, it transpires that a significant number can be traced back to design issues, either physical design of the workplace or job design. This paper looks at how such issues can affect the likelihood of violence in the workplace and discusses methods of reducing this risk. A background of the CIRAS taxonomy and methodology will be presented, along with examples of relevant reports and suggestions for improvements in the future.

## Introduction

In 2002/03 it was estimated that workers in England and Wales experienced 849,000 incidents of violence in the workplace. The Health and Safety Executive (HSE) defines work-related violence (WRV) as, "Any incident in which a person is abused, threatened or assaulted in circumstances relating to their work". It is estimated that on average 1.2% of the working population experienced physical assault and 1.7% were threatened at work during the year 2002/03 (TSSA website). In the public transport sector, this figure rises to 2.8%.

The Railway Group Safety Plan 2003/04 states that "88% of staff say they have experienced assault in the past six months" and that many of these incidents go un-reported. The cost of such incidents is not only physical, but also psychological and financial. The threat of assault can lead to anxiety and stress that in turn can result in low staff morale and high staff turnover.

The TUC state that certain jobs have higher risks, inherent within the design of tasks they are expected to carry out. For example, jobs such as those that involve handling money, provide care, advice or information, work with violent people, deal with complaints, work alone or work unsocial hours.

Certain work places also have higher risks inherent within their design. For example, if you work on your own, away from other staff, in the community, in a workplace that is badly lit, not on your own employer's premises or in multi-occupied premises (TUC website).

## Methodology

CIRAS is the UK Rail Industry's Confidential Incident Reporting and Analysis System. It offers staff an alternative route through which to report their safety concerns.

The system was set up to capture safety-critical information that may go unreported through company channels for fear of recrimination.

Trained CIRAS analysts conduct short interviews with staff when they make reports, in order to establish the nature of the concern and the issues surrounding it. The issue is written up in a de-identified fashion and fed back to the relevant company for comment. The details of the concern and the transcript of the interview are entered onto the national database and analysed by the CIRAS Core Facility.

The CIRAS Core analysis method enables the assignment of up to three actual, near miss and/ or potential events to each safety issue that is reported by staff. The events in the taxonomy were developed in consultation with RSSB. Using the taxonomy, each safety issue is coded by trained CIRAS analysts based on the information in the report and on the company response(s) provided to that report.

An event is an undesirable occurrence related to railway operations (e.g. a derailment or a collision). These can be actual, potential or near miss events. In addition, operational failures can be assigned to indicate that an error or violation has been or could have been made by front line operational personnel. Operational failures are only assigned when stated explicitly by the reporter or commented upon in the company response. They may be actual failures (have happened) or potential failures (could have happened).

Up to three causal factors can also be coded for each report. Causal factors (precursors to the event), such as assault/ abusive behaviour, can also be assigned to describe the problem and help us understand why the event occurred. The causal factors analysed in this report are "Malicious Act – Assault" and "Welfare/OHS problems – Inadequate Security".

## Reports received by CIRAS

### *Overview of data*

In the period June 2000 to June 2004, a total of 3972 reports were received by CIRAS from all sectors, including LUL. A search of the CIRAS database found 238 reports that related to violence. Figure 1, below, shows the root causes associated with the causal factors of "Malicious Act –Assault" and "Welfare/OHS problem – Inadequate Personal Safety (Threat of Assault)". Of these reports, 120 of them had "Design" as the root cause. These reports are analysed in the following section.

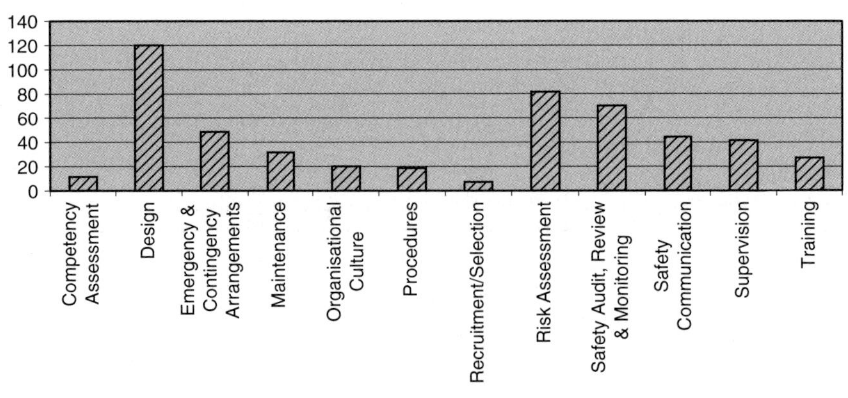


**Figure 1. Root causes relating to assault.**

*Design issues*

Figure 2 shows the job categories of the staff that made reports regarding risk of assault and design issues. Figure 2 shows that Train Crew made the most reports to CIRAS (n = 40), followed closely by Station Staff (n = 36).This is to be expected as it is these job groups that deal with the public most frequently and therefore they are more likely to encounter violent situations.

*Safety consequences*

Figure 3 shows the safety consequences associated with violence problems. It is possible to have more than one event associated with each report, for example one reporter was concerned that certain trains were overcrowding and that this could lead to passengers being caught in the doors and also passengers becoming violent with staff. Figure 3 shows that reporters believe major (n = 57) and minor injuries (n = 92) are a possible safety consequence of violence. Considering that those reports relate to Malicious Acts and Welfare/OHS Problems it follows that the main concern of reporters would be an injury. However, as Figure 3 shows, these are not the only potential ramifications of assault. For example, there are 8 cases of "person struck/crushed by train". This could occur when an attack breaks out on a platform and someone is pushed into the path of an oncoming train.

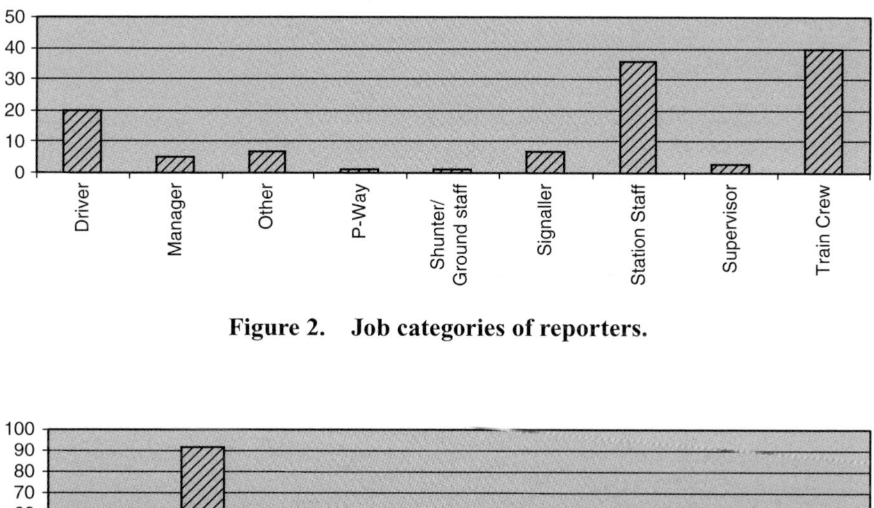


**Figure 2. Job categories of reporters.**

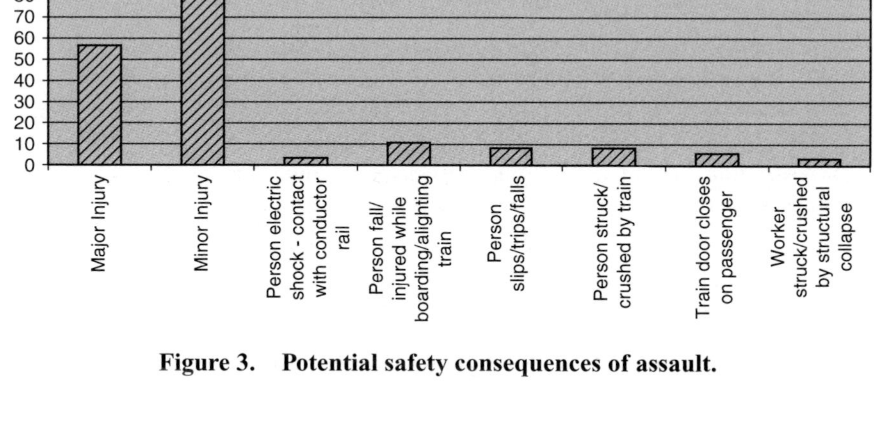


**Figure 3. Potential safety consequences of assault.**

## Job design

Reports that relate to job design and assault often also relate to rostering. Numerous concerns have been raised to CIRAS regarding staff working alone at specific times of the day, such as early morning or late at night, as can be seen in the extract from a report synopsis below.

> *"Concern has been raised regarding the security of staff whilst carrying out station inspections at unmanned stations. The reporter believes as these inspections are carried out after peak hours (often about 7pm) there is the risk of assault on staff by members of the public."*

This report also highlights the further risks of working alone. Other issues raised relating to job design and violence include concerns over handling money.

> *"…Well it's the threat of it, because we do come under, quite a few threats just doing our regular duties as revenue inspectors, but it's erh, it's just the threat of carrying money, whatever some, the attacker obviously doesn't know, but if they see you there for a few hours they will presume that you have money on you"*

Both of these issues would hopefully be overcome by highlighting problem areas and altering the job design. For example, by altering the time that the inspections need to be carried out or by sending staff to carry out these inspections in groups, the risk of assault could be reduced.

## Physical design

The risk of assault due to the physical design of the workplace (be it a station, a train or a worksite) can be seen in the following examples.

> *"…a guard reports that this service is often dangerously overcrowded and, as a result, passengers are frequently left at station platforms on route. Recently on this service a passenger became verbally abusive and threatening to staff in relation to the poor provision of service information and the overcrowded conditions. The guard making this report suggests that by putting a 4-car set on this service, train overcrowding and passenger pressure on Staff could be reduced."*

> *"It is stated that Drivers must use a public pathway that is often frequented by gangs of youths, and in an area where there have been serious attacks on members of public and also on a member of staff. It is said that Drivers are seriously concerned about using this pathway and would like action to be taken quickly to provide a different access route."*

> *"A CIRAS report has highlighted that for over three years this station has been without a passenger lift to all platforms…when disabled passengers are unable to gain access to the platforms, staff are often verbally abused, with suggestions of discrimination in relation to the poor provision disabled access facilities."*

Overcrowding, lack of security and inadequate facilities are just some of the concerns raised by staff that can result in assaults against them.

## Reducing the risk

More than half of the reports made to CIRAS that relate to assault have design as a root cause, be it job design or physical design. Often when making such reports, reporters themselves suggest solutions for mitigating risks and often believe that it is the cost of implementing such changes means that nothing is done.

Alterations to job design, such as removing the need for train crew to carry cash, increasing the number of staff working during certain periods and ensuring staff are trained in conflict resolution, may reduce the risk of assault.

Similarly, simple changes to actual physical design and layout can make a huge difference in the probability of bring attacked. For example, ensuring that walkways are well lit, that CCTV is installed at places of high risk and that the physical environment on trains and in stations is at a comfortable level (HSE Website).

By involving Ergonomists and Human Factors Consultants when first considering the redesign of a station, train or worksite and indeed when designing jobs or shift patterns, it would be hoped that some of this risk could be mitigated.

## References

*Railway Group Safety Plan 2003/04*. Railway Safety.
www.hse.gov.uk/violence
www.tssa.org.uk/advice/hs/hs09.htm
www.tuc.org.uk/h_and_s/index.cfm?mins=30

# VALIDATION OF MINIMUM READING TIME REQUIREMENTS FOR RAILWAY SIGNAL READING

**Mark Newman[1], W. Ian Hamilton[1] & Ian Maxwell[2]**

[1]*Human Engineering Limited, Shore House, 68 Westbury Hill, Westbury-on-Trym, Bristol, BS9 3AA*
[2]*Rail Safety and Standards Board, Evergree House, 160 Euston Road, London NW1 2DX*

Human Engineering Limited was commissioned to validate and extend the concept of the Minimum Reading Time for railway signals, a vital measure in rail safety, specifying how much time is required by drivers to read and respond to signals. This work involved an assessment of the behaviours that comprise the signal reading task via Cognitive Task Analysis, and an examination of the hazards that are associated with this task via survey and performance modelling. For some hazards, it was found that additional time could be added to the signal reading process as a form of mitigation. For other hazards, the analysis showed that additional time alone could not help drivers, and alternative means of mitigation were required. The results were used to provide a tool for use in the rail industry that is supported both by human factors theory and evidence from experience.

## Introduction

The signal reading task is a vital part of the train driver's role. Scanning for and reading signals on approach requires a driver to maintain his or her vigilance throughout a journey. This task takes a finite amount of time, and it is therefore vital to ensure that drivers have sufficient time to locate, read, interpret and then respond to each signal, before it is passed.

The Minimum Reading Time (MRT) is currently specified as "a seven second time window starting eight seconds travelling time before the signal and finishing one second before the signal" (Railway Group Standard, 2001). However, some signals may require longer reading times. A number of hazards, such as *the signal is viewed against a dark or complex background*, have been identified that affect the signal reading process. Additional reading time has been suggested as a means of mitigating the impact of several of these hazards. This information has been used to develop the current MRT checklist, which is used when signals are first positioned, and also during accident investigations.

Unfortunately, neither the MRT nor the additional time specified for different hazards appears to be based on human factors evidence or theory, but instead is an approximation based on operational experience. Given the rarity of accidents on the railways it could be argued that these approximations are reasonable. However, following the accident at Ladbroke Grove, great emphasis has been given to the application of human factors to questions of rail safety.

Human Engineering Limited was commissioned by the Rail Safety and Standards Board to validate and extend the concept of the MRT, by looking at:

- The variability in time required to complete signal reading;
- The hazards associated with signal reading; and
- How these hazards could be mitigated.

## Method

Three distinct lines of work were used:

- A literature review, to determine what hazards should be considered;
- A Cognitive Task Analysis of driver behaviour for the signal reading task and for dealing with each hazard;
- A survey of driver attitudes to the hazards identified from the literature review.

### *Literature review*

The literature review was used to determine what hazards should be considered. It examined empirical research, both real world and simulator studies of driver performance, and other research into issues that can affect drivers, such as the speed at which items such as signals on a gantry can be searched. The review was also used to identify the hazards described in Railway Group Standards and other tools and checklists used in signal assessment.

The aim of this assessment was to identify what hazards should go forward for assessment in the driver survey and in the MRT worksheet, and (where possible) to identify experimental evidence to show the size of the effect of these hazards.

### *Cognitive Task Analysis*

The Cognitive Task Analysis (CTA) was used to make a prediction of how long it takes for drivers to perform the normal signal reading task, and allowed the effects of different hazards to be calculated.

CTA is a type of cognitive modelling in which a task, such as reading a signal, is broken down into a series of behaviours, such as "scan for signal" and "interpret meaning". These behaviours are described at a sufficiently fine grain of detail that estimates can be made of the processing time required for them to be performed. For example, the time required to respond to the onset of a light by pressing a button is the sum of the time required for:

- Perceptual processing: to see that the light is on;
- Cognitive processing: to decide to act; and
- Motor processing: to press the button.

Estimates for the time required for basic activities such as these have been derived from evidence from a large number of laboratory studies, and have been used to develop the keystroke level model (Card et al., 1983). This allows the prediction of how much time it takes to complete a task by slow, median and fast operators. These estimates include a variety of individual and environmental differences. This means that, for example, *fast man* times are the fastest times reported for any individual in any of the studies (e.g. to perform a visual processing behaviour). It can be assumed that the activity was performed under optimal environmental conditions. The *slow man* times are, similarly, the results for the worst person working in the worst conditions. This distribution of scores has a significant positive skew. Given the age and level of skill of drivers, it was thought that the worst case figures would be unrealistically slow for the purposes of this study. A modified score was therefore calculated for the slowest drivers, based on a more symmetrical distribution of task times. This was subsequently confirmed by observation of empirical data.

### *Driver survey*

The driver survey was aimed at identifying which hazards the drivers rated as being the most severe. An initial evaluation identified a variety of different signal types. The

**Table 1. Behaviours involved in signal reading.**

| Behaviour | Fast driver | Median driver | Slow driver |
|---|---|---|---|
| Monitoring of track ahead | 0.170 | 0.470 | 0.770 |
| Detection & recognition of signal | 0.220 | 0.570 | 0.920 |
| Association of signal with line | 0.302 | 0.617 | 0.932 |
| Hear AWS alarm | 0.130 | 0.310 | 0.490 |
| Respond to AWS | 0.305 | 0.436 | 0.566 |
| Interpretation of signal meaning | 0.100 | 0.240 | 0.380 |
| Selection of (goal) action required | 0.192 | 0.430 | 0.668 |
| Manipulate power control | 0.467 | 0.731 | 0.996 |
| Manipulate brake control | 0.467 | 0.731 | 0.996 |
| Strategy review | 0.142 | 0.290 | 0.438 |
| Total | 2.495 | 4.826 | 7.156 |

Key: Times are in seconds.

impact of hazards on each signal type could, in principle, be assessed separately. However, it was considered that this would not be practical – because of both resource limitations and problems of usability of the end product.

Resource limitations were a factor because assessing different signals would require longer questionnaires. These take longer to complete, which will typically reduce the number returned, or require different respondents to answer questions on different signal types. Both these problems reduce the statistical power of the study. Assuming that there were differences between signals, the result of this survey could be an MRT form which would be far longer and more detailed than currently is the case – and so far less usable.

The severity of different hazards was, therefore, assessed in relation to a four aspect colour light signal. The design for the questionnaire finally selected was one that asked drivers to assess the severity of hazards on a 4 point scale, which ranged from "this is not a hazard" (marked as 1) to "this is a severe hazard" (marked as 4).

## Results

The CTA used to identify the series of behaviours that make up the signal reading task, was based on the recognise-act cycle and research conducted by Human Engineering Limited (2003). An estimate of the time required to complete each behaviour was calculated, using the keystroke- level model (Card et al., 1983), see Table 1.

Over 650 responses from the driver survey were analysed. These results were used principally to identify the mean rating of hazard severity, which allowed those hazards that the drivers perceived to be the most severe to be identified. This process allowed the contents of the MRT checklist to be verified in workshops with subject matter experts, to ensure that no hazard was dropped, needlessly, from the list. The top rated hazards are shown in Table 2.

Models of the impact of the hazards allowed three different types of solution to be identified.

(1) Some hazards require additional tasks to be performed by the driver during the signal reading process. For example, when there are parallel signals on a curve the normal behaviour of line association becomes far more difficult; drivers may instead need to recall which is the correct signal and then use an additional search strategy to locate it. For such hazards, an increased reading time can be seen as a suitable form of mitigation.

**Table 2. Hazards rated as "high" severity.**

| Hazard | Mean rating |
|---|---|
| There are abnormal railhead conditions (such as grease or leaves) | 3.54 |
| The signal is viewed against a complex background – such as street-lights, housing, or lots of gantries | 3.50 |
| You are at the end of a long shift, which may lead to a loss of concentration while driving | 3.35 |
| There are poor driving conditions: fog or heavy rain reduces visibility | 3.25 |
| The signal is near strong lighting, such as station lighting – which risks diminishing the apparent intensity of the signal light | 3.25 |
| Obstructions mean that parallel signals are not all in view together, 8 seconds in rear of the signal | 3.11 |
| Parallel signals are on a curve: other signals may be more prominent than your signal | 3.11 |
| When the AWS horn sounds, another signal is more conspicuous than your signal – may lead to you reading the wrong signal as your own | 3.07 |

(2) Other hazards were found to be difficult to mitigate with additional reading time alone. For example, where the light from a signal is diminished by nearby lighting (or where the light from the signal itself is so bright that the message it conveys is difficult to distinguish), additional reading time would be unlikely to be of help. Instead, alternative mitigation measures are required, such as shielding the signal from external light sources or using a more appropriate signal lamp intensity.

(3) Finally, a set of hazards was identified that are difficult to mitigate at all. For example, when signals are not placed in parallel on a bend, they can appear to swap positions as the train approaches, making it very difficult for the driver to identify the relevant signal. If one of a group of parallel signals is obscured on approach, a driver may read across to the wrong signal. These are extremely dangerous situations, that need to be removed by repositioning signals or objects obscuring them, wherever possible.

## Conclusion

From Table 1, the time required for signal reading can be seen to be approximately 7.2 seconds. This is an extremely conservative model of signal reading, assuming for example that the behaviours are performed in series, rather than in parallel. However, this approach is justified, given the safety critical nature of the activity and the fact that it will frequently be performed in sub-optimal conditions, such as when there is fog or glare from the sun.

This is slightly less than the current standard, in which the MRT is 8 seconds; nevertheless, it is proposed that the longer reading time be used. While the higher time is not required for completion of the signal reading task, it will be of value in allowing for variations in driver strategy and in recovery from error. Any hazards associated with a particular signal would require that even more time be added.

However, since the individual behaviours involved in the signal reading task have been specified, it is possible to reduce the reading time if some of these behaviours are not required. For example, the behaviour of *associate signal with line* will not be performed if there is a completely unambiguous association between signal and line. This may happen where there is only one line signalled in the direction of travel. The reading time required for such signals can be reduced by approximately 1 second.

If trains have a combined power and brake controller, the time required to complete the planned actions can be reduced. An additional 1 second reduction in reading time is possible, if the trains using the track *all* have combined controllers. This means that, under certain circumstances, the reading time for a signal may be as low as 6 seconds.

This study has, in fact, supported many of the suggestions made in the existing MRT checklist. However, some changes have been suggested to bring the checklist into line with the results of empirical studies and performance models. The revised checklist is currently being reviewed by RSSB before use in the field.

## References

Card, S.K., Moran, T.P., and Newell, A., 1983. *The Psychology of Human-Computer Interaction*, Lawrence Earlbaum Associates, Hilldale, New Jersey.

Human Engineering Limited, 2003, *Cognitive Task Analysis for Signal Sighting Hazards.* Report No.HEL/RSSB/03862/RT1, Issue 1, 8th October 2003.

Railway Group Standard, 2001, *Signal Positioning and Visibility*. Report No. GK/RT0037, Issue Four, October 2001.

# DEVELOPMENT AND VALIDATION OF THE HMRI SAFETY CULTURE INSPECTION MODEL AND TOOLKIT

**Joanna Foulkes[1], Nicola Stapley[1], Surbjit Laroya[1], Kim Bowling[1] & Claire Dickinson[2]**

*[1]Human Engineering Limited, Shore House, 68 Westbury Hill, Westbury-on-Trym, Bristol BS9 3AA*
*[2]HMRI, Grove House, Skerton Road, Manchester M16 0RB*

There is an ongoing programme of work within HMRI to validate the implementation of recommendations made from public inquiries into rail accidents. A safety culture assessment toolkit was developed for use by HMRI Inspectors to address one of the themes identified within the recommendations. The aim of the toolkit was to provide a pragmatic approach for the assessment of safety culture within the rail industry. The toolkit focuses on assessing performance based on 5 key indicators known to influence safety culture. An underlying theme embedded within the toolkit is that it is based on verifiable, evidence-based measures of safety culture derived from an extensive literature review and key stakeholder experience. The toolkit was tested through a programme of stakeholder site visits, during which information concerning the strengths and weaknesses associated with the safety culture of a participating organisation were captured and recorded.

## Introduction

### *Background*

One of the themes identified within the extensive set of recommendations from recent public inquiries, that investigated and reported on rail accidents, is that of safety culture. Her Majesty's Railway Inspectorate (HMRI) has an ongoing programme of work to validate the implementation of the recommendations and it was identified that the validation of these recommendations would benefit from clear programmes of work that inspectors could undertake.

### *Aim of the toolkit*

The aim of this project was to develop a pragmatic tool to aid HMRI inspectors validate the implementation of the safety culture recommendations within rail organisations. The toolkit focuses on a number of indicators that are known to influence safety culture:

- Leadership;
- Two-way communication on health and safety matters;
- Involvement of staff (including staff representatives) in decision-making;
- The existence of a learning organisation;
- The prevailing attitude towards blame.

### *What really happens?*

The safety culture assessment toolkit was developed to capture, as far as is practicable, the activities actually being undertaken within an organisation, rather than focusing upon the perceptions of staff. Cooper (2000) developed a framework that distinguishes between three interrelated aspects of safety culture:

- Psychological aspects – how people feel, e.g., about safety and safety management systems;
- Behavioural aspects – what people do, e.g., safety related activities and actions of employees and management;
- Situational aspects – what the organisation has in place, e.g., the organisation's policies, management systems and operating procedures.

It is acknowledged that these three aspects are not mutually exclusive, but intertwined and thus difficult to disentangle. Some existing safety culture assessment frameworks have focused upon the subjective and psychological aspects (safety climate), but the focus of the HMRI safety culture assessment toolkit was directed towards what happens in the company (a reality check). The HMRI safety culture assessment toolkit therefore provides an objective methodology for safety culture inspections, by concentrating on the behavioural and situational aspects of an organisation to assess what actually happens in relation to health and safety matters.

## Method

### *Literature review*

In order to develop the toolkit, an extensive but focused literature review was undertaken. The review was conducted to identify the prevalent features of the 5 indicators that could be used to form the toolkit. The review was carried out using a variety of sources including research papers related to the 5 indicators, standards and other guidance material, and tools and approaches used to measure safety culture within rail and other industries.

### *Stakeholder meetings*

Meetings with stakeholders (i.e., HMRI inspectors and representatives from railway organisations) were undertaken to identify best practice approaches that are used within the industry and to capture the user requirements of HMRI inspectors, in order to develop a usable toolkit that is tailored to fulfil the HMRI inspectors' needs.

### *Visits*

A pilot study was conducted at one rail organisation, on a preliminary version of the toolkit, which was followed by 6 further visits to a range of different rail organisations to validate the final version of the toolkit. The visits consisted of a series of interviews, with a range of employees from a vertical slice within the organisation, and site walk-abouts. The organisations varied in size and function and each visit was completed within two days.

## Development of the inspection model

A number of methods and approaches were selected and adapted, from the review of literature and current safety culture assessment methods employed, to structure the HMRI safety culture inspection model and toolkit.

The Health and Safety Executive (HSE) POPMAR Model for Successful Health and Safety Management (1997) and Regulation 5: Health and Safety Arrangements of The

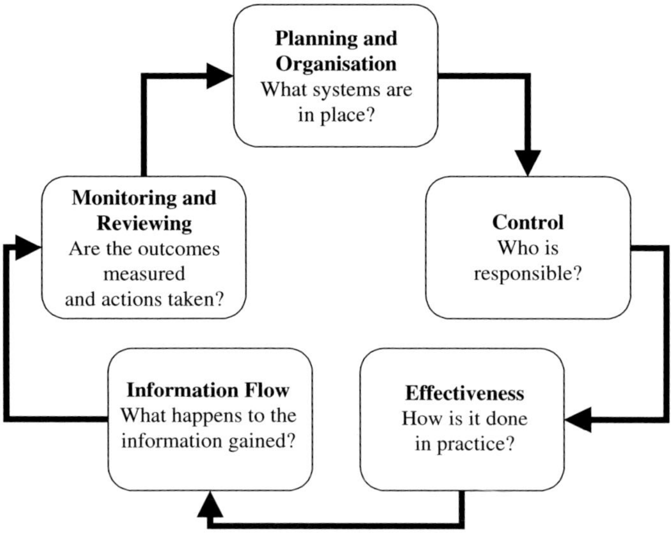


**Figure 1. Safety culture model.**

Management of Health and Safety at Work Regulations 1999, provide a framework for management of health and safety. The stages within the framework incorporate: policy, organisation, planning and implementing, measuring, auditing and reviewing, and each of the stages must be accomplished to achieve successful health and safety management.

The HSE framework was applied to the safety culture toolkit to structure its contents and form a model of safety culture, which outlines the steps that should be taken to ensure there are effective arrangements in place for the management and implementation of a successful health and safety strategy. The steps are outlined in Figure 1.

## Development of the inspection toolkit

An underlying theme embedded within the toolkit is that it is based on verifiable, evidence-based measures of safety culture derived from the extensive literature review and key stakeholder experience. The toolkit was developed to assess performance based on the 5 key indicators and comprised the following:

- Question set and evidence review;
- Pocket card;
- Overview diagram;
- Inspector's guide;
- Data collection pro forma.

### *Question set and evidence review*

Question prompts were developed to provide the inspector with a logical and consistent approach to investigate the performance of safety culture within an organisation. The questions were organised to encompass all stages of the health and safety management model, and structured in order that they could be used to interview all levels of personnel within an organisation.

The question set was divided into six scenarios, which were designed as a mechanism, or context, for eliciting the required information regarding the company's safety culture. The scenarios were developed from a product of the interviews with subject matter experts (e.g., HMRI inspectors and other rail stakeholders), a review of incidents within the HSE Railway Enforcement Database, Confidential Incident Reporting and Analysis System reports, and findings from the safety culture literature review. Scenarios that demonstrated direct relationships to a number of the 5 indicators were selected as the most suitable and are as follows:

- Safety management;
- Safety concern;
- Change management;
- Transfer of information about shift duties;
- Time-critical and degraded situations;
- Incident management.

Each scenario had relevant assessment criteria assigned that contained positive and negative evidence points. The assessment criteria comprised facets of an organisation, based on the 5 safety culture indicators, which could be measured to determine its safety culture. The evidence points consisted of features that could be used to evaluate the progress of the organisation against each of the assessment criteria and, where possible, were derived from the literature, standards, or guidelines. The evidence points were developed to aid an inspector to assess whether the data gathered indicated a satisfactory, or unsatisfactory, safety culture.

*Pocket card*

A pocket card was developed for inspectors that provides an overview of the structure of the safety culture assessment methodology in a pocket-sized format. It was designed to be used by inspectors as a memory prompt during an inspection situation.

*Overview diagram*

An overview diagram was produced to provide a summarised version of the 5 safety culture indicators in relation to the health and safety management framework. It was designed to provide quick, clear information to inspectors regarding the questioning approach during interviews and data collection objectives.

*Inspector's guide*

A guidance document on how to use the toolkit was produced to support inspectors during safety culture inspections. The guide comprises the following sections:

- Introduction to the inspector's guide;
- What is the safety culture inspection toolkit?
- How should I prepare for the inspection?
- How should safety culture be assessed during the inspection?
- How should I report the findings of the inspection?
- How should I monitor and review progress?

*Data collection pro forma*

A pro forma was developed to aid inspectors collect, organise and report the data collected.

## Validation of the toolkit

The safety culture inspection toolkit was tested through a programme of stakeholder site visits at various rail organisations. The data gathered were analysed and the performance of each organisation was presented in a report that identified the organisations' strengths and weaknesses in relation to the 5 safety culture indicators.

The content of the toolkit was validated through the series of stakeholder visits by enabling the users to capture and assess information regarding the status of each organisation's safety culture. The usability of the toolkit was also tested during the series of visits. Usability criteria were assessed at each visit and the toolkit's structure and format was refined.

## Summary

The safety culture inspection toolkit was developed to support the ongoing HMRI programme of work to validate the implementation of safety culture recommendations from rail public inquiries. The toolkit provides inspectors with a focused, systematic approach to assess specific areas that influence safety culture for any type of rail organisation. Collective findings from a number of inspections will enable a comparison to be made between organisations, industry trends to be developed and safety culture improvement of the rail industry to be tracked.

## References

HSC. 2001. *The Ladbroke Grove Rail Inquiry. Part 2 Report.* The Rt Hon Lord Cullen PC. HSE Books.

HSC. 2003. The Rail Public Inquiries. *Report on Overall Progress as of April 2003 on the Remaining Recommendations from the Rail Public Inquiries.* HSE Books.

HSE. 1999. *The Management of Health and Safety at Work Regulations 1999, Approved Code of Practice and Guidance*, Regulation 5: Health and Safety Arrangements. HSE Books.

HSE. 1997. *Successful Health and Safety Management.* HSG 65. HSE Books.

M.D. Cooper. (2000). Towards a model of safety culture. *Safety Science*, 36, 111–136.

# USABILITY OF PEDESTRIAN CROSSINGS: FURTHER RESULTS FROM FIELDWORK

**S. Robertson & R. Thoreau**

*Centre for Transport Studies, University College London,*
*Gower St., London WC1E 6BT*

The USAPED project, described in in Robertson and Thoreau (2004) and Robertson Thoreau and Allsop (2004), has worked to identify the difficulties associated with crossing the road with a view to improving usability for all road users. The study aims to develop a framework to assess the usability of pedestrian crossing places, and as part of that process examined what makes a crossing place easy to use. For this study only zebra crossings and informal crossing places were used. The study employed a combination of user workshops and observational studies to help determine what makes a crossing or crossing place easy to use and how the different types of road user interact.

## Introduction

This study examines the perceptions and behaviours of road users when navigating pedestrian crossings. The aim is that through this examination researchers will be able to gather information as to what is important in making a crossing easy or difficult to use. The details of this project have been discussed previously in Robertson and Thoreau (2004) and Robertson Thoreau and Allsop (2004).

The objectives for the project relevant to this paper are:

- Identify what is perceived by road users as being important in making a crossing place easy to use.
- Relate the behaviour of road users to the factors perceived by them to be important.
- In terms of these and other factors define the relevant elements of usability of the crossing place.
- Identify and quantify observables that could measure or act as proxies for good or poor usability at a sample of crossing places.

## The sample and methodology

User workshops were used to ascertain what factors effected usability of crossing places. Observational studies of crossing places in the vicinity of the user workshops were used to augment the workshop findings and to obtain further information about the use of crossing places.

The workshops and filming were undertaken in a total of five local authorities (four London boroughs, central, inner and outer London as well as one shire county). In each local authority a pair of sites, one refuge and one zebra were filmed. Two days filming was completed at each of the ten sites, with over a total of 200 hours filming completed.

The observational study aimed to determine how and to what extent factors and issues relating to use of crossing places identified in the workshops operate in practice to estimate delays incurred by crossing pedestrians.

*User workshops*

A series of ten workshops were arranged so that they could be followed by observational studies at site in the vicinity of where the sample of users was drawn. Due to lack of response in one of the boroughs no specific workshop was able to be held there, however an additional workshop of younger adults was undertaken instead.

For each workshop participants were self-selected via leaflet drops and posters in the local area. They were asked to complete a questionnaire prior to the workshop. During the workshop two paper based activities in the form of questionnaires were undertaken and a short film shown various examples of crossings and crossing behaviours was shown. Discussions on issues relating to crossings were held after the paper based activities and after the film had been shown.

*Fieldwork*

In each borough, two sites were chosen; a refuge and a zebra. These pairs of sites were within about half a mile of each other. Sites were chosen so they would be similar to each other in design and location. No split zebra crossings were used. Filming was carried out in consultation with the local authorities who gave approval for filming in each borough and facilitated liaison with the local police, lighting authorities and members of public.

Two cameras were used at each site, allowing vehicles and pedestrians approaching from both sides to be observed. Cameras were mounted onto lampposts in the vicinity of the crossing. A camera mounted on a bracket attached to a lighting or sign column was connected via a cable transmitting video signal and 12 V power to a box containing a digital recording system secured to the column. The time starting and finishing filming depended mainly on the light available with a maximum recording period of about 11 hours.

## The results

The findings reported here are preliminary and will be reported on in more depth in due course. The sample size in each workshop was between three and eleven. The mean age of participants was 55.5years, with 35%) of the total sample over 65. Of the total sample 43% were female and 57% were male. The participants who had had some kind of formal training regarding rules of the road accounted for 80%, of those with formal training 94% had full drivers licenses whereas the remaining participants had either a motorcycle license or had done a cycling proficiency test.

*Walking speeds*

The results from the workshops have confirmed the findings presented in Robertson and Thoreau (2004). From the questionnaires, an average of 5.4 trips per person per day on days which participants walked were reported.

Of those trips an average of 3.5 are made on foot. The average walking trip was 770 m. The average walking journey took 13.5 minutes and an estimated average journey speed of 0.91 metres per second. The average number of road crossings per walking trip was found, in this sample, to be 2.40. This corresponded to crossing a road every 320 metres. No differences were found between the genders for self reported journey speed, distance walked or time spent walking.

**Table 1. Sample.**

| Area | Total workshop participants in each area: Total number of Participants | Number of females | Number with drivers licenses | Observational study: Month filmed | Number of crossing events |
|---|---|---|---|---|---|
| Richmond | 10 | 4 | 9 | December | Zebra 71<br>Refuge 39 |
| Bromley | 15 | 9 | 10 | February | Zebra 75<br>Refuge 11 |
| Wandsworth | 6 | 1 | 4 | March | Zebra 80<br>Refuge 39 |
| Hertfordshire | 5 | 2 | 5 | April | Zebra 20<br>Refuge 20 |
| Kensington | – | – | – | March and May | Zebra 224<br>Refuge 30 |
| General workshop | 4 | 1 | 2 | – | – |
| Totals | 40 | 17 | 30 | | |

**Table 2. Comparison of walking and crossing speeds at Zebra and refuge.**

| | First half crossing speed m/s | | Second half crossing speed m/s $t = 7.75$, df 649, $p < 0.01$ | | Total average crossing speed including wait m/s | | Average walking speed while crossing excluding wait in centre of refuge m/s $t = 4.20$, df 649, $p < 0.01$ | |
|---|---|---|---|---|---|---|---|---|
| | mean | S.D. | mean | S.D. | mean | S.D. | mean | S.D. |
| Zebra (n = 470) | 1.57 | 0.47 | 1.43 | 0.44 | 1.45 | 0.37 | 1.46 | 0.37 |
| Refuge (n = 181) | 1.61 | 0.64 | 1.79 | 0.69 | 1.39 | 0.51 | 1.61 | 0.51 |

There were differences between the walking speeds of pedestrians at a refuge and at a zebra. These are shown in Table 2. People crossing at a refuge walked significantly faster on the second half of the crossing when compared the second half of a zebra.

There were qualitative differences in the pattern of walking speeds between crossing types. At a zebra, people slowed on the second half whereas on the refuge they speeded up on the second half relative to their speed on the first half of the crossing.

A significant difference was found between males and females in walking speed on the second half of the crossing (zebra and refuge combined). Men walked significantly faster than women did (male 1.59 m/s and female 1.46 m/s respectively) ($t = 3.083$, df = 641).

## *Perceptions of crossing places*

In general the perception of zebra crossings and of refuge crossings expressed during the workshops were positive. Zebra crossings were thought by some participants to be simple and efficient. *"They are so clear, really visible the markings on the road and the Belisha beacon. …quite simple in a way and not offensive. … they work very well."* The size, particularly width of a refuge was thought to be an important factor in its ease of use. Wide refuges were preferred – so that they could fit more than one person at a time – or one person with a pushchair or bicycle. Visibility of crossing, vehicle or crosser was one of the most commonly mentioned factors considered to be important across all types of crossings from both pedestrian and driver perspectives. In the workshops both pedestrians and motorists described that they perceived the actions of the other with some uncertainty and in some cases suspicion. Views toward drivers were much harsher than views toward pedestrians despite most participants being drivers as well as pedestrians. People who were pedestrians only also noted, at the end of the workshop that they had not realized the difficulties and complexity of the task facing the driver. There appeared to be a more lenient view taken of pedestrians breaking rules of behaviour than there was for motorists. One concern of motorists was that it could be difficult to judge when a pedestrian was about to cross, however the observational study showed that 91% of the observed pedestrians approached the kerb directly.

### *Knowledge of rules of the road*

From the qualitative information obtained at the user workshops there were strong suggestions of gaps in knowledge on how crossings and crossing places operate. While most users had a broad grasp of rights of way, a substantial minority were unclear about the formal, specific rules for using pedestrian crossings and crossing places. There was also a lack of knowledge about operation of some new informal crossing places. Participants in a number of the workshops expressed the wish that there was more public education regarding rules of the road.

### *Communication*

Communication between road users was viewed very positively by almost all participants of the workshops, with the use of an acknowledgement as being a positive thing that could encourage a more co-operative approach by road users. There appeared to be widespread use of communication between users despite it being officially discouraged in the context of crossings. Three methods of communication to other road users were mentioned frequently; motorists flashing their lights, hand waving and eye contact. However there was no agreed communication protocol nor consensus about what constituted an indication of intention to cross at a zebra crossing. Eye contact was the main form that pedestrians said they used to communicate with motorists. Eye contact with drivers was perceived by pedestrians as likely to make drivers give way to pedestrians.

The observational study could only identify light flashing or some types of hand signal, 4% of pedestrians communicated to the driver on the first half of the road and 3% on the second half of the road. These percentages are small since the workshop findings suggested communication was used widely and seen as positive by pedestrians and motorists.

### *Active choice to use crossings*

During the workshops participants were asked to say how far they thought they would walk out of their way to cross the road at a crossing place. This was found to be associated with a number of factors including age, level of traffic, speed of traffic, encumberment, and whether accompanied. In the context of the workshop the exact numbers and distances could not be quantified by the respondents. Of those people observed to cross the road, 39% walked out of their way to use a crossing. Significantly ($X^2 = 4.05$, df 1 $p < 0.05$), men were more likely to go out of their way to use a crossing place than women, with 43% of observed men walking out of their way to use a crossing place compared to 35% of observed women who did so. Of those people we observed about 1 in 5 people started their crossing away from the facility.

## Conclusions

The key points to emerge from the work to date are that:

Users need more information about how crossings and crossing places are supposed to work. A substantial minority of users will walk out of their way to use a crossing or crossing place. This suggests that these facilities are fulfilling a need for these users. Communication between road users at crossings and crossing places while officially discouraged was perceived to be a good thing and was reported to be widely used, though there was no fully agreed protocol.

Final analysis has yet to be completed. The results will be disseminated in due course and will be available from the Centre for Transport Studies, UCL (usaped@transport.ucl.ac.uk).

## References

DETR 1999, *The Highway Code*, The Stationery Office (London)

Ward, H., Cave, J., Morrison, A., Allsop, R., Evans, A., Kuiper, C. and Willumsen, L. 1994, *Pedestrian Activity and Accident Risk,* AA Foundation for Road Safety Research, (Basingstoke)

Robertson, S. and Thoreau, R. 2004, Usability of pedestrian crossings: Some initial findings. In P. McCabe (ed.) Contemporary Ergonomics 2004, (CRC Press London), 253–257

Robertson, S., Thoreau R. and Allsop R. 2004, Usability of pedestrian crossings places. Paper presented at ICTTP 2004, Available on line at: http://www.psychology.nottingham.ac.uk/IAAPdiv13/ICTTP2004papers2/Children%20&%20Adolescents/Robertson.pdf (Accessed 22 December 2004)

# CONTROL ROOM DESIGN: NEW THINKING

**Jonathan Bainbridge[1], Gary Davis[2] & Glyn Roberts[3]**

[1]*HF Consultant, Davis Associates Ltd, Wyllyotts Place, Potters Bar, EN6 2JD*
[2]*Managing Director, Davis Associates Ltd, Wyllyotts Place, Potters Bar, EN6 2JD*
[3]*HF Delivery Manager, Tube Lines Limited, Canary Wharf, London E14 4HD*

Under the London Underground Public Private Partnership (PPP) scheme, Tube Lines Ltd is modernising the Jubilee, Northern and Piccadilly lines. To accommodate new signalling and communications equipment, two of the key deliverables of the project are the building of a new Northern Line Service Control Centre and the refurbishment of the existing Jubilee Line Service Control Centre. Tube Lines have developed concept control room layouts and an overview display for these control centres that are distinctly different from traditional control room arrangements commonly found in London Underground control centres and other control facilities around the world. This paper presents the design concepts, describes the design rationale behind them and outlines a mock-up exercise conducted to introduce stakeholders to the concepts.

## Introduction

Under the London Underground PPP scheme, Tube Lines Limited is tripling investment and committing over £4,400 million over the next 7.5 years to modernise the Jubilee, Northern and Piccadilly lines. A major part of the Jubilee and Northern Line Upgrade Project (JNUP) is the replacement of the Northern and Jubilee lines "Fixed Block" signalling control system with a new Transmission Based Train Control "Moving Block" system, as well as new communications systems. To accommodate this new equipment, two key deliverables of JNUP are the refurbishment of the Jubilee Line Service Control Centre at Neasden and the building of a new Northern Line Service Control Centre at Highgate. During the concept design phase it was necessary to develop an initial design concept for a control desk, overview display and control room layout so that:

- The control room dimensions for the new Northern Line control centre could be defined allowing the architects to draw up initial building plans.
- It could be determined whether or not the control desk arrangement and overview display designed for the Northern Line control room could also be installed in the existing Jubilee Line control room.
- It could be determined whether or not Tube Lines could design a suitable control room layout to facilitate the operating structure that London Underground desired.

## Traditional control room arrangements

Traditional control rooms have banks of control desks arranged in rows with operators sitting face-to-back looking at a large line overview display at the front of the room (as shown in Figure 1). Historically this layout exists because of:

- Technical limitations of the overview display – the overview display had to be a large filament lamp or LED panel (for example in the Northern Line Control Room at Cobourg Street) or more recently a rear projection display (for example in the Jubilee Line Service Control Centre at Neasden).

- Technology and operating systems could not support the operators' tasks sufficiently enough to facilitate control of the whole railway from one control desk. There are also distinct limitations to human reach and viewing angles. Operators were therefore allocated an area of control and the control desks were arranged along the length of the overview display opposite these control areas.
- Hierarchical operating structure and physical constraints dictated that supervisors were positioned at the rear of the room looking over less senior personnel in front.

## Operating structure

Early in the concept design phase, London Underground provided Tube Lines with an Operational Concept. This outlined how London Underground planned to operate the future railway. Rather than use a traditional hierarchical operating structure, their vision was for a two-tier command and control structure. This operating structure encompasses the scope from tactical command (undertaken by a Duty Line Control Manager – DLCM) to execute control (undertaken by Service Duty Managers – SDM).

Tube Lines believes that two-tier command and control is a team working environment and one in which information is shared between operators who are mutli-trained rather than fixed to particular job roles. For these reasons, it was felt that a traditional control room ("class room") arrangement (as shown in Figure 1), or one with operators sitting back-to-back (as on the Central Line) was inappropriate.

## Concept control desk

Before the overall "foot-print" size of the desk arrangement could be defined, Tube Lines had to first design a control desk to accommodate the new signalling and communications equipment (as well as equipment provided by London Underground's other suppliers). Advances in technology and operating system designs now enable one operator to control the whole railway from one control desk (during off peak normal operating

**Figure 1. Example of a "traditional" control room layout.**

conditions). This, together with the flexibility of two-tier command and control and multi-skilled SDMs, required that any operator could sit at any of the control desks and perform his/her tasks. Therefore, each of the six control desks has the same equipment.

## Concept control room arrangement

Advances in overview display technology, operating systems and operational strategy, have enabled Tube Lines to design a concept control room and overview display that, compared to more traditional control room layouts, better supports the operators' operational requirements. The control room concept arranges the desks in a circle with operators sitting face-to-face and looking at an overview display suspended from the ceiling in the centre (see Figure 2). This is distinctly different from traditional control room arrangements commonly found in London Underground's control centres and other control facilities around the world.

### *Advantages of the concept control room arrangement*

The circular control desk arrangement is one of those identified in the international standard Ergonomics Design of Control Centres (ISO 11064-3, 1999) and in Human Factors for Engineers (Harvey and Sandom, 2004) and is currently in use in a similar format in the air traffic control tower at Heathrow Airport. At Heathrow, operators rely on communication between different job roles in the control tower to safely control aircraft. The arrangement allows operators to quickly and efficiently communicate between "Take Off, Landing and Taxi-ing Controllers", and so react to situations.

Tube Lines analysis and design reviews identified that the circular arrangement has the following advantages over a more traditional layout:

- More efficient sharing of information: because operators have direct eye contact with each other making it easier for them to catch people's attention, to see facial and bodily gestures and to talk face-to-face (ISO 11064-3, 1999).
- Better team working: because control desks are grouped to maximise the opportunity for team working or social interaction (Harvey and Sandom, 2004).
- Better team structure: the DLCM sits within the circle with the SDMs so the "them and us" attitude that can exist in some traditional control rooms can be minimised.

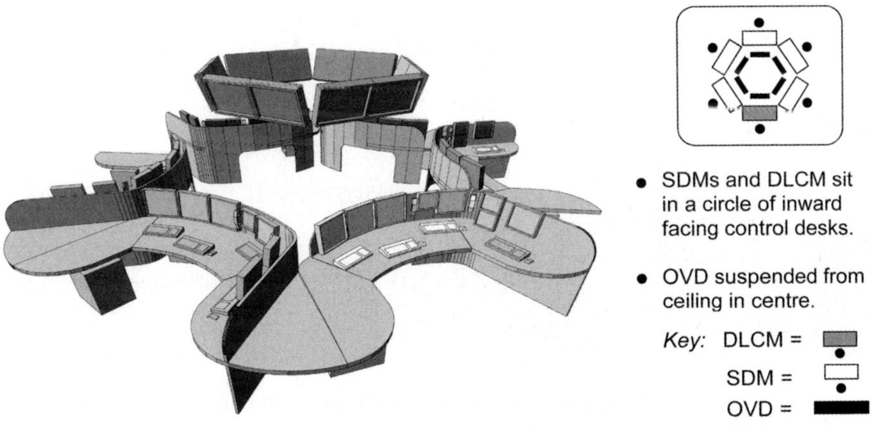


**Figure 2. The concept layout of inward facing control desks and suspended line overview display (OVD).**

- Operating structure flexibility: layout allows operators to sit at any desk. This also gives London Underground the flexibility to adopt an operating structure other than two-tier command and control.

However, ISO 11064 (1999) identifies two particular disadvantages of the circular control desk arrangement compared to a more traditional layout. These are worse noise interference and the difficulty for operators in sharing off-workstation displays (for example a line overview display).

*The overview display*

The need for a large line overview display is supported by preliminary analysis of operators' tasks. It is currently recognised that there are significant operational advantages if operators can see the whole line at a glance without the need to select it on their desktop monitors, in this particular case, service pattern recognition. To overcome the difficulty that ISO 11064 (1999) identifies in sharing off-workstation displays, Tube Lines have produced an innovative design solution for the line overview display. In line with appropriate recommendations on viewing angles and viewing distances, LCD screens are suspended from the ceiling in the centre of the circle of control desks (as shown in Figure 2). Two LCD screens are shared per pair of control desks. These screens are placed suitably high enough to enable the operators to have a clear unobstructed view across the circle of control desks even when standing.

## Control room mock-up exercise

In-line with good human factors practice and London Underground's Human Factors Integration standard (2-01018-003, 2003), Tube Lines wanted to involve end-users and other stakeholders throughout the design process. It was therefore decided during the concept design phase to produce a simple mock-up of the new control desk, room layout and overview display concepts and to invite stakeholders and end-users to view and comment (see Figure 3). The mock-up was conducted at a number of locations to maximize the potential for a number of different stakeholders to attend. The mock-up consisted of an MDF control desk with representative foam-board monitors arranged on the desk. Five standard desks (six control desks in total) were arranged around the circle as per the concept and mock LCD screens hung from the ceiling.

**Figure 3. (Left image) Control desk mock-up with concept equipment arrangement. (Right image) A visitor trying the control desk for size and looking at a small overview display placed behind the desk to simulate approximate viewing angles.**

Over 100 stakeholders and end-users visited the mock-up and over 40 of these completed a questionnaire to provide feedback on the concepts. This exercise allowed Tube Lines to gain feedback on the concept of a circular control desk as well as the equipment arrangement on the control desk it self. The mock-up also illustrated the concepts to London Underground much better than a CAD model ever could. In addition to this, because the control room arrangement and overview display concept are distinctly different from traditional control room arrangements, it was important that London Underground gained a better understanding of the concepts and its potential impact upon overall building design requirements as soon as possible in the design process. This would prevent Tube Lines conducting potentially abortive design work early on in the design process only to discover later on that London Underground were not happy with the design direction during more detailed design work.

## The future

Tube Lines believe that the circular control room arrangement will support the operators' tasks within a two-tier command and control structure much better than in a control room with a traditional classroom layout. ISO 11064 (1999) identifies that this layout has numerous advantages over a more traditional layout. In August 2004, the concept design phase of JNUP was completed. During the next stage of the project (preliminary design phase), the following human factors work is planned to progress the design concepts:

- Detailed requirements definition of overview display symbol and text indications.
- Detailed task analysis and task synthesis of control room operators' tasks.
- "Day in the life" exercises with London Underground representatives and operators to investigate how the new control room arrangement functions during different operational scenarios.
- Further mock-up exercises to further evaluate the control room arrangement and overview display and to ensure stakeholder and operator "buy-in" and feedback throughout the project.

## Reference

Harvey R.S. and Sandom C. (Editors) (2004), *Human Factors for Engineers*, (The Institution of Electrical Engineers, Stevenage)

International Organisation for Standardisation (1999), *Ergonomic Design of Control Centres – Part 3: Control Room Layout (ISO 11064-3)* (International Organisation for Standardisation, Geneva)

London Underground (2003), *Engineering Standard – Integration of Human Factors into Systems Development (2-01018-003)* (London Underground, London)

# IS EXTINCTION BEHIND INTERSECTION COLLISIONS?

**Tay Wilson & Gulaid Egeh**

*Psychology Department, Laurentian University,
Ramsey Lake Road, Sudbury, Ontario P3E 2C6, Canada
tel (705) 675-1151 fax (705) 675-4889*

The mystery of how drivers, after searching an intersection arm can emerge into an immediate collision, might be explained in part by the phenomenon of extinction, a limited attentional capacity in which multiple stimuli cannot be processed simultaneously to conscious awareness. The stimulus not processed is generally farthest away on the contra-lateral side to brain lesion patients and in some normal cases just the leftmost stimulus. Two sites were studied. One was deemed more likely to induce extinction consistent driving behaviour on the grounds of road geometry and site nature. Failure to stop, rolling stops and full stops were tabulated by type of signal and turn. Overall at the presumed high extinction site 5% of drivers failed to stop versus 0% at the other site. The ratio of rolling stop non signallers to rolling stop signallers overall and in turns to the right was higher at the presumed low extinction site than at the presumed high extinction site.

## Introduction

One mystery in intersection collisions under clear conditions in which drivers emerge from a stop sign controlled intersection directly into the path of an approaching vehicle is that drivers frequently report, in good faith, having looked for approaching traffic and seen the road as clear.

It is suggested here that the phenomenon of *extinction* may be responsible for some of these collisions. Extinction in brain injured patients is observed as the failure to see the member of a pair of targets which is contralaterally farthermost from the site of the brain injury. It is thought to be a pathologically limited attentional capacity in which multiple stimuli cannot be processed simultaneously to conscious awareness (Vaishnavi et al. 2000). Vuilleumier and Rafal (2000) suggest that allocation of attention to stimuli can be modulated by task demand and further, location information may be critical for attentional mechanisms subserved by the parietal cortex. Extinction can also be produced in healthy subjects (Meador et al.). Ray et al. (2001) report relatively more extinction in healthy subjects on the left side in the tactile mode. Oliveri et al. (1999) found that left frontal transcranial magnetic stimulation can reduce contralesional extinction leading to the suggestion that extinctions due to right brain damage may be dependent on a breakdown in balance of hemispheric rivalry in directional spatial attention to contralateral hemispace so that the unaffected hemisphere generates an unopposed orienting response to the side of the lesion. See also Persinger's (1999) vectorial hemisphericity explanation of "sensed presence" as demons or gods in terms of relative stimulation of right- left hemispheres and concomitant limbic affect.

In view of these findings, it was decided to study intersection entering at two stop sign controlled sites with different apparent likelihood of inducing extinction consistent driver behaviour. The Laurentian Cancer Hospital site was chosen as likely to induce extinction for two reasons. First, many drivers would be experiencing the stress of being a cancer outpatient or of visiting a seriously ill patient. Second, the intersection being scalloped to the left was not at right angles with the result of requiring entering drivers to look somewhat behind themselves thus moving road furniture such as lamp posts into the left visual field in a more central position than that of any approaching vehicle (remember it is the left most visual object which suffers extinction.). This site was further a candidate for extinction behaviour because recently two serious collisions had occurred in which the protagonist drivers had allegedly reported that they had looked and seen the road as clear of traffic. The neighbouring Nepahwin site was chosen as a lower extinction candidate intersection on the grounds of its right angle geometry and the lower likelihood of immediate driver stress because it involved only a small residential community.

## Method

Two intersections controlled by stop signs were selected in Sudbury, Canada: the joining road from Laurentian Hospital to Ramsey Lake Road and the joining road from Nepahwin to Paris Street. Ramsey Lake road is the only road leading to and from the University. At the observed Laurentian Hospital stop, the approaching driver is faced with two lanes of Ramsey Lake traffic moving left to right and a further three lanes of traffic filtering into appropriate turn lanes moving right to left. Because the road at this point is "scalloped" the approaching driver must scan somewhat backwards beyond the right angle to search for oncoming traffic. In addition, the driver approaching the stop sign must negotiate a small declivity. The Nepahwin stop also involves a small declivity but the sight line to traffic is at right angles to the stop line. The approaching driver is faced with two lanes of left to right traffic and then a further two lanes of right to left traffic. Drivers must turn left or right.

Drivers were observed over a series of 16 approximately one hour sessions at the Laurentian Hospital site and 18 similar sessions at the Nepahwin site at various times varying from 7:45 a.m. to 18:30 p.m. from Feb. to May, 2003 spanning late winter to late spring weather. Traffic counts were taken for five minutes before and after each observation session. Over the observation period means and standard deviations of traffic pro-rated per hour on the Laurentian Hospital road, Ramsey Lake Road, Nepahwin Road and Paris Street were respectively (108, 69, 668, 256, 74, 24, 1451, 413).

Drivers approaching either stop sign were classified as having made a full stop at the appropriate stop line, having made a full stop past the appropriate stop line, having made a rolling stop or having failed completely to stop. In addition, use of signal lights and turning direction was recorded.

## Results

At Laurentian Hospital, 358 drivers were observed. Two hundred and two drivers (57%) made a complete stop of which only 46 (22% of full stoppers) managed to stop at or behind the stop line, 137 made a rolling stop (38%) and 19 (5%) failed to stop at all. Two hundred and ninety drivers signalled their further intentions, 261 turned left across traffic, and 82 turned right. At Nepahwin, 191 drivers were observed. One hundred and fourteen made a full stop of which 69 managed to stop at or behind the stop line, 77 made a rolling stop and no one failed to stop at all. One hundred and forty one drivers signalled their further intentions, 85 turned left across traffic, and 106 turned right.

Various combinations of stopping type, signalling, and turning were examined with the following results. As expected, the ratio of signalling to not signalling right turners

(135/53) was significantly lower than the ratio of signalling to not signalling left turners (294/46; chi square = 17.1 $p < .01$). Significantly more (165 versus 56) drivers made a rolling stop while signalling than made a rolling stop while not signalling (chi square = 58, $p < .01$). Significantly more (161 versus 37) drivers stopped past the stop line while signalling than did so without signalling (chi square = 77, $p < .01$).

Various combinations of stopping type, signalling and turning by *location* were then examined with the following results. Significantly more (16 versus 0) drivers failed to stop at all at Laurentian Hospital relative to Nepahwin (chi square = 16, $p < .01$). The ratio of signalling right turners to signalling left turners at Nepahwin (69/72) was significantly higher than those at Laurentian Hospital (66/222, chi square = 29.7, $p < .01$). The ratio of rolling stop non signallers to rolling stop signallers at Nepahwin (28/49) was significantly higher than those at Laurentian Hospital (18/116), chi square = 15.1, $p < .01$). For those turning right, the ratio of rolling stop not signallers to rolling stop signallers at Nepahwin (21/27) was significantly higher than that at Laurentian Hospital (4/27, chi square = 8.2 $p < .01$).

No significant differences were noted by location by location in signalling for those stopping past the stop line compared with those stopping at the stop line, or by signalling or not signalling by direction of turn, or by rolling or full stop by turning direction, or by those turning left making a rolling stop and either signalling or not signalling.

## Discussion

One would expect extinction phenomena to be strongest at the Laurentian Hospital site for two reasons. First, more people driving on that road would be expected to be experiencing strong emotion because of recent visits to cancer patients who were unlikely to recover than on the residential Nepahwin road. This in part is the logic behind slower speed limits near hospitals. The finding of 16 failures to stop at Laurentian versus 0 at Nepahwin is at least consistent with this surmise. Second, because of the scalloping of the entry road it is more difficult to track oncoming vehicles and more likely that a potentially dangerously close vehicle would be in the far left visual field while some roadside furniture is in the relatively close left visual field. The expectation under both provisos would be that less precise/safe driving practices would be exhibited at the Laurentian site. This was not borne out by the other two significant findings that failure to signal combined with rolling stops overall and to the right were more prevalent at the presumed low extinction site. More research particularly at presumably high stress sites such as hospitals is indicated.

## References


Meador, K.J., Ray, P.G., Day, L.J. and Loring, D.W. 2001, *Neurology,* **56**(8), 1044–7.

Olivieri, M., Rosssini, R.M., Traversa, R., Cicinelli, P., Filippi, M.M., Pasqualetti, P., Tomaiulo, F. and Caltagironde, C. 1999, *Brain*, **122**(9), 1731–1739.

Persinger, M.A. 1999, Near-death experiences and ecstasy: A product of the organization of the human brain. In *Mind myths* (Wiley, New York), chapter 5.

Vaishnavi, S., Calhoun, J., Southwood, M.H. and Chatterjee, A. 2000, Sensory and response interference by ipsilesional stimul in tactile extinction. *Cortex,* **36**, 81–92.

Vuilleumier, P.O. and Rafal, R.D. 2000, A systematic study of visual extinction. *Brain*, **123:** 1263–1279.

# VISUALISING USER EXPERIENCES THROUGH VIDEO

**Hayley Dixon[1], Gary Davis[1] & Hilary Cottam[2]**

*[1]Davis Associates Limited, Wyllyotts Place, Potters Bar, Hertfordshire EN6 2JD*
*[2]Design Council, 34 Bow Street, London WC2E 7DL*

Shadowing user journeys and capturing them through video is a powerful and effective way of communicating rich insights of psychological and physiological user experiences. This method enables many people to see the ups and downs of a personal journey that would not otherwise be experienced or observed. It is also particularly effective in creating empathy with end-users. On behalf of the Design Council, Davis Associates conducted user research to generate stimulus material for use at a later stage of an urban transport project. The research brief defined by the Design Council required a diverse range of individuals and their different journeys to be shadowed and captured through the medium of video. The journeys selected included various travel modes, cross modal journeys and various scenarios for people travelling in and around the Paddington area in London. The user research delivered six brief, yet rich and insightful movies, documenting the diversity of urban journey experiences of very different users.

## Introduction

On behalf of the Design Council (DC), Davis Associates (DA) conducted user research to generate stimulus material for use at a later stage of an urban transport project, demonstrating how user-centred design can improve the quality of life for those living and travelling in and around urban environments. The objectives of the user research were:

- to research urban journey experiences from the users' perspective;
- capture a variety of issues associated with living and travelling in urban environments; and
- provide suitable stimulus material to demonstrate a user-centred approach to transport design.

The research, focused upon the area of Paddington Basin, West London, was carried out over a period of 2 months, from the selection of users to the compilation of video material, using both shadowing of user journeys and post-journey interviews. This paper outlines the approach undertaken for capturing and visualising user's urban journey experiences.

## Using video to visualise user experiences

A person's journey through an urban environment can be one of many psychological and physiological experiences. Capturing user insights and communicating this dynamic situation to others, while retaining aspects such as noise, pace of movement, conflict with the environment and many other tangible issues cannot be achieved by the more

**Figure 1. Trial shadowing – "That was lucky! I haven't missed my bus".**

traditional user research methods, such as questionnaires or interviews. In particular, these traditional methods fail to create empathy with end-users and ultimately lack the inspiration to drive appropriate changes in urban and transport design.

Video analysis is an established and practical design tool that has been widely used to support the capture and subsequent analysis of human behaviour, for stimulating thought and driving changes in design. It provides a visual record of observed users, capturing their everyday activities and experiences, their decisions, thoughts and the environment and context within which they live, work or socialise. Above all, it is a valuable tool for providing a visual representation of end-users, enabling others to see with their own eyes, and empathise with the personal journeys of others. Those outside the human factors and design world, such as policy makers and brand managers, can be influenced by the use of video to support them in visualising user experiences and in understanding high-level issues relevant to end-users.

## Testing the process

Prior to undertaking the user research, a half-day user journey (figure 1) was undertaken as a pilot trial to determine the audio-visual requirements and to evaluate alternative camera techniques. This included assessing appropriate camera angles, capturing the participant's speech and facial expressions close-up, and from further back, capturing the wider context of their journey. In addition to the video capture, a post-journey interview was carried out.

The combination of wide angle video capture and post-journey interviews proved to be an effective method of capturing a comprehensive view of a user's journey experience. Each method alone would fall short of fully exploring and conveying the dynamic and complex nature of a person's journey profile.

## Recruitment of users

Before commencing participant recruitment, DA familiarised themselves with the transport links and key destinations in the Paddington Basin area. This included a review of the variety of travel modes used, the alternative journeys undertaken and the urban spaces that people lived, worked and passed through.

The recruitment of appropriate users, who were diverse in age, mobility, physical characteristics and the journey undertaken, was key to the success of delivering a realistic view of the depth and breath of urban experiences. The DC provided the following criteria for consideration during the recruitment process (see figure 2).

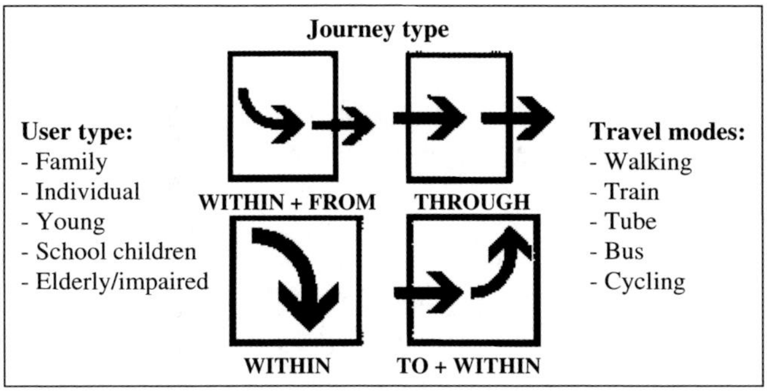


**Figure 2. Selection criteria for users.**

**Table 1. Users selected for shadowing and interview.**

| Participant | Participant details | Travel mode |
|---|---|---|
| Caroline | Mother of two children. Lives in Paddington and accompanies children daily to and from school | Bus & by-foot |
| John | Police officer, working in Victoria, London. Lives in Reading and commutes via Paddington rail station | National rail and bicycle |
| Sev | Young female worker. Lives in London and travels to Paddington to meet with friends during the evening | Tube & by-foot |
| Eileen | Elderly lady who lives in South London. Travels to Paddington on a regular basis for hospital visits | Bus, tube & by-foot |
| Reg | Elderly gentlemen, lives in Paddington and makes frequent trips to shops and hospital | By-foot |
| Natalie | Student living in East London with her mother. Travels on a daily basis to college in Paddington | By-foot & tube |

To support the discussion of preferred users with the DC, DA formulated image boards showing the ideal mix of participants based on the selection criteria in figure 2. Various recruitment methods were employed, such as approaching people in the streets, which proved most effective, and posting messages in community areas, including the local hospital, schools, colleges and transport interchanges to ensure a wide audience of users were reached. Recruiting very specific participants within a small geographical area proved to be a particularly time consuming activity.

From the responses received, DA discussed with the DC how well the potential participants represented the ideal mix of users shown on the image boards. As a result, the following six users were selected (table 1).

## Shadowing journeys

Each user was shadowed on a two-way journey from a discreet distance that minimised the influence on their behaviour throughout. The user brief prior to each journey also emphasised the importance of completing their journey as normal and to avoid any interaction with the DA team.

**Figure 3. Video clips from the user's urban journey.**

Two members of the DA team were each equipped with a video camera to record the journey undertaken, as well as the use of a mini-camera with microphone attached to the upper body of the participant to capture the journey directly from their perspective. However, certain journey types were more challenging to capture, and required alternative methods to ensure adequate video footage was obtained. John's cycle journey from Paddington rail station to his place of work in Victoria required an alternative method of capture in addition to the mini-camera attached to his cycle helmet. A DA team member followed John on another bicycle with a video camera attached to the handle-bars, capturing a complete journey route in both directions, across roads, major junctions and dedicated cycle paths. Sev's journey to meet with a friend for dinner during the evening also presented technical challenges for obtaining quality video footage. Moving between artificially lit transport hubs and poorly lit streets during the hours of darkness required regular adjustments to the recording equipment mid-journey.

*Still images*

In addition to the video footage, a diverse range of still images of the area in and around Paddington, and of the six participants during the shadowing of user journeys were captured by a professional photographer. A catalogue of high quality images were produced, which were included as part of the user experience stories.

## Post-journey interviews

Post-journey interviews were completed with each of the six participants after their return journey. The interviews were held either at their home or place of work to provide a comfortable and familiar environment so they were as relaxed as possible. Each user was encouraged to talk about their journey stages in detail and directed towards exploring both the physical aspects of the journey as well as reflecting on their thoughts and emotions. Differences between the observed journey made and previous journeys were also explored, considering the impact of other factors, such as poor weather conditions and delays to journeys.

The interview put into context many of the issues observed during the journeys, in particular, Reg's feelings on using a poorly lit underpass and Eileen's alternative use of travel modes for her return journey. This combined method of shadowing and post-journey interview supported the analysis of user behaviour and the identification of underlying issues that would not otherwise have been uncovered using just one of these methods.

## Bringing together the urban journey experience

With six user journeys of various lengths in time, there were approximately 20 hours of video footage to be edited down to six, five to six minute movies, while retaining the character and tone of the participant's journey. A semi-structured analysis of the video material focused on the identification of high-level themes that represented the physical aspects of the journey, and the range of emotions and feelings experienced by the different participants. Several emergent themes were identified from reviewing the journeys and interviews collectively, including, conflict with others, isolation, stress, fear, enjoyment, physical effort and errors.

Each identified theme was represented by at least one user experience, using a combination of video and audio from the journey and interview. High quality still images were also integrated into each movie to maximise the impact of the participant's journey experiences.

### *Editing time*

For each journey, the compilation of video, audio and still images were fairly time consuming. However in this case, the audience of the final movies were part of a project illustrating the value of a user-centred process, and therefore the video footage dictated a highly polished finish, more akin to a television documentary. This extra refinement would be unnecessary if DA were using video to feed user experience insights into a design team, resulting in a reduction in editing time.

## Summary

Using video to capture and visualise the experiences of people on journeys and other dynamic situations is a valuable tool that should be considered alongside traditional user experience research tools. Although the cost of using video can be greater in terms of equipment and time, situations where the audience need to be engaged and influenced, or when the experiences of end-users are set within a dynamic context, the value of visualising user experiences through video outweighs the cost and technical challenges.

# DEVELOPING A DESIGN FOR THE NEW TICKET OFFICES AT KING'S CROSS UNDERGROUND STATION

**Ian Rowe & Duncan Smith**

*Rossmore Group, Haseley Business Centre, Warwick CV35 7LU*

This paper discuses the development of human factors integration into the new ticket offices at King's Cross underground station. Including the examination of issues associated with compliance to many published standards, the engagement of a wide range of stakeholder groups and the operational and commercial needs of the company. Particular attention is paid to the primary research carried out in order to meet the objective of "inclusive design" and therefore meet the requirements of the Disability Discrimination Act (DDA). This paper includes details of the original "drivers" for carrying out the work, methodologies employed, results and conclusions.

The design work of the ticket offices is nearing completion and will shortly be in the process of being fabricated. As such this paper should be considered as a case study covering a completed project with a practical outcome.

## Introduction

The design development of the new King's Cross underground station is one of the first to incorporate the mandatory standard LUL-2-01018-003-A1 – Human Factors integration into system development. This standard was developed by London Underground (LU) as a response to the lessons learnt from the Jubilee Line extension project. This standard covers all areas within the station that are operated by LU staff including control rooms, tickets offices, etc. LU also specified that the design must be DDA compliant and be "inclusive" in its approach.

## Motivations

London Underground introduced a new systems integration standard in 2001 entitled "Human Factors Integration into System Design" that was introduced as a category 1 standard. The application of this standard is mandatory on all design whether it be new build or refurbishment of existing environments. Other drivers for the HF integration work included the Transport for London directive that states that all new design should be "inclusive" in its nature and the statutory requirement of the DDA 1994. It was felt that in order to provide assurances to all stakeholders and compliance with regulation, an holistic approach would need to be taken that included high levels of engagement from all stakeholders including:

- Operations staff
- Union representatives
- LU managers
- Disabled customers

- Able bodied customers
- Equipment suppliers/maintainers

## Methods

The methods used to ensure that Human Factors considerations were integrated into the design of the new ticket office are set out below. All methods were fully documented to provide the necessary assurance for regulatory bodies such as HMRI and compliance with London Underground's own standards.

## Literature search

Various publications were reviewed to establish available standards and to ensure that published information based on previous experience was taken into account. The documents that were considered appropriate and reviewed along with a summary of their contents are included in Table 1 below. Particular attention was given to the height of the serving counter both on the customer side and the employee side. This was of critical importance to make the "reasonable adjustments" required by the DDA. It can be seen that the guidance is disparate and as such it was necessary to conduct primary research to test usability of different counter heights with a broad population of both able bodied and disabled users.

Conclusions drawn from this review are as follows:

- Whilst 760 mm is a common height within both publications, the SRA guidance specifies kitchen worktops and café table heights higher than that of a ticket counter. (Generally task duration and complexity is greater in the case of the latter – see task analysis results.)
- Operable parts for parking meters in the SRA guidance is outside the extended reach specified in BS8300.
- BS8300 suggests that purchasing tickets could be performed at an increased height due to the short duration expected in transaction time and the infrequency of the task.

## Research of existing environments

As part of a standard methodology, research was conducted into existing operational environments to understand how these are operated, what works well and what does not.

**Table 1. Summary of guidance.**

| Standard or guidance | Summary of information |
|---|---|
| BS8300 | General counter heights for wheelchairs users should be 760 mm. Tickets may be purchased at an extended reach of 1170 mm. |
| SRA Train and station services for disabled passengers. A code of practice. | Ticket counters should be 760 mm.<br>Shared kitchen areas – Worktop height 850 mm.<br>Café table height 800 mm.<br>Parking meters must not have operable parts above 1200 mm.<br>Lift control buttons should be between 900 and 1100 mm. |

Site visits during normal operating times were organised to a total of three existing LU ticket offices and two mainline ticket offices. During the site visits interviews with operations staff and customers were conducted and digital stills/video footage taken.

*Observations – existing LU ticket office, Jubilee Line Extension*

The Jubilee Line ticket office visited is considered to be the new LU "standard" and features an "open plan" approach. This uses large glass panels and modern feel brushed steel and blue glass sign frieze. Results from this visit are summarised as follows:

- The design works fairly well in general.
- The design had been the subject of a "mock up" and as a result the equipment was reasonably placed and fitted adequately.
- There was adequate space for performing all tasks including paperwork, etc.
- The cross glass communication performance was felt to be poor.

*Observations – refurbished ticket office 1, Piccadilly Line*

This ticket office had been recently redeveloped using the new "open plan" approach as per the Jubilee Line office. However, due to changes that have been made in the way that these offices perform their accounting function and the discontinuation of certain pieces of equipment, operators felt less happy with the performance of the system. Results from this visit are summarised as follows:

- The design does not work well with the revised machine specifications (keyboard too large for space allocated, ticket issuing machine position causes difficulty for operator).
- Redeveloped cash tray generally configured in a non-preferred way.
- The cross glass communication performance was felt to be poor.

*Observations – refurbished ticket office 2, Piccadilly Line*

This ticket office had been recently redeveloped using the new "open plan" approach as per the previous environments. In this case, however, the operators had decided to make physical modifications to the layout and furniture in order to make their working environment more comfortable and efficient. The results of the visit are summarised as follows:

- Counter height is too low – only suitable for seated operation.
- Equipment does not fit onto the work surfaces (as a result users have fabricated additional surfaces to lay over the original ones in order to increase capacity).
- The cross glass communication performance was felt to be poor.
- There is no facility for the housing of the regularly used fares chart and other paperwork.

*Observations – mainline ticket office, Chesterfield Station*

A group representing the disabled community as a "best practice" example identified this ticket office. The design featured a dual height counter at one counter position and way finding tactile flooring for use by visually impaired persons. Both operators and customers were interviewed. The results are summarised as follows:

- The low counter is never used.
- The regular wheelchair users interviewed prefer to use the standard height counter because:
- They are able to reach and perform the ticket purchasing function relatively easily.

- They do not like drawing attention to their disability by waiting for the low counter to be free.
- The low counter does not work well with the "Disney" queuing system.
- Operationally the low counter needs to be manned all the time. (A seated only operator position.)

*Observations – mainline ticket office, Liverpool Central*

This ticket office features an adjustable height counter. The visit is summarised as follows:

- The height adjustability is appropriate for configuring the height for maximum comfort of the operator.
- The height adjustment process takes a finite time and is felt inappropriate for customers since the adjustment time is a significant proportion of the transaction time. That is, Customers do not want to wait for the counter height to be adjusted delaying the start of their transaction for 10 seconds, when the normal transaction time is normally less than 20 seconds
- Operators do not use the height adjustment facility for convenience of customers.

## Existing environment research conclusions

From the research into existing ticket office environments the following conclusions were reached:

- The design of all environments researched had lacked user input and as a result all had major opportunities for improvement.
- There was little evidence that lessons had been learned from the earlier developments and considered for newer designs.
- Efforts made in the design to accommodate wheelchair users were not used in practice.

## Process mapping and task analysis

Having established that it would be necessary to conduct fitting trials a task analysis would be required prior to the commencement of the fitting trial. In order to understand the nature of the tasks being performed a process mapping exercise was conducted.

## Process mapping

A total of 17 operations staff were involved in defining the processes that occur in the ticket office. This exercise enabled the consultants to gain a good understanding of all processes conducted and enabled the operations staff to map and discuss the processes. This exercise also encouraged a dialogue between different operations staff and highlighted the slight differences in the way staff undertake their tasks. A total of 12 high level processes were identified.

## Task analysis

Video footage was taken at three different ticket office locations during normal operating times. From this careful analysis was conducted of tasks and movement. This analysis identified a hierarchy of the most frequently used equipment and how other reference materials were used. e.g. timetables, fares, etc. On the customer side the tasks and behaviours of customers were also analysed. It was established that the average transaction time for ticket purchase is 17 seconds.

## Anthropometrics analysis

The anthropometrics data for 5th percentile females to 95th percentile males (UK) were overlaid on a scaled layout diagram and zones of convenient reach identified and mapped. This was performed for both seating and standing operator positions.

## Fitting trials

A full size mock up was manufactured from MDF and Perspex in order to perform fitting trials for both operators and customers.

The counter height on this was made adjustable using commercially available car jacks such that satisfaction rating could be obtained by the same users but with different counter heights.

Fitting trials were conducted at a London Underground office near to the operating environment at King's Cross station (for operators and able bodied customers) and subsequently the mock up was moved to the spinal injuries unit at Stoke Mandeville Hospital where wheelchair users were invited to take part in the trial as customers. The actual equipment that will be used in the final design was used during the trial. Future technology such as "oyster card" swipes were also employed during the trial.

## Results

### *Operators*

17 operators were engaged in the trial and overall a satisfaction rating of 94% was achieved at counter heights of 950 mm. This allows both comfortable standing or seating serving position.

Equipment positioning was reviewed and included the ability for operators to change all equipment positioning during the trial. From this a "best" fit layout was derived with various pieces of equipment employed in different positions at the workstation dependent on operator choice.

### *Able bodied customers*

A total of 40 able-bodied customers were engaged in the trial. From these 100% satisfaction levels were recorded with a counter height of between 760 mm and 1000 mm.

### *Disabled customers*

A total of 19 wheelchair users agreed to take part in the trial. All of these subjects had spinal injuries and a number had other disabilities in relation to the use of their arms and hands. From the sample 95% of the users reported high usability for the recommended counter height of 950 mm.

*Acoustic trials*

In all visits to existing LU ticket offices, the performance of cross glass communication was identified as a major opportunity for improvement. Acoustic analysis was conducted at the existing King's Cross ticket office in order to establish the performance of the system. The mock up was then used to trial alternative microphone and speaker positioning and new technology used to provide enhancement of system performance.

## Conclusions from fitting trials

There is no published standard covering recommended heights for disabled users other than those in wheelchairs. To this end it is felt likely that users with other muscular-skeletal conditions affecting the back and spine and those with sight impairment would find the traditional disabled counter height of 760 mm uncomfortable to use. This part of the population as a group is many times greater than populations of wheelchair users. Making a single counter height that can be used by the majority of the population has the following advantages:

- Preferred by majority of users.
- Satisfies "inclusivity" requirement by not discriminating against any user group.
- Operationally more efficient.

## Overall conclusions

In conclusion this study has, in certain respects, challenged the standard and guidance available and has indicated that an alternative strategy that takes into consideration the requirements of the wider population in a truly inclusive manner as well as the operational and commercial requirements of London Underground and the usability of the design by the staff, has been achieved.

## Integration into system design

All elements of the research carried out were fully documented and presented to representatives of all stakeholder groups. Comments were collected from these groups and the final report was issued with recommendations to the designers such that the final design will include the results from the integration work.

## Application review

A review of the final design will be performed at some time after this ticket office design is commissioned in 2007. This will include further operator and users engagement that will serve as a conduit to feedback and understand how effective the integration process has been.

# GENERAL ERGONOMICS

# CAN THE WORDING OF GM LABELS AFFECT PERCEPTIONS OF GM FOODS?

**Alexandra Costello, Elizabeth Hellier & Judy Edworthy**

*School of Psychology, University of Plymouth, North Hill, Plymouth PL4 8AA*

In the UK, mandatory labels are required for GM food products. However, little is known of the impact that GM label variation will have on public perception of these novel food products. Previous warning label research suggests that linguistic variations affect perception. Previous research by the research team here has found that the manipulation of physical characteristics, such as label colour and shape, have a significant effect on hazard perception and purchase intention for foods with labels indicating GM content. This experiment examines the effect of five within subject linguistic variables: signal word, process, item, personal pronoun and content on hazard perception and purchase intention. 71 participants took part in the experiment. ANOVA revealed significant main effects for all five variables on hazard perception. Only process, item and content, however, had significant effects on purchase intention. The results suggest that the findings from warning label research generalize across to GM labels.

## Introduction

Mandatory labels are currently required for Genetically Modified (GM) food products in the European Union. The Food Standards Agency guidelines suggest that the wording "This product contains genetically modified organisms" should be used to indicate GM content on food products in the UK (e.g. Food Standards Agency). They also suggest the wording "This product is produced from Genetically Modified (e.g.) soya" could be used. However, there is also no current legislation to mandate for the wording of GM free labels and consequently a range of labels exist such as "Contains no GM ingredients". Little is known as to the effect that these or any alternative wording to indicate GM content, will have on consumer perception of food products.

Previous warning label research suggests that linguistic variation on labels affects not only the perceived risk from the product, but also behaviour (e.g. Edworthy and Adams, 1996). This warning label research suggests that a range of linguistic variations have an effect on hazard perception, such as the presence or absence of a signal word (e.g. Wogalter and Silver, 1990) and the presence or absence of a personal pronoun (e.g.Hellier et al., 2001).

The strength of signal words, placed on the top of warning labels, have been calibrated into a list with words such as "Deadly" conveying more hazard than "Urgent", "Important" and "Notice" respectively (e.g. Hellier et al., 2000). The use of personal pronoun such as the wording "You should be aware that this product…." Vs "this product…." on pesticide labels has also been demonstrated to increase hazard perception (e.g. Hellier et al., 2001). There is, therefore, some indication in the previous research that linguistic variation on labels affects the perception of the product and consequently behaviour.

Previous research by this research team has also found that the manipulation of physical characteristics such as colour and shape have a significant effect on hazard perception and purchase intention for GM food labels (e.g. Costello et al., 2002, 2003). The results of these two studies have therefore demonstrated the generalisation of the findings from warning label research to this new area of GM label research. It was therefore hypothesized that the effect of linguistic variation on warning labels may transfer to labels indicating GM content.

This experiment examines the effect of linguistic variation in GM labels on measures of hazard perception and purchase intention. Five factors were examined namely Signal Word, Personal Pronoun, Process, Item and GM content. The signal word manipulations were no signal word, "notice" or "important" on the label indicating GM content. The personal pronoun manipulations were the presence or absence of the wording "You should be aware that" on the label indicating GM content. Process was an indication of the type of product and stated either "this product contains GM" or "this product is produced from GM". Item was one of two alternative descriptions of the GM content and stated either "GM ingredients" or "GM organisms". Content indicated whether the product was GM or non-GM. These manipulations were chosen on the basis of previous research or current wordings used on GM and non-GM food products in the UK. It was expected that each of the five linguistic variations would affect hazard perception and consequently purchase intention.

## Method

### *Design*

There were five within subject variables, signal word (none, notice, important) process (contains GM, produced from GM), item, (GM organisms, GM ingredients) personal pronoun (none, *you* should be aware that) content (GM, no GM).

### *Participants*

71 participants aged 18 to 51 took part in the experiment, 20 were male and 51 female. All were undergraduate students at the University of Plymouth and took part in exchange for a coursework credit or were paid £3.

### *Stimuli and materials*

Factorial combinations of all levels of signal word, process, item, personal pronoun and content produced 48 different labels. The labels were scanned onto identical pictures of a food product. A picture of the food product with no label acted as the control. Each label was of identical size with a white background and black Times New Roman text of size 10 font.

### *Procedure*

Participants were presented a folder containing 49 photographs of the food product each with a different label, in one of 5 different randomised orders. When viewing the pictures, participants were told that each product was the same price and quality and they should assume in principle that they would buy the product. The participants were asked "How likely would you be to buy the product?" and directed to indicate the number that best represented their answer on a scale which ran from 0 to 8 where 0 represented the answer "Not at all" and 8 represented "Extremely likely". Participants were also asked "How much hazard is indicated by the label?" and indicated their answer on a scale which ran from 0 to 8, where the number 0 represented the answer "No hazard" and the number 8 represented "Extreme Hazard".

## Results

### *Hazard perception*

The mean score for hazard perception for the control (no label) condition was 1.10. Repeated measures analysis of variance revealed significant main effects for all 5 within subject variables.

*Signal word*
Main effect of signal word F (2,140) = 17.89, $p < 0.01$. The hazard perception mean score for "no signal word" was 2.34 followed by 2.66 for "notice" and 2.86 for "important".

*Process*
Main effect of process F (1,70) = 8.71, $p < 0.01$. The mean score for "contains GM" was 2.58 and 2.67 for "produced from GM".

*Item*
Main effect of item, F (1,70) = 15.36 $p < 0.01$. The mean score for the use of the word "ingredient" was 2.42 and 2.82 for the word "organism".

*Personal pronoun*
Main effect of personal pronoun F (1,70) = 16.44, $p < 0.01$. The mean score for the use of the phrase "you should be aware that" was 2.70 and 2.54 for no use of personal pronoun.

*Content*
Main effect of content F (1,70) = 155.16, $p < 0.01$. The mean score for GM content was 4.11 and 1.13 for no GM content.

Tests of between-subjects effects revealed no significant effect of either gender for measures of hazard perception.

### *Purchase intention*

The mean purchase intention score for the control (no label) condition was 4.87. Repeated measures ANOVA revealed significant main effects on purchase intention for Process, Item and Content, but not Signal word or Personal pronoun.

*Process*
Main effect of F (1,70) = 10.86, $p < 0.01$. The mean score for process was 3.56 for "contains GM" and 3.48 for "produced from GM".

*Item*
Main effect of F (1,70) = 17.33, $p < 0.01$. The mean score for the use of the word "ingredient" was 3.73 and 3.28 for "organism".

*Content*
Main effect of F (1,70) = 120.24, $p < 0.01$. The mean score for GM was 2.17 and 4.84 for non-GM.

Tests of between-subjects effects revealed no significant effect of either gender for measures of purchase intention.

### *Hazard perception and purchase intention*

A strong inverse relationship was found between the mean score for hazard perception and purchase intention for the 49 different labels with a Pearson correlation of −.982, $p < 0.01$.

## Discussion

All five factors were found to have a significant effect on hazard perception. The signal word "Important" resulted in a higher mean score for hazard perception than "Notice" or no signal word respectively. This finding is in the direction suggested by Hellier et al.'s (2000) arousal strength ranking of signal words. The wording "Produced from GM" yielded a higher mean score for hazard perception than the wording "Contains GM" and the wording "GM organism" yielded higher mean hazard perception score then the wording "GM ingredient". It is argued here that the reasons for this finding require further investigation. The use of the personal pronoun in the form of the wording "You should be aware that" increased hazard perception compared to the absence of the wording. This was in line with the findings of Hellier et al., (2001). Finally, GM content had the greatest effect on hazard perception. Those labels indicating that the product was GM yielded a much higher mean score for hazard perception than those indicating that the product was not GM.

Only "Process", "Item" and GM content were found to have significant effects on purchase intention. These results were in the reverse direction to those found for hazard perception. The wording "Produced from GM" decreased purchase intention relative to "Contains GM" and the wording "GM organism" decreased purchase intention relative to "GM ingredient". GM content again had the greatest effect on the mean scores for purchase intention. Those label indicating that the product was GM greatly decreased purchase intention.

Whilst significant main effects on hazard perception were found for all 5 factors, only process, item and content had main effects on purchase intention. What explanation could we have for this difference? One possible explanation is the nature of these linguistic variations. These variations were whether the product was "produced from" or "contained", GM or non-GM "ingredients" or "organisms". Hence it is possible that these linguistic variations were interpreted by participants as describing possible differences in the nature of the product with some differences being significantly less desirable than others. However the signal word and personal pronoun variations although affecting hazard perception did not significantly affect purchase intention. These variations although attracting attention to the label and its informational content, may not necessarily have changed the impression of the product itself.

The strong correlation found between hazard perception and purchase intention of –0.982 supports the previous findings by Costello et al. (2002, 2003). That is, as the perceived hazard from the label on the product increased, the perceived likelihood of purchasing the product reduced. In effect then, we would suggest that those wishing to increase the purchase of their products, choose low hazard label designs and wordings.

The mean score for hazard perception for the control (no label) condition was 1.10, for purchase intention it was 4.87. Having no indication of GM content on a product therefore results in lower hazard perception and higher purchase intention than the majority of the labels tested here. This is even when the labels clearly state the product to be non GM. This has been a consistent finding by this research team here and we would argue that non GM labeling may be counter-productive at increasing the sales of a food product, unless great care is taken over the design and linguistic content of the label to avoid the increased hazard perception of labeled products.

These results indicate that the findings from warning label research generalise across to GM informational labels as previously demonstrated by Costello et al., (2002, 2003). It is therefore suggested that the relevant regulatory bodies may be wise to take into consideration that changing the linguistic content of GM informational labels may possibly alter hazard perception and/or purchase intention for food products. It is therefore not just a matter of providing a label to inform of GM content, but providing the most appropriate label for consumers in terms of design and linguistic content. It is also important to emphasise that significant results were found for the effect of linguistic manipulations for labels indicating both GM and non-GM content. More research is therefore needed to determine the effect of such manipulations on other types of informational label.

## References

Costello, A.J., Hellier, E.J., Edworthy, J., and Coulson, N. 2002, Can food label design characteristics affect perceptions of genetically modified foods? In P.T. McCabe (ed.) *Contemporary Ergonomics 2002*, (Taylor and Francis, London), 442–446

Costello, A.J., Hellier, E.J., and Edworthy, J. 2003, GM informational labels. Does shape matter? In P.T. McCabe (ed.) *Contemporary Ergonomics 2003*, (Taylor and Francis, London), 507–512

Edworthy, J., and Adams A.S. 1996, *Warning design: a research perspective*, (Taylor and Francis, London)

Food Standards Agency 2004, GM and Novel Foods, www.food.gov.uk/gmfoods (Accessed 27/08/2004)

Hellier, E.J., Edworthy J., Lambell, N., Grey, C., Aldrich, K., and Lee, A. 2001, *The effectiveness of labeling of pesticides*, (Health and Safety Executive, report number 3917/R67.137)

Hellier, E.J, Wright, D.B., Edworthy, J., and Newstead, S. 2000, On the stability of the arousal strength of warning signal words, *Applied Cognitive Psychology,* **14**, 577–592

Wogalter, M.S., and Silver N.C. 1990, Arousal strength of signal words, *Forensic Reports*, **3**, 407–420

# AUTHOR INDEX

# SUBJECT INDEX